Food Science and Technology

Food Science and Technology

Edited by **Lisa Jordan**

SYRAWOOD
PUBLISHING HOUSE
New York

Published by Syrawood Publishing House,
750 Third Avenue, 9th Floor,
New York, NY 10017, USA
www.syrawoodpublishinghouse.com

Food Science and Technology
Edited by Lisa Jordan

International Standard Book Number: 978-1-68286-255-1 (Hardback)

Printed in China.

Contents

Preface

Food science and technology is an applied branch of science dealing with food processing, production and consumption processes and techniques. It is an interdisciplinary field that aims to understand the nutritional value of food and involves research and development for technological advancements in mass production of food products. The book emphasizes on various techniques to enhance product quality and improve efficiency of processed food products. It provides a detailed account of emerging technologies in food processing and preservation, food engineering and biotechnology applications, etc. This book is helpful for students, researchers and academicians interested in food science.

This book is the end result of constructive efforts and intensive research done by experts in this field. The aim of this book is to enlighten the readers with recent information in this area of research. The information provided in this profound book would serve as a valuable reference to students and researchers in this field.

At the end, I would like to thank all the authors for devoting their precious time and providing their valuable contributions to this book. I would also like to express my gratitude to my fellow colleagues who encouraged me throughout the process.

Editor

Influence of smoking method on quality of traditional smoked Bonga shad (*Ethmalosa frimbriata*) fish from Lagos State, Nigeria

Adeyeye, S. A. O[1] , Oyewole, O. B.[1], Obadina, A. O.[1] and Omemu, A. M.[2]

[1]Department of Food Science and Technology, Federal University of Agriculture, Abeokuta, Nigeria.
[2]Department of Hospitality and Tourism Management, Federal University of Agriculture, Abeokuta, Nigeria.

This study was carried out to assess the influence of smoking method on quality attributes of traditional smoked bonga shad fish. Fresh bonga shad fish (100 samples) and smoked bonga shad fish (100 samples) were collected from 20 different processing centres and the fresh samples were smoked with convectional smoking kiln as the control. Laboratory analyses were conducted and each batch was assessed for; proximate and quality analyses. The results obtained show significant variations ($p<0.05$) for all the proximate composition and quality indices of the smoked bonga shad fish samples. The mean moisture content of fresh bonga shad ranged from 72.96 - 76.89% and that of the smoked bonga shad and control ranged from 10.89 - 14.38% and 8.56 - 10.12%, respectively. The mean protein content of 15.18 - 16.95% was recorded for fresh bonga shad samples and 51.86 - 60.24% and 58.86 - 64.84% for samples of smoked bonga shad. The fat content was determined at 6.46 - 8.84%, while the smoked bonga shad and the control samples were found to have fat content at 16.13 - 20.84% and 12.87 - 17.34%, respectively. The mean pH, thiobarbituric acid value (TBA), total volatile base- nitrogen (TVB-N), TMA, trimethylamine value (TMA) and Free fatty acids values (FFA) values of the smoked bonga shad were lower than the control and the fresh samples. However, all values are within the range of legislative standard.

Key words: Bonga shad, smoking, traditional, proximate, quality indices.

INTRODUCTION

In most sub-saharan Africa countries, traditional fishing is practiced in almost all rivers, lakes, ponds, and seas. It represents an important part of total fish captures and is an important sector in the national strategies of fight against poverty, food security and food safety (Ahmed et al., 2011). Fish is an important dietary component of people all around the world and represents a relatively cheap and accessible source of high quality protein for poorer households (Ikutegbe and Sikoki, 2014). In West Africa, fish has been reported to provide 40–70% of the protein intake of the population (Béné and Heck, 2005; Ikutegbe and Sikoki, 2014) and it is a critical source of dietary protein that is not readily available in the carbohydrate-based staple foods of the populations. In Nigeria, fish has an edge over meat because it is cheaper and relatively more abundant (Eyo, 2001) and constitutes about 40% of the animal protein intake (Eyo, 2001; Abolagba and Melle, 2008). Fish can be consumed in several forms; fresh, dried, frozen, fermented, and brined depending on the consumer preference. In a study by Mafimisebi (2012), it was discovered that majority of the Nigerian people had a preference for fresh fish; however limitations such as the low keeping quality of fish after harvested and the tendency of fish to spoil quickly make this very difficult. This results in a higher reported consumption of smoke-dried fish, which has relatively longer shelf life (Mafimisebi, 2012).

Fish is highly perishable and a considerable effort has been directed to extend the shelf-life of fish using presservation and processing techniques, such as refrigeration, freezing, canning, smoking, salting, and drying (Nwachukwu and Madubuko, 2013). Besides, some of these techniques can also be used to enhance the value of fish, such as smoked fish.

The smoking of fish smoldering dates back to civilization (Eyo, 2001). Smoking method mostly imparts a desirable flavour and inhibits the growth of microorganisms (Swastawati, 2008). It has been used for centuries in food preservation, and is still widely used for this purpose among several communities in the third world where up to 70% of the catch is smoked for preservation (Omojowo and Raji, 2010). Consumers are rediscovering the good taste of smoked seafood, including smoked catfish. To satisfy the consumer demand, it is necessary to produce good quality and safe smoked seafood products (Omojowo and Raji, 2010).

There is insufficient data on the quality characteristics of traditional smoked wild fish from Nigeria despite the fact that it constitutes a substantial portion of fish available for Nigerians. This study is carried out to investigate the influence of smoking method on quality of traditional smoked bonga shad from Lagos State.

MATERIALS AND METHODS

Fish used

Fresh bonga shad fish (100 samples) and smoked bonga shad fish (100 samples) were collected from 20 different processing centres from fishing communities of Badagry and Epe Local Government Areas of Lagos State, Nigeria. The fresh bonga shad fish samples were freshly harvested. The fresh samples were taken to the Institute of Food Security, Environmental Resources and Agricultural Research (IFSERAR) laboratory, Federal University of Agriculture, Abeokuta for smoking.

Area of study

Using a current geopolitical map of Nigeria, Lagos State (Figure 1) lies to the south-western part of Nigeria and has boundaries with Ogun State both in the north and east. It is bordered on the west by the Republic of Benin and in the south, stretches for 180 km along the coast of the Atlantic Ocean. It therefore has 22.5% of Nigeria's coastline and occupies an area of 3,577 sq km land mass with about 786.94 sq km (22%) of it being lagoons and creeks. The state is endowed with marine, brackish and fresh water ecological zones with varying fish species that provide productive fishing opportunity for fishermen. Two local government areas (Badagry and Epe Local Government) were covered because they are highly densed fish processing centers. They were selected for the study and hazard analyses of the products.

Sampling procedure

Fresh samples (100) and smoked bonga shad samples (100) were collected from each of the 10 processing centres from each of the two local government areas (Badagry and Epe Local Government) by purposive sampling in sterile containers (Ziploc).

All freshly harvested bonga shad samples were kept on ice during transportation to the laboratory and smoked on the same day. Smoked bonga shad fish samples were analyzed immediately.

Fish smoking process

Smoked fish was prepared following the method as described by Crapo (2011) with modifications. Bonga shad fish samples were carefully cleaned to remove slime, blood and harmful bacteria. The fish were eviscerated, leaving the skin on the fish. The fish were cut into uniform pieces (fillet) so that no parts will get overheated.

The bonga shad fish were smoked to 80°C internal temperature for 18 h. The kiln temperature was adjusted as needed throughout this smoking period to maintain the 80°C internal temperature. Hands, utensils and work surfaces were cleaned when transferring fish from smoker to oven to cool down to avoid cross-contamination. Smoking was done for 24 h until the fish is fully dried.

Proximate analysis

The following proximate analyses were carried out on fresh bonga shad and the smoked bonga shad samples collected from the processing centres. The moisture content of the fresh bonga shad and smoked bonga shad were determined by the oven-drying method. Protein contents of the bonga shad were extracted and fractionated by the method of AOAC (2000) method. The crude fat, crude fibre and ash content of the fresh bonga shad and smoked bonga shad were determined by AOAC (2000) method.

Physico-chemical analysis

Kent pH meter (Kent Ind. Measurement Ltd., survey) model 7020 equipped with a glass electrode was used to measure the pH of the flesh, employing 10 g of fish homogenized in 10 ml of distilled water. Triplicate determinations were made in all cases. The pH meter was calibrated using pH 4.0 and pH 7.0 buffers. The total volatile base- nitrogen, trimethylamine value (TMA), thio-barbituric acid value, peroxide value and free fatty acid value of the fresh fish and smoked fish were determined by AOAC (2000) method. All chemicals used in this study were of the analytical grade unless stated otherwise.

Data analysis

All data analyses were done in triplicates. The data obtained were

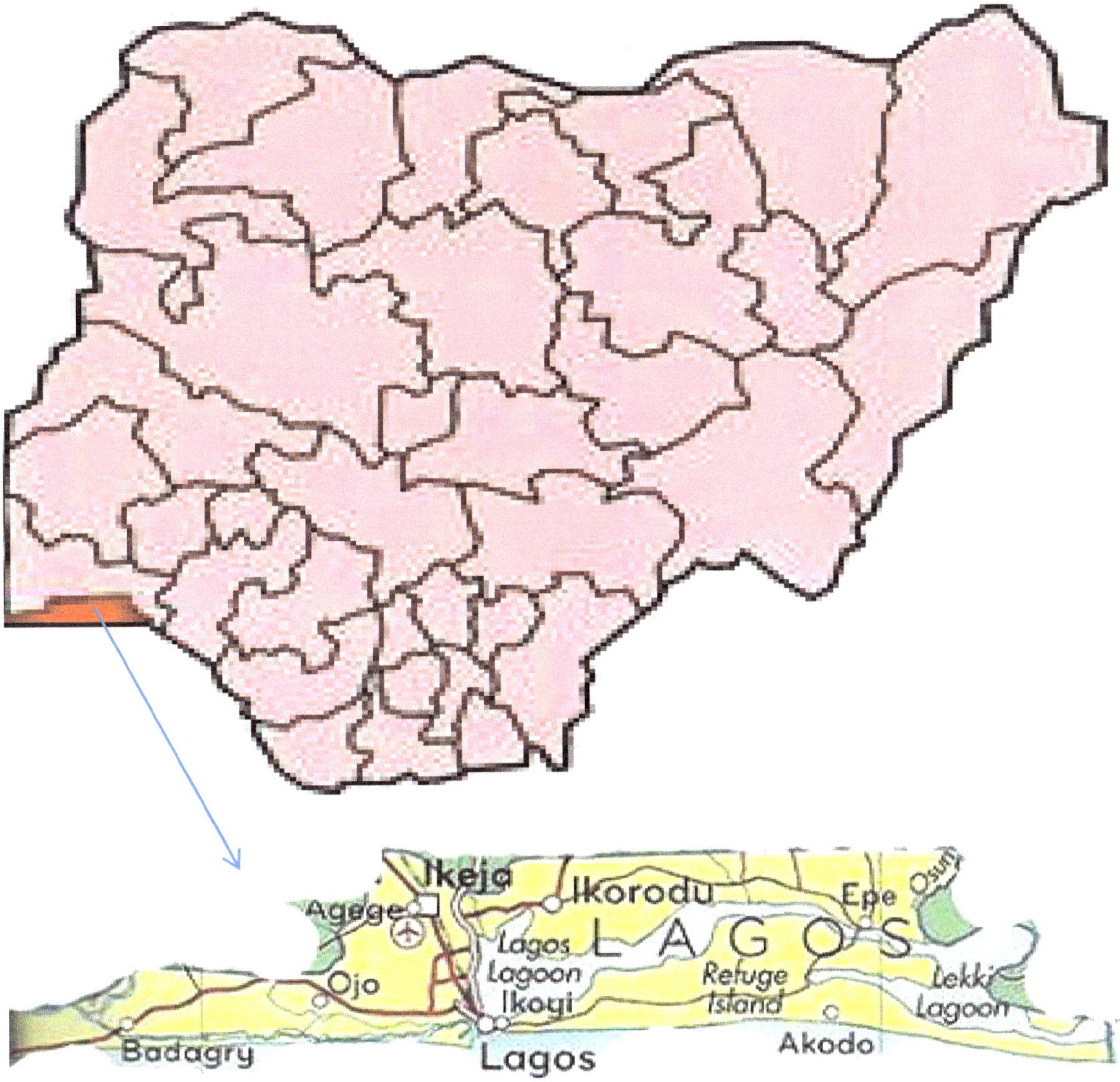

Figure 1. Map of Lagos State showing the 20 Local Government Areas

subjected to descriptive statistics using IBM SPSS version 21.0 software. One way analysis of variance (ANOVA) was done using Duncan's Multiple Range Test ($p<0.05$) to study the difference between means.

RESULTS AND DISCUSION

Moisture content of fresh bonga shad samples ranged from 72.96 - 76.89% (Table 1) and that of smoked bonga shad samples obtained using local drum kiln and conventional smoke kiln ranged from 11.22 - 14.64% and 8.64 - 10.32% (Table 2). In contrary to protein, fat, and ash, the moisture content of fresh bonga shad samples decreased sharply after the smoking process. This decrease was due to loss of water during smoking. Moisture content of fish is of great importance as it promotes microbial growth. In the present study, the protein content of the fresh bonga shad samples ranged from 15.18 - 16.95% and that of the smoked bonga shad samples ranged from 51.86 - 60.24% and 58.86 - 64.84%. There was an inverse relationship between the moisture and protein content in the smoked bonga shad. Fat content of fresh bonga shad samples ranged from 6.46% - 8.84% and that of smoked bonga shad samples ranged from 16.13 - 20.84% and 12.87 - 17.34%. Crude fibre content of fresh bonga shad samples ranged from0.21 - 0.37% and crude fibre content of smoked bonga shad samples ranged from 2.14 - 4.32% and 2.26 - 4.56%. Ash content of fresh bonga shad samples ranged from 0.12 - 0.16% and ash content of smoked bonga shad samples ranged from 1.18 - 1.46% and 1.24 - 1.79% . The increase in mineral content, ash and crude fibre can be attributed to an increase in the dry matter content per unit weight following sample dehydration

Table 1. Proximate composition of fresh Bonga shad (*Ethmalosa frimbriata*) samples from 20 different processing centres.

Processing centres	Moisture %	Protein %	Fat %	Crude fibre %	Ash %	Carbohydrate %
Agbalata	74.81_{a}	15.18_{a}	7.81_{cdefg}	0.29_{abc}	0.15_{a}	0.92_{ab}
Ajido	75.68_{a}	15.96_{a}	5.73_{a}	0.26_{abc}	0.12_{a}	1.04_{ab}
Asakpo	76.54_{a}	15.79_{a}	7.86_{defg}	0.3_{4bc}	0.14_{a}	1.01_{ab}
Boguru	73.86_{a}	16.21_{a}	7.90_{fg}	0.21_{a}	0.12_{a}	1.05_{ab}
Fvanoveh	75.87_{a}	16.95_{a}	8.84_{fg}	0.26_{abc}	0.12_{a}	0.94_{ab}
Gberefun	74.69_{a}	15.72_{a}	8.45_{efg}	0.23_{ab}	0.13_{a}	0.99_{ab}
Gbetrome	74.75_{a}	16.78_{a}	8.87_{g}	0.33_{bc}	0.12_{a}	1.01_{ab}
Ilaje	75.91a	16.26_{a}	7.49_{bcde}	0.31_{abc}	0.14_{a}	1.16_{ab}
Kofegameh	74.62_{a}	15.93_{a}	6.73_{abc}	0.24_{ab}	0.14_{a}	1.08_{ab}
Pako	76.72_{a}	16.11_{a}	7.94_{defg}	0.27_{abc}	0.16_{a}	0.95_{ab}
Afuye	74.95_{a}	$16._{89a}$	8.84_{fg}	0.25_{abc}	0.14_{a}	0.92_{ab}
BodinYawa	73.89_{a}	16.61_{a}	6.93_{bcd}	0.31_{abc}	0.12_{a}	1.06_{ab}
Idale	74.63_{a}	16.96_{a}	7.89_{bcd}	0.21_{a}	0.13_{a}	0.95_{ab}
Igbodun	75.36_{a}	16.83_{a}	8.43_{efg}	0.23_{ab}	0.14_{a}	1.04_{ab}
Ilogun	76.52_{a}	15.68_{a}	7.68_{cdef}	0.30_{abc}	0.12_{a}	1.24_{b}
Mejona	74.93_{a}	16.89_{a}	7.74_{cdef}	0.24_{ab}	0.15_{a}	1.12_{ab}
Oluwo	75.84_{a}	16.48_{a}	6.46_{ab}	0.37_{c}	0.14_{a}	1.01_{ab}
Okorisan	77.69_{a}	16.91_{a}	7.90_{defg}	0.34_{bc}	0.12_{a}	0.9_{a}
Orita	76.92_{a}	16.86_{a}	7.82_{defg}	0.31_{abc}	0.14_{a}	1.23_{b}
Orogoro	75.45_{a}	16.62_{a}	7.96_{defg}	0.28_{abc}	0.13_{a}	1.09_{ab}

Data are means of 3 replicates. Data with the same subscript are not significantly different at $p<0.05$.

Table 2. Proximate composition of smoked Bonga shad (*Ethmalosa frimbriata*) from 20 different processing centres using local drum kiln and convective smoking kiln.

Processing	Moisture %		Protein %		Fat %		Crude fibre %		Ash %		Carbohydrate %	
centres	Local	Convect	Local	Convect	Local	Convect	Local	Convect	Local	Convect	Local	Convect
Agbalata	13.37gh	9.23e	56.18h	63.18m	18.09g	13.31c	2.91e	2.98f	1.46hi	1.79j	7.99g	9.51hi
Ajido	13.84i	9.56f	59.06p	64.06p	16.61c	12.98b	2.63c	2.87e	1.18a	1.33ab	6.68b	9.20f
Asakpo	14.31k	10.12h	54.39e	61.39h	18.15gh	14.43f	3.24g	3.56i	1.36efgh	1.54efg	8.55h	8.96e
Boguru	12.59e	8.78bc	57.21k	63.21n	18.20gh	14.60g	2.14a	2.43b	1.24abcd	1.61ghi	8.62h	9.37g
Fvanoveh	12.11d	8.64a	60.24q	64.84q	17.13e	12.87a	2.65c	2.78d	1.19a	1.24a	6.68b	9.63jk
Gberefun	14.26jk	10.06h	53.72c	60.72d	20.58k	14.15d	2.38b	2.59c	1.32cdef	1.68i	7.74e	10.80n

Table 2. Contd

Gbetrome	11.83c	8.89c	55.48g	60.48c	19.27j	15.82l	3.32gh	3.46h	1.21ab	1.43bcde	8.89j	9.59ij
Ilaje	12.67e	8.68ab	53.26n	63.86o	18.15gh	13.36c	3.09f	3.32g	1.40efgh	1.62ghi	6.43a	9.16f
Kofegameh	13.28g	9.09d	57.63l	62.63l	16.43b	14.53g	2.43b	2.57c	1.44ghi	1.73i	8.79i	9.45gh
Pako	11.54b	8.83c	54.11d	61.11g	18.72i	16.87o	2.72cd	2.91ef	1.61j	1.47cde	11.30m	8.81d
Afuye	12.04d	8.78bc	51.86a	58.86a	20.84l	17.34p	4.32j	4.56l	1.42fghi	1.73i	9.52k	8.73d
Bodin Yawa	14.16j	10.32i	58.21mn	61.61i	16.13a	14.28e	3.10f	3.34g	1.18a	1.65hi	7.22c	8.80d
Idale	12.10d	8.65a	52.96b	60.96f	16.84d	16.76n	2.18a	2.26a	1.25abcd	1.54efg	14.67n	9.83l
Igbodun	11.33a	9.68g	58.13m	62.13k	18.25h	16.35m	2.25a	2.41b	1.42fghi	1.38bcd	8.62h	8.05a
Ilogun	13.28g	9.08d	57.18k	60.18b	16.46b	17.82h	3.04f	3.32g	1.16a	1.51def	8.62h	8.01a
Mejona	13.41h	9.23e	54.89f	62.89m	17.73f	15.13j	2.46b	2.64c	1.51i	1.69i	10.00l	8.42c
Oluwo	12.94l	8.88c	58.42o	61.72j	16.12a	14.82h	3.79j	3.91j	1.30bcde	1.43bcde	7.43d	10.24m
Okorisan	11.22i	9.56f	56.94j	60.94f	19.44j	15.58k	3.41h	3.57j	1.23abc	1.36abc	7.76f	8.99e
Orita	14.64	10.23i	53.83c	60.83e	18.64i	13.31c	4.11i	4.23k	1.41fghi	1.71i	7.37d	9.69k
Orogoro	13.89	9.71g	56.18h	63.18m	18.15gh	15.89l	2.82de	2.96ef	1.34defg	1.63ghi	7.46d	8.17b

Local = Local drum kiln; Convect = Convective smoking kiln. Data are means of 3 replicates. Data with the same subscript are not significantly different at $p<0.05$.

and during the smoking process. In this study carbohydrate content is given by difference that is the percentage of water, protein, fat and ash subtracted from 100. The carbohydrate content of fresh unsmoked bonga shad samples ranged from 0.92 - 1.24% and carbohydrate content of smoked bonga shad and control samples ranged from 6.43 - 10.00% and 8.01 - 10.24%. Carbohydrate content of smoked bonga shad samples is low because it is a proteinous food.

The quality indices of the fresh and smoked bonga shad were studied (Tables 3 and 4). Fats undergo changes during storage which result in production of an unpleasant taste and odour which is commonly referred to as rancidity. The peroxide value (PV) results are similar in pattern to TBA. In this study PV of fresh bonga shad samples was 6.11 – 8.59 mg Eqperoxide/kg and 6.22 – 10.41 mg Eqperoxide/kg and 7.12 – 9.86 mgEq.peroxide/kg for smoked bonga shad and control samples. These values are below the recommended value of between 20 and 40 mgEq.peroxide/kg for rancid taste to begin. Free fatty acids values (FFA) of fresh unsmoked bonga shad samples was 0.86 - 1.13% while that of smoked bonga shad and control samples ranged from 1.05 - 1.26% and 1.00% - 1.19%. These values are very low and below the threshold for rancidity detection in smoked fish. The thiobar-bituric acid value (TBA) is used to assess the degree of fish spoilage especially in fatty fish. The TBA test measures a secondary product of lipid oxidation, malonaldehyde. The TBA values of fresh bonga shad samples ranged from 0.84 – 1.11 mgMol/kg and thio-barbituric acid value (TBA) of smoked bonga shad samples and control samples ranged from 1.01 – 1.23 mgMol/kg and 1.01 – 1.15 mgMol/kg. The TBA (1.00 to 1.15 mg TBA/kg) did not exceed 1 to 2 mg TBA/kg which was well within acceptable limits. The sensory threshold level for detecting rancidity in fresh meat was reported to be between 1 and 2 TBA (Calhoun et al., 1999). The increased TBA values in the smoked fish probably originated from the breakdown of oxidation products, mainly malonaldehyde, during smoking due to the high temperature (Goktepe and Moody, 1998). Beltran and Moral (1991) reported that high TBA values are correlated with the degree of oxidation of fats in hot smoked sardines.

The legislative standard for TVB-N include: 20 mgN/100 g for fresh fish, 30 mgN/100 g stale fish and 40 mgN/100 g for fish that is unfit for human consumption but can be used for animal feed (da Silva, 2002). In this study, the total volatile base-nitrogen (TVB-N) of fresh bonga shad ranged from 13.38 – 15.57 mgN/kg and TVB-N of smoked bonga shad and control samples ranged from 16.43 – 19.36 mgN/kg and 15.49 – 18.83mgN/kg. These values are within the range of legislative

Table 3. Quality indices of fresh Bonga shad (*Ethmalosa frimbriata*) samples from 20 different processing centres.

Processing centres	Peroxide value (PV)(mEq. peroxide/kg)	Free fatty acid (FFA) %	Thiobarbituric acid (TBA) (mg Mol/kg)	Total volatile base-nitrogen (TVB-N) (mgN/kg)	Trimethyl amine value (TMA) (mgN/kg)	pH
Agbalata	6.11$_{a}$	1.03$_{bcdefg}$	0.96$_{abcde}$	13.89$_{e}$	2.42$_{ghij}$	6.96$_{abcd}$
Ajido	6.57$_{ab}$	1.08$_{efgh}$	1.00$_{bcde}$	13.56$_{bc}$	1.98$_{abc}$	7.03$_{bcdef}$
Asakpo	8.04$_{def}$	1.00$_{bcdef}$	1.19$_{g}$	15.38$_{k}$	2.31$_{efghi}$	7.18$_{fgh}$
Boguru	6.38$_{a}$	0.95$_{abc}$	1.07$_{efg}$	13.49$_{ab}$	2.49$_{ijk}$	6.91$_{ab}$
Fvanuveh	7.20$_{bc}$	1.06$_{defg}$	1.11$_{fg}$	15.73$_{l}$	2.26$_{efgh}$	6.93$_{abc}$
Gberefun	7.63$_{cde}$	1.01$_{bcdef}$	0.89$_{abc}$	14.63$_{gh}$	2.03$_{abcd}$	7.11$_{defgh}$
Gbetrome	8.13$_{def}$	1.09$_{efgh}$	0.94$_{abcde}$	14.92$_{i}$	2.51$_{jk}$	6.93$_{abc}$
Ilaje	8.00$_{def}$	1.11$_{fgh}$	1.13$_{efg}$	13.71$_{cd}$	2.43$_{hijk}$	6.82$_{a}$
Kofegameh	6.75$_{ab}$	0.93$_{abc}$	0.86$_{ab}$	13.47$_{ab}$	2.56$_{k}$	6.86$_{a}$
Pako	8.27$_{def}$	0.98$_{bcde}$	1.04$_{def}$	14.78$_{hi}$	2.12$_{abcde}$	7.07$_{bcdef}$
Afuye	7.18$_{bc}$	1.04$_{cdefg}$	0.95$_{abcde}$	15.17$_{j}$	1.94$_{a}$	7.10$_{defgh}$
BodinYawa	6.46$_{a}$	1.07$_{efgh}$	1.02$_{cdef}$	13.79$_{de}$	2.23$_{defg}$	7.13$_{efgh}$
Idale	6.31$_{a}$	1.05$_{defg}$	1.00$_{bcdef}$	13.73$_{de}$	2.16$_{bcde}$	7.24$_{h}$
Igbodun	8.35$_{ef}$	1.18$_{h}$	0.85$_{a}$	15.57$_{l}$	2.11$_{abcde}$	7.00$_{bcde}$
Ilogun	8.49$_{ef}$	0.87$_{a}$	0.97$_{abcdef}$	14.45$_{f}$	1.96$_{ab}$	7.14$_{efgh}$
Mejona	7.59$_{cd}$	0.92$_{ab}$	0.84$_{a}$	15.63$_{l}$	1.93$_{a}$	6.81$_{a}$
Oluwo	8.06$_{def}$	0.86$_{a}$	1.08$_{efg}$	13.85$_{de}$	1.99abc	7.20$_{gh}$
Okorisan	7.61$_{cd}$	1.06$_{defg}$	1.05$_{defg}$	14.91$_{i}$	2.17$_{cdef}$	7.09$_{cdefg}$
Orita	6.38$_{a}$	1.13$_{gh}$	0.91$_{abcd}$	14.57$_{fg}$	2.36$_{fghi}$	7.01$_{bcdef}$
Orogoro	8.59$_{cd}$	1.02$_{bcdef}$	0.94$_{abcdef}$	13.38$_{a}$	2.13$_{abcde}$	7.13$_{efgh}$

Data are means of 3 replicates. Data with the same subscript are not significantly different at $p<0.05$.

Table 4. Quality indices of smoked Bonga shad (*Ethmalosa frimbriata*) from 20 different processing centres using local drum kiln and convective smoking kiln.

Processing centres	Peroxide value (PV)(mEq,peroxide/kg)		Free fatty acid(FFA) %		Thiobarbituric acid (TBA) (mg Mol/kg)		Total volatile base-Nitrogen (TVB-N) (mgN/100g)		Trimethyl amine value (TMA)(mgN/kg)		pH	
	Local	Convect	Local	Convect	Local	Local	Local	Convect	Local	Convect	Local	Convect
Agbalata	9.05$_{ghi}$	8.92$_{n}$	1.13$_{abc}$	1.10$_{b}$	1.07$_{ab}$	1.02$_{a}$	18.33$_{g}$	16.62$_{c}$	2.61$_{i}$	2.49$_{kl}$	6.41$_{defg}$	6.56$_{a}$
Ajido	8.96$_{gh}$	7.78$_{g}$	1.18$_{cd}$	1.12$_{b}$	1.12$_{ab}$	1.09$_{ab}$	19.06$_{i}$	18.31$_{k}$	2.36$_{def}$	2.18$_{cde}$	6.38$_{cdef}$	6.72$_{bcde}$
Asakpo	9.01$_{ghi}$	8.21$_{hi}$	1.26$_{d}$	1.19$_{b}$	1.18$_{ab}$	1.15$_{bc}$	17.54$_{c}$	16.96$_{d}$	2.56$_{ij}$	2.37$_{ij}$	6.62$_{ij}$	6.93$_{i}$
Boguru	9.13$_{i}$	8.46$_{l}$	1.05$_{ab}$	1.03$_{b}$	1.04$_{ab}$	1.02$_{a}$	17.31$_{b}$	16.63$_{c}$	2.62$_{jk}$	2.51$_{lm}$	6.29$_{abcd}$	6.65$_{abc}$
Fvanuveh	9.94$_{k}$	8.63$_{m}$	1.09$_{abc}$	1.04$_{b}$	1.06$_{ab}$	1.04$_{ab}$	19.11$_{i}$	17.23$_{e}$	2.49$_{hi}$	2.30$_{ghi}$	6.58$_{hij}$	6.90$_{hi}$

Table 4. Contd

Gberefun	9.09_{hi}	8.32_{k}	1.06_{ab}	1.03_{b}	1.02_{ab}	1.00_{a}	17.29_{b}	16.36_{b}	2.27_{abcd}	2.10_{abc}	6.53_{fghi}	6.61_{ab}
Gbetrome	8.77_{f}	7.61_{e}	1.04_{a}	1.01_{b}	1.01_{a}	1.00_{a}	19.36_{j}	17.62_{hi}	2.73_{jk}	2.59_{mn}	6.37_{bcde}	6.72_{def}
Ilaje	8.92_{fg}	7.68_{ef}	1.19_{cd}	1.15_{b}	1.11_{ab}	1.10_{abc}	19.18_{i}	18.23_{k}	2.64_{i}	2.52_{lm}	6.43_{defgh}	6.84_{ghi}
Kofegameh	9.11_{hi}	8.31_{jk}	1.16_{bcd}	1.11_{b}	1.13_{ab}	1.11_{abc}	18.24_{fg}	17.67_{i}	2.81_{l}	2.61_{n}	6.71_{j}	6.63_{abc}
Pako	8.98_{ghi}	8.95_{n}	1.10_{abc}	1.07_{b}	1.08_{ab}	1.06_{ab}	17.96_{d}	16.82_{c}	2.35_{cdef}	2.19_{cde}	6.24_{abc}	6.91_{i}
Afuye	7.99_{d}	7.73_{fg}	1.08_{abc}	1.06_{b}	1.02_{a}	1.01_{a}	17.28_{b}	16.58_{c}	2.32_{cde}	2.14_{bcd}	6.32_{abcde}	6.65_{abc}
BodinYawa	10.41_{l}	9.86_{o}	1.12_{abc}	1.08_{b}	1.05_{ab}	1.03_{ab}	16.43_{a}	15.49_{a}	2.47_{ghi}	2.34_{hij}	6.42_{defgh}	6.71_{bcde}
Idale	6.22_{a}	6.09_{a}	1.06_{ab}	1.02_{b}	1.01_{a}	1.01_{a}	17.91_{d}	16.68_{c}	2.38_{defg}	2.18_{cde}	6.21_{a}	6.64_{abc}
Igbodun	8.43_{e}	8.23_{ij}	1.13_{abc}	1.12_{b}	1.07_{ab}	1.04_{ab}	19.46_{j}	18.34_{k}	2.31_{bcde}	2.22_{def}	6.58_{hij}	6.77_{def}
Ilogun	9.61_{j}	8.47_{l}	1.08_{abc}	1.05_{b}	1.03_{ab}	1.01_{a}	18.21_{efg}	17.53_{gh}	2.17_{a}	2.0_{4a}	6.41_{defg}	6.93_{i}
Mejona	9.59_{j}	8.14_{h}	1.12_{abc}	1.10_{a}	1.08_{ab}	1.06_{ab}	19.06_{i}	18.12_{j}	2.20_{ab}	2.08_{ab}	6.22_{ab}	6.72_{bcde}
Oluwo	8.01_{d}	8.41_{l}	1.10_{abc}	1.05_{b}	1.23_{b}	1.21_{c}	18.19_{ef}	17.32_{f}	2.23_{abc}	2.05_{ab}	6.56_{ghij}	6.80_{efg}
Okorisan	7.69_{c}	7.32_{c}	1.02_{a}	1.00_{b}	1.14_{ab}	1.13_{abc}	18.09_{e}	17.46_{g}	2.42_{efgh}	2.24_{ef}	6.23_{abc}	6.68_{bcd}
Orita	6.66_{b}	7.12_{b}	1.08_{abc}	1.03_{b}	1.01_{a}	1.01_{a}	18.79_{h}	16.31_{b}	2.57_{ij}	2.41_{jk}	6.20_{a}	6.74_{cdef}
Orogoro	7.51_{c}	7.45_{d}	1.11_{abc}	1.10_{b}	1.06_{ab}	1.05_{ab}	19.14_{i}	18.83_{l}	2.35_{cdef}	2.26_{efg}	6.46_{efgh}	6.92_{i}

Local = Local drum kiln; Convect = Convective smoking kiln. Data are means of 3 replicates. Data with the same subscript are not significantly different at $p<0.05$.

standard for TVB-N which is 20 mgN/100 g for fresh fish. This suggests that the level of protein decomposition or breakdown in all the samples is low.

Trimethylamine (TMA) is a reduction product of trimethylamine oxide during spoilage and ammonia is mainly formed as a product of protein break-down. Trimethylamine (TMA) is one of the volatile amines plus ammonia which can be used as an index of spoilage (da Silva, 2002). In this study, the trimethylamine value (TMA) for fresh unsmoked bonga shad samples ranged from 1.93 –2.56 mgN/kg and 2.17 – 2.81mgN/kg and 2.05 – 2.61 mgN/kg for smoked bonga shad and control samples. The trimethylamine value (TMA) of 1.93 –2.56mgN/kg for fresh bonga shad samples and 2.17 – 2.81 mgN/kg and 2.05 – 2.61 mgN/kg for smoked bonga shad samples are within the range of $<$ 3 mgN/100g for fresh fish, $>$8 mgN/100g for spoiled fish and $\geq$ 5 mgN/100g for doubtful quality specified U.S.F.D.A (da Silva, et al., 2008). pH is the most critical factors affecting microbial growth and spoilage of foods. The pH value of fresh unsmoked bonga shad samples ranged from 6.81 – 7.24 and pH value of smoked bonga shad samples ranged from 6.20 – 6.71 and 6.56 – 6.93. The pH values of the fresh bonga shad samples was high compared to smoked bonga and control samples; this may be due to biochemical reac-tions and enzyme action as a result of delay in reaching the shore from the sea because most of the fishermen had no cooling system in their boats or canoes. However, the pH in fish tissues drops due to smoking.

Conclusion

This research work revealed that in contrary to protein, fat, and ash, the moisture content of fresh bonga shad samples decreased sharply after the smoking process.

This decrease had been found to be due to loss of water during smoking. Protein content in smoked bonga shad fish and control samples has been found to increase due to an increase in the dry matter content per unit weight following sample dehydration during smoking compared to fresh samples this is important- as so many people in West Africa eat smoked dried fish and are involved in producing smoked dried fish- has a high protein content lends support to national strategies of fight against poverty, food shortage and malnutrition. There was an inverse relationship between the moisture and protein content in the smoked bonga shad fish. There were increases in mineral content, ash, crude fibre, TBA and pH values. Convective smoking kiln did not significantly affect pH and proximate composition of smoked fish. It, however, significantly

($p<0.05$) reduced the moisture content of smoked fish and the quality indices such as FFA, TBA and PV.

Conflict of interests

The authors did not declare any conflict of interest.

REFERENCES

Abolagba OJ, Melle OO (2008). Chemical composition and keeping qualities of a Scaly Fish Tilapia (*Oreochromis niloticus*) Smoked with two Energy Sources. Afr. J. Gen. Agric. KLOBEX 4(2):113-117.

Ahmed A, Dodo A, Bouba AM, Clement S, Dzudie T (2011) Influence of Traditional Drying and Smoke-Drying on the Quality of Three Fish Species (*Tilapia nilotica, Silurus glanis* and *Arius parkii*) from Lagdo Lake, Cameroon. J. Anim. Vet. Adv. 10(3):301-306.

AOAC International (2000). Official Methods of Analysis, 20th ed. AOAC International, Gaithersburg, MD.

Beltran A, Moral A (1991). Changes in fatty acid composition of fresh and frozen sardine (*Sardinella pilchardus*) during smoking. Food Chem. 42:99-109.

Béné C, Heck S (2005). Fish and food security in Africa. World Fish Centre 28(3/4):8.

Calhoun CM, Gaebler DM, Mandingo RW (1999) Storage stability of ground pork containing meat from an advanced meat recovery system. J. Food Sci. 64:69-75.

Crapo C (2011) Smoking Fish at Home. The University of Alaska, Fairbanks Cooperative Extension Service programs. pp.1- 4

da Silva LVA (2002). Hazard Analysis Critical Control Point (HACCP), Microbial Safety, and Shelf Life of Smoked Blue Catfish *(Ictalurus furcatus*). A Master of Science in Food Science Thesis, the Graduate Faculty of the Louisiana State University and Agricultural and Mechanical College.

da Silva LVA, Prinyawiwatkul W, King JM, No HK, Bankston JD Jr, Ge B (2008). Effect of preservatives on microbial safety and quality of smoked blue catfish (Ictalurus furcatus) steaks during room-temperature storage. Food Microbiol. 25(8):958-63.

Eyo AA (2001). Fish processing technology in the tropics. University of Ilorin press, Nigeria. 403pp. ISSN 978 1770457

Goktepe I, Moody MW (1998). Effect of modified atmosphere package on the quality of smoked catfish. J. Muscle Foods 9:375-389.

Ikutegbe V, Sikoki F (2014). Microbiological and biochemical spoilage of smoke-dried fishes sold in West African open markets. Food Chem. 161:332-336.

Mafimisebi T (2012). Comparative analysis of fresh and dried fish consumption in rural and urban households in Ondo State, Nigeria. In: Visible possibilities: The economics of sustainable fisheries, aquaculture and seafood trade: Proceedings of the sixteenth biennial conference of the international institute of fisheries economics and trade, July 16–20, Dar es Salaam, Tanzania. Tanzania Proceedings. International Institute of Fisheries Economics & Trade (IIFET), Corvalli.

Nwachukwu VN, Madubuko CU (2013). Microflora Associated With Processing and Storage of the White Catfish (*Chrysichthys nigrodigitatus*). J. Fish. Aquat. Sci. 8:108-114.

Omojowo FS, Raji, A (2010). Assessment of the use of sodium benzoate on the safety and shelf life of smoked tilapia. New York Sci. J. 3:6.

Swastawati F (2008). Quality and safety of smoked catfish (Aries talassinus) using paddy chaff and coconut shell liquid smoke. J. Coast. Dev. 12(1):47-55.

2

Physical properties of high quality maize (Swam 1 variety) seeds (*Zea mays*) as affected by moisture levels

Sobukola, O. P.*, Kajihausa, O. E., Onwuka, V. I. and Esan, T. A.

Department of Food Science and Technology, University of Agriculture, P.M.B. 2240, Abeokuta, Ogun State, Nigeria.

A study was conducted to evaluate the physical properties that affect equipment design, processing, storage and transportation of high quality protein maize (SWAM 1) seeds as a function of moisture content varying from 9.38 to 32.7% (db). The length, width, thickness and the geometric diameter increased linearly from 9.80 to 10.55, 8.60 to 9.06, 4.00 to 4.75 and 6.85 to 7.69 mm, respectively. The sphericity index, seed volume, seed surface area and thousand seed mass also increased linearly from 69.89 to 72.85, 99.36 to 138.56 mm, 124.55 to 157.76 mm^2 and 240.36 to 303.71 g, respectively. Bulk density, true density and porosity decreased linearly from 1.109 to 1.057 g/m^3 1.365 to 1.176 g/m^3 and 18.75 to 10.12%, respectively. Static coefficient of friction was found to increase on plywood, galvanized iron, aluminium and stainless steel surfaces and it increased logarithmically from 0.55 to 0.91; 0.52 to 0.81, 0.49 to 0.70, and 0.46 to 0.68, respectively. Angle of repose increased linearly on plywood, galvanized iron, aluminium and stainless steel surfaces from 18.91 to 29.05, 17.00 to 26.96, 15.93 to 23.98 and 15.55 to 22.19°, respectively.

Key words: High quality maize (SWAM 1), moisture content, physical properties.

INTRODUCTION

Maize (*Zea mays* L.) which is also referred to as corn is the most important cereal next to wheat and rice and it is widely distributed throughout the world (Ahn, 1993). It belongs to the tribe Tripsceae and family Gramineae. The stalk, leaves, silk, cob and kernels have a commercial value, with that of the kernel being the largest (Kochhar, 1981). Many areas of Africa are very suited to the crop and it has largely replaced traditional grains such as sorghum and millet (Ahn, 1993). In industrialized countries, maize is largely used as livestock feed and as raw material for industrial products, while in developing countries, it is mainly used for human consumption. Maize is an important source of carbohydrates, protein, iron, vitamin B and minerals. Africans consume maize as a starchy base in a wide variety of porridges, pastes, grits and beer. Green maize (fresh on the cob) is eaten parched, baked, roasted or boiled; and is playing an important role in filling the hunger gap after the dry season. The major varieties of corn are pod, flint, dent, sweet, pop, flour and waxy (Staackmann and Matz, 1977) and is a major staple food in Nigeria where it is utilized by the food industry in the manufacture of weaning foods.

The knowledge of physical properties constitutes important and essential engineering data in the design of machines, storage structure and processes. The value of this basic information is not only important to engineers but also to food scientist, processors and other scientists who may want to exploit these properties and find new uses (Işik and Ünal, 2007). Determination of physical properties as a function of moisture content are important to design equipment for handling, conveying, separation, drying, aeration, storing and processing (Sobukola and Onwuka, 2011). The size and shape are, for instance, important in their separation from undesirable materials and in the development of sizing and grading machinery

*Corresponding author. E-mail: sobukolaop@unaab.edu.ng.

(Mohsenin, 1970). The shape of the material is important for an analytical prediction of its drying behavior (Işik and Ünal, 2007). Bulk density, true density and porosity are major consideration in designing the drying and aeration and storage systems, as these properties affect resistance to air flow of the mass (Amin et al., 2004).

Information on the coefficient of friction is used in designing equipment for solid flow, storage systems, seed harvesting and handling systems (Amin et al., 2004) while angle of repose is used to study the flowability of seeds and other powdery materials on various surfaces. The angle repose is important in designing equipment for mass flow and structures for storage (Isik and Unal, 2007). These properties have been studied for various crops such as African breadfruit seeds (Omobuwajo et al., 1999); sorrel seeds (Omobuwajo et al., 2000); bambara groundnut (Baryeh, 2001); groundnut kernels (Olajide and Igbeka, 2003); hazel nut (Ozdemir and Akinci, 2004); rape seed (Calisir et al., 2005); green wheat (Al-Mahansneh and Rababah, 2007); white speckled red kidney bean grains (Isik and Unal, 2007); and recently locust bean seed (Sobukola and Onwuka, 2011). High quality maize such as SWAM 1 is a useful industrial raw material especially in the production of weaning foods, bakery products and its starch is used for industrial application. Hence, there is need for detailed information on the physical properties of these maize seeds within the moisture content range for storage and processing. Therefore, the objective of this study was to investigate some physical properties of maize hybrid (SWAM 1) seeds at different moisture contents. The properties investigated include principal dimensions, effective diameter, one thousand seed mass, sphericity, bulk and true densities, porosity, angle of repose and coefficient of static friction.

MATERIALS AND METHODS

Materials and preliminary operation

The high quality maize seeds recently developed (SWAM 1 variety) were obtained from Research and Development Centre (RESDEC) now Institute of Food Security, Environmental Resources and Agricultural Research (IFSERAR) of the University of Agriculture, Abeokuta, Nigeria. They were cleaned manually to remove all unwanted materials such as chaffs, cobs, stones, insects and damaged or unhealthy seeds. The initial moisture content was determined by oven drying method at 105 ± 1°C for 24 h (AOAC, 1980) and was found to be 9.38% (db). In order to obtain maize seed of different moisture levels (9.38, 16.92, 19.93, 24.70, 27.89 and 32.70%) applicable during processing, the method reported by Sobukola and Onwuka (2011) was used. After conditioning, the seeds were kept in polyethylene bags and placed in the refrigerator at 5°C for equilibration to occur. Similar methods have been used by Karababa and Coşkuner (2007). Before starting a test, the required quantity of seed was taken from the refrigerator and allowed to equilibrate to the room temperature for about 2 h (Singh and Goswami, 1996; Cetin, 2007). All measurements (except principal dimensions) were obtained in 10 replicates at the six moisture levels, while principal dimensions were obtained at 100 replicates at the different moisture levels.

Determination of physical properties

The thousand seed mass of the seeds were determined at the six moisture levels. It was determined by direct weighing of clean 1000 seeds of SWAM 1 at each moisture level by means of an electronic balance (Mettler, model AE 240, Switzerland) with an accuracy of 0.001 g (Aviara et al., 1999).

The three linear dimensions namely length L, width W and thickness T, of the seeds at the different moisture levels were measured using a micrometer screw gauge (Model 436-25M, Starret, Brazil) to an accuracy of 0.01 mm. Effective diameter, (De), sphericity, (Φ), kernel volume (V) and surface area (S) were calculated as a function of the linear dimensions.

The effective diameter (De) of the seed at the six different moisture levels was calculated using the following formula (Jain and Bal, 1997):

$$De = (LWT)^{1/3} \tag{1}$$

According to Mohsenin (1970), the degree of sphericity (Φ) was calculated using the expression below:

$$\Phi = \frac{(LWT)^{1/3}100}{L} \tag{2}$$

Jain and Bal (1997) have reported that kernel volume, V and kernel surface area S can be estimated using relationship (3) and (4):

$$V = \frac{\pi B^2 L^2}{6(2L-3)} \tag{3}$$

$$S = \frac{\pi B L^2}{2L-B} \tag{4}$$

Where

$$B = (WT)^{0.5} \tag{5}$$

The bulk density is the ratio of the mass sample of the kernels to its total volume. It was determined by filling a 500 ml container with seeds from a height of 150 mm, striking off the top level without the seeds being compacted in any way and then the content was weighed (Singh and Goswami 1996). Bulk density was then calculated as a ratio between the kernel weight and the volume of the cylinder as shown in equation 6:

$$\rho_b = \frac{Kernelweight}{500ml} \tag{6}$$

The true density (ρ_t) was determined using toluene displacement method. Toluene was used in place of water because it is absorbed by seeds to a lesser extent. Also, its surface tension is low, so that it fills even shallow dips in a seed and its dissolution power is low (Aydin, 2002). 500 ml of toluene was placed in a 1000 ml graduated measuring cylinder. 300 g of seed were then immersed in toluene. The amount of displacement was recorded. True density (ρ_t) was obtained as the ratio of kernel weight to volume of displaced toluene (Aviara et al., 1999; Ogunjimi et al., 2002; Omobuwajo et al., 1999).

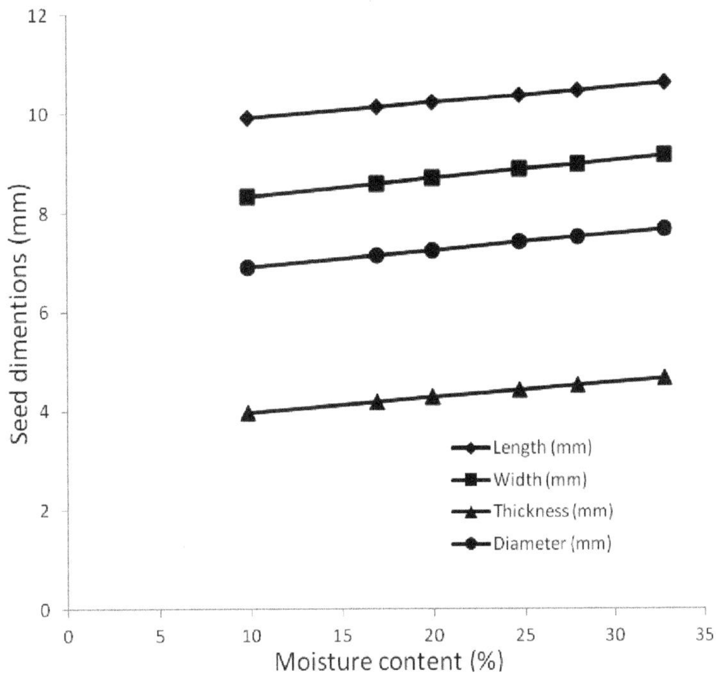

Figure 1. Effect of variation in moisture content on seed dimensions of SWAM 1 maize.

$$\rho_t = \frac{30g}{V_2 - V_1} \quad (7)$$

Where, V_2 = Final volume and V_1= initial volume.

Porosity (ε) was calculated using the relationship between bulk and true densities according to Mohsenin (1970) as follows:

$$\varepsilon = \frac{\rho_t - \rho_b}{\rho_t} \times 100 \quad (8)$$

Where, ρ_t = true density; ρ_b = bulk density.

The static coefficient of friction which is a dimensionless quantity used to calculate the force of friction was determined on four surfaces: plywood, galvanized iron, aluminium and stainless steel. These materials are common materials used for handling and processing of agricultural seeds and construction of storage and drying bins. This was measured using a hollow metal cylinder 50 mm in diameter and 50 mm high and opened at both ends. It was filled with seeds at the desired moisture content and placed on an adjustable tilting table such that the metal cylinder did not touch the table surface. The tilting surface was raised gradually by means of a screw device until the cylinder just started to slide down. The angle of the surface was read from a scale and coefficient of friction (μ) was taken as the tangent of this angle (Dutta et al., 1988; Joshi et al., 1988; Singh and Goswami, 1996).

$$\mu = \text{Tan } \alpha \quad (9)$$

Where μ = coefficient of the static friction and α = the angle of tilt in degrees.

The dynamic angle of repose or the emptying angle was determined on four surfaces namely plywood, galvanized iron, stainless steel and aluminium sheet. This was determined by using a topless and bottomless cylinder of 30 cm diameter and 50 cm height. The cylinder was placed at the centre of a raised circular plate having a diameter of 70 cm and was filled with the seeds. The cylinder was raised slowly until it formed a cone on a circular plate. The height of the cone was measured carefully with a meter rule while the base radius traced out was also measured with a divider (Maduka and Faborode, 1990; Owolarafe et al., 2007). The angle of repose (θ) was estimated using the relationship below:

$$\theta = \frac{Tan^{-1}(2H)}{D} \quad (10)$$

Where H = height of the cone; D = diameter of the cone; θ = angle of repose.

RESULTS AND DISCUSSION

The variations of axial dimensions (L, W and T) and De of SWAM 1 maize seeds with respect to moisture levels are shown in Figure 1. All dimensions increased linearly as moisture levels increased. The increase was from 9.8 to 10.5 mm (7.14% increase), 8.2 to 9.06 mm (10.49% increase), 4.00 to 4.75 mm (18.75% increase) and 6.85 to 7.69 mm (12.2% increases) for L, W, T and De, respectively when moisture levels increase from 9.38 to 32.70%. The linear relationship between the axial dimensions (L, W and T), De and moisture levels (M_c) are shown below:

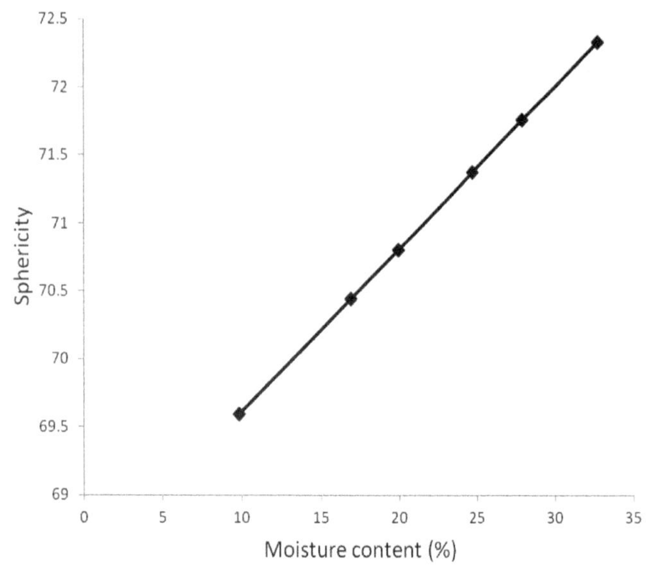

Figure 2. Effect of variation in moisture content on sphericity of SWAM 1 maize.

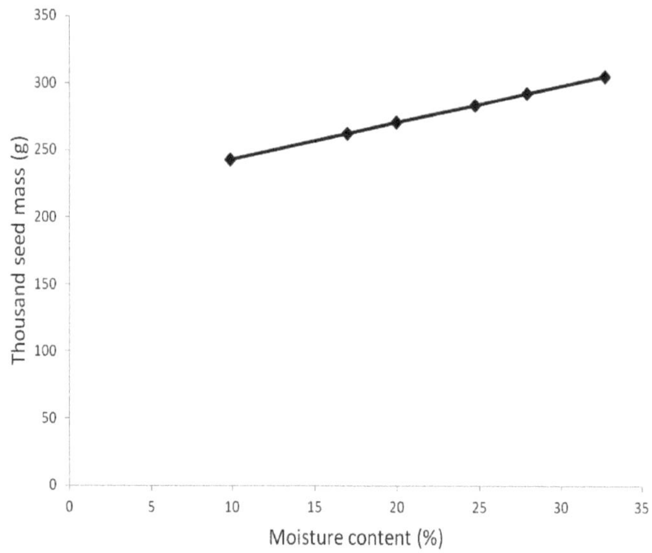

Figure 3. Effect of variation in moisture content on thousand seed mass (g) of SWAM 1 maize.

$$L = 0.0309\ M_c + 9.6105\ (R^2 = 0.9202) \tag{11}$$

$$W = 0.0364\ M_{c\,+}7.975\quad (R^2\ 0.8934) \tag{12}$$

$$T = 0.0301\ M_c + 3.6732\ (R^2 = 0.8896) \tag{13}$$

$$De = 0.0342\ M_c + 6.5603\qquad (R^2 = 0.923) \tag{14}$$

Where, *Mc* is the moisture level (%)

The increase in linear dimension may be due to the fact that as moisture content increases, these seeds tend to swell due to moisture absorption. The effective diameter of maize seeds was found to be higher than the thickness of the seeds. Similar findings were reported for millet (Jain and Bal, 1997), sakiz faba beans (Haciseferoğullari et al., 2003); popcorn kernels (Karababa et al., 2003) and sweet corn (Karababa and Coşkuner, 2007).

The sphericity index (Φ), variation with moisture levels of maize seed is shown in Figure 2. The values of Φ for different moisture levels increased linearly from 69.898 to 72.852 (4.24% increase) as moisture level increased from 9.38 to 32.70%. The Φ of maize is higher than the values reported for grain wheat (Karababa and Coşkuner, 2007); barbunia bean seed (Cetin, 2007) and safflower (Baumler et al., 2006). Such high values suggested that these high quality maize might be expected to roll rather slide on the surface and this is a property quite important in the design of grain hoppers (Al-Mahasneh and Rababah, 2007). It indicated maize seeds can be treated as sphere for an analytical prediction of their drying behavior. The mathematical relationship between moisture level and Φ is given below:

$$\Phi = 0.1197M_c + 68.418\ (R^2 = 0.8432) \tag{15}$$

The one thousand maize seed mass was found to increase linearly from 240.36 to 303.71 g (26.36% increases) as moisture level increased from 9.38 to 32.70% as shown in Figure 3. The linear relationship between a thousand mass and moisture level is expressed as Equation 16.

$$M_{1000} = 2.7376M_c + 216.04\ (R^2 = 0.9845) \tag{16}$$

Deshpande et al. (1993), Öğüt (1998), Baryeh (2001), Mahapatra et al. (2002), Dursun and Dursun (2005) and Cetin (2007) have also found similar relationship to be true for soybeans, white lupin, bambara groundnut, sorghum, caper seed and barbunia beans, respectively.

The variation of kernel volume (V) and surface area (S) with seed moisture content for SWAM 1 seeds are displayed in Figures 4 and 5, respectively. The volume of seeds increased linearly from 99.36 to 158.56 mm^3 (39.45% increase) with increase in moisture level from 9.38 to 32.70%. The linear relationship between V and moisture content is shown in Equation 17.

$$V = 1.6076\ M_c + 84.976\ (R^2 = 0.9729) \tag{17}$$

Similar results have been reported for popcorn (Karababa, 2006); safflower (Baumler et al., 2006); millet (Baryeh, 2002); rape seed (Calisir et al., 2005); sweet corn (Karababa and Coşkuner, 2007) and white lupin (Öğüt, 1998). The kernel surface area increased linearly with increase in moisture content of the seeds. There was an increase from 124.58 to 157.76mm^2 (26.60% increase)

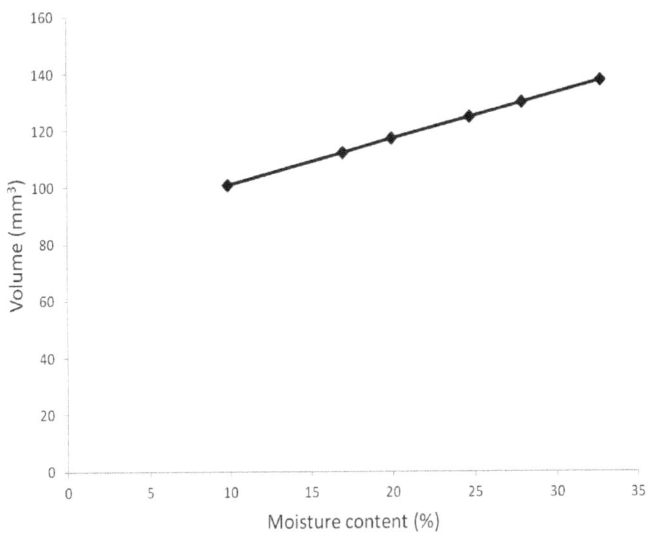

Figure 4. Effect of variation in moisture content on volume of SWAM 1 maize.

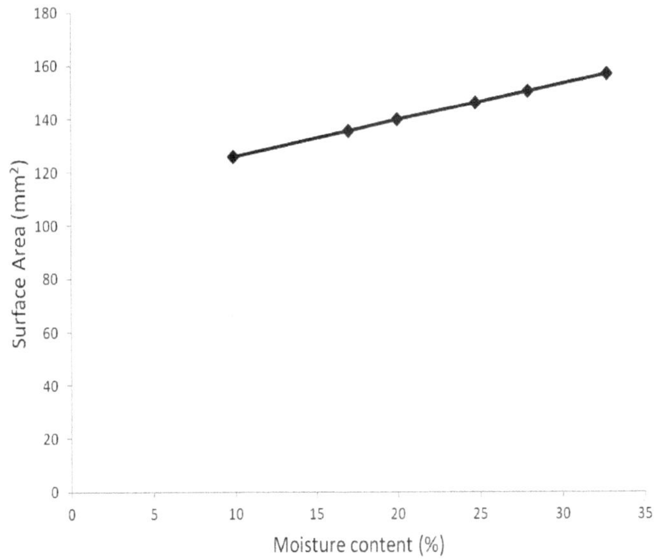

Figure 5. Effect of variation in moisture content on surface area of SWAM 1 maize.

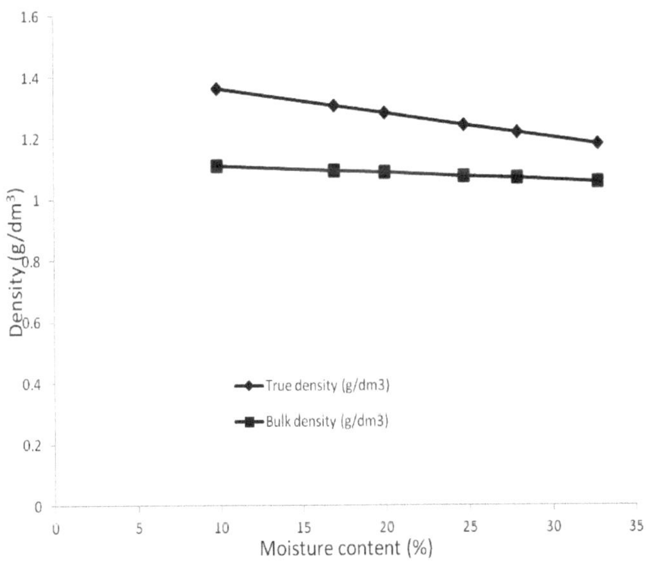

Figure 6. Effect of variation in moisture content on density of SWAM 1 maize.

as moisture content increased from 9.38 to 32.70%. The linear relationship between the surface area and the moisture content of maize seeds can be represented by the equation below:

$$S = 1.3612\ M_c + 112.58\ \ (R^2 = 0.9731) \qquad (18)$$

Similar relationship was reported for popcorn (Karababa, 2006); sweet corn (Karababa and Coskuner, 2007) and millet (Baryeh, 2002). The ratio between volume and surface area is usually called the characteristic length. Characteristic length has important role in the case of irregular shape objects. Some of its applications include determination of projected area of particles moving in turbulent air stream, which can be useful in designing grain cleaners, separators and pneumatic conveyors. As the ratio between surface area and volume increases, the rate of heat and mass transfer from kernel increases, which affects several unit operations such as drying, cooling and heating.

The variations of bulk density, (ρ_b) and true density (ρ_t) of maize seeds with moisture levels are shown in Figure 6. The bulk densities were observed to decrease linearly from 1109 to 1057 kg/mm^3 (4.69% increase) as moisture level increased from 9.38 to 32.70%. The linear relationship between bulk density and moisture level can be expressed by Equation 19.

$$\rho_b = -0.0023\ M_c + 1.1328\ (R^2 = 0.9425) \qquad (19)$$

This decrease was due to the higher rate of increase in volume relative to the increase in weight. Similar relationships have been reported for white lupin (Öğüt, 1998), gram seed (Chowdhury et al., 2001), millet (Baryeh, 2002), chicken pea (Konaki et al., 2002), rape seed (Calisir et al., 2005), safflower (Baumler et al., 2006) and sweet corn (Karababa and Coşkuner, 2007). However, linear increasing trends were reported for sunflower (Gupta and Das, 1997), coffee (Chandrasekar and Viswanathan, 1999), cashew nut (Balasubramanian, 2001) and pistachio (Kashaninejad et al., 2006).

The true density of maize decreased linearly from 1365 to 1176 kg/m^3 (13.85% decrease) with increasing moisture

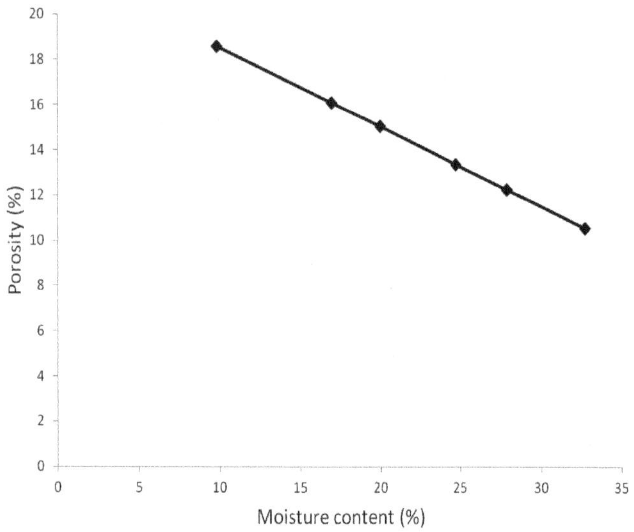

Figure 7. Effect of variation in moisture content on porosity of SWAM 1 maize.

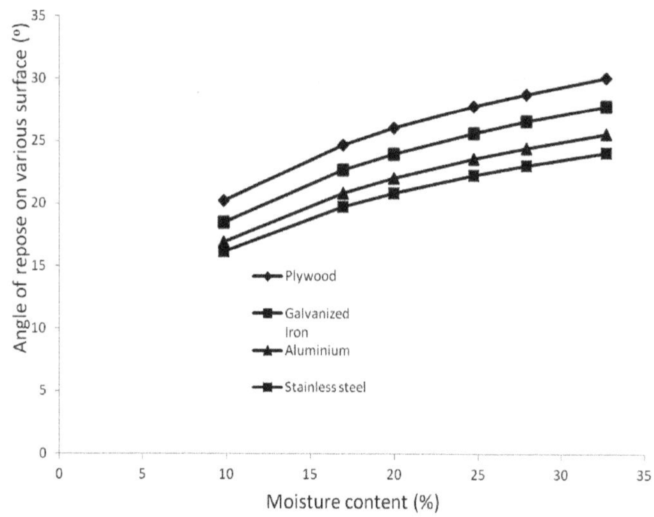

Figure 8. Effect of variation in moisture content on angle of repose on various surfaces of SWAM 1 maize.

content from 9.38 to 32.70%. The linear relationship can be represented by Equation 20:

$$\rho_t = -0.0079\ M_c + 1.4384\ (R^2 = 0.9944) \qquad (20)$$

This decrease in true density may be due to higher rate of increase in seed volume relative to weight increase. Similar results were reported for gram seed (Chowdhury et al., 2001), chicken pea (Konaki et al., 2002), hazel nut (Aydin, 2002), popcorn (Karababa, 2006) and sweet corn (Karababa and Coşkuner, 2007). However, a linear increasing trend was reported for sunflower (Gupta and Das, 1997), white lupin (Öğut, 1998), coffee (Chandrasekar and Viswanathan, 1999), cashew nut (Balasubramanian, 2001), millet (Baryeh, 2002) and cacao (Bart-Plange and Baryeh, 2003). Thus, these seeds have lower weight increase in comparison with volume increase as their moisture content increases. At all moisture levels, it was observed that true density was greater than bulk density. Similar relationships have been reported for sweet corn (Karababa and Coşkuner, 2007) and barbunia beans (Cetin, 2007).

Porosity (ε) was calculated as a function of the bulk density and true density of the seeds. The porosity was found to decrease linearly from 18.75 to 10.12% (46.03%) with increase in moisture level from 9.38 to 32.70% as shown in Figure 7. A linear relationship between ε and varying moisture level was obtained by Equation 21.

$$\varepsilon = -0.3511\ M_c + 22.042\ (R^2 = 0.9783) \qquad (21)$$

Higher porosity provides better aeration and water vapour diffusion during deep bed drying. Similar trend was reported for hazel nut (Aydin, 2002), gram (Dutta et al., 1988), sunflower (Gupta and Das, 1997) and popcorn (Karababa, 2006). However, porosity was reported to increase with increase in moisture content for millet (Baryeh, 2002), rape seed (Calisir et al., 2005) and soybeans (Desphande et al., 1993).

The angle of repose, (θ), of maize seeds increased logarithmically from 18.91 to 29.05° (53.62% increase); 17.00 to 29.96° (58.59% increase); 15.95 to 23.98° (50.53% increase) and 15.15 to 22.19° (39.87% increase) on plywood, galvanized iron, aluminium and stainless steel sheet surfaces with increase in moisture content from 9.38 to 32.70% (Figure 8). The logarithmic relationship between θ and moisture content on plywood (PW), galvanized iron (GFe), aluminium (Al) and stainless steel sheet (SS) can be represented by the following equations:

$$\theta_{PW} = 8.2006\ \ln(M_{c)} + 1.5173\ (R^2 = 0.8967) \qquad (22)$$

$$\theta_{GFe} = 7.732\ \ln(M_{c)} + 0.8625\ (R^2 = 0.8403) \qquad (23)$$

$$\theta_{Al} = 7.2013\ \ln(M_{c)} + 0.5049\ (R^2 = 0.9141) \qquad (24)$$

$$\theta_{SS} = 6.6015\ \ln\ (M_{c)} + 1.109\ (R^2 = 0.8991) \qquad (25)$$

At higher moisture content, seeds tend to stick together resulting in better stability and less flowability and hence, resulting in increase in the angle of repose. Similar trends have been observed for chick pea (Konak et al., 2002); okra (Sahoo and Srivastava, 2002); green gram (Nimkar and Chattopadhyay, 2001); canola and sunflower meal pellets (White and Jayas, 2001) and arecanut kernels (Kaleemullah and Gunasekar, 2002). At all moisture levels,

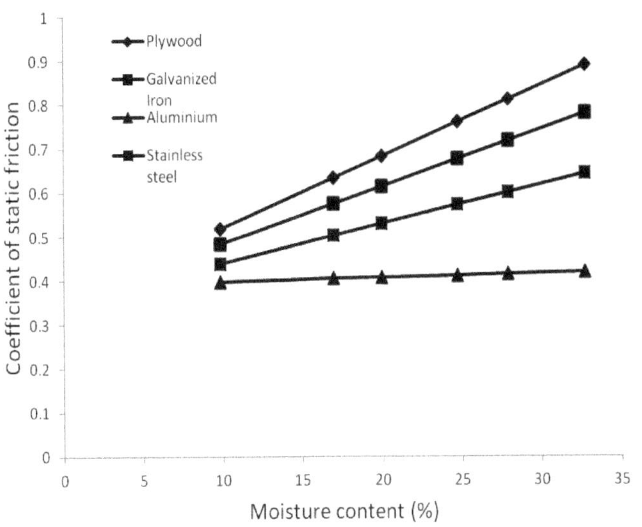

Figure 9. Effect of variation in moisture content on coefficient of static friction of SWAM 1 maize.

the highest angle of repose was on plywood followed by galvanized iron and then aluminium sheets. The least angle of repose was on stainless steel and this may be because of smoother and more polished surface of the stainless steel sheet relative to other materials used, thus causing the seeds to form flatter piles. Consequently, the data on frictional properties will be useful in hopper design for gravity flow, since the angle of inclination of the hopper walls should be greater than angle of repose to ensure continuous flow of the material.

Variations of static coefficient of friction (µ) of high quality maize seeds with moisture level for different surfaces are displayed in Figure 9. The static coefficient of friction on the four surfaces (plywood, galvanized iron, aluminium and stainless steel) increased linearly with increase in moisture level. It increased from 0.55 to 0.91 (65.45% increase); 0.52 to 0.81 (55.78% increase); 0.49 to 0.70 (42.86% increase) and 0.48 to 0.68 (41.67% increase) on plywood, galvanized iron, aluminium and stainless steel surfaces, respectively as moisture level increased from 9.38 to 32.70%. The mathematical relationship between µ and moisture level on plywood (PW), galvanized iron (GFe), aluminium (Al) and stainless steel sheets (SS) can be represented by the following equations:

$$\mu_{PW} = 0.0162M_c + 0.3599 \ (R^2 = 0.9580) \quad (26)$$

$$\mu_{GFe} = 0.0131M_c + 0.3549 \ (R^2 = 0.9135) \quad (27)$$

$$\mu_{Al} = 0.00094M_c + 0.3885 \ (R^2 = 0.9619) \quad (28)$$

$$\mu_{SS} = 0.009M_c + 0.3500 \ (R^2 = 0.8713) \quad (29)$$

The increase in µ with increase in moisture level may be due to the adhesion between the seed and the material surfaces at higher moisture levels. At all moisture levels, the least static coefficient was on stainless steel. This may be because of the smoother and more polished surface of the stainless steel relative to other materials used. The static coefficient of friction at all moisture level was highest on plywood followed by galvanized iron and aluminium. This can be attributed to the surface roughness which is largest on plywood. This order has been reported for pumpkin seeds (Joshi et al., 1988); cumin seeds (Singh and Goswami, 1996); sunflower (Gupta and Das, 1997), bambara groundnuts (Baryeh, 2001) and sweet corn (Karababa and Coşkuner, 2007). However, Amin et al. (2004) reported that no variation existed between plywood and galvanized iron for pulse grains. Tunde-Akintunde and Olajide (2005) also reported low static coefficient of friction values on plywood for soybean as compared to results obtained for maize and locust bean seeds in this work. Tsang-Mui-Chung et al. (1984), Dutta et al. (1988) for gram seed, Joshi et al. (1988) for pumpkin seeds, Carman (1996) for lentil seeds, Öğüt (1998) for white lupin and Aydin (2002) for hazel nuts reported that as moisture level increased, static coefficient of friction increased. This is because, the seeds may become rough and static coefficient of friction is increased.

Conclusions

Axial dimensions (length, width and thickness), geometric mean diameter, mass of 1000 seeds, surface area, volume and sphericity of high quality maize seeds (SWAM 1) was observed to increase linearly with increase in moisture levels. The bulk density, true density and porosity however decreased linearly with increase in moisture levels. Static coefficient of friction of the seeds on all surfaces increased linearly, while the angle of repose increased logarithmically with increase in moisture contents.

ACKNOWLEDGEMENT

We are indebted to the Institute of Food Security, Environmental Resources and Agricultural Research (IFSERAR) for the supply of the newly developed hybrid maize.

REFERENCES

Ahn PM (1993). Tropical soils and fertilizer use: Intermediate Tropical Agricultural series. Longman Nigeria Ltd. p. 36.

Al-Mahasneh MA, Rababah TM (2007). Effect of moisture content on some physical properties of green wheat. J. Food Eng. 79:1467 - 1473.

Amin MN, Hossain MA, Roy KC (2004). Effect of moisture content on some physical properties of lentil seeds. J. Food Eng. 65: 83-87.

AOAC (1980). Official methods of analysis (13th ed.). Arlington, VA:

Association of Official Analytical Chemists.
Aviara NA, Gwandzang MI, Haque MA (1999). Physical properties of guna seeds. J. Agric. Eng. Res. 73(2):105-111.
Aydin C (2002). Physical properties of hazel nuts. Bios. Eng. 82:297-303.
Balasubramanian D (2001). Physical properties of raw cashew nut. J. Agric. Eng. Res. 78:291-297.
Bart-Plange A, Baryeh EA (2003). The physical properties of category B cocoa beans. J. Food Eng. 60:219-227.
Baryeh EA (2001). Physical properties of Bambara groundnuts. J. Food Eng. 47:321-326.
Baryeh EA (2002). Physical properties of millet. J. Food Eng. 69(10):61-66.
Baumler E, Cuniberti A, Nolasco SM, Riccobene IC (2006). Moisture dependent physical and compression properties of safflower seed. J. Food Eng. 72:134-140.
Calisir S, Marakoglu T, Ogut H, Ozturk O (2005). Physical properties of rapeseed (*Brassica napus oleifera* L.). J. Food Eng. 69:61-66.
Carman K (1996). Some physical properties of lentil seeds. J. Agric. Eng. Res. 63(2):87- 92.
Cetin M (2007). Physical properties of barbunia beans (*Phaseolus vulgaris* L. cv. 'Barbunis') seed. J. Food Eng. 80: 353-358.
Chandrasekar V, Viswanathan R (1999). Physical and thermal properties of coffee. J. Agric. Eng. Res. 73: 227-234.
Chowdhury MMI, Sarker RI, Bala BK, Hossain MA (2001). Physical properties of gram as a function of moisture content. Inter. J. Food Prop. 4: 297-310.
Deshpande SD, Bal S, Ojha TP (1993). Physical properties of soybean. J. Agric. Eng. Res. 56: 89 - 98.
Dursun E, Dursun I (2005). Some physical properties of caper seed. Biosyst. Eng. 92(2):237-245.
Dutta SK, Nema VK, Bhardwaj RJ (1988). Physical properties of gram. J. Agric. Eng. Res.39:259-268.
Gupta RK, Das SK (1997). Physical properties of sunflower seeds. J. Agric. Eng. Res. 66:1-8.
Haciseferoğullari H, Gezer I, Bahtiyarca Y, Mengeş HO (2003). Determination of some chemical and physical properties of Sikiz faba bean (*Vicia faba* L. Var Major). J. Food Eng. 60:475-479.
Isik E, Ünal H (2007). Moisture-dependent physical properties of white speckled red kidney bean grains. J. Food Eng. 82: 209-216.
Jain RK, Bal S (1997). Physical properties of pearl millet. J. Agric. Eng. Res. 66:85-91.
Joshi DC, Das SK, Mukherjee RK (1988). Physical properties of gram. J. Agric. Eng. Res. 39:259-268.
Kaleemullah S, Gunasekar JJ (2002). Moisture dependent physical properties of arecanuts. Biosys. Eng. 82(3):331-338.
Karababa E (2006). Physical properties of popcorn kernels. J. Food Eng. 72:100-107.
Karababa E, Coşkuner Y (2007). Moisture content dependent physical properties of dry sweet corn kernels. Int. J. Food Prop. 10:549-560.
Kashaninejad MA, Mortazavi A, Safekordi A, Tabil LG (2006). Some physical properties of pistachio (*Pistacia vera* L.) nuts and its kernel. J. Food Eng. 72:30-38.
Kochhar SL (1981). Tropical Crops: A textbook of economic botany. Macmillan London. pp. 91-100.
Konak M, Carman K, Aydin C (2002). Physical properties of chick pea seeds. Biosys. Eng. 82(1):73-78.
Maduka JN, Faborode MO (1990). Some physical properties of cocoa pods in relation to processing. Ife J. Tech. 2:1-7.
Mahapatra AK, Patrick C, Diau P (2002). Effect of moisture content on some physical properties of sorghum (Mahube variety) seeds. Agric. Eng. J. 11(1):1-10.
Mohsenin NN (1970). Physical properties of plant and animal material. New York: Gordon and Breach Science Publishers.
Nimkar PM, Chattopadhyay PK (2001). Some physical properties of green gram. J. Agric. Eng. Res. 80(2):183-189.
Ogunjimi LAO, Aviara NA, Aregbesola OA (2002). Some engineering properties of locust bean seed. J. Food Eng. 55(2):95-99.
Öğüt H (1998). Some physical properties of white lupin. J. Agric. Eng. Res. 56:273-277.
Olajide JO, Igbeka JC (2003). Some physical properties of groundnut kernels. J. Food Eng. 58:201-204.
Omobuwajo TO, Akande EA, Sanni LA (1999). Selected physical properties of guna seeds. J. Agric. Eng. Res. 73(2): 105-111.
Omobuwajo TO, Sanni LA, Balami YA (2000). Physical properties of sorrel (*Hibiscus sabdariffa*) seeds. J. Food Eng. 45(1):37-41.
Owolarafe OK, Olabige MT, Faborode MO (2007). Physical and mechanical properties of two varieties of fresh oil palm fruits. J. Food Eng. 78:1228-1232.
Ozdemir F, Akinci I (2004). Physical and nutritional properties of four major commercial Turkish hazelnut varieties. J. Food Eng. 63:341-347.
Sahoo PK, Srivastava AP (2002). Physical properties of okra seeds. Biosys. Eng. 83(4):441-448.
Singh KK, Goswami TK (1996). Physical properties of cumin seed. J. Agric. Eng. Res. 64(2):93-98.
Sobukola OP, Onwuka VI (2011). Effect of moisture content on some physical properties of locust bean seed (P*arkia fillicoidea* L.). J. Food Process Eng. 30:1946-1964.
Staackmann JM, Matz SA (1977). Elemnts of Food Technology. In: (Dresrosier, N.M. ed.). AVI publishing company Inc., Westport, Connecticut. pp. 148-154.
Tsang-Mui-Chung M, Verma LR, Wright ME (1984). A device for friction measurement of grains. Trans. ASAE 27:1938-1941.
Tunde- Akintunde TY, Olajide JO (2005). Mass-volume-area related and mechanical properties of soybean as a function of moisture and variety. Int. J. Food Prop. 8:449-456.
White NDG, Jayas DS (2001). Physical properties of canola and sunflower meal pellets. Canadian Biosys. Eng. 43:349-352.

Analysis of quality attributes of *Hibiscus sabdariffa* (*zobo*) drinks blended with aqueous extract of ginger and garlic

Adesokan I. A., Abiola O. P., Adigun M. O. and Anifowose O. A.

Department of Science Laboratory Technology, Faculty of Science, The Polytechnic, Ibadan, Oyo State, Nigeria.

Various types of highly valued food and medicinal products are produced from *Hibiscus sabdariffa* in different parts of the world. In Nigeria, the calyces of this plant are processed into a refreshing non-alcoholic beverage known as *zobo*. In this work, *zobo* drink was prepared using hot water extraction method and the juice obtained was blended separately with aqueous extracts of ginger and garlic. Physico-chemical, microbiological and sensory properties of the various blends of *zobo* were investigated. The results of this study showed that the *zobo* samples contained different concentrations of ascorbic acid which ranged between 22.5 and 35.8 mg/100 g. The pH ranged between 3.94 and 7.67, while the total sugar was between 2.5 and 3.56 mg/100 g. The samples also contained varying concentrations of Mg^{2+}, K^{+}, Ca^{2+}, Fe^{2+}, Na^{+} and Zn^{2+}. The total viable counts of the samples during storage showed that the microbial loads increased steadily and ranged between 1.2×10^{3} and 3.7×10^{7} (CFU/ml). The samples containing ginger and garlic had the lowest microbial count of 2.7×10^{4} CFU/ml on the seventh day of storage, while the sample containing no extract had the highest microbial load of 3.7×10^{7} CFU/ml within the same period of time. The sensory attributes of *zobo* beverage degenerate during the seven days of storage. However, the samples containing ginger and/or garlic were better rated in terms of the parameter tested. The results of this study indicate that incorporation of ginger and garlic extracts into *zobo* drink could be an effective means of improving quality attributes of this drink.

Key words: *Hibiscus sabdariffa*, sensory properties, zobo drink, microbial count.

INTRODUCTION

Hibiscus sabdarffa belongs to the super order Malvaceae and it is believed to originate from East Africa (Ilondu and Iloh, 2007). *H. sabdariffa* plants are cultivated and consumed as vegetable and tea, whereas other *Hibiscus* varieties are planted for the fibres they produce. It is called different names like Roselle and Sorrel in English and it is locally called *zobo* and *Isapa* in Nigeria (Adebayo-Taye and Samuel, 2000).

Many parts of Roselle including seeds, leaves, fruits and roots, are used in various foods. Among them, the fleshy red calyces are the most popular (Yadeng et al., 2005). Roselle is used in many folk medicines. In Nigeria, the dry red calyces are processed into a refreshing non-alcoholic beverage known as *zobo*. Despite the fact that the popularity of *zobo* juice is increasing, one of its greatest limitation for large scale production is that it has a very short shelf life of 24 h if not refrigerated (Omemu et al., 2006). Therefore, there is urgent need to explore various preservation methods that could be employed to extend the shelf life of this product.

Dougheri et al. (2007) employed some chemical preservatives to improve the shelf life of *zobo* drink. They reported that only samples treated with benzoic acid remained organoleptically attractive after 14 days of storage. However, the problem with the use of chemical preservatives in food is that they tend to have adverse effects on the health of consumers (Adesokan et al., 2010). Therefore, naturally occurring plant extracts with proven antimicrobial properties will be preferred in food preservation (Kolapo et al., 2007). The use of natural preservatives has become more popular as compared to synthetic antimicrobials and antioxidants (Aliu et al., 2007).

Ginger (*Zingiber officinale*) and garlic (*Allium sativum*) are two plants traditionally used as spices in food preparation but have been demonstrated to have both antioxidant and antimicrobial activities (Kolapo et al., 2007). Therefore, the objective of this present study was to evaluate the physico-chemical and microbiological qualities of *zobo* drinks blended with ginger and garlic extracts.

MATERIALS AND METHODS

Collection of samples

Dry calyces of *H. sabdariffa* and fresh wet form of ginger rhizomes and garlic bulbs were obtained from local Bodija market in Ibadan, Nigeria. The samples were manually cleaned by handpicking stones and other unwanted debris. The samples were then thoroughly washed separately using sterile de-ionized water and allowed to air dry under the sun (40°C).

Preparation of ginger and garlic extracts

One hundred grams of each of the spices were separately chopped into small pieces with a clean stainless steel knife. The chopped spices were then blended into 50 ml de-ionized water with a Kenwood blender having stainless steel blades until smooth pastes were obtained. The pastes were diluted further with 100 ml de-ionized water and filtered using a clean Muslin cloth. The resulting extracts were stored in clean bottles inside a refrigerator (5°C) until needed for use in *zobo* preparation.

Preparation of *zobo* drinks

For hot extraction, 175 g of already cleaned calyces of *H. sabdariffa* was added to 300 ml hot boiling water and was left to stand for 15 min. The hot, red-coloured aqueous extract was filtered with the white muslin cloth into a plastic bowl and tightly covered. The resulting *zobo* juice was separately blended with ginger, garlic, ginger/garlic mixtures at different concentration of 2.50 and 10%. The unblended *zobo* juice served as control.

Physico-chemical analysis

The pH of the *zobo* samples was measured by using a standardized pH meter (Micron pH meter). Appropriate procedure of A.O.A.C. (1990) was used to determine the metallic ion contents. Flame photometry was used to determined sodium, potassium and iron and Buck scientific atomic absorption spectrophotometer (model 200A) was used to determine calcium, magnesium and zinc. Crude protein and total sugar content was also estimated (A.O.A.C., 1990).

Microbiological analysis

The various *zobo* drink samples were serially diluted in sterile distilled water and appropriate dilutions were plated on nutrient agar, potato dextrose agar (PDA) and MacConkey agar for total viable, fungal and coliform counts, respectively. The nutrient and MacConkey agar plates were incubated at 35°C for 24 h, while the PDA were incubated at 28°C for 72 h and were supplemented with streptomycin to inhibit the growth of bacteria (Adesokan, 2005).

Sensory analysis

The blended *zobo* samples were analyzed by a 10 member panel that are familiar with the product which comprises of the students of the Polytechnic, Ibadan. The samples were scored based on a nine point hedonic scale where 1 equals extremely like and 9 equals extremely dislike. The data obtained were subjected to analysis of variance (ANOVA) and the means were separated by Duncan multiple range test (Kolapo et al., 2007).

RESULTS

The physico-chemical analyses of the various blends of *zobo* beverages are presented in Table 1. The result showed that the ascorbic acid content (mg/100 g) increased gradually and its values ranged between 22.5 in pure *zobo* juice (Sample A) and 35.8 in *zobo*/ginger/garlic mixture (Sample D). The pH values increased steadily and ranged between 3.94 in sample A and 7.67 in sample D. The total sugar also ranged between 2.5 and 3.56% in sample D. The result also indicated that the *zobo* samples blended with spices had higher crude protein contents than non-spiced *zobo* beverage and it values ranged between 9.1 (Sample A) and 62% (Sample D).

The various blends of *zobo* drink contained varying concentrations of Mg^{2+}, K^{+}, Ca^{2+}, Fe^{2+}, Na^{+} and Zn^{2+}. Another finding of this study is that the *zobo* beverages extracted with hot water had higher amount of nutrients than the sample extracted with cold water (result not shown).

The total viable counts (TVC) (cfu/ml) of the *zobo* samples are presented in Table 2. This result showed that there was a progressive increase in TVC with storage time. However, the *zobo* samples containing ginger or garlic had lower ($P \leq 0.05$) TVC than *zobo* samples containing no spice. Sample A had the highest count of 3.7×10^7 cfu/ml, while sample D had the lowest count of 2.6×10^4 cfu/ml by the seventh day of storage. The organoleptic properties of the beverages are presented in Table 3 and the result indicated that sample D had the highest overall acceptability of 2.3, while sample A had the lowest value of 1.9.

Table 1. Physico-chemical properties of different blends of *zobo* drink.

Parameter	Sample A*	Sample B	Sample C	Sample D
Ascorbic acid (mg/100 g)	22.5**	25.0	23.5	35.8
pH	3.94	6.63	7.05	7.67
Total sugar (mg/100 ml)	2.5	3.2	2.95	3.56
Crude protein (%)	9.1	57	40	62
Mg^{2+} (mg/100 ml)	243	276	203	225
K^{+} (mg/10 0ml)	2410	2615	2673	2450
Ca^{2+} (mg/100 ml)	1756	2100	1905	1895
Fe^{2+} (mg/100 ml)	39.2	32.1	28.9	35.3
Na^{+} (mg/100 ml)	7.2	6.8	8.7	6.9
Zn^{2+} (mg/100 ml)	7.7	12.5	9.5	11.2

*Sample A, Unspiced *zobo* beverage; sample B, *zobo* with ginger; sample C, *zobo* with garlic; sample D, *zobo* with ginger and garlic. **Values are means of three determinations.

Table 2. Total viable counts (cfu/ml) of the different *zobo* blends during storage.

Samples	Day 1	Day 4	Day 7
Sample A*	$2.6x10^{4**}$	$3.5x10^{5}$	$3.7x10^{7}$
Sample B	$3.0x10^{3}$	$3.7x10^{3}$	$5.5x10^{6}$
Sample C	$1.5x10^{3}$	$1.9x10^{3}$	$2.5x10^{5}$
Sample D	$1.2x10^{3}$	$1.3x10^{3}$	$2.6x10^{4}$

*Sample codes are as stated in Table 1; **values are means of three replicates.

Table 3. Sensory evaluation of different blends of *zobo* drink.

Sample	Appearance		Taste		Texture		Colour		Aroma		Overall acceptability	
	D1	D7	D1	D7	D1	D7	D1	D7	D1	D7	D1	D7
Sample A	5.0*	2.3	4.5	2.5	4.5	1.0	5.2	2.2	5.3	1.5	4.9	1.9
Sample B	5.0	2.1	5.5	2.6	4.4	1.4	5.0	2.4	5.7	2.1	5.12	2.12
Sample C	5.0	2.4	3.3	1.3	4.3	2.3	4.9	2.6	3.1	1.9	4.1	2.1
Sample D	5.0	2.6	4.2	2.2	4.1	2.1	4.7	2.9	3.4	1.7	4.3	2.3

*Values are means of three replicates.

DISCUSSION

This work demonstrated that the ascorbic acid contents of *zobo* drinks supplemented with garlic and ginger are higher than that of ordinary *zobo* juice. This means that the vitamin C status of *zobo* beverages can thus be improved by blending with these spices. Babalola et al. (2001) reported the presence of ascorbic acid in various varieties of *H. sabdariffa*. The pH of the *zobo* samples blended with ginger and garlic are higher than that of the *zobo* juice which contain no extract. This implies that *zobo* juice is slightly acidic and should not be consumed without snack or on an empty stomach. A recent study revealed that the pH of the fruit flavoured *zobo* drinks had a low value which ranged between pH 2.19 and 3.62 (Fasoyiro et al., 2005).

The crude protein content of non-spiced *zobo* juice was 9.1% but higher values were obtained for the samples containing spices. This means that the spices are rich in protein. The crude protein value is similar to the one obtained by Olayemi et al. (2011). The *zobo* drinks evaluated in this work contained various metallic ions like Mg^{2+}, K^{+}, Ca^{2+}, Fe^{2+}, Na^{+} and Zn^{2+}. Olayemi et al. (2011) also reported various concentrations of these metallic ions in three different varieties of *zobo* subjected to the same preparation condition. In addition to this Builder et al. (2010), reported the presence of certain phytochemicals like alkaloids, saponin, tannin, phenols, sterols, etc.

The spices used in this study were able to reduce the microbial loads of *zobo* samples during the seven days storage. However, Ogiehor et al. (2008) extended the shelf life of *zobo* for 42 days using 0.2% ginger extract and refrigeration. This method of storage under refrigeration is not practicable in present day Nigeria because of epileptic public power supply and the use of power generators is not economical. The sensory attributes of the *zobo* samples containing spices were also enhanced. The use of extracts of ginger and garlic in *zobo* preparation should therefore be encouraged as the sensory properties and shelf life of resulting product will be enhanced.

REFERENCES

A.O.A.C. (1990). Official methods of Analysis (15th Ed.). Association of Official Analytical Chemist, Washington D.C.

Adebayo-Tayo BC, Samuel UA (2000). Microbial quality and proximate composition of dried *Hibiscus sabdariffa* calyces in Uyo, Eastern Nigeria. Malayas. J. Microbiol. 5(1):13-18.

Adesokan IA (2005). Characterization of killer yeasts from Nigerian traditional fermented alcoholic beverages. M.Sc. Thesis, Department of Microbiology, University of Ibadan, Nigeria.

Adesokan IA, Abiola IP, Ogundiya MO (2010). Influence of ginger on sensory properties and shelf - life of Ogi, a Nigerian traditional fermented food. Afr. J. Biotechnol. 9(12):1803-1808.

Aliu J, Gun IU, Mustapha A (2007). Effects of Plant extracts on microbial growth, colour change and lipid oxidation in cooked beef. Food Microbiol. 24:7-14.

Babalola SO, Babalola AO, Aworh OC (2001). Compositional attributes of the calyces of Roselle (*Hibiscus sabdariffa L.*). J. Food Technol. Afr. 6(4):133-134.

Builder PF, Ezeobi CR, Tanfa FD, Buildes MI (2010). Assessment of the intrinsic and stability properties of the freeze-dried and formulated extract of *Hibiscus sabdariffa* Linn. (Malvaceae). Afr. J. Pharm. Pharmacol. 4(6):304-313.

Dougheri JH, Alabi G, Elmahmood AM (2007). Effect of some chemical preservatives on the shelf-life of zobo drink. Afr. J. Microbiol. Res. 2:037-041.

Fasoyiro SB, Ashaye O.A, Adeola A, Samuel FO (2005). Chemical and storability of fruit-flavoured (*Hibiscus sabdariffa*) drinks. World J. Agric. Sci. 1(2):165-168.

Ilondu EM, Iloh AC (2007). Inhibition of three fungal isolates from sorrel drink (Zobo) using hurdle technique. World J. Agric. Sci. 3(3):339-343.

Kolapo AL, Popoola TOS, Sanni MO, Afolabi RO (2007). Preservation of soybean daddawa condiment with dichloromethane extract of ginger. Res. J. Microbiol. 6:13-18.

Ogiehor IS, Nwafor OE, Owhe-uUreghe UB (2008). Changes in the quality of zobo beverages produced from *Hibiscus sabdariffa* (Lin Roscelle) and the effects of extract of ginger alone or in combination with refrigeration. Afr. J. Biotechnol. 7(8):1176-1180.

Olayemi F, Adedayo R, Muhummad R, Bamishaiye E (2011). The nutritional quality of three varieties of zobo (*Hibiscus sabadariffa*) subjected to the same preparation condition. Plant Food Hum. Nutr. 6:10-15.

Omemu AM, Edema MO, Atayese AO, Obadina AO (2006). A Survey of the microflora of *Hibiscus sabdariffa* (Roselle) and resulting 'zobo' juice. Afr. J. Biotechnol. 5(3):254-259.

Yadeng Q, Chin KL, Malekian F, Berhane M, Gager J (2005). Biological characteristics nutritional and medicinal values of Roselle, *Hibiscus sabdariffa*. Circular-Urban Forestry Natural Resources and Environmental No. 604.

Quality assessment of acha-based biscuit improved with bambara nut and unripe plantain

Agu H. O.[1] , Ezeh G. C.[1] and Jideani A. I. O.[2]

[1]Department of Food Science and Technology, Federal Polytechnic, Bauchi, P.M.B. 0231 Bauchi, Bauchi State, Nigeria.
[2]Department of Food Science and Technology, University of Venda, Thohoyandou, 0950 Limpopo province, South Africa.

Five value added biscuit products were produced from three different raw materials, namely acha (*Digitaria exilis* Staph), bambara nut and unripe plantain, at different proportions: 100:0:0% (ACH105), 80:10:10% (ACH801), 70:20:10% (ACH702), 60:30:10% (ACH603) and 50:40:10% (ACH504), respectively. The raw materials were cleaned, sorted, dried and milled into flour and used to produce biscuits. Sensory parameters evaluated were taste, texture, colour, flavour, crispness and general acceptability. Result showed that biscuit products ACH801 and ACH702 were acceptable ($p < 0.05$). Physical parameters determined were spread ratio, weight and break strength. The proximate composition for 80:10:10% acha, bambara nut and unripe plantain flour biscuit are 6.20% protein, 2.04% moisture, 20.12% fat, 2.28% ash, 2.27% crude fibre, 69.36% carbohydrate. The energy content is 483.32 Kcal/100 g. Shelf life studies on the bacterial, mould and yeast were also carried out and the results obtained showed a snack product containing nutrients required for diabetic patient with a projected shelf life of one year.

Key words: Quality, biscuit, acha, *Digitaria exilis*, bambara, unripe plantain.

INTRODUCTION

Biscuit may be regarded as a form of confectionary, dried to very low moisture content. Consumption of whole grains is an excellent source of dietary fibre and nutraceutical that are of benefit in the management of obesity and diseases such as diabetes (Jideani and Jideani, 2011). The consumption of cereal foods such as biscuit has become very popular in Nigeria especially among children. Most of these cereal foods are poor in protein content and protein quality (Alobo, 2001). Enrichment of cereal-based foods with other protein sources such as oil seeds and has received considerable attention (Ayo et al., 2007; Dhingra and Jood, 2002; Elkhalifa and El-Tinay, 2002; Ayo and Gaffa, 2002; Ayo and Olawale, 2003). Legumes are high in lysine, an essential limiting amino acid in most cereals (Alobo, 2001).

The production of good quality biscuit would depend on selecting the correct flour for each type and appropriate processes involving steps such as mixing, aeration and fermentation, machining including laminating, baking, cooling and packaging (Okaka and Okaka, 2005). The introduction of composite flour into the bakery world has brought about different changes into baked products

Some of the oldest popular cereal grains of *Digitaria* spp., *Digitaria exilis* (acha) and *Digitaria iburua*, (iburu), also known as fonio or hungry rice are indigenous grains of West Africa (Jideani, 2012). Acha and iburu proteins have composition similar to that of white rice (Temple and Bassa, 1991; Jideani and Jideani, 2011), but having relatively higher sulphur amino acids (methionine and cystine) content (de Lumen et al., 1993; Lasekan, 1994; Jideani et al., 1994). These and other attributes of acha and iburu show the uniqueness of the grains and their potential in contributing significantly to whole grain diets. Acha grain can also be grounded into flour to produce biscuit (Jideani, 2012).

However, there is sufficient evidence showing that higher whole grains diet when compared with refined grain diet are beneficial for treating or managing several health problems (Jones, 2009). Recent research has shown that acha helps diabetic patients to recover due to its low glycemic index (Balde et al., 2008). The *in vitro* starch digestibility and glycemic property of acha, iburu and maize porridge has been investigated (Jideani and Podgorski, 2009). The study showed that the total starch (TS) for maize, acha and iburu flours were 45.3, 43.6 and 41.5%, respectively. The resistant starch (RS) was 2.9, 2.1 and 1.2, respectively for maize, acha and iburu flours and the digestible starches (DS) 43.7, 41.4 and 40.0%. The authors concluded that acha and iburu may have potential in a low GI food as porridge from both grains had low estimated value of 40. Low-GI diets may improve both glycemic control and cardiovascular risk factors for patients with type 2 diabetes (Jenkins et al., 2008).

Bambara nut (*Vigna subterranea*) is an important legume grain in semi arid Africa. It is resistant to high temperature and suitable for marginal soil where other leguminous crops cannot be grown (Baryeh, 2001). Bambara nut is an important source of high protein value for poorer people in Africa who cannot afford expensive animal protein (Baryeh, 2001; Stading, 2006).

Plantain is the common name for herbaceous plants of the genus *Musa*. Plantains are classified formally as *Musa acuminate* and *Musa balbisiana* depending on their genomic constitution. It provides more than 25% of the carbohydrate requirements for over 70 million people and tends to be firmer and lower in sugar content. Plantains are commonly cooked or otherwise processed and are used either when green or unripe (starchy) or over ripe (sweet) (Oke et al., 1998). An average plantain has about 220 calories and is a good source of potassium and dietary fiber (Randy et al., 2007). It is rich in carbohydrate, dietary fiber, iron, vitamins and minerals. It is ideal for diabetics, children and pregnant women and can also be a good supplement for marasmus patients. Plantain contain small amount of serotonin which has the ability to dilate the arteries and improve blood circulation. Its regular consumption helps to cure anemia and maintain a healthy heart (USDA, 2010). A diet of unripe plantain is filling and can also be a good inclusion in a weight loss diet plan (Oke et al., 1998).

Many countries have produced composite product using other cereals like rice, sorghum, acha, bambara nut, maize and starchy root. Biscuit may be classified either by the degree of enrichment and processing or by the method adopted in shaping them (Okaka and Okaka, 2005). Wheat flour blended with acha flour up to 30% has shown no significant difference in terms of sensory qualities (Ayo and Nkama, 2003). Acha-wheat flour has been fortified with soybean flour to improve the nutritional value when used for production of biscuit (Ayo et al., 2007). Acha-malted soybean has been used for the production of bread and biscuits (Ayo et al., 2014). Acha-based biscuit improved with bambara nut and unripe plantain is a good and cheap source of nutrient than wheat flour used by the baking industry.

Despite the inherent potentials of acha, bambara and unripe plantain, little has been done to incorporate them in most food formulations. The objectives of this research work are (i) production and packaging of acha biscuit, improved with bambara nut and unripe plantain at different proportions, (ii) evaluation of the physical properties of the biscuits, (iii) determination of acceptability of the biscuits and (iv) evaluation of the chemical composition of the acha-based biscuits and microbiological properties of the best two biscuits.

MATERIALS AND METHOD

Source of material

Acha grain (*D. exilis*), bambara nut (*V. subterranea*), unripe plantain (*M. paradisiaca*), Royal baking powder, evaporated peak milk, Simans baking fat, Dangote granulated sugar, and salt were purchased from Wunti Market Bauchi, Nigeria. The equipment and chemicals used were of analytical grade.

Preparation of acha and bambara nut flours

Acha (*D. exilis*) grains were washed with clean water several times in a container in order to remove tiny stones and dust as well as foreign materials. The bambara nut (*V. subterranea*) were dry clean to remove shells and other foreign material such as stones in the grain, Both acha and bambara nut were drained and sun-dried, milled using hammer mill (CD1155-3-1317) and sieved with the aid of a 425 µm sieve (Endecotts Ltd, London, England). The flour was packaged in polyethylene bag and stored at low temperature as acha or bambara flour

Preparation of unripe plantain flour

Unripe matured plantain (*M. paradisiaca*) was washed with clean water and peeled with stain less knife to separate skin from fruit. The peeled fruit was sliced to equal size and dried under the sun. The drying, milling, sieving and storage were as in acha and bambara nut flours.

Table 1. Recipe (%) for acha-based biscuit samples.

Sample code	Acha flour	Bambara nut flour	Unripe plantain flour	Fat	Egg	Sugar	Salt	Baking powder
ACH105	100	0	0	49	30	45	0.6	3.6
ACH801	80	10	10	49	30	45	0.6	3.6
ACH702	70	20	10	49	30	45	0.6	3.6
ACH603	60	30	10	49	30	45	0.6	3.6
ACH504	50	40	10	49	30	45	0.6	3.6

Source: Department of Food Technology, Kaduna Polytechnic, Kaduna, Nigeria (2004).

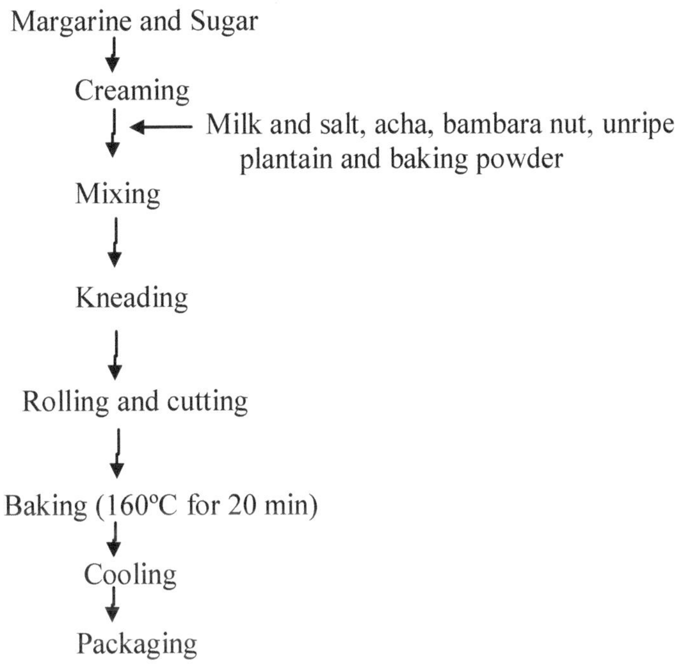

Figure 1. Production of acha, bambara nut and unripe plantain biscuit.

Production of acha, bambara and unripe plantain flour biscuit

The essential ingredients for the production of composite biscuit and their various proportions are shown in Table 1. The acha, bambara nut and unripe plantain flours were used at various proportions while the milk, fat, baking powder, sugar and salt were the same for all the proportions. The recipe for the composite biscuit was chosen based on the Department of Food Technology Kaduna Polytechnic, Kaduna, Nigeria (2004) manual with slight moderations. The flow chart for production of composite biscuit from acha, bambara nut and unripe plantain is shown in Figure 1. After weighing with an electric weighing balance (Sauter, RC 8021 model), the fat was manually mixed vigorously with sugar for 10 min to form a cream. The acha, bambara nut and unripe plantain flours at different levels as shown in the recipe in Table 1 was added with the other ingredients like salt, baking powder. The mixing was done properly and the method of Okaka (1997) was used to produce the biscuits.

Physical analysis

For spread ratio, two rows of five well formed biscuits were made and the height measured. They were also arranged horizontally edge to edge and the sum of the diameters measured. The spread ratio was calculated as diameter divided by height (Gomez et al., 1997). Break strength was determined using break strength device (Okaka and Isieh, 1990). Biscuit sample of 0.4 cm thickness was placed centrally between two parallel metal bars 2 cm apart and weights were applied until the biscuit snapped. The least weight that caused the breaking of the biscuit was regarded as the break strength of the biscuit.

Sensory evaluation of the biscuit samples

The biscuits were subjected to sensory evaluation using twenty panelists from the Federal Polytechnic, Bauchi, Nigeria based on their familiarity with the product. The products, appropriately coded and of the same size and temperature (29 ± 3°C) were placed in white plastic plates. The panelist rinsed their mouths with bottled water after tasting each sample (Larmond, 1977) served to the in different boots under the florescent light. A nine-point Hedonic scale with one (1) representing "extremely dislike" and nine (9) "extremely like" was used. The qualities assessed were color, texture, taste, flavor, crispness and general acceptability (Akinjayeju, 2009).

Although, the panelists were not trained, their selection was based on basic requirements of a panelist, such as availability for

Table 2. Physical quality of acha-based biscuit samples.

Sample code	Acha flour (%)	Bambara nut flour (%)	Unripe plantain flour (%)	Weight (g)	Spread ratio	Break strength (g)
ACH105	100	0	0	10.0	9.15	900
ACH801	80	10	10	10.0	9.30	600
ACH702	70	20	10	9.50	9.50	700
ACH603	60	30	10	9.50	10.32	700
ACH504	50	40	10	9.20	11.22	850

the entire period of evaluation, interest, willingness to serve, good health (not suffering from colds), not allergic or sensitive to the products evaluated (Penfield and Campbell, 1990).

Proximate composition

The moisture, protein, fat, ash and crude fibre contents were determined according to AOAC (2000). The total carbohydrate (CHO) was determined by difference: CHO = 100 – (% moisture + % protein + % fat + % ash). Food energy (kcal/100 g) was calculated according to the method of Marero et al. (1988) using the factor: [(4 x % Protein) + (4 x % Carbohydrate) + (9 x % Fat)]

Microbiological analysis

The pour plate method was adopted as described by Jideani and Jideani (2006). Nutrient agar and potato dextrose agar were prepared as specified by the manufacture for bacteria and fungi respectively. Serial dilution was carried out on suspension of the biscuit samples using quarter strength peptone water before plating on the media. Incubation was done for 24 h at 37°C for bacteria, and 3 to 4 days for yeast and mould at room temperature of 25°C. The colonies were counted using an electronic colony counter (Gallenkamp, 443 300 66087, UK).

Statistical analysis

The results obtained were subjected to analysis of variance (ANOVA) and Duncan multiple range test (Duncan, 1955) was used to separate means where significant differences existed. The software used for the statistical analysis was MINITAB ver. 16.

RESULTS AND DISCUSSION

The physical analysis of the biscuits is shown in Table 2. The weight of the biscuits ranged from 9.2 - 10.0 g. The control sample ACH105 (100:0:0% acha, bambara nut and unripe plantain flour biscuit) had the highest weight. The weight of the biscuits decreased with increase in bambara nut flour with sample ACH801 (80:10:10% acha, bambara nut and unripe plantain flour biscuit) having the highest weight and sample ACH504 (50:40:10% acha, bambara nut and unripe plantain flour biscuit) having the lowest. The decrease in weight could be due to the increase in the fat content of the blended bambara nut flour, as fat is lighter in weight (Ayo et al., 2007). These results were similar to those for bambara groundnut-maize flour (Akpapunam and Darbe, 1994), millet-sesame flour (Alobo, 2001) and amaranth-wheat flour (Ayo, 2001).

The spread ratio of the biscuit increased from 9.30 - 11.22. The increase is an indication of the binding properties of the flour and of the texture of the biscuits. The increase in the fats content (Table 4) could also affect the spread ratio (Ayo et al., 2007). The break strength decreased from 900 - 600 g with increase in percentage of bambara nut flour. The decrease could be due to the increase in the percentage of fats (19.70-20.12%) with increase in percentage of bambara nut flour added, diluting the protein and carbohydrate level which are the principal compounds responsible for hardness in biscuits (Okaka and Isieh, 1990).

The mean scores for the sensory evaluation of the biscuits are shown in Table 3. There was no significant difference ($p < 0.05$) in crispness and flavour of the biscuits while significant differences ($p < 0.05$) existed in taste, texture, colour and overall acceptability. The biscuits compared favourably with 100% acha biscuit which significantly differed ($p < 0.05$) from the other biscuits except in flavor and crispness.

There was a general decrease in the mean scores of all the parameters with increase in bambara nut and unripe plantain flours except for texture which increased. Sample ACH801 (80:10:10% acha, bambara nut and unripe plantain flour biscuit) had the highest mean value in most of the parameters monitored while sample ACH105 (100:0:0%) acha, bambara nut and unripe plantain flour biscuit) had the lowest mean value in all the parameters with the exception of texture. The crust texture was related to the external appearance of the biscuit top, which is the smoothness or roughness of the crust (Ayo et al., 2007). The texture reduced from 7.30 for 100% acha flour biscuit to 6.85 for 80:10:10% acha, bambara nut and unripe plantain flour biscuit and gradually increased as the proportion of bambara nut and unripe plantain flour increased.

There was a general decrease in the overall acceptability of the biscuits with decrease in acha flour and increase in bambara nut flour and unripe plantain flour. Sample ACH801 (80:10:10% acha, bambara nut and unripe plantain flour biscuit) had the highest mean

Table 3. Sensory evaluation of the biscuit samples.

Sample code	Proportion of flours (%)			Taste	Crispness	Flavour	Texture	Colour	Overall Acceptability
	Acha	Bambara nut	Unripe plantain						
ACH 105	100	0	0	$6.80^{a} \pm 1.39$	$7.15^{a} \pm 1.53$	$6.35^{a} \pm 1.63$	$7.30^{a} \pm 1.46$	$6.95^{a} \pm 1.43$	$6.90^{ab} \pm 1.58$
ACH 801	80	10	10	$7.30^{a} \pm 1.63$	$7.20^{a} \pm 1.80$	$7.25^{a} \pm 1.48$	$6.85^{a} \pm 1.95$	$7.50^{a} \pm 1.61$	$8.00^{a} \pm 0.85$
ACH 702	70	20	10	$7.00^{a} \pm 1.65$	$7.05^{a} \pm 2.01$	$6.85^{a} \pm 1.42$	$7.05^{a} \pm 1.28$	$7.40^{a} \pm 1.67$	$7.50^{ab} \pm 1.49$
ACH 603	60	30	10	$6.95^{a} \pm 1.82$	$7.50^{a} \pm 1.64$	$6.85^{a} \pm 1.63$	$7.25^{a} \pm 1.21$	$7.30^{a} \pm 1.56$	$6.40^{b} \pm 1.78$
ACH 504	50	40	10	$6.35^{a} \pm 1.60$	$7.05^{a} \pm 2.04$	$6.65^{a} \pm 1.76$	$7.35^{a} \pm 1.46$	$7.25^{a} \pm 1.83$	$6.80^{ab} \pm 1.79$

Values are mean ± standard deviation of twenty panelists. Means within each column not followed by the same superscript are significantly different ($p < 0.05$) from each other using Duncan multiple range test.

Table 4. Proximate (%) and energy compositions of acha based biscuit improved with bambara nut and unripe plantain.

Parameter	Samples				
	ACH 105	ACH 801	ACH 702	ACH 603	ACH 504
Acha	100	80	70	60	50
Bambara nut	0	10	20	30	40
Unripe Plantain	0	10	10	10	10
Moisture	$2.06^{a} \pm 0.05$	$2.04^{a} \pm 0.10$	$2.03^{a} \pm 0.00$	$2.08^{a} \pm 0.01$	$2.07^{a} \pm 0.09$
Protein	$5.30^{d} \pm 0.13$	$6.20^{c} \pm 0.02$	$6.43^{c} \pm 0.11$	$7.00^{b} \pm 0.15$	$7.51^{a} \pm 0.03$
Fat	$19.86^{a} \pm 0.14$	$20.12^{a} \pm 0.02$	$20.10^{a} \pm 0.07$	$20.08^{a} \pm 0.06$	$19.70^{a} \pm 0.57$
Ash	$2.31^{d} \pm 0.02$	$2.28^{d} \pm 0.14$	$2.67^{c} \pm 0.06$	$3.41^{a} \pm 0.02$	$3.09^{b} \pm 0.05$
Crude fibre	$3.56^{a} \pm 0.05$	$2.27^{ab} \pm 0.02$	$2.43^{a} \pm 0.06$	$2.08^{b} \pm 0.04$	$2.40^{a} \pm 0.13$
Carbohydrate	$71.47^{a} \pm 0.15$	$69.36^{ab} \pm 0.15$	$68.77^{abc} \pm 0.06$	$67.43^{bc} \pm 0.06$	$67.63^{bc} \pm 1.02$
Energy (Kcal/100 g)	$485.82^{a} \pm 1.22$	$483.32^{a} \pm 0.56$	$481.70^{ab} \pm 0.83$	$478.44^{b} \pm 0.21$	$477.86^{b} \pm 1.75$

Values are mean ± standard deviation. Any mean not followed by the same letter on each column are significantly different ($p < 0.05$).

value of 8.0 in overall acceptability which made the biscuit most acceptable (Figure 2). This is an indication that the improvement was acceptable to the consumers. Therefore, the two best biscuits based on the overall acceptability of sensory properties were ACH801 and ACH702 (Table 3).

The proximate and energy compositions of acha biscuit improved with bambara nut and unripe plantain flours are shown in Table 4. The moisture content (%) of the acha based biscuit increased from 2.03-2.08 with addition of bambara nut and unripe plantain. The low moisture content of the biscuit will require a unique packaging material to prevent reabsorption of moisture. The protein content (%) of the biscuit increased from 5.30 - 7.51 with increase in bambara nut and unripe plantain. Sample ACH105 had no addition of bambara nut hence had lowest protein content. Bambara nut is a rich source of protein. Legume was used and they are known to be high in both fat and protein (Dhringa and Jood, 2002). The fat

Figure 2. Acha based biscuit (80:10:10% acha, bambara nut and unripe plantain flours) in polyethylene bag and paperboard.

content (%) ranged from o 19.70 - 20.12 with increase in bambara nut and because of added baking fat. The baking fat could be reduced in subsequent baking as to elongate the shelf life of the biscuits. The ash content (%) increased from 2.28 - 3.09 with increase in bambara nut and added unripe plantain flour which is a rich source of mineral.

Ash content indicates the presence of mineral matter in food. Ash is a non organic compound containing the mineral content of food. It aids in the metabolism of other compound such as protein fat and carbohydrate (Okaka and Ene, 2005). The crude fibre (%) decreased from 3.56-2.08 with increase in bambara nut and unripe plantain flour. Also, decrease was observed in the carbohydrate and energy contents.

The microbial analysis for first week showed no growth in bacteria, mould and yeast. This could be because the product is freshly produced and packaged but growth began at the second week with 1.0 x 10^4 cfu/g for sample ACH 801 and at fourth week it increased to 3.5 x 10^4 cfu/g (Table 5). This increase could be due to make up of the product which contained protein. From the result, it could be concluded that mould and yeast are the possible spoilage organism associated with dry product. Bacteria growth in the product was as a result of handling during processing.

Conclusion

High quality biscuit was produced from acha, bambara nut and unripe plantain. The addition of bambara nut and unripe plantain improved the physical, proximate and sensory quality of acha biscuit. Nutritionally, the protein

Table 5. Microbial count of the two acceptable biscuit.

Sample	No. of viable count		
	Weeks	Bacteria (cfu/g)	Mould/yeast (cfu/g)
ACH801 (80% acha,10% bambara nut & 10% unripe plantain)	1st	Nil	Nil
	2nd	1.0×10^4	1.0×10^4
	3rd	2.0×10^4	3.0×10^4
	4th	2.1×10^4	4.0×10^4
ACH702 (70% acha, 20% bambara nut & 10% unripe plantain)	1st	Nil	Nil
	2nd	1.0×10^4	2.0×10^4
	3rd	2.0×10^4	3.0×10^4
	4th	3.5×10^4	5.0×10^4

content increased from 5.30 - 7.51%, fat from 19.86 - 20.12%, ash 2.28 - 3.09%, crude fibre 2.27 - 2.39% at different proportions. The sensory qualities were significantly different with addition of unripe plantain and bambara nut flours. For ACH801 (80: 10: 10% acha, bambara nut and unripe plantain flours), the biscuit was generally acceptable. The production of biscuit from acha could reduce the use of wheat importation for baking. The authors recommend acha based biscuit improved with bambara nut and unripe plantain for both children and diabetic patients.

Conflict of Interests

The author(s) have not declared any conflict of interests.

ACKNOWLEDGEMENT

The authors acknowledge the Federal Polytechnic, Bauchi Nigeria that supported this work.

REFERENCES

Akinjayeju O (2009). Quality control for the industry: A statistical approach. Concept publications, Lagos, pp. 229-273.

Akpapunam MA, Darbe JW (1994). ChemicalComposition and functional Properties of blendsof maize bambara groundnut flours for Cookieproduction. Plant Food Hum. Nutr. 46:147-155.

Alobo AP (2001). Effect of sesame seed on millet biscuit. Plant Food Hum. Nutr. (2): 195-202.

AOAC (2000) Official method of analysis of the AOAC (W. Horwitz Editor) 16th Edition Association of Official Analytical Chemists. Washington D.C.

Ayo JA, Gaffa T (2002). Effect of undefated soybean flour on the protein content and sensory quality of "Kunnu zaki". Niger. Food J. 20:7-9.

Ayo JA, Olawale O (2003). Effect of defatted groundnut concentrate on the physico-chemical and sensory quality of "fura". Nutr. Food Sci. 33(4):175-176.

Ayo JA (2001). The effect of Amaranth grain flour on the quality of bread. Int. J. Food Properties 4(2):341-352.

Ayo JA, Nkama I (2003). Effect of acha (*Digitaria exilis* Staph) grain flours on the physical and sensory quality of biscuit. Nutr. Food Sci. 33:125-130.

Ayo JA, Ayo VA, Nkama I, Adewori R (2007). Physicochemical, in-vitro digestibility and organoleptic evaluation of acha wheat biscuit supplemented with soybean flour. Niger. Food J. 25(1):77-89.

Ayo JA, Ayo VA, Popoola C, Omosebi M, Joseph M (2014). Production and evaluation of malted soybean-acha composite flour bread and biscuits. Afr. J. Food Sci. Technol. 5(1):21-28.

Balde NM, Besancon S, Sidibe TA (2008). Glycemic index fonio (*Digitaria exilis*) interested in feeding in diabetic subjects in West Africa. Diabetes Metab. 34(3):190-192.

Baryeh EA (2001). Physical properties of bambara groundnuts. J. Food Eng. 47(4): 321-326.

de Lumen O, Thompson S, Odegard JW (1993). Sulphur amino acid-rich proteins in acha (*Digitaria exilis*), a promising underutilized African cereal. J. Agric. Food Chem. 41:1045-1047.

Department of Food Technology Kaduna Polytechnic Kaduna-Nigeria (2004). Practical Manual on Food Technology, Nutrition and Dietetics for Schools and Industries (2nd Edition), National Science and Technology Forum, Kaduna Polytechnic. pp. 19-21.

Dhingra S, Jood S (2002). Organoleptic and nutritional evaluation of wheat bread supplemented with soybean and barley flour. Food Chem. 77:479-488.

Duncan DB (1955). New multiple range and multiple F tests. Biometrics 11:1-11.

Elkhalifa AE, El-Tinay AH (2002) Effect of cysteine on bakery products from wheat – sorghum blends. Food Chem. 77:133-137.

Gomez MI, Obliana AB, Martins DF, Madzavamuse M, Monyo ES (1997). Manual of laboratory procedures for quality evaluation of sorghum and pearl millet. Technical Manual No. 2 ICRISAT, Patencheru 502, 324, Andhra Pradesh, India.

Jenkins DJ, Kendall CW, McKeown-Eyssen G, Josse RG, Silverberg J, Booth GL, Vidgen E, Josse AR, Nguyen TH, Corrigan S, Banach MS, Ares S, Mitchell S, Emam A, Augustin LS, Parker TL, Leiter LA (2008). Effect of a low-gylcermic index or a high-cereal fiber diet on type 2 diabetes: a randomised trial. J. Amer. Med. Assoc. 300(23):2742-2753.

Jideani IA (2012). *Digitaria exilis* (acha/fonio), *D. iburua* (iburu/fonio) and *Eluesine coracana* (tamba/finger millet) – non-conventional cereal grains with potentials. Sci. Res. Essays 7(45): 3834 -3843

Jideani IA, Jideani VA (2011). Developments on the cereal grains *Digitaria exilis* (acha) and *Digitaria iburua* (iburu). J. Food Sci. Technol. 48(3):251-259.

Jideani IA, Owusu RK, Muller HG (1994). Proteins of acha (*Digitaria exilis* Stapf): solubility fractionation, gel filtration and electrophoresis of protein fractions. Food Chem. 51:51–59.

Jideani VA, Jideani IA (2006). Laboratory manual of food bacteriology. Amana printing and Advertising Ltd, Kaduna. pp. 90-112.

Jideani VA, Podgorski SC (2009). In-vitro starch digestibility and glycemic property of acha, *Iburu* and maize porridge. Cereal Foods World (Suppl.) 54: A48.

Jones JM (2009). The second C & E spring meeting and the international whole grain global summit. Cereal Foods World 54:132-

135.
Larmond E (1977). Laboratory methods for sensory evaluation of foods. Publication No.1637. Canada Department of Agriculture, Ottawa
Lasekan DO (1994). Chemical composition of acha (*Digitaria exilis*) flour. J. Food Sci. Agric. 14:177-179.
Marero LM, Payumo EM, Librando EC, Lainez WN, Gopez MD, Homma S (1988). Technology of using food formulations prepared form germinated cereals and legumes. J. Food Sci. 53 (5):1391-1395, 1455.
Okaka JC (1997). Cereal and legume: storage and processing technology. Data and micro systems publishers Ogui-Enugu, pp. 116-131.
Okaka JC, Ene GL (2005). Food microbiology: method in food safety control. OCJANCO Academic publishers Enugu. pp. 262.
Okaka JC, Isieh MI (1990). Development and quality evaluation of cowpea-wheat biscuit. Niger. Food J. 8:56-60
Okaka JC, Okaka ANC (2005). Food composition and spoilage, shelf life extension OCJANCO Academic publishers Enugu. pp. 48-68.
Oke OL, Redhead J, Hussain MA (1998). Roots, tubers, plantains bananas in humans nutrition. Rome, Italy: Food and Agriculture Organisation of the United Nations (FAO) and the Information Network On post – Harvest Operations (INOPH). pp.198.
Penfield MP, Campbell AM (1990). Experimental food science, 3rd edn. Academy Press, Inc. San Diego, New York.
Randy CP, Kepler AK, Daniells J, Nelson SC (2007). Banana & Plantain -an overview with emphasis on the pacific Island cultivars. Species Profiles for Pacific Island Agroforestry (www.traditionaltree.org). Retrieved 5/7/2012.
Stading M (2006). Introduction to traditional grains. www.sik.sse/traditionalgrains/review. Accessed 25th February, 2011
Temple VJ, Bassa JD (1991). Proximate chemical composition of acha (*D. exilis*) grain. J. Sci. Food Agric. 56:561-563.
USDA (2010). Nutrient Database.Com. United States Department of Agriculture Retrieved July 5th 2012.

Comparative studies of physico-chemical properties of yam (*Discorea rotundata)*, cocoyam (*Collocasia taro*), breadfruit (*Artocapus artilis*) and plantain (*Musa parasidiaca*) instant flours

Oladeji, Babatunde Stephen, Akanbi, Charles Taiwo and Gbadamosi, Saka Olasunkanmi

Department of Food Science and Technology, Obafemi Awolowo University, Ile-Ife, Nigeria.

Cocoyam, breadfruit and plantain usage is similar to that of white yam. They offer similar and sometimes better nutrition advantage to yam though with less prestigious value. This study is aimed at producing instant flour of yam, cocoyam, breadfruit and plantain and to compare their physico-chemical properties. The instant flours were prepared separately with slight modifications. Proximate analysis, starch, vitamin C, physico-chemical and pasting properties of the flours were determined. Significant differences existed in the protein content of the flours ($p<0.05$) with cocoyam flour having the highest value (8.9%). The vitamin C contents of the flours ranged from 2.4 to 7.3 mg/100 g. The water binding and swelling capacities ranged 160 to 268 and 2.72 to 3.99%, respectively. While the pasting temperature and peak viscosity ranged from 63 to 93°C and 60 to 400 BU, respectively. The instant flours (*fufu* form) of cocoyam, breadfruit and plantain compared well with that of yam and offer nutritional and economic advantages to both household and industrial production.

Key words: Starch staples, instant flour, nutritional composition, pasting properties.

INTRODUCTION

Yam (*Discorea rotundata* Poi) is an important staple food in many tropical countries (Akanbi et al., 1996). The tuber is a tropical root crop that grows extensively in West Africa and to a lesser extent in other tropical areas (Anon., 2007). The yam tuber is prepared for consumption in a variety of ways including boiling, frying and baking. One of the most important traditional culinary preparations of yam in West Africa is pounded yam (Ihekoronye and Ngoddy, 1985; Asiedu, 1989). Cocoyam, an edible aroid originated in India and South East Asia, is presently cultivated in many tropical and subtropical countries (Anon., 2007; AUSAID, 1995). It belongs to the family *Araceae* that comprises many underground food crops grown in tropical and subtropical countries with taro as one of the most important species (Opara, 2007).

Breadfruit (*Artocarpus altilis*) which belongs to the family of *Moracaera* is a tree and fruit native to the Malary Peninsula and Western pacific Island (Loos et al., 1981). It has been widely planted in the tropical regions including Nigeria. Breadfruit is extremely cheap in Nigeria as two mature fruits each weighing approximately 1.5 kg can be obtained for the equivalent of US 25 cents and can feed four adults (Omobuwajo, 2003). The plantain (*Musa paradisiaca)* resembles banana but are longer in length, have a thicker skin, and contain more starch. They are also a major staple food in Africa, Latin America and Asia. They are usually cooked and not eaten raw unless they are very ripe (IITA, 2012). Plantain tend to be firmer and lower in sugar content than the desert banana and are used when green or under ripe (and therefore starchy) or overripe (sweet). It is a staple food in the tropical regions of the world, treated in much the same way as potatoes and with similar neutral flavour and texture when unripe (Elevitch, 2006).

In the growing region, Nigeria for example, cocoyam, breadfruit and plantain usage is similar to that of white yam. They are used in the production of stiff porridge or their instant flours *(fufu)* form which is eaten with soup when yam is not in season or by those who could not afford the price of yam. They offer similar and sometimes better nutrition advantage to yam though, with lesser prestigious value.

There is limited scientific information on the post-harvest characteristics of these starch staples, which perhaps contributes to their very limited application of improved post-harvest technologies to maintain quality and improve marketing potential (Opara, 2007). Therefore, there is need to compare the physico-chemical properties of their instant flours. The objectives of the study were therefore to produce instant flours of yam, cocoyam, breadfruit and plantain and to compare the physico-chemical properties of the flours.

METHODOLOGY

Materials

The materials used for the research include: yam tubers (*D. rotundata*), cocoyam corms (*Colocasia esculenta* or taro), breadfruit (*A. artilis*) and plantain (*M. parasidiaca*). These materials were purchased at a local market in Ile-Ife, Nigeria.

Processing of the instant flours

Yam tubers were peeled, diced into cubes with dimension 1.0×1.5×1.5 cm. They were sulphited with 40 ppm of Na_2SO_3 and washed, steamed for eight minutes and dried in cabinet drier at 60°C for twenty hours. The dried yam cubes were milled with attrition mill and packaged. Cocoyam flour, breadfruit flour and plantain flour were similarly prepared with appropriate modifications according to the method of Oladeji and Akanbi (2011).

Chemical analyses of instant flours

Moisture, protein, crude fat, crude fibre and ash contents of the flours were determined by the standard official methods (AOAC, 1990), while carbohydrate was determined by difference. The starch analysis was done by the modification of Lane and Eynon process (James, 1995). Ascorbic acid was determined by dyestuff titration. Sample (5 g) was digested by 0.4% oxalic acid. The aliquot was titrated against dyestuff, which was previously standardized by standard ascorbic acid solution, and the ascorbic acid content was calculated using the following expression.

$$\text{Vitamin C (mg/100 g} = \frac{\text{Titre} \times 0.606 \times 100}{\text{Wt of sample}} \quad (1)$$

Physicochemical properties

Bulk density

Bulk density was determined according to the method described by Okaka and Potter (1979). A 15 g sample was put into a 100 mL graduated cylinder. The cylinder was tapped forty (40) times and the bulk density was calculated as weight per unit volume (g/cm^3)

Water binding capacity

Water binding capacity was determined according to the procedure described by Medcalf and Gilles (1965). The sample (2 g) was transferred into a weighed centrifuge tube to which 30 mL of distilled water was added. The bottle was tarred and agitated in a shaker for one hour. It was then centrifuged at 500 rpm for 10 min. The supernatant was decanted immediately and the bottle was tipped up and allowed to drain for 10 mins more. The bottle was weighed and the amount of water held by the flour was determined using the formula:

$$\text{Water binding capacity} = \frac{\text{g bond water} \times 100}{\text{Wt. of the sample (2 g)}} \quad (2)$$

Swelling capacity

The swelling capacity of the samples was determined by modification of Leach et al (1959) method. Each sample (5 g) was dispersed in 40 mL distilled water. The resultant slurry was heated at a temperature of 70°C for thirty minutes in water bath, cooled to room temperature and centrifuged at 2300 rpm for thirty minutes. The supernatant liquid was decanted and the centrifuge tube was placed in a hot air oven to dry for 25 minutes at 50°C. The residue was then weighted (W_2). Centrifuge tubes containing samples alone were weighed prior to adding distilled water (W_1).

$$\text{Swelling capacity} = \frac{\text{W (2g)} - \text{W (g)}}{\text{Wt of sample (g)}} \quad (3)$$

pH value

The pH was measured by preparing 10% (w/vol) suspension of each sample in distilled water; this was achieved by dissolving 10 g of the sample in 100 mL of distilled water. Each sample was mixed thoroughly and filtered. The pH of the filtrate was measured using pH meter (Combo pH and EC, model H19812, Hanna, England.

Pasting characteristics

The pasting characteristics of the flours were evaluated using a Brabender visco-amylograph. Flour slurry, containing 8% solids (w/w, dry basis), was heated from 30 to 90°C at a rate of 1.5°C /min, held at 95°C for 15 min, and cooled at the same rate to 50°C (Shuey and Tripples, 1982). The pasting performance was automatically recorded on the graduated sheet of he amylograph. The pasting temperature, peak viscosities, viscosity at 95°C, stability, cooking time and setback viscosities were read off the amylograph.

Statistical analysis

The sensory and objective data were subjected to analysis of variance to find out whether a significant difference existed between the samples or not. Least significant difference (LSD) test was calculated for the parameters to determine the level of significance

Table 1. Proximate composition of the flours.

Sample	Composition (%)					
	Moisture	Crude protein	Crude Fibre	Crude fat	Ash	Carbohydrate
Yam	5.6±0.40 ^{b}	4.8±0.03 ^{b}	0.2±0.01^{b}	1.0±0.50 ^{b}	1.5±0.00^{c}	87.0
Cocoyam	6.3±0.06^{a}	8.9±0.34^{a}	1.2±0.14^{a}	1.1±0.06^{a}	3.5±0.05^{a}	78.7
Breadfruit	5.6±0.14^{b}	5.7±0.08^{ab}	1.1±0.02^{a}	1.7±0.03^{a}	3.0±0.07 ^{b}	82.9
Plantain	4.9±0.01^{c}	5.1±0.08 ^{b}	1.4±0.01^{a}	1.5±0.03 ^{b}	2.5±0.04 ^{c}	84.7

Values are means of triplicate sample ±SD. Values with different superscript in the same column are significantly different at p<0.05.

Table 2. Starch and ascorbic acid contents of the flours (mg/100 g).

Sample	Composition	
	Starch	Ascorbic Acid
Yam flour	75.3±0.35^{b}	6.1±0.12 ^{b}
Cocoyam flour	78.5±0.70^{a}	2.4±0.95^{c}
Breadfruit flour	69.2±0.28^{c}	7.3±0.51^{a}
Plantain flour	78.3±0.42^{ab}	3.0±0.98^{c}

Values are means of triplicate sample ±SD. Values with different superscript in the same column are significantly different at p<0.05.

at 95% confidence interval as described by Akindele (2004).

RESULTS AND DISCUSSION

Proximate composition of the flours

The proximate compositions of instant flours of yam, cocoyam, breadfruit and plantain (Table 1) show that significant differences existed (p<0.05) in the percentage protein contents of cocoyam flour and that of others, and cocoyam having the highest value (8.9%), followed by breadfruit (5.7%), while yam has the least value of 4.8%. The value obtained for cocoyam was higher than that reported by Ajewole (2004) for *Zanthosoma* variety, due to possible variation in chemical composition of different varieties. Also, a lower value of protein content (3.16%) for yam flour (Amala) was reported by Jimoh and Oatiloye (2009). This may be due to the browning reaction involved in the production of "amala" which did not require any sulphiting to control browning reaction.

Significant differences existed in the ash contents of all the flour samples at p<0.05, with cocoyam having the highest value (3.5%), followed by breadfruit. Though, the ash content of cocoyam was lower than 4.0% reported by Ajewole (2003), that of breadfruit was close to 2.7% reported by Mayaki et al. (2003), the high value of ash recorded for cocoyam and breadfruit is an indication that they are rich in mineral elements. The flours are also rich in crude fibre (0.2 to 1.4%) indicating their usefulness in the production of roughage and bulk and in contributing to a healthy condition of the intestine when consumed (AUSAID, 1995). The range of 1.0 to 1.7% of crude fat recorded for the instant flour is desirable in the products as fat constitutes the largest percentage of calorie needed by the body (Smith and Ojofeitimi, 1995).

Starch and Ascorbic acid contents of instant flours

The starch contents of the flours ranged between 69.2 and 78.5% (Table 2). This is an expected range for starch staples (Smith and Ojofehintimi, 1995). Cocoyam had the highest starch content of 78.5% followed by plantain (78.3%) (Table 2), and breadfruit with the least (69.2%). The starch result of breadfruit was a bit lower than 71.1% which was reported by Mayaki et al. (2003) and higher than 68% reported by Loos et al. (1981). This may be due to difference in preparation operations which may have resulted in greater loss of starch.

Moreover, breadfruit had the highest vitamin C content of 7.3 mg/100g followed by yam 6.1 mg/100 g while cocoyam and plantain were found to be poor in terms of vitamin C content; 2.4 and 3.0 mg/100 g were recorded for cocoyam and plantain flours, respectively. This suggests that plantain and cocoyam are better cooked with their peel to retain their water soluble vitamins such as vitamin C which might have been partially removed with the peels.

Mineral contents of the instant flours

Yam had the highest iron content (2.37 mg/100 g) but its zinc value was the lowest (0.035 mg/100 g) (Table 3). Breadfruit seemed to be richer in mineral elements than cocoyam and plantain from the result in Table 3. The mineral composition of breadfruit however agrees with

Table 3. Mineral elements content (mg/100g) of the flours.

Sample	Calcium	Iron	Magnesium	Zinc
Yam flour	0.015±0.00	2.37±0.39	0.174±0.05	0.035±0.01
Cocoyam Flour	0.044±0.01	0.24±0.11	0.167±0.04	0.057±0.01
Breadfruit flour	0.011±0.00	0.36±0.02	0.332±0.02	0.072±0.02
Plantain Flour	0.055± 0.03	0.21±0.04	0.225± 0.01	0.081± 0.01

Values are means of triplicate sample ± SD.

Table 4. Physicochemical properties of the instant flour samples

Sample	Bulk density (g/ml)	Water Binding Capacity (%)	Swelling Capacity (%)	pH
Yam flour	$0.65±0.12^{b}$	$160.2±0.28^{b}$	$2.72±0.04^{c}$	$5.88±0.03^{b}$
Cocoyam Flour	$0.71±0.03^{ab}$	$163.1±0.14^{b}$	$3.89±0.13^{b}$	$5.23±0.01^{ab}$
Breadfruit flour	$0.56±0.14^{c}$	$268.4±0.57^{a}$	$3.99±0.15^{a}$	$5.84±0.07^{a}$
Plantain Flour	$0.71±0.14^{ab}$	$236.4±0.57^{ab}$	$3.22±0.06^{c}$	$5.74±0.03^{c}$

Values are means of dublicate sample ±SD. Values with different superscript in the same column are significantly different at p<0.05.

the report of Itiola (1991) and that of plantain, though not the same value but followed almost the same trend with the report of Ahenkora et al. (1997).

Physico-chemical properties

Bulk density

The bulk density is generally affected by the particle size and the true density of the matter in flour. Bulk densities of flours from cocoyam and plantain were the same (0.71 g/ml). This value was greater than the bulk densities of yam and breadfruits flours with 0.65 and 0.56 g/ml, respectively (Table 4). This suggests that the particle size of cocoyam and plantain flours are smaller compare to that of yam and breadfruit with breadfruit flour having the biggest particle size. The small particle size of cocoyam flour is in connection with taro's smaller starch grains (1 to 4 µm) compared to the large grain of tannia (17 to 20 µm) (Opara, 2007). He also noted that this makes taro suitable for several products especially as food for potentially allergic infants and persons with gastro intestinal disorder. Bulk density is important for determining package requirement, materials handling, and application in wet processing in the food industry (Kulkarni et al., 1996).

Water binding capacity

Breadfruit flour had the highest water binding capacity, that was comparable (p<0.05) to the water binding capacity of plantain flour (Table 4). The reason for high water binding capacity recorded for breadfruit flour would have been due to higher content of undamaged starch granules in the flour (Mayaki et al., 2003). However, there was no significant difference (p<0.05) in the water binding capacity of yam and cocoyam flour samples. Little wonder, cocoyam is the closest alternative to yam especially in term of pounding for food. Both the crude fibre content of plantain flour (Table 1) and its starch content (Table 2) were high. The high water binding capacity of plantain may be due to its high crude fibre (Mayaki et al., 2003) and starch contents.

Swelling capacity

The swelling capacity, does not follow the same trend with starch contents of the flours but followed the same trend with the bulk densities (Table 4). This suggests that particle size of the starch granules and damages done to starch by milling operation had great effect on the swelling capacity of the flour. Breadfruit flour has the highest swelling capacity (3.99) followed by cocoyam flour (3.89), while yam flour recorded the lowest value (2.71) and significant differences existed in the swelling capacity of all the flour samples at p<0.05.

Pasting performance of instant flours

Amylographic plot generated by computer shows the pasting performance of yam, cocoyam, breadfruit and plantain instant flour samples (Figure 1). The apparent temperatures (Tp) of the flours ranged from (63 to 93°C). Yam and plantain instant flours had the lowest and highest pasting temperatures, respectively (Table 5). The peak viscosities; 150, 60 and 190 BU for cocoyam, bread fruit and plantain, respectively were low compared to that of yam flour (400BU), an indication that the starch

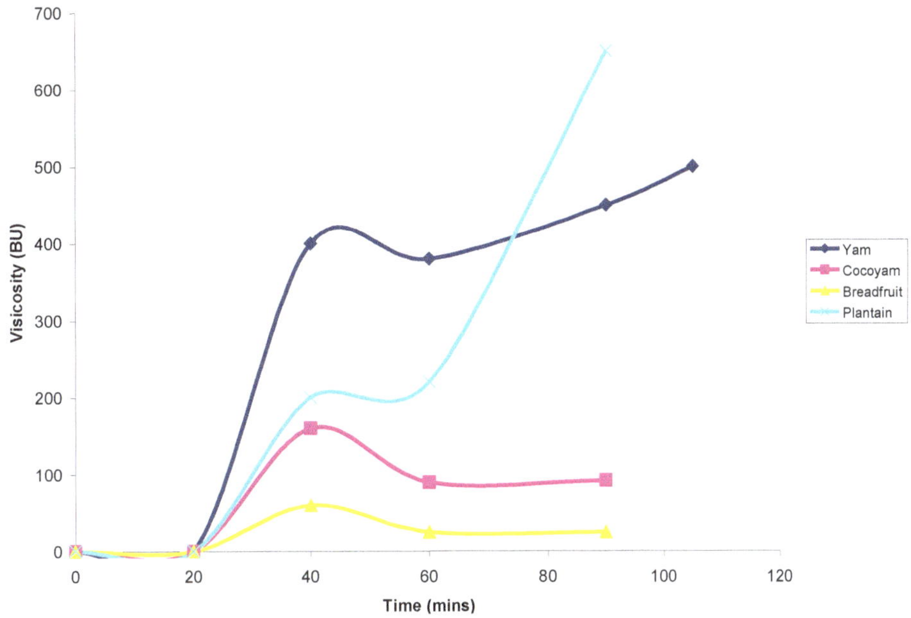

Figure 1. Pasting profile of yam, cocoyam, breadfruit and plantain.

Table 5. Pasting characteristics of the instant flours.

Parameter	Yam flour	Cocoyam flour	Breadfruit flour	Plantain flour
Tp (°C)	63	63	93	81
Mg (Mins)	22	22	42	34
Tvp (°C)	93	93	96	95
Vp (BU)	400	150	60	190
Mn (,in)	42	42	44	42
VI (BU)	390	150	60	190
Vr (BU)	480	80	25	360
Vc (BU)	540	80	20	640
Mn-Mg (Mins)	20	20	2	8
Vp-Vr (BU)	-80	70	35	-170
Vc-Vp (BU)	140	-70	-40	450
Vc-Vr (BU)	60	0	-5	280

Tp, Pasting temperature; Mn, time at peak viscosity; Mg, gelatinization time; VI, viscosity at 95°C; Tvp, peak temperature; Vr, viscotity at 95°C after 15 mins; Vp, peak viscosity; Vc, viscosity at 50°C; Vp-Vr, stability cooking; Vc-Vp, set-back value; Vc-Vr, gelatinization index.

structure of yam flour is stronger than other starch staples (Svanberg, 1987). The lowest peak viscosity recorded for breadfruit flour may be due to its high fat content (Table 1) which might have caused a buffering effect on the gelling properties of its starch (Oluwamukomi et al., 2005).

After a 15 mins hold at 95°C, viscosities of cocoyam and breadfruit flour sample were reduced while that of yam and plantain increased considerably. This is a pointer to stability (Vp -Vr) of cooking cocoyam flour and breadfruit flours seems to be more stable compared to yam and plantain (Table 5). The high Vp reflects fragility of the swollen granules, which first swell and break down under continuous mechanical shearing condition of brabender visco-amylograph (Oluwamukomi et al., 2005). However, when cooled to 50°C, there were significant increases in the viscosities of yam and plantain flours compared to that of cocoyam and breadfruit. On the other hand the low viscosities obtained on cooling cocoyam and breadfruit pastes may be due to early breakdown of their starch structures thereby preventing formation of a stronger cohesive force or network.

Conclusion

The investigation concluded that all the starch staples could be used to produce acceptable instant flour for reconstitution in boiling water and eaten with soup. Though, yam instant flour showed a better pasting performance than others, the pasting characteristics of cocoyam, breadfruit and plantain instant flours compared relatively well with that of yam and can be used in it stead. Not only that they have better nutritional advantage over yam flour, a combination of plantain with either cocoyam or breadfruit will reduce cost of production and improve nutritional qualities of the instant flour samples. The products will still remain cheaper than whole yam flour which will offer an economic advantage to both industrialists and consumers.

REFERENCES

Ajewole AI (1994). Enrichment of Pregelatinized Cocoyam Flour with melon seeds protein Concentrate as an Instant Food Gel. JAT 5(2): 28-32

Akanbi CT, Gureje PO, Adeyemi IA (1996). Effect of heat-moisture pre-treatment on physical characteristic of dehydrated yam. J. Food Eng. 28(45):45-54.

Akindele SO (2004). Basic Experimental Design in Agricultural Research, Montem Paperbacks, Akure. pp. 41-64.

Ahenkora K, Kyei EKM, Banful B (1997): Nutritional composition of false horn Apantupa plantain during ripening and processing. Afr. Crop Sci. J. 5(2): 243-248.

AOAC (1990). Official Method of Analysis, 15th edition. Association of Official Analytical Chemists, Washington

Asiedu JJ (1989): Processing Tropical Crops – A Technological Approach. Macmillan Publishers. p. 265.

AUSAID (1995). Use of local Pacific foods. Australia Agency for International Development, 9th South Pacific leaflets. South Pacific Commission, New Caledonia.

Elevitch C (2006). Musa Species (Banana and Plantain) specie profile for pacific Island Agroforestry. Available on www.trditionaltree.org

IITA (2012). Banana and Plantain. International Institute of Tropical Agriculture. Available on http://www.iita.org/banana-and-plaintain

Ihekoronye AI, Ngoddy PO (1985). Integrated Food Science and Technology for the tropics. Macmillan Publishers Ltd, London

Itiola AO (1991). Compressional characteristics of three starches and the mechanical properties of their tablets Pharm. World J. 8:91-94.

James CS (1995). Analytical Chemistry of Foods. Blackie Academic & Professional London and New York. pp. 37-38.

Jimoh KO, Olatidoye OP (2009). Evaluation of Physicochemical and Rehological characteristics of soybean fortified yam flour. J. Appl. Biosci. 13:703-706.

Kulkarni KD, Govinden N, Kulkarni D 2008. Production and use of raw potato flour in Muaritian traditional foods. J. Sci. Food Agric. 17(2):162-168

Leach HW, McCowen LD, Schoch TT (1959). Structure of the starch granule (1) Swelling and solubility patterns of various starches. Cereal Chem. 36:534-539

Loos PJ, Hood LF, Graham HD (1995). Isolation and Characterization of starch from breadfruit. Cereal Chem. 58, 282-286.

Mayaki OM, Akingbala JO, Baccus-Taylor GSH, Thomas S (2003). Evaluation of breadfruit (*Artocarpus communis*) in traditional stiff porridge foods. J. Food Agric. Environ. 1(2): 54-59

Medcalf DG, Gilles RA (1965). Wheat starches: comparison of physiochemical properties. Cereals Chem. 42:558-66.

Okaka JC, Potter NN (1979). Physiochemical and functional properties of Cowpea flours, J. Food Sci. 44:1237-42.

Oladeji B, Akanbi CT (2011). Physico-chemical and sensory evaluation of "poundmix" from yam, cocoyam, breadfruit and plantain blends. Nutr. Food Sci. 41(6):430-436

Oluwamukomi MO, Eleyinmi AF, Enujiugha VN (2005). Effect of soy supplementation and its stage of inclusion on quality of Ogi – a fermented maize meal. J. Food Chem. 91:651-657

Omobuwajo TO (2003). Compositional characteristic and sensory quality of biscuit, prawn cracker and fried chips produced from breadfruit. Innov. Food Sci. Emerg. Technol. 4:219-225

Opara LU (2007). Yam: Post Harvest Operation. Massey University, Palmerston North, New Zealand

Smith IF, Ojofeitimi EO (1995). Nutrition and Diet Therapy for Health Care Professionals in Africa. Y. Books, Ibadan Nigeria. pp. 13-14.

Shue WC, Tripples KH (1982). The Amylograph Handbook, American Association of Cereal Chemist, St Paul, MN

Svanberg U (1987). Dietary bulk in weaning foods. Department of Food Science, Chalmers University of Technology, p. 3.

Effect of fermentation on physico-chemical, textural properties and yoghurt bacteria in mango soy fortified yoghurt

Ashu Walia[1], H. N. Mishra[2] and Pradyuman Kumar[1]

[1]Department of Food Engineering and Technology, Sant Longowal Institute of Engineering and Technology, Longowal - 148106 India.
[2]Department of Agricultural and Food Engineering, Post Harvest Technology Centre, Indian Institute of Technology, Kharagpur 721302 India.

Mango soy fortified yoghurt (MSFY) samples were prepared by using blends of 78.3% milk containing 3% fat, 14.5% soymilk containing 8.2% total solids and 7.2% mango pulp containing 18% total solids. The physico-chemical, textural properties and yoghurt bacteria counts of MSFY were analyzed at an interval of 30 min during fermentation upto 270 min. The fermentation time had a positive effect on acidity but a negative effect on pH, total solids and sugars in MSFY. Quadratic model were well fitted into different physico-chemical properties of MSFY during fermentation. Textural profile data of MSFY during fermentation were: hardness (46.44 - 154.6 g), cohesiveness (0.4133 - 0.8137), adhesiveness (-2.57 to -59.77 g.s), springiness (1.037 - 1.416) and gumminess (26.22 - 93.80 g). The *Streptococcus thermophilus* counts increased from 1.3 × 10^8 to 1.75 × 10^9 cfu/ml and *Lactobacillus bulgaricus* from 0.9 × 10^8 to 1.17 × 10^9 cfu/ml. Specific growth rate of *S. thermophilus* is higher than that of *L. bulgaricus* during MSFY fermentation.

Key words: Yoghurt, mango, soymilk, fermentation, texture, physico-chemical properties.

INTRODUCTION

Milk, for thousands of years, has been transformed through microbial fermentation into various food products with high nutritional value (Tamime and Robinson, 1999). Yoghurt is a coagulated milk product obtained by the lactic acid fermentation through the action of *Streptococcus thermophilus and Lactobacillus delbrueckii* spp. *bulgaricus*. The product is accepted by consumers due to its flavor and aroma, mainly attributed to acetaldehyde and texture. Viability and activity of yoghurt bacteria are important commercial considerations so that they survive throughout shelf life, transit through acidic conditions in the stomach as well as enzymes and bile salts in the small intestine.

Fortification of yoghurt with soy solids and mango pulp not only increases its nutrient content including minerals and vitamins but also adds to its health promoting value. Soymilk based yoghurt would offer several distinct nutritional advantages over milk yoghurt to the consumer, that is reduced level of cholesterol, saturated fat and lactose (Lee et al., 1990). Soybean is a good source of niacin, riboflavin, iron, potassium, calcium, magnesium and phosphorus, with several fat and water soluble vitamin B-complex (Zucker and Zucker, 1943). Addition of mango pulp adds a variety for consumer to select as they both add to the nutritional value of the yoghurt. Mango pulp contains a high proportion of sugar, mostly monosaccharide that is easily adsorbed by the digestive system.

Bonezar et al. (2002) studied the effects of certain factors on the properties of ewe's milk and observed that the overall properties of yoghurt, such as acidity level, the production of aroma compounds (diacetyl, aceteldehyde) as well as the nutritional value are important traits of the product. These aspects are influenced by the chemical composition of the milk base, processed conditions, the added flavors and the activity of starter culture during the incubation period. There have been many investigations involving optimization of yoghurt texture. These studies have demonstrated that the total solids and fat levels in the milk, heat treatment of the milk prior to inoculation, homogenization, presence of stabilizers (Kumar and Mishra, 2004) and incubation conditions will affect the body of the final product (Schellhaass and Morris, 1985).

Buono et al. (1990) studied the carbohydrate utilization and growth kinetics in the production of yoghurt from soymilk. Utilization of soymilk as an ingredient for human food necessitates a reduction in the concentration of the carbohydrates that can cause flatulence and a reduction in the objectionable beany flavor typically associated with soy products. The fermentation process undergoes several distinct phases based on the changes in pH and acidity and sugar content. The first stage has a rapid decrease of pH and an increase of acidity. These changes are accompanied by a decrease of reducing sugars after the initial lag phase. The next stage shows a gradual drop in pH, a further increase in acidity and a rapid disappearance of reducing sugars. The final stage of fermentation proceeds with slight changes in pH, acidity and reducing sugars.

In the present paper, the effect of fermentation time on physico-chemical, textural and growth kinetics of yoghurt bacteria in Mango soy fortified yoghurt (MSFY) is discussed.

MATERIALS AND METHODS

Pure freeze dried cultures of *S. thermophilus* NCDC 074 and *L. delbrueckii* spp. *bulgaricus* NCDC 009 were procured from National Dairy Research Institute, Karnal, India. These cultures were maintained in 12% skim milk (SM) media and autoclaved at 15 psi for 20 min. Toned milk used for yoghurt preparation was procured from local market and has been standardized to a fat content of 3%. Soymilk was prepared in the laboratory from whole soybeans using the procedure described by Bourne et al. (1976). A canned mango pulp (Kaytis brand) containing 18% total solid was purchased from local market. These were used to prepare experimental samples.

MSFY preparation

The blend of 78.3% milk (3% fat), 14.5% soymilk (8.2% total solids) were heated to 95°C for 5 min, cooled to 45°C and 7.2% mango pulp (18% total solids) was added with high speed stirring. The blends were inoculated with 2% (v/v, *S. thermophilus* and *L. bulgaricus* in 1:1 ratio) of 24 h old culture. The blend was incubated at 42 ± 2°C in a plastic cup (100 ml). Mango soy fortified yoghurt (MSFY) was prepared by as per method (Kumar and Mishra, 2003a). Samples were kept for 2 h at 8°C before analysis.

Addition of inoculums

Two cultures *S. thermophilus* and *L. bulgaricus* were used for studying the effect of fermentation time on physico-chemical properties and growth kinetics of Mango Soy Fortified Yoghurt (MSFY). The physicochemical properties of MSFY such as pH, acidity, total solids, sugars were analyzed at an interval of 30 min during fermentation up to 270 min. Microbiological and textural properties were also studied for each 30 min after the culture had been allowed to grow for 270 min.

The experiments were carried out in triplicates. All physico-chemical and textural properties were fitted into linear and quadratic equations to see the effect of fermentation time. The effect of fermentation time on microbial growth was described by the exponential phase. The exponential phase may be described by the equation:

$$dx / dt = \mu x \qquad (1)$$

Where:

x = microbial counts (cfu/ml)
t = time (min)
μ = specific growth rate (min^{-1})

Integrating Equation (1) gives:

$$x_t = x_0 e^{\mu t} \qquad (2)$$

Where:

x_0 = original microbial counts (cfu/ml)
x_t = microbial counts after the time interval, t (min)
t = Time (min)
e = base of the natural logarithm.

Taking natural logarithms, Equation (2) becomes:

$$\ln x_t = \ln x_0 + \mu t \qquad (3)$$

Thus, a plot of natural logarithm of microbial counts of ST and LB against time should yield a straight line, the slope of which would be the value of specific growth rate. During the exponential phase, nutrients are in excess and the organism is growing at its maximum specific growth rate, μ_{max}, for the prevailing conditions (Stanbury et al., 1995).

Analysis of physicochemical characteristics

Physicochemical properties of MSFY such as pH, titrable acidity (IDF, 1991a) and total solids (IDF, 1991b) were determined. Reducing and non reducing sugars were estimated by Lane and Eynon Method (Ranganna, 1996).

Texture profile analysis

In the recent past, instrumental texture analysis (TPA) has been applied as a useful method to evaluate mechanical properties in a

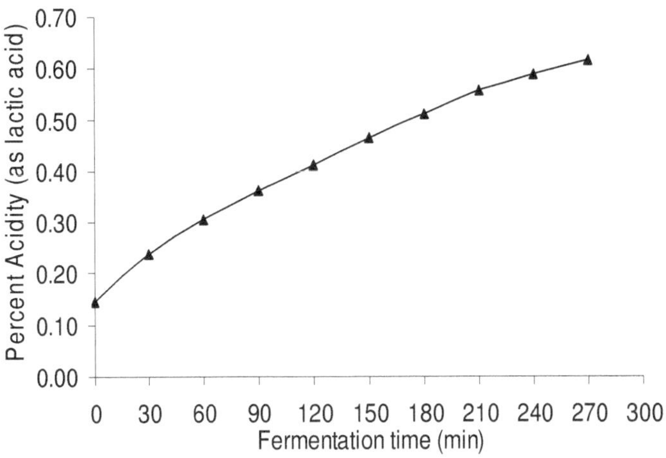

Figure 1a. Effect of fermentation time on per cent acidity (as lactic acid).

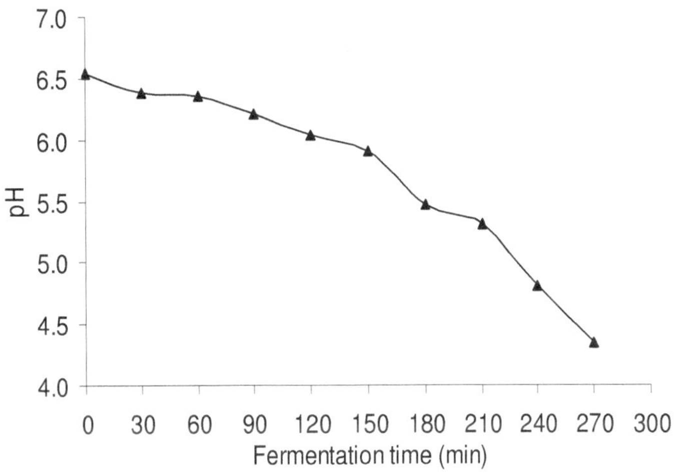

Figure 1b. Effect of fermentation time on pH.

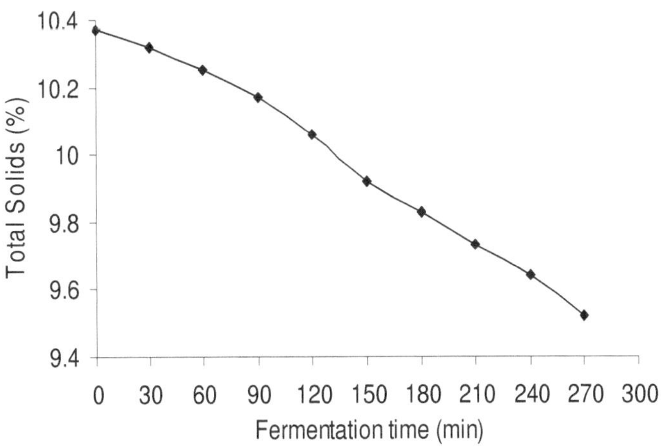

Figure 1c. Effect of fermentation time on total solids (%).

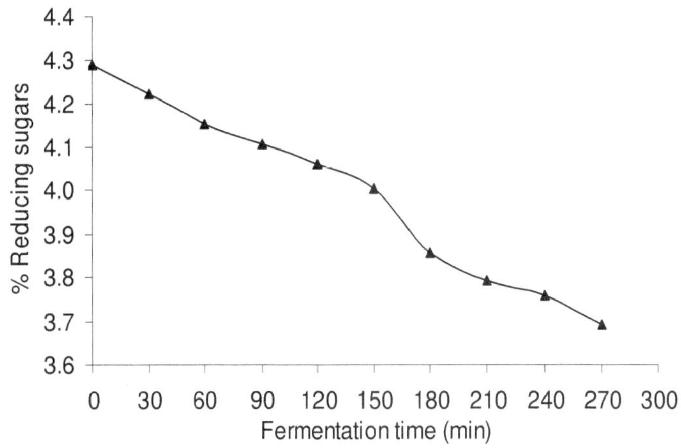

Figure 1d. Effect of fermentation time on reducing sugars (%).

wide range of foods (Pons and Fiszman, 1996). TPA compresses a piece product twice imitating the conditions in the mouth (Szczesniak et al., 1963). Five basic textural parameters may be obtained in only one test (hardness, cohesiveness, springiness, adhesiveness and gumminess), extracted from the resulting force-time curves. Texture profile analysis was carried out by the method described by Kumar and Mishra (2003b). The data obtained from typical texture profile curve was used for the calculation of the textural parameters (Bourne, 1978; Bourne, 1982) as follows:

Hardness, g = maximum force of the first compression

$$\text{Cohesiveness} = \frac{\text{Area under } 2^{nd} \text{ compression (Area 2)}}{\text{Area under } 1^{st} \text{ compression (Area 1)}}$$

Adhesiveness, g.s = negative area in the graph (Area 3).

Springiness = Length 2 / Length 1

Gumminess, g = Hardness × Cohesiveness

Enumeration of *S. thermophilus* and *L. bulgaricus*

International Dairy Federation methods (1988) were used for the enumeration of *S. thermophilus* and *L. bulgaricus*. The *S. thermophilus* agar (M948), *L. delbrueckii* subsp. *bulgaricus* agar (M927) produced from Hi Media Laboratories Limited, Mumbai, were used. The samples were diluted in peptone (1%) in water. The agar plates of *S. thermophilus* were incubated aerobically at 37°C for 48 h while the plates of *L. bulgaricus* were incubated at 37°C for 72 h under anaerobic conditions. After incubation, typical colonies of *S. thermophilus* and *L. bulgaricus* were counted. Plates with 30 - 300 colonies were used in counting and the counts were averaged for three replicates and expressed as c.f.u/ml.

The significance of fermentation time was determined through Single factor analysis (ANOVA) of the data obtained in three experimental runs.

RESULTS AND DISCUSSION

The results of different physico-chemical parameters that is, acidity, pH, total solids, reducing sugars and total sugars are shown in Figures 1a to 1e. The fermentation time

Table 1. Linear and quadratic models of different physico-chemical properties of MSFY.

Parameters	Linear model ($y = mx + c$)			Quadratic model ($y = ax^2 + bx + c$)			
	m	c	R^2	a	b	c	R^2
Acidity	0.0017	0.191	0.979	-3×10^{-6}	0.0026	0.1569	0.998
pH	-0.0077	6.7781	0.922	-3×10^{-5}	-4×10^{-5}	6.4699	0.994
Total solids	-0.0033	10.422	0.992	-3×10^{-6}	-0.0025	10.393	0.996
Reducing sugars	-0.0023	4.3002	0.985	-1×10^{-6}	-0.002	4.2895	0.986
Total sugars	-0.0022	4.939	0.994	-1×10^{-6}	-0.0019	4.9272	0.995

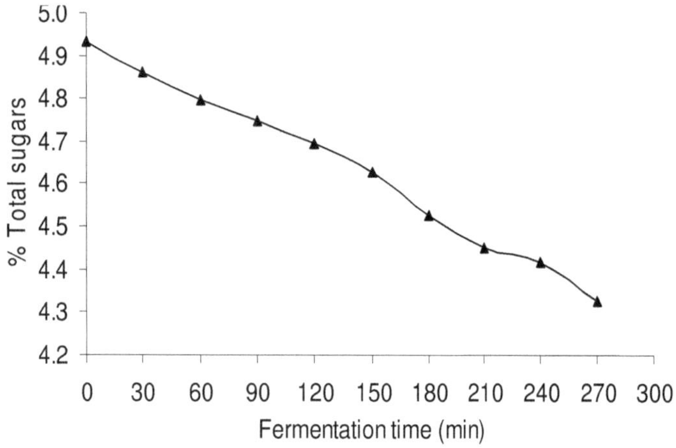

Figure 1e. Effect of fermentation time on total sugars (%) of MSFY.

had a positive effect on acidity but a negative effect on pH, total solids, reducing and total sugars in MSFY (Table 1). The acidity of MSFY as % lactic acid increased from 0.15 at 0 min to 0.62 at 270 min of fermentation and total solids decreased from 10.37 to 9.52. The pH decreased from 6.53 to 4.34 and total sugars from 4.93 to 4.33 at these fermentation times.

Different culture combination of *S. thermophilus* (HST, 197) and *L. bulgaricus* (RTS, Yb) were studied for total acid production in cow milk and buffalo milk from 2 - 10 h of incubation. It was observed that with the increase in incubation time, there was an increase in acid production in the culture combinations of different strains in both types of milk. Earlier reports by Macquot and Hurrel (1970) and Tramer (1973) have indicated that with the increase in fermentation time, there was an increase in acid production in the culture combinations of different strains in both cow and buffalo milk from 10 to 12 h of incubation. In yoghurt, lactose utilization proceeds as long as the external pH allows metabolic activity of the lactic acid bacteria. When lactic acid is produced in yoghurt and pH has reached 4.0, *L. bulgaricus* cease their metabolic activity and lactose is no further utilized. By rerouting metabolism of the lactic acid bacteria towards production of more pH-neutral components, further conversion of lactose is possible (de Vos, 1996). Goodenough and Kleyn (1976) investigated qualitative and quantitative changes in sugars of yoghurt during its manufacture and reported average lactose concentration of 8.5% in yoghurt mix which dropped to 5.75% with a concomitant increase of 1.20% galactose. Glucose was detected only in trace quantities. Commercial yoghurts had an average of 4.06% lactose, 1.85% galactose, 0.05% glucose.

The dependency of response variables on coefficients associated with linear and quadratic equations and their significance with coefficient of determination (R^2) of physicochemical properties is shown in Table 1. High coefficient of determination (R^2) explained the goodness of fit in the experimental data. The quadratic model was found to be suitable than corresponding linear model for each physico-chemical property. The fermentation time had significant ($P < 0.01$) effects after analyzing the data for different experimental runs through Single factor analysis (ANOVA) (Table 2) on all physico-chemical characteristics of yoghurt sample during fermentation.

Textural characteristics

Yoghurt rheology is described not only in terms of viscosity, viscosity loss and its recovery, but is also described in terms of how hard, brittle, elastic and cohesive it is. Textural characteristics and rheological properties of coagulated dairy products are affected by their structural characteristics. The structural arrangement of the network determines the textural characteristics of these coagulated products and is affected by factors such as composition and manufacturing processes (Rawson and Marshall, 1997). Force measuring instruments like the TA-XT2 texture analyzer allow food materials to be evaluated in such terms and have been used to assess yoghurt texture (Toba et al., 1990; Kaur et al., 2009).

TPA curves of MSFY were obtained during fermentation at 120,150,180, 210, 240 and 270 min fermentation time. TPA curves at 0, 30, 60, 90 min were not possible as the force required by the probe to mea-sure TPA curves was not attained at these fermentation times. The results of different TPA curves are summarized in Table 3.

Hardness (force necessary to attain a given deformation) is a commonly evaluated parameter when

Table 2. Single factor analysis (ANOVA) for physico-chemical properties and *S. thermophilus* (ST) and *L. bulgaricus* (LB) counts of MSFY.

Physico-chemical and microbiological properties	Source of variation	Sum of squares (SS)	Degree of Freedom (df)	Mean sum of square (MSS)	Calculated F - value	P-value	F-critical
Acidity	Fermentation time	90557.98	1	90557.98	21.95	1.84×10^{-4}	4.41
	Error	74250.2	18	4125.01			
	Total	165808.2	19				
pH	Fermentation time	83550.65	1	83550.65	20.25	2.77×10^{-4}	4.41
	Error	74254.83	18	4125.27			
	Total	157805.5	19				
Total solids	Fermentation time	78148.75	1	78148.75	18.94	4.32×10^{-4}	4.41
	Error	74250.8	18	4125.04			
	Total	152399.6	19				
Reducing sugars	Fermentation time	85812.42	1	85812.42	20.80	2.42×10^{-4}	4.41
	Error	74250.39	18	4125.02			
	Total	160062.8	19				
Total sugars	Fermentation time	84970.82	1	84970.82	20.60	2.54×10^{-4}	4.41
	Error	74250.37	18	4125.02			
	Total	159221.2	19				
ST	Fermentation time	25063.2	1	25063.2	5.24	3.44×10^{-2}	4.41
	Error	86108.27	18	4783.793			
	Total	111171.5	19				
LB	Fermentation time	32213.69	1	32213.69	6.62	1.92×10^{-2}	4.41
	Error	87613.51	18	4867.417			
	Total	119827.2	19				

Table 3. Textural profile data of TPA curves at different fermentation time of MSFY.

Fermentation time (min)	Hardness (g)	Cohesiveness	Adhesiveness (g.s)	Springiness	Gumminess (g)
120	46.44	0.5646	-2.57	1.126	26.22
150	63.6	0.4133	-5.48	1.078	26.28
180	66.9	0.4906	-5.49	1.416	32.82
210	76.6	0.3674	-22.71	1.067	28.14
240	93.1	0.8137	-43.30	1.066	75.76
270	154.6	0.6067	-59.77	1.037	93.80

determining yoghurt texture. Hardness of yoghurt increases with the increase in fermentation time. The hardness of the yoghurt increases with the duration of heating and has been attributed to sulphydral groups (Vijayananda et al., 1989). During the fermentation period of milk, gel is produced to make yoghurt, thus increasing the hardness. In semi solid like products, cohesiveness represents how well the product withstands a second deformation relative to how it behaved under the first deformation. An important characteristic of TPA curve is the negative area between the first and second compression cycle which corresponds to adhesiveness of product. Adhesiveness is the force necessary to remove the material that adheres to the mouth during eating. Springiness is the rate at which the sample returns to its original shape when the deforming force is removed. The textural data of MSFY were hardness (46.44 to 154.6 g), cohesiveness (0.4133 to 0.8137), adhesiveness (-2.57 to -59.77 g.s), springiness (1.037 to 1.416) and gumminess (26.22 to 93.80 g) during fermentation of MSFY. Hardness, gumminess, springiness, adhesiveness and cohesiveness were significantly ($P < 0.01$) affected by

Table 4. Analysis of variance (ANOVA) for textural properties of MSFY.

Properties of MSFY	Source of variation	Sum of Squares (SS)	Degree of freedom (df)	Mean sum of square (MSS)	Calculated F - value	P-value	F-critical
	Fermentation time	46720.13	1	46720.13	25.03	5.35×10^{-4}	4.96
Hardness	Error	18663.72	10	1866.37			
	Total	65383.85	11				
	Fermentation time	113440.9	1	113440.90	72.03	6.99×10^{-6}	4.96
Cohesiveness	Error	15750.13	10	1575.01			
	Total	129191	11				
	Fermentation time	142859.9	1	142859.9	77.03	5.18×10^{-6}	4.96
Adhesiveness	Error	18544.85	10	1854.485			
	Total	161404.8	11				
	Fermentation time	112754.8	1	112754.8	71.59	7.18×10^{-5}	4.96
Springiness	Error	15750.1	10	1575.01			
	Total	128504.9	11				
	Fermentation time	72732.75	1	72732.75	40.53	8.18×10^{-5}	4.96
Gumminess	Error	17945.5	10	1794.55			
	Total	90678.25	11				

fermentation time (Table 4).

Yoghurt bacteria counts

During incubation, as the starter culture grows, a positive interaction is generally observed between *S. thermophilus* and *L. bulgaricus* mixed cultures. The *S. thermophilus* and the *L. bulgaricus* are inoculated simultaneously at 1:1 composition, for instance and remain present throughout the production of yoghurt, as well as in the final product. When both bacteria grow in association, the times for milk coagulation are faster than if either of them is grown separately. The *S. thermophilus* grow and, while they grow, they produce formic acid that, in turn, stimulates the growth of the *L. bulgaricus.* The activity of the latter on casein induces the presence of free amino acids, which, in turn, stimulate the growth of the former (Ginovert et al., 2002).

Production of MSFY involves batch culture fermentation. There were marked changes in starter counts with the fermentation time. Data were plotted as the logarithm of the microbial counts of ST and LB as shown in Figure 2a and 2b. Batch growth curves are prepared based on measurement of viable cell counts (VCC) for *L. bulgaricus* and *S. thermophilus.* As shown in Figure 2, both organisms displayed relatively straight line plots. The *S. thermophilus* and *L. bulgaricus* counts increased from 0 to 270 min of fermentation. The *S. thermophilus* and *L. bulgaricus* counts varied from 1.3×10^8 to 1.75×10^9 cfu / ml and 0.9×10^8 to 1.17×10^9 cfu / ml, respectively.

The fermentation time had a significant effect ($P < 0.05$) on starter counts (Table 2). F_{cal} value (5.24) is greater than F_{tab} value (4.41) which signifies that fermentation time is significantly affecting the number of Streptococci cells at $P < 0.05$. The Single factor analysis for *L. bulgaricus* counts (Table 2) also showed the significant effect of fermentation time on *lactobacilli* cells at $P < 0.05$.

A plot of natural logarithm of microbial counts of ST and LB against time yield a straight line (Figure 2), the slope of which was the specific growth rate (μ_{max}) for each microorganism.

Variations in *S. thermophilus* cell numbers

The equation for the increase in the number of *streptococci* cells in MSFY, 270 min after inoculation:

$$ln(ST) = 0.0092 \times t + 19.145 \ (R^2 = 0.9012)$$

Slope = μ_{max} = 0.0092 where μ_{max} is the maximum specific growth rate of *streptococci.*

Variations in *L. bulgaricus* cell numbers

The equation for the increase in the number of *lactobacilli*

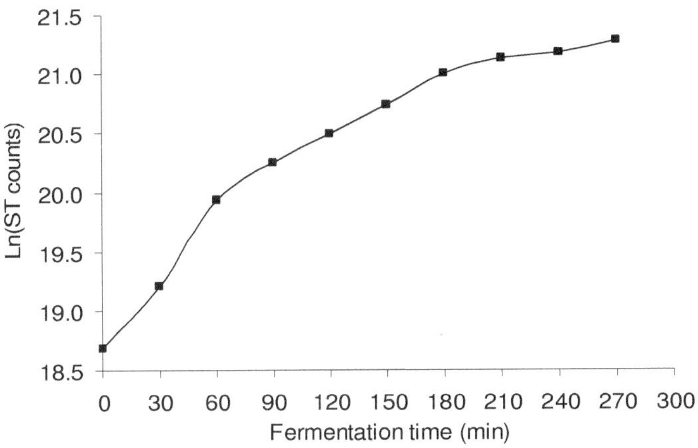

Figure 2a. Growth curves of yoghurt bacteria *Streptococcus thermophilus* during fermentation of MSFY.

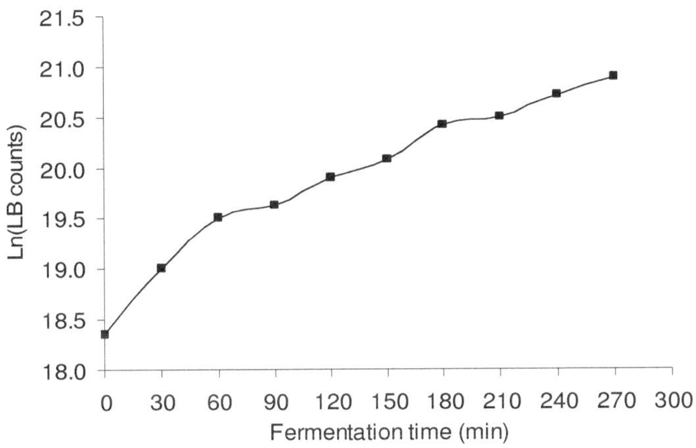

Figure 2b. Growth curves of yoghurt bacteria *Lactobacillus bulgaricus* during fermentation of MSFY.

cells in MSFY, 270 min after inoculation:

$ln(LB) = 0.0085 \times t + 18.745$ ($R^2 = 0.9475$)

Slope = $\mu_{max} = 0.0085$ where μ_{max} is the maximum specific growth rate of *lactobacilli*.

Therefore, μ_{max} for *streptococci* was larger than that for *lactobacilli* which indicated that *S. thermophilus* grows faster than *L. bulgaricus*. This result was in agreement with the earlier findings of Ginovert et al. (2002). They studied the simulation modeling of bacterial growth and noticed a faster increase for both *S. thermophilus* and *L. bulgaricus* in the mixed culture, a consequence of their symbiotic relation.

Conclusion

All the physico-chemical, textural properties and yoghurt bacteria counts were analyzed at an interval of 30 min after the two cultures that is ST and LB had been allowed to grow for 270 min. The acidity (as % lactic acid) increased whereas pH, total solids and sugars decreased with the increase in fermentation time. The value of acidity of MSFY as % lactic acid ranged from 0.15 at 0 min to 0.62 at 270 min of fermentation. The decrease in pH was from 6.53 to 4.34 and reduction in total sugars was 4.93 to 4.33 while reducing sugars decreased from 4.29 to 3.69 at these fermentation times. The TA-XT2 Texture Analyzer was used to measure the textural characteristics of MSFY. The TPA curves were obtained at 120 - 270 min of fermentation. TPA curves at 0 - 90 min were not possible as the force required by the probe to measure TPA curves was not attained at these fermentation times. The *S. thermophilus* and *L. bulgaricus* counts increased from 0 to 270 min of fermentation. The *S. thermophilus* counts increased from 1.3×10^8 to 1.75×10^9 cfu/ml and *L. bulgaricus* from 0.9×10^8 to 1.17×10^9 cfu /ml.

REFERENCES

Bonezar G, Wszolek M, Siuta, A (2002). The effects of certain factors on the properties of yoghurt made from ewe`s milk. Food Chem. 79: 85-91.

Bourne MC (1978). Texture profile analysis. Food Technol. 32(7): 62-72.

Bourne MC (1982). Food Texture and Viscosity: Concept and Measurement. Academic Press, New York.

Bourne MC, Escuela EE, Benzon J (1976). Effect of sodium alkali and salt on pH and flavor of soymilk. J. Food Sci. 41: 62- 66.

Buono M, Larry E, Fung D, Jeon IJ (1990). Carbohydrate Utilization and Growth Kinetics in the Production of Yoghurt from Soymilk Part I: Experimental Methods. J. Food Proc. Preserv. 13: 135-153.

De Vos WM (1996). Metabolic engineering of sugar catabolism in lactic acid bacteria. Anotonie van Leeuwenhoek 70: 223-242.

Ginovert M, Lopez D, Valls J, Silbert M (2002). Simulation modeling of bacterial growth in yoghurt. Int. J. Food Microbiol. 73: 415- 425.

Goodenough ER, Kleyn DA (1976). Qualitatative and quantitative changes in carbohydrates during the manufacture of yoghurt. J. Dairy Sci. 59: 45-47.

International Dairy Federation (1988). Yoghurt: Enumeration of characteristics organisms-colony count technique at 37ºC. IDF Standard No. 117 A Brussels.

International Dairy Federation (1991a). Yoghurt- determination of titrable acidity- potentiometric method. IDF Standard No. 150 Brussels.

International Dairy Federation (1991b). Yoghurt- determination of total solids IDF Standard No. 151 Brussels.

Kaur H, Mishra HN, Kumar P (2009). Textural properties of mango soy fortified probiotic yoghurt: optimization of inoculum level of yoghurt and probiotic culture. Int. J. Food Sci. Technol. 44(2): 415-424.

Kumar P, Mishra HN (2003a). Optimization of Mango soy fortified yoghurt formulation using response surface methodology. Int. J. Food Prop. 6(3): 499-517.

Kumar P, Mishra HN (2003b). Effect of mango pulp and soymilk fortification on textural profile of set yoghurt made from buffalo milk. J. Texture Stud. 34(3): 249-269.

Kumar P, Mishra HN (2004). Mango soy fortified set yoghurt: effect of stabilizer addition on physico-chemical, sensory and textural properties. Food Chem. 87: 501-507.

Lee SY, Morr CV, Seo A (1990). Comparison of milk based and soymilk based yoghurt. J. Food Sci. 55(2): 532-536.

Macquot G, Hurrel C (1970). The Selection and use of some micro-orgs for the manufacture of fermented and acidified milk products. J. Soc. Dairy Technol. 23: 130.
Pons M, Fiszman SM (1996). Instrumental texture profile analysis with particular reference to gelled systems. J. Texture Stud. 27: 597-624.
Ranganna S (1996). Handbook of analysis and quality control for fruit and vegetable products. Tata McGraw-Hill Publishing Co. Ltd.
Rawson HL, Marshall VM (1997). Effect of ropy strains of *Lactobacillus delbrueckii* ssp. bulgaricus and *Streptococcus thermophilus* on rheology of stirred yoghurt. Int. J. Food Sci. Technol. 32: 213- 220.
Schellhaass SM, Morris HA (1985). Rheological and Scanning Electron Microscopic Examination of skim milk gels by fermenting with ropy and non-ropy strains of lactic acid bacteria. Food Microstruc. 4: 279-287.
Stanbury PF, Whitaker A, Ha SJ (1995). Microbial growth kinetics. Principles of Fermentation Technology, 2nd edition, pp. 13-16.
Szczesniak AS, Brandt MA, Friedman HH (1963). Development of standard rating scales for mechanical parameters of texture and correlation between the objective and sensory methods of texture evaluation. J. Food Sci. 28: 397-403.
Tamime AK, Robinson RK (1999). Background to manufacturing practices. Yoghurt Science and Technology. Woodhead Publishing Ltd. and CRC Press LLC, Boca Raton, Florida, pp. 11-619.
Toba T, Nakajima H, Tobitani A, Adachi S (1990). Scanning electron microscope and texture studies on characteristics consistency of Nordic ropy sour milk. Int. J. Food Microbiol. 11: 313- 320.
Tramer J (1973). Yoghurt Cultures. Nature 211: 204.
Vijayananda P, Mittal BK, Kulshreshtha M (1989). Instrumental Textural Profile Analysis of soybean curds prepared by acid and salt coagulation. J. Food Sci. Technol. 26(4): 223- 224.
Zucker TF, Zucker L (1943). Nutritive value of cotton, peanut and soy seeds. Ind. Eng. Chem. 35: 868- 871.

7

Resistance to antimicrobials and acid and bile tolerance of *Bacillus* spp isolated from *Bikalga*, fermented seeds of *Hibiscus sabdariffa*

Clarisse S. Compaoré[1,2] , Lars B. Jensen[3], Bréhima Diawara[1], Georges A. Ouédraogo[2], Mogens Jakobsen[4] and Labia I. I. Ouoba[5,6]

[1]Département Technologie Alimentaire (DTA/IRSAT/CNRST), Ouagadougou 03 BP 7047, Burkina Faso.
[2]Institut du Développement Rural/Université Polytechnique de Bobo, BAMSB, 01 BP 10 91 Bobo-Dioulasso, Burkina Faso.
[3]Technical University of Denmark (DTU), National Food Institute, Division of Food Microbiology, Mørkhøj Bygade 19, 2860 Søborg, Denmark.
[4]Department of Food Science, Faculty of Science, University of Copenhagen, Rolighedsvej 30, 1958 Frederiksberg C, Denmark.
[5]London Metropolitan University, FLSC/SHS, Microbiology Research Unit, 166-220 Holloway Road, London N7 8DB, United Kingdom.
[6] Independent Senior Research Scientist-Consultant, London, UK.

In the aim of selecting starter cultures, thirteen species of *Bacillus* spp. including six *Bacillus subtilis* ssp. *subtilis*, four *Bacillus licheniformis* and three *Bacillus amyloliquefaciens* ssp. *plantarum* isolated from traditional *Bikalga* were investigated. The study included, for all isolates, genes, determination of minimal inhibitory concentration (MIC) for 24 antimicrobials and detection of resistance by PCR using specific primers. The isolates were also examined for their resistance to pH 2.5 and their tolerance to 0.3% bile over 4 h. Results showed that most studied isolates, in particular *B. subtilis* ssp. *subtilis* G2, H4, C6, I7 and *B. amyloliquefaciens* ssp. *plantarum* A4, I8, G3 were susceptible to most antimicrobials tested while all *B. licheniformis* isolates showed high resistance level. The resistance observed towards the antimicrobials (chloramphenicol, erythromycin, kanamycin, penicillin, streptomycin and trimethoprim) in this study may be intrinsic as no positive amplicon was observed for the most prevalent resistance genes investigated (*catp*IP501, *erm*(A), *erm*(B), *erm*(C), *aph*(3")-I, *aph*(3")-III, *ant*(2")-I, *blaZ*, *aadA*, *aadE*, *StrA*, *StrB*, *dfr*(A)). Furthermore, based on their good survival in pH 2.5 and in 0.3% bile all the tested isolates may be able to resist passage through the gastro-intestinal tract conditions. Regarding these results, isolates G2, C6, I7, H4, A4, I8 and G3 may be useful as starter cultures to optimize *Hibiscus sabdariffa* seeds fermentation into *Bikalga*.

Key words: *Bikalga, Bacillus*, antimicrobial resistance, acid resistance, bile tolerance, starter cultures.

INTRODUCTION

Bikalga is a traditional food condiment obtained by an alkaline fermentation of *Hibiscus sabdariffa* seeds. It is generally used as a seasoning condiment in most staples in Burkina Faso and is also known in other African countries under different names such as *Dawadawa bosto* in Niger, *Datou* in Mali, *Furundu* in Sudan and *Mbuja* in Cameroun (Yagoub et al., 2004). The main steps of *Bikalga* production process include cleaning of the seeds, cooking and a two-stage spontaneous fermentation (Parkouda et al., 2008). The nutritional value of *Bikalga* has been studied showing that it is a rich source of proteins, lipids, carbohydrates, essential amino acids, fatty acids and vitamins (Bengaly et al., 2006; Parkouda et al., 2008; Yagoub et al., 2004). Members of the *B. subtilis* group were shown to be the main microorganisms involved in the fermentation of *H. sabdarifa* seeds into *Bikalga* (Ouoba et al., 2008). The nutritional value of *Bikalga* as well as other similar African traditional fermented foods is believed to be due to the fermentation process and metabolic activities of the associated microorganisms (Bengaly 2001; Mohamadou et al., 2010). However, the production of such products relies on spontaneous fermentation with uncontrolled processes and hazardous starters, leading to a product of varying hygienic, nutritional and organoleptic quality. Therefore, more emphasis has since been put on the mastering of starter cultures for use in a more standardized process to produce African fermented products. In this aim, in previous studies, we have pre-selected potential starter cultures from *Bikalga* predominant *Bacillus* spp. based on their capacity to inhibit pathogens and spoilage microorganisms via bacteriocin and lipopeptide antibiotics production (Compaoré et al., 2013 a, b).

However, in addition to antimicrobial properties, another required property of starter cultures is that they are safe for human consumption (Ammor et al., 2007). Such safety includes that they do not harbor acquired and transferable antimicrobial resistance elements. Indeed, antimicrobial resistance is a worldwide public health problem that continue to grow and bacteria used as starter cultures for the production of foods could be a source of spread of antimicrobial resistances, which might be transferred to commensal or pathogenic bacteria (Danielsen and Wind, 2003; Ammor et al., 2007). Furthermore, in order to survive and establish within the human Gastro Intestinal Tract (GIT), some of the desirable properties of starter cultures/probiotics include their ability to resist the acidity (pH 2.5-pH 3.5) of the stomach and the exposure to bile in the upper part of the intestine (Holzapfel et al., 1998; Huang and Adams, 2004).

The aim of the present study was to determine the antimicrobial resistance profile of the predominant *Bacillus* spp. isolated from *Bikalga* and to explore their capacity to survive in acidic pH and to tolerate bile salts, in order to select suitable starter cultures for a controlled fermentation of *H. sabdariffa* seeds.

MATERIALS AND METHODS

Bacterial strains

Thirteen (13) strains of *Bacillus* including six strains of *B. subtilis* ssp. *subtilis* (F1, C3, C6, H4, G2, I7), four strains of *B. licheniformis* (E3, F9, J3, E5) and three strains of *B. amyloliquefaciens* ssp. *plantarum* (A4, I8, G3) isolated from different productions of *Bikalga* were investigated. The *Bacillus* isolates were maintained as stock cultures at -80°C in Brain Heart Infusion (BHI, broth, Oxoid CM1135 Basingstoke, Hampshire, England) supplemented with 20% (v/v) glycerol. The strains were sub-cultured in 10 ml BHI broth at 37°C for 24 h before the cells were used.

Resistance of bacteria to antimicrobials

Determination of the susceptibility of the studied bacteria to antimicrobials: determination of Minimal Inhibitory Concentrations (MIC)

For 24 antimicrobials (Table 1), MICs were determined by the microwell and agar dilution methods using BHI broth (Oxoid CM1135) and BHI agar (Oxoid CM1136). This was done according to standards set by Clinical and Laboratory Standard Institute (CLSI), Wayne, USA. Breakpoint values towards antimicrobials for *Bacillus* spp, were used as described by European Food Safety Agency (EFSA, 2008). For antimicrobials not included in the EFSA (2008) list, breakpoints for closely related Gram-positive species were used tentatively (Table 1).

Detection of resistance genes by PCR

For some antimicrobials to which the tested bacteria showed reduced susceptibility, PCR were conducted to screen the isolates for the presence of resistance genes. Amplification of genes associated with resistance to chloramphenicol (*catp*IP501), erythromycin (*erm*(A), *erm*(B), *erm*(C)), kanamycin (*aph*(3")-I, *aph*(3")-III, *ant*(2")-I), penicillin (*blaZ*), streptomycin (*aadA, aadE, StrA, StrB*), and trimethoprim (*dfr(A)*) was done accordingly to Jensen et al. (1999). All PCR were performed using the following temperature program: 94°C for 3 min, 25 or 35 cycles of 94°C for 1 min, 45–65°C, according to annealing temperature for the individual primers (Table 2) and extension at 72°C for 1 min. A final extension step at 72°C for 10 min ended the PCR protocol. The PCR products (10 µl) were subjected to electrophoresis on 1.5% agarose gels (120 V, 2 h) and the products were visualized by staining with ethidium bromide.

Screening of the *Bacillus* isolates for acid resistance and bile tolerance

Preparation of inocula

From BHI agar plates incubated for 24 h at 37°C, the *Bacillus* strains were sub-cultured under agitation for 18 h at 37°C in 10 ml BHI broth, pH 7. The cultures were centrifuged at 5000 g, 4°C for 10 min and the pellet re-suspended in 5 ml of sterile saline solution (Becton 211677; Sparks, MD, USA), pH 7.0. The number of cells was estimated by microscopy using a counting chamber (Neubauer, Wertheim, Germany) and dilutions were made in sterile saline to obtain an inoculum concentration of approximately 10^6 CFU/ml.

Acid resistance

The method described by Klingberg et al. (2005) was used. The survival of the bacteria in acidic pH was examined in BHI broth adjusted with hydrochloric acid (HCl) 1 N to obtain a final pH of 2.5. For each *Bacillus* isolate, 100 µl of inoculum were added into 10 ml of BHI broth, pH 2.5 and incubated at 37°C in a rotary shaker at 120 cycles per min. Samples were taken at various times (0, 1, 2, 3

Table 1. Antimicrobial breakpoints and resistance genes investigated.

Antimicrobial	Proposed breakpoints (μg/ml)	Source	Resistance genes investigated
Avilamycin	≥ 16	CLSI (*Enterococcus)*	
Bacitracin	≥ 16	Jensen et al. (2001)	
Chloramphenicol	> 8	EFSA (2008)	*catp*IP501
Ceftiofur	≥8	CLSI (*Staphylococcus*)	
Ciprofloxacin	≥ 4	Jensen et al. (2001)	
Daptomycin	≥ 8	CLSI (*Enterococcus)*	
Erythromycin	> 4	EFSA (2008)	*erm*(A), *erm*(B), *erm*(C)
Flavomycin	≥ 16	CLSI (*Enterococcus*)	
Florfenicol	≥ 32	CLSI (*Staphylococcus*)	
Gentamicin	> 4	EFSA (2008)	
Kanamycin	> 8	EFSA (2008)	*aph*(3")-I, *aph*(3")-III, *ant*(2")-I
linezolid	≥ 4	CLSI (*Staphylococcus*)	
Penicillin	≥ 0.25	Luna et al. (2007)	*blaZ*
Salynomycin	≥ 16	CLSI (*Enterococcus*)	
Spectinomycin	≥128	CLSI (*Staphylococcus*)	
Streptomycin	> 8	EFSA (2008)	*aadA, aadE, StrA, StrB,*
Sulphamethoxazol	≥ 256	CLSI (*Staphylococcus*)	
synercid	> 4	EFSA (2008)	
Tetracyclin	> 8	EFSA (2008)	
Tiamulin	≥ 32	CLSI (*Staphylococcus*)	
Tigecycline	> 0.5	Luna et al. (2007)	
Trimethoprim	≥ 16	CLSI (*Staphylococcus*)	*dfr(A)*
TMP+SMX	≥4	Luna et al. (2007)	
Vancomycin	> 4	EFSA (2008)	

and 4 h), serially 10-fold diluted and plated in duplicate onto BHI agar. The plates were incubated at 37°C for 24 h and bacterial colonies were counted. The number of bacteria was calculated according to the standard of ISO 7218 (2007).

Tolerance to bile salts

The bile tolerance was examined using the method described by Klingberg et al. (2005). For each *Bacillus* isolate, 100 μl of inoculum were added into 10 ml of BHI broth, pH 7 containing 0.3% (w/v) oxgall bile (Sigma-Aldrich 30209037, Steinheim, Germany) and incubated at 37°C in a rotary shaker at 120 cycles per min. Samples were taken at various times (0, 1, 2, 3 and 4 h), serially 10-fold diluted and plated in duplicate onto BHI agar. The plates were incubated at 37°C for 24 h and bacterial colonies were counted. The number of *Bacillus* was calculated according to the standard of ISO 7218 (2007).

To determine bile tolerance of the *Bacillus* isolates after pre-exposure to low pH, 100 μl of each *Bacillus* inoculum were first added into 10 ml of BHI broth at pH 2.5. After incubation for 3 h at 37°C, cells were harvested by centrifugation (5000 g, 10 min),re-suspended into 10 ml of BHI broth containing 0.3% (w/v) oxgall bile and incubated at 37°C in a rotary shaker. Bacterial growth was checked after 24 h of incubation at 37 °C (from the beginning of the first incubation) by plate counting on BHI agar.

For the experiments described above, cells growth for each isolate in 10 ml BHI broth, pH 7 was monitored simultaneously as a positive control. The experiments were conducted in duplicate on two separate occasions.

RESULTS AND DISCUSSION

Antimicrobial resistance

The antimicrobial susceptibility of the studied *Bacillus* isolates was variable according to the *Bacillus* isolate and the antimicrobial tested (Table 3). For all tested bacteria, no resistance to avilamycin, ceftiofur, fluoroquinolones (ciprofloxacin), glycopeptides, florfenicol, gentamycin, linezolid, salynomycin, spectinomycin, kanamycin (except *B. subtilis* ssp. *subtilis* F1), tetracyclin, trimethoprim (except *B. subtilis* ssp. *subtilis* C6), trimetoprim + sulfametoxazole (except *B. subtilis* ssp. *subtilis* C3), and tigecycline was observed. However, reduced susceptibility towards flavomycin was found for all isolates with the exception of *B. subtilis* ssp. *subtilis* C3. All *B. licheniformis* showed reduced susceptibility towards chloramphenicol, daptomycin, β-lactams (penicillin) and streptomycin while all *B. amyloliquefaciens* ssp. *plantarum* were susceptible to the same antimicrobials (Table 3). The reduced susceptibility of all *B. licheniformis* toward chloramphenicol correlate with the findings of Sorokulova et al. (2008) who reported that the East European probiotic strain *B. licheniformis* 31 (BL31) was resistant to chloramphenicol. Similar to *B. subtilis* PY79, a laboratory strain derived from the 168 type strain and *B.*

Table 2. Primers used.

Resistance gene	Primer	Annealing temperature (°C)
*catp*IP501	5'-GGATATGAAATTTATCCCTC-3' 5'-CAATCATCTACCCTATGAAT-3'	47
erm(A)	5'-AAGCGGTAAAACCCCTCTGAG-3' 5'-TCAAAGCCTGTCGGAATTGG-3'	55
erm(B)	5'-CATTTAACGACGAAACTGGC-3' 5'-GGAACATCTGTGGTATGGCG-3'	52
erm(C)	5'-CAAACCCGTATTCCACGAGG-3' 5'-ATCTTTGAAATCGGCTCAGG-3'	48
aph(3")-I	5'-AACGTCTTGCTCGAGGCCGCG-3' 5'-GGCAAGATCCTGGTATCGGTCTGCG-3'	68
aph(3")-III	5'-GCCGATGTGGATTGCGAAAA-3' 5'-GCTTGATCCCCAGTAAGTCA-3'	52
ant(2")-I	5'-GGGCGCGTCATGGAGGAGTT-3' 5'-TATCGCGACCTGAAAGCGGC-3'	67
blaZ	5'-CAGTTCACATGCCAAAGAG-3' 5'-TACACTCTTGGCGGTTTC-3'	54
aadA	5'-ATCCTTCGGCGCGATTTTG-3' 5'-GCAGCGCAATGACATTCTTG-3'	56
aadE	5'-ATGGAATTATTCCCACCTGA-3' 5'-TCAAAACCCCTATTAAAGCC-3'	50
StrA	5'-CTTGGTGATAACGGCAATTC-3' 5'-CCAATCGCAGATAGAAGGC-3'	55
StrB	5'-ATCGTCAAGGGATTGAAACC-3' 5'-GGATCGTAGAACATATTGGC-3'	56
dfr(A)	5'-CCTTGGCACTTACCAAATG-3' 5'-CTGAAGATTCGACTTCCC-3'	50

subtilis Natto obtained from the Japanese soybean staple Natto (Hong et al., 2008) as well as the East European probiotic *B. subtilis* VKPM B2335 (Sorokulova et al., 2008), most *B. subtilis* ssp. *subtilis* (G2, H4, C6 and I7) investigated in the present study were susceptible to most antimicrobials tested including those highlighted by EFSA (2008). Knowledge on the antimicrobial resistance of *B. subtilis* group species is limited and therefore little is known on the population distributions of susceptibility for these compounds.

In the present study, no positive amplicons were obtained when the isolates were screened for the presence of the most prevalent genes associated with resistance to chloramphenicol (*catp*IP501), erythromycin (*erm*(A), *erm*(B), *erm*(C)), kanamycin (*aph*(3")-I, *aph*(3")-III, *ant*(2")-I), penicillin (*blaZ*), streptomycin (*aadA, aadE, StrA, StrB*), and trimethoprim (*dfr(A)*). This result suggests that the reduced susceptibility towards chloramphenicol, erythromycin, kanamycin, penicillin, streptomycin and trimethoprim may be intrinsic or natural. This finding is very important, since antimicrobial resistance plasmids are of special interest from the safety point of view, because they may be transferred to other strains including pathogens. Indeed, intrinsic resistance is considered to present a minimal risk for spread whereas acquired resistance mediated by mobile genetic elements like plasmids and transposons is considered to have a high risk for spread (European Commission, 2001). Given that the studied *B. subtilis* ssp. *subtilis* and *B. amyloliquefaciens* ssp. *plantarum* are for the most part susceptible to the antimicrobials tested and that they do not carry the most prevalent resistance genes for chloramphenicol, erythromycin, kanamycin, penicillin, streptomycin and trimethoprim, they can be considered as potentially safe for use as starter cultures. However, additional studies will be required to determine the exact nature of the antimicrobial resistance observed in this study.

Acid resistance and bile tolerance of the *Bacillus* isolates

As seen in Table 4, all 13 *Bacillus* isolates studied were able to survive in BHI broth (pH 2.5) following 4 h of incubation at 37°C. There was no sensitive variation in viable cell numbers (about 10^4 CFU/ml) from 0 to 4 h.

Table 3. Minimal Inhibitory Concentrations (MIC) and antimicrobial susceptibility of *Bikalga Bacillus* spp.

Antimicrobial	MIC (µg/ml)*												
	B. amyloliquefaciens ssp. *plantarum*			*B. subtilis* ssp. *subtilis*						*B. licheniformis*			
	A4	I8	G3	G2	H4	F1	C3	C6	I7	J3	E5	E3	F9
Avilamycin	<2 s	<2 s	<2 s	<2 s	4 s	<2 s	<2 s	<2 s	<2 s	<2 s	<2 s	<2 s	<2 s
Bacitracin	>64 r	>64 r	>64 r	>64 r	8 s	>64 r	4 s	4 s	>64 r	>64 r	>64 r	>64 r	>64 r
Chloramphenicol	4 s	4 s	<2 s	8 s	4 s	8 s	16 r	8 s	4 s	>64 r	32 r	32 r	16 r
Ceftiofur	0.25 s	0.25 s	0.25 s	1 s	0.5 s	2 s	<0.12 s	0.5 s	0.5 s	0.5 s	1 s	1 s	1 s
Ciprofloxacin	<0.12 s	<0.12 s	<0.12 s	<0.12 s	0.5 s	1 s	<0.12 s	<0.12 s	<0.12 s	<0.12 s	<0.12 s	<0.12 s	<0.12 s
Daptomycin	4 s	4 s	2 s	2 s	2 s	4 s	2 s	4 s	4 s	16 r	>16 r	16 r	>16 r
Erythromycin	<0.12 s	<0.12 s	<0.12s	<0.12 s	0.25s	<0.12 s	<0.12 s	<0.12 s	<0.12s	0.25 s	>32 r	>32 r	0.5 s
Flavomycin	>32 r	>32 r	>32 r	>32 r	>32 r	>32 r	8 s	>32 r	>32 r	>32 r	>32 r	>32 r	>32 r
Florfenicol	<1 s	<1 s	<1 s	<1 s	<1 s	<1 s	<1 s	2 s	<1 s	2 s	2 s	2 s	1 s
Gentamycin	≤2 s	≤2 s	≤2 s	≤2 s	≤2 s	4 s	≤2 s	≤2 s	≤2 s	≤2 s	≤2 s	≤2 s	≤2 s
Kanamycin	≤2 s	≤2 s	≤2 s	≤2 s	≤2 s	>64 r	≤2 s	≤2 s	≤2 s	4 s	4 s	4 s	4 s
Linezolid	<1 s	<1 s	<1 s	<1 s	<1 s	<1 s	<1 s	<1 s	<1 s	<1 s	<1 s	<1 s	<1 s
Penicillin	<0.06 s	<0.06 s	<0.06 s	<0.06 s	<0.06 s	<0.06 s	<0.06 s	<0.06 s	<0.06 s	0.5 r	0.25 r	0.25 r	0.25 r
Salynomycin	<2 s	<2 s	<2 s	<2 s	<2 s	<2 s	<2 s	<2 s	<2 s	<2 s	<2 s	<2 s	<2 s
spectinomycin	64 s	64 s	64 s	64 s	64 s	32 s	32 s	32 s	32 s	32 s	64 s	32 s	64 s
Streptomycin	8 s	8 s	4 s	4 s	4 s	8 s	4 s	4 s	4 s	64 r	16 r	16 r	32 r
Sulphamethoxazol	64 s	128 s	64 s	512 r	512 r	256 r	16 s	>512 r	256 r	16 s	256 r	32 s	64 s
Synercid	8 r	8 r	8 r	2 s	<0.5 s	>16 r	2 s	<0.5 s	<0.5 s	4 s	4 s	4 s	2 s
Tetracyclin	8 s	8 s	8 s	<0.5 s	<0.5 s	4 s	<0.5 s	1 s	2 s	<0.5 s	8 s	<0.5 s	2 s
Tiamulin	>32 r	>32 r	>32 r	32 r	<0.25 s	>32 r	>32 r	<0.25 s	0.5 s	32 r	>32 r	32 r	0.5 s
Tigecycline	0.06 s	0.06 s	0.06 s	0.03 s	0.03 s	0.06 s	0.03 s	0.03 s	0.03 s	<0.015 s	0.03 s	<0.015 s	<0.015 s
Trimethoprim	<1 s	<1 s	<1 s	<1 s	<1 s	<1 s	<1 s	32 r	<1 s	<1 s	<1 s	<1 s	<1 s
TMP+SMX	<0.25 s	<0.25 s	<0.25 s	0.5 s	<0.25 s	<0.25 s	8 r	2 s	<0.25 s	<0.25 s	<0.25 s	<0.25 s	<0.25 s
Vancomycin	<2 s	<2 s	<2 s	<2 s	<2 s	<2 s	<2 s	<2 s	<2 s	<2 s	<2 s	<2 s	<2 s

* s, sensitive; i, intermediary; r, resistant; according to the proposed breakpoints mentioned in Table 1.

Viable counts from 1, 2 and 3 h were identical to viable counts at 4 h (data not shown). This indicates that these strains may be able to survive the acidic conditions of the stomach. The pH value (2.5) used in the present study for the selection of potential starter cultures has been shown to be very selective and even though it is not the most common pH value of the human stomach it assures the isolation of very acid-tolerant strains (Pennachia et al., 2004). Interestingly, the vegetative cells of all *Bacillus* spp. showed excellent resistance to 0.3% bile. A growth was even observed from 10^4 CFU/ml at 0 h up to 10^6 CFU/ml after 4 h of incubation at 37°C (Table 4). After pre-exposure to BHI broth pH 2.5 for 3 h, all *Bacillus* isolates were

Table 4. Survival of *Bacillus* spp. in low pH and tolerance to bile.

Strain		Viable count (CFU/ml)								
		pH 7		pH 2.5		Bile		Bile/ pre-exposure to low pH		
		0 h	4 h	0 h	4 h	0 h	4 h	0 h	3 h/pH2.5	24 h/Bile
B. amyloliquefaciens ssp. *plantarum*	A4	1.3×10^5	1.2×10^7	2.3×10^4	1.8×10^4	2.2×10^4	3.0×10^6	5.0×10^4	4.2×10^4	7.8×10^7
	I8	8.3×10^4	1.7×10^7	8.6×10^4	7.4×10^4	1.9×10^4	4.2×10^6	6.2×10^4	2.5×10^4	1.5×10^8
	G3	9.4×10^4	2.1×10^7	1.2×10^4	1.0×10^4	6.5×10^4	6.2×10^6	5.4×10^4	1.6×10^4	6.5×10^7
B. subtilis ssp. *subtilis*	H4	2.0×10^4	8.9×10^6	2.6×10^4	2.3×10^4	3.6×10^4	2.3×10^6	1.1×10^4	6.5×10^3	1.4×10^7
	F1	5.5×10^4	9.6×10^6	6.0×10^4	4.4×10^4	5.4×10^4	5.6×10^6	5.3×10^4	3.4×10^4	1.3×10^8
	C3	4.2×10^4	7.7×10^6	2.4×10^4	1.8×10^4	1.2×10^4	8.0×10^5	1.5×10^4	4.0×10^3	5.4×10^6
	C6	4.5×10^4	1.1×10^6	1.4×10^4	1.3×10^4	2.1×10^4	4.5×10^5	5.5×10^4	3.4×10^4	1.0×10^6
	G2	5.1×10^4	4.9×10^6	8.6×10^4	7.7×10^4	5.7×10^4	5.0×10^5	3.7×10^4	2.0×10^4	7.5×10^7
	I7	4.3×10^4	4.7×10^6	1.4×10^4	1.2×10^4	3.0×10^4	3.1×10^6	2.0×10^4	1.9×10^4	8.2×10^7
B. licheniformis	E3	2.7×10^5	1.8×10^7	4.0×10^4	1.5×10^4	3.5×10^4	7.5×10^5	4.5×10^4	1.2×10^4	9.2×10^7
	F9	8.0×10^4	5.9×10^6	1.9×10^4	7.7×10^3	5.1×10^4	4.2×10^5	1.6×10^4	1.1×10^4	5.1×10^7
	J3	9.0×10^4	3.2×10^6	4.0×10^4	2.8×10^4	4.9×10^4	1.7×10^5	3.8×10^4	1.0×10^4	2.1×10^8
	E5	6.8×10^4	3.7×10^6	5.1×10^4	3.4×10^4	5.4×10^4	1.6×10^6	3.2×10^4	2.5×10^4	1.1×10^8

able to grow in BHI broth containing 0.3% oxgall bile with viable cell counts reaching up to 10^8 CFU/ml after 24 h incubation at 37°C (Table 4). The concentration of bile (0.3%) used has been recommended to be suitable for the selection of probiotics (Goldin and Gorbach, 1992) while other authors reported that this concentration is discriminatory (Chateau et al., 1994; Papamanoli et al., 2003). The fact that our isolates showed full resistance to acidic pH and bile is not in agreement with the findings of Barbosa et al. (2005). Indeed, these authors reported that vegetative cells of different *Bacillus* species (including *B. subtilis* and *B. licheniformis*) isolated from chicken fecal materials were unable to survive the simulated gastro-intestinal tract conditions. However, they observed that spores of the same *Bacillus* species showed excellent tolerance to bile salts and simulated gastric conditions. In contrast, some *Bacillus* spp. spores were reported to be susceptible to these conditions (Duc et al., 2004; Guo et al., 2006). *B. amyloliquefaciens* ssp. *plantarum* A4, I8 and G3 and *B. subtilis* ssp. *subtilis* H4 G2, C6 and I7 showed susceptibility to most antimicrobials tested and full resistance to simulated gastro-intestinal tract conditions making them promising starter cultures candidates for *H. sabdariffa* fermentation into *Bikalga*. However, to complete the selection of the starter cultures, further studies including for example proteolytic, lipolytic and organoleptic properties need to be addressed. It will also be interesting to investigate the probiotic potential of these isolates.

ACKNOWLEDGMENT

This work was supported by a Danish International Development Agency (DANIDA) funded project, London Metropolitan University/FLSC/SHS/MRU, UK and the Technical University of Denmark/National Food Institute. They are gratefully acknowledged.

REFERENCES

Ammor MS, Belen Florez, Mayo B (2007). Antibiotic resistance in non-enterococcal lactic acid bacteria and bifidobacteria. Food Microbiol. 24:559-570.

Barbosa MT, Serra RC, La Ragione MR, Woodward JM, Henriques AO (2005). Screening for *Bacillus* isolates in the Broiler Gastrointestinal Tract. Appl. Environ. Microbiol. 71:968-978.

Bengaly MD (2001). Microbiological Study and nutritional value of traditional protein-rich condiment, obtained by natural fermentation of Hibiscus sabdariffa seeds. PhD Thesis, University of Ouagadougou, p. 116.

Bengaly MD, Béré A, Traoré A (2006). The chemical composition of bikalga, a traditional fermented roselle (*Hibiscus sabdariffa* L.) seeds condiment. Part II: evaluation of mineral, total polyphenols and phytic acid content, predicting the iron bioavailability. Elec. J. Food Plant Chem. 1:7-11.

Chateau N, Deschamps AM, Sassi AH (1994). Heterogeneity of bile salts resistance in the Lactobacillus isolates of a probiotic consortium. Lett. Appl. Microbiol. 18:42-44.

Compaoré CS, Nielsen DS, Ouoba LII, Berner TS, Nielsen KF, Sawadogo-Lingani H, Diawara B, Ouédraogo GA, Jakobsen M, Thorsen L (2013a). Co-production of surfactin and a novel bacteriocin by *Bacillus subtilis* subsp. *subtilis* H4 isolated from *Bikalga*, an African alkaline *Hibiscus sabdariffa* seeds fermented condiment. Int. J. Food Microbiol. 162:297-307.

Compaoré CS, Nielsen DS, Sawadogo-Lingani H, Berner TS, Nielsen KF, Adimpong DB, Diawara B, Ouédraogo GA, Jakobsen M, Thorsen L (2013b). *Bacillus amyloliquefaciens* ssp. *plantarum* strains as potential protective starter cultures for the production of *Bikalga*, an alkaline fermented food. J. Appl. Microbiol. 115 :133-146.

Danielsen M, Wind A (2003). Susceptibility of *Lactobacillus* spp. to antimicrobial agents. Int. J. Food Microbiol. 82:1-11.

Duc LH, Hong HA, Barbosa TM, Henriques AO, Cutting SM (2004). Characterization of *Bacillus* probiotics available for human use. Appl. Environ. Microbiol. 70:2161-2171.

EFSA (2008). Technical guidance prepared by the Panel on Additives and Products or Substances used in Animal Feed (FEEDAP) on the update of criteria used in the assessment of bacterial resistance to antibiotics of human or veterinary importance. J. EFSA. 732:1-15.

European Commission (2001). Opinion of the Scientific Committee on Animal Nutrition on the Criteria for Assessing the Safety of Microorganisms Resistant to Antibiotics of Human Clinical and Veterinary Importance. [http://www.europa.eu.int/comm/food/fs/sc/scan/out64_en.pdf].

Goldin B, Gorbach S (1992). Probiotics for humans. In R. Fuller (Ed.), *Probiotics: the scientific basis* (pp. 355-376). London: Chapman and Hall

Guo X, Li D, Lu W, Piao X, Chen X (2006). Screening of *Bacillus* strains as potential probiotics and subsequent confirmation of the in vivo effectiveness of *Bacillus subtilis* MA139 in pigs. Antonie van Leeuwenhoek. 90:139-146.

Holzapfel WH, Haberer P, Snel J, Schillinger U, Huis in't Veld JHJ (1998). Overview of gut flora and probiotics. Int. J. Food Microbiol. 41:85-101.

Hong HA, Huang JM, Khaneja R, Hiep LV, Urdaci MC, Cutting SM (2008). The safety of *Bacillus subtilis* and *Bacillus indicus* as food probiotics. J. Appl. Microbiol. 105:510-520.

Huang Y, Adams MC (2004). In vitro assessment of the upper gastrointestinal tolerance of potential probiotic dairy propionibacteria. Int. J. Food Microbiol. 91:253-260.

International Organization for Standardization: ISO 7218 (2007). Food Microbiology-General Requirements and Recommendations. pp. 66.

Jensen BL, Baloda S, Boye M, Aarestrup MF (2001). Antimicrobial resistance among *Pseudomonas* spp. and the *Bacillus cereus* group isolated from Danish agricultural soil. Environ. Int. 26:581-587.

Jensen BL, Frimodt-Møller N, Aarestrup FM (1999). Presence of erm gene classes in Gram-positive bacteria of animal and human origin in Denmark. FEMS Microbiol. Lett. 170:151-158.

Klingberg TD, Axelsson L, Naterstad K, Elsser D, Budde BB (2005). Identification of potential probiotic starter cultures for Scandinavian-type fermented sausages. Int. J. Food Microbiol. 105:419-431.

Luna VA, King DS, Gulledge J, Cannons AC, Amuso PT, Cattani J (2007). Susceptibility of *Bacillus anthracis*, *Bacillus cereus*, *Bacillus mycoides*, *Bacillus pseudomycoides* and *Bacillus thuringiensis* to 24 antimicrobials using Sensititre® automated microbroth dilution and Etest® agar gradient diffusion methods. J. Ant. Chem. 60:555-567.

Mohamadou BA, Mbofung CM, Thouvenot D (2010). Characterization of some atypical lactic acid bacteria associated with the fermentation of *Hibiscus sabdariffa* seeds. Afr. J. Microbiol. Res. 4(24):2655-2660.

Ouoba LII, Parkouda C, Diawara B, Scotti C, Varnam AH (2008). Identification of *Bacillus* spp from bikalga, fermented seeds of *Hibiscus sabdariffa*: phenotypic and genotypic characterization. J. Appl. Microbiol. 104:122-131.

Papamanoli E, Tzanetakis N, Litopoulou-Tzanetaki E, Kotzekidou P (2003). Characterization of lactic acid bacteria isolated from a Greek dry-fermented sausage in respect of their technological and probiotic properties. Meat Sci. 65:859-867.

Parkouda C, Diawara B, Ouoba LII (2008). Technology and physico-chemical characteristics of bikalga, alkaline fermented seeds of *Hibiscus sabdariffa*. Afr. J. Biotechnol. 7:916-922.

Pennachia C, Ercolini D, Blaiotta G, Pepe O, Mauriello G, Villani F (2004). Selection of *Lactobacillus* strains from fermented sausages for their potential used as probiotics. Meat Sci. 67:309-317.

Sorokulova IB, Pinchuk IV, Denayrolles M, Osipova IG, Huang JM, Cutting SM, Urdaci MC (2008). The safety of two *Bacillus* probiotic strains for human use. Digestive Diseases Sci. 53:954-963.

Yagoub AEGA, Mohamed BE, Ahmed AHR, El Tinay AH (2004). Study on *furundu*, a traditional Sudanese fermented Roselle (*Hibiscus sabdariffa* L.) seed: Effect on in vitro protein digestibility, chemical composition, and functional properties of the total proteins. J. Agric. Food Chem. 52:6143-6150.

8

The microorganisms and compounds influencing the organoleptic properties of Ugba (fermented *Pentaclethra macrophylla* Benth. seeds)

N. O. Kabuo*, J. O. Uzuegbu, C. N. Ubbaonu and E. U. Onyeka

Department of Food Science and Technology, Federal University of Technology, Owerri, Nigeria.

The microorganisms and compounds influencing the organoleptic properties of Ugba were studied. Ugba was produced by the fermentation of boiled and shredded seeds of African oil bean. The pure cultures of microorganisms responsible for Ugba fermentation were isolated from Ugba samples, produced by traditional method and were used in singles and in combination to ferment African oil bean shreds. The Ugba so fermented were analyzed organoleptically and sensory scores statistically analyzed using analysis of variance at *P<0.5.* Compounds in the samples of oil extracted from unfermented African oil bean shreds and Ugba samples fermented for 72 h with the microorganisms (in singles and in combinations) were analyzed by gas chromatograph. The result showed that the following microorganisms were involved in the fermentation of Ugba: *Staphylococcus saprophyticus, Bacillus pumilus, Bacillus subtilis* and *Bacillus licheniformis*. The organoleptic tests showed that in all parameters tested, the sample fermented by mixed starter culture of *B. subtilis* and *B. licheniformis* was generally liked, which implied 'overall best'. Compounds found in unfermented Ugba were nineteen in number. Out of this number, eleven were found in unfermented samples, while eight were in the fermented Ugba. The best sample contained the following compounds: ethanol, ethyl stearate, ethyl oleate, ethyl linoteate, ethyl phenol and ethyl octanoate (their percentage concentration ranged from 2.69% for ethyl octanoate to 67.85% for ethanol). These compounds influenced the perceived Ugba flavor but had no direct influence on color and texture of Ugba sample.

Key words: African oil bean seed, Ugba, fermentation, microorganisms, starter culture, organoleptic characteristics, flavor compounds.

INTRODUCTION

Fermentation is a chemical change in food brought about by enzymes from living microorganisms (Uzogara et al., 1990). Fermented foods are prepared from plant and animal materials by processes in which microorganisms play important role in modifying the substrate physically, naturally and sensorily (Njoku and Okemadu, 1989). Some desirable changes in foods (especially legume based) which occur during fermentation include changes in texture and organoleptic characteristics (flavors, aroma and appearance or consistency); especially elimination of beany flavors, improvement in digestibility, enhancement in the quality of the product, improved safety (absence of toxins and partial and/or complete elimination of nutritional factors), increased nutritional quality and reduced cooking time (Wang and Hesseltine, 1981; Beuchart, 1984). Many of the food fermentations are natural and/or controlled fermentation consisting of different species and genera of yeast, fungi and/or bacteria (Kuboye, 1985; Reddy et al., 1986). These microorganisms can cause desirable changes in various foods, which distinguish them from the ones that are responsible for undesirable changes, including bad flavor and spoilage (Kuboye, 1985). Flavor components are the primary contributors to product acceptability either desirable or undesirable, if fermentation is uncontrolled (Sathe et al., 1986). Identification and characterization of

*Corresponding author. E-mail: ookabuo@yahoo.com.

flavor compounds responsible for the final flavor may help in development of synthetic flavor, which may reduce product cost and improve product reproducibility. There is a relationship between flavor compounds and organoleptic qualities of fermented foods (Nunomura and Sasaki, 1986; Okafor and Uzuegbu, 1987).

Ugba, a fermented product of African oil bean seed is one of the common fermented legumes predominantly consumed by the Ibos and other smaller ethnic groups of South Eastern Nigeria. Many researches have been carried out on Ugba, which included nutrients and other biochemical changes in Ugba associated with microorganisms during fermentation (Enujiugha, 2003; Kolawole and Okonkwo, 1985; Njoku and Okemadu 1989). These works indicated that enzymes bring about the major changes in the principal chemical components observed during the natural fermentation of Ugba. In the starter culture evaluation, the microorganisms involved in Ugba fermentation were related to its organoleptic properties (aroma, taste, texture and overall acceptability) (Mbajunwa et al., 1998).

The organoleptic properties of Ugba have only been associated with microorganisms and biochemical changes during fermentation, but no effort has so far been made to relate these microorganisms and organoleptic properties to the compounds produced during Ugba fermentation. There is a relation between flavor compounds and organoleptic properties (Sathe et al., 1986; Nunomura and Sasaki, 1986) which may be used to control the manufacturing process of Ugba and improve reproducibility.

The objectives of the work are:

i. Isolation and identification of microorganisms involved in fermentation of African oil bean seed.
ii. Preparation of starter cultures which were used in production of Ugba.
iii. Identification and characterization of flavor compounds produced by these microorganisms when used to ferment African oil bean shreds in singles and in combinations.

At the end of this work, microorganisms/their combinations, which contribute to the organoleptic properties of Ugba and compounds produced, which contribute to Ugba flavor will be known. This will not only help to produce pure starter cultures of microorganisms responsible for organoleptic properties of Ugba, but will also help to produce Ugba with consistent organoleptic properties. The data from this investigation will be a step forward to production of synthetic Ugba flavors.

MATERIALS AND METHODS

The African oil bean seeds and the *Alchornea laxiflora* Benth. leaves (Akwukwo Ugba – the popular leaves for wrapping Ugba) were obtained from Mbaise, Imo State Nigeria. The seeds were sorted and washed in order to remove spoilt seeds, dust and extraneous materials from wholesome seeds.

Production of Ugba (fermented African oil bean seeds)

2 kg of raw African oil bean seeds were sorted to remove rotten seeds and extraneous materials and washed with clean water. The seeds were put in the pot and covered with water and boiled with occasional stirring for 45 min. The heating was discontinued and the seeds were removed in batches and defueled while hot. After dehulling, a local vegetable shredder (called Nkwoo in Igbo – a perforated piece of metal, which, when the seed is made to run over it at a certain angle, the seed came out in shreds) was used to shred the seeds. Then the shreds (shredded seeds) were poured into a covered pot of boiling water and stirred at 5 min intervals, for 30 min. The boiled shreds were poured into sterile sieve to drain out the hot liquor. Water was sprayed on the shreds to completely remove the hot liquor and to cool the shreds. Then the shreds were washed three times, drained of wash water, and steeped in distilled water in a pot and covered. The shreds were steeped for 10 h. At the end of steeping, the shreds were vigorously stirred and poured into a sterile sieve (which has been autoclaved at a temperature of 121°C and pressure of 16 psi) to completely drain the steep water from the shreds. Then the shreds were poured into a sterile sieve lined with steam heated and cooled akwukwo Ugba (*A. laxiflora* Benth. leaves) and covered with the leaves, and kept in a warm environment (temperature 34°C) to initiate fermentation.

After 5 h, the fermenting shreds were aseptically taken, weighed into 50 g portions and wrapped in sterile akwukwo Ugba. The wraps were packed in sieves lined and covered with the same leaves, put in sterile pot, kept in a warm environment and fermented for 3 days (72 h). Figure 1 shows the flow diagram of Ugba production via traditional procedure.

The unfermented sample and samples, fermented from 24 up to 72 h were aseptically taken with sterile spoon (the spoon was sterilized by dipping in ethanol and flamed on a Bunsen flame) and ground in a sterile porcelain mortar, and diluted with 45 ml 0.1% normal saline. This was serially diluted and plated out (using NA) using pours plate method. The plates were incubated at 30°C for 24 to 48 h. Thereafter, colony forming units per gramme (cfu/g) was calculated (Harrigan and McCance, 1976). Representative colonies were isolated, purified by repeated streaken. The cultures were identified using biochemical methods according to Bergey's Manual of Determinative Bacteriology (Buchanan and Gibbon, 1974). The Biological tests carried out include gram staining, sugar fermentation test, catalase tests and starch hydrolysis test.

The bacteria isolates for fermentation of Ugba were prepared by the aforementioned methods (Mgbajunwa et al., 1998; Buchanan and Gibbson, 1974). The starter cultures were prepared as follows: The subcultures of the bacteria isolates were harvested with sterile water in screw-cap bottles and thoroughly mixed. The experimental procedure for Ugba preparation was followed as in traditional process except that the shreds of African oil bean samples were autoclaved for 121°C for 5 min at 16 psi and the pure cultures were inoculated into the autoclaved samples (Figure 1), via experimental procedure. Ten milliliters of the cultures containing 7.2 to 11.4 x 10^{10} cful/mol was used to inoculate 100 g sterile African oil bean shreds in singles and in combinations, and aseptically wrapped in sterile akwukwo Ugba sterilized over a steam. The wrapped samples were kept in warm environment and fermented for 72 h. At the end of the 72 h fermentation, the samples were pasteurized using steam and allowed to cool.

The oils extracted from unfermented African oil bean shreds and 72 h samples fermented with the starter cultures were used in determination of the compounds in Ugba. Diethyl ether was used as solvent because of its low boiling point (34°C) and the soxhlet extraction method was used to extract the oils. These oils were put in vials and stored in a refrigerator for analysis. A Shimadu Gas chromatograph instrument (model 17A) was used to determine the compounds in the Ugba oil samples using Nitrogen at a pressure of 16 psi as the carrier gas. The system was automated with its own

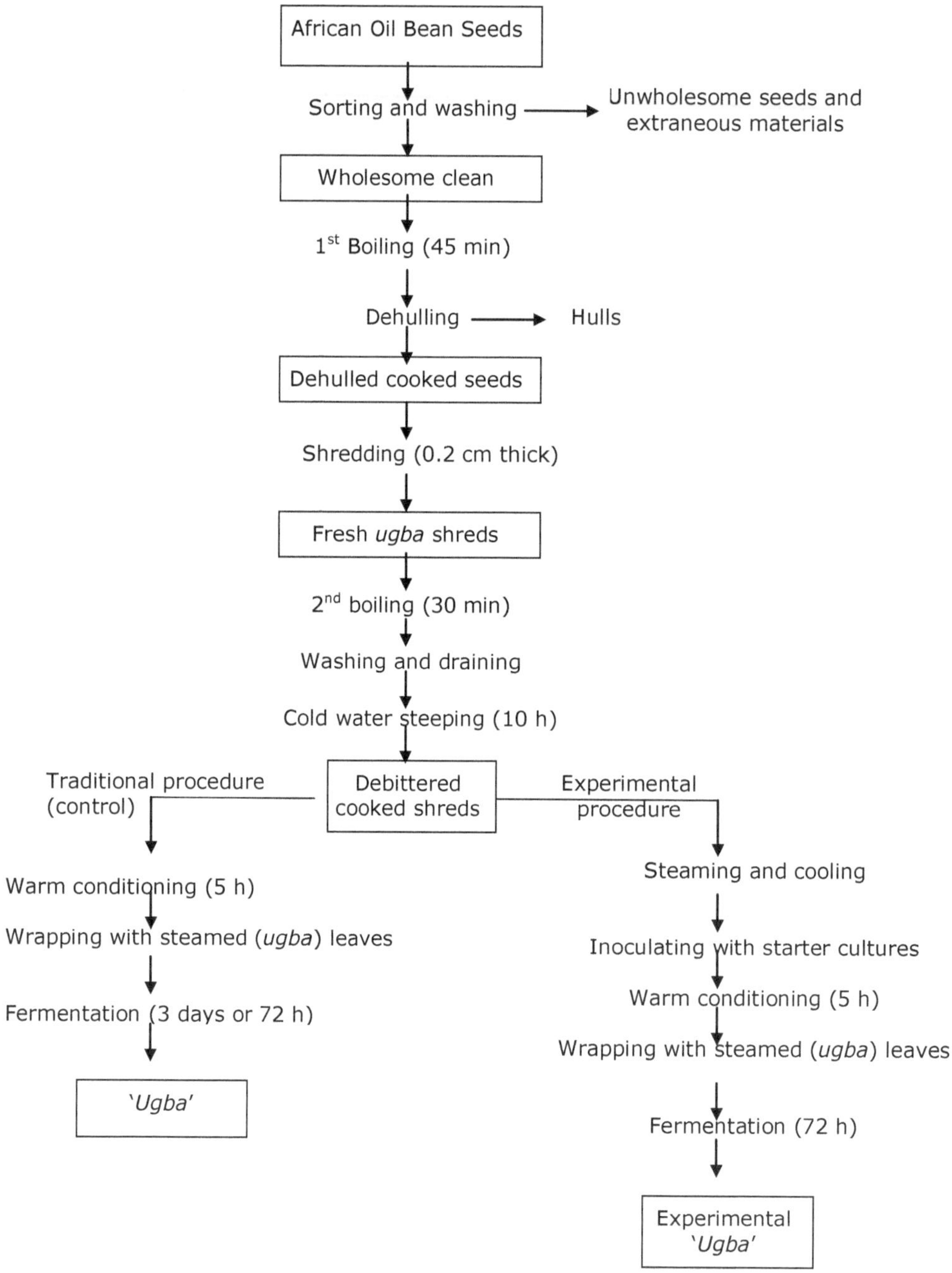

Figure 1. Flow diagram of 'Ugba' production.

workstation. Computer software was connected to the Gas chromatograph (GC), via the communication bus module. The instrument was set at an injector and detector temperatures of 240°C with programmed initial oven temperature of 50°C, increasing at 5°C per min for 20 min, and then 30°C per min for 20 min. The sample size injected into the system was 0.01µl. The internal standard used was Isopropanol. The calculations of the peak areas and retention times of compounds were carried out by the computer integrator coupled to the GC. The retention times of the compounds are matched with those of pure standards on the chromatograms obtained, to identify the various compounds in Ugba oil samples. The amount (in percentages) of each compound in a sample was calculated from the fractions of its peak area in the total peak areas of all the identified compounds.

Modified assessment methods described by Okafor and Uzuegbu (1987), Ihekoronye and Ngoddy (1985) were used in the sensory evaluation of Ugba samples. Organoleptic quality of the 3 days fermented and pasteurized Ugba samples were evaluated subjectively by a set of 30 panelists drawn from the University community, made up of staff and students. The selection of the panelists from the university environment was for practical convenience. The panelists were untrained and made up of males and females who indicated having eaten Ugba as a meal. All the samples were presented at the same time and the identities of the samples were not revealed to the panelists. Each panelist was provided with sufficient privacy to ensure that his/her results

Table 1. Properties and bacterial isolates identified from fermenting African oil bean shreds (control – as presently consumed).

Isolates	Colony morphology	Cell characteristics	Size (mm)	Gram reaction	Biochemical tests							Probable organism
					Catalase Test	Glucose	Starch	Sucrose	Fructose	Maltose	Mannitol	
A	Whitish (creamy) convex, entire and smooth	Cocci in clusters	0.5	+ve cocci	+	+	+	+	-	+	+	*Staphyloccus saprophyticus*
B	Creamy convex	Long rods	0.5	+ve rods	+	+	-	+	+	+	+	*Bacillus pumilus*
C	Creamy, pin point and submerged in agar (when streaked – rough on agar surfaces	Rods in singles	0.1	+ve rods	+	+	+	+	+	-	+	*Bacillus subtilis*
D	Whitish, flat entire	Rods in clusters	0.2	+ve rods	+	+	+	+	+	-	+	*Bacillus licheniformis*

+ Positive to fermentation, acid or gas forming; - Negative to fermentation, acid or gas forming.

would be arrived at, independently and without being influenced by other panelists. The quality parameters of Ugba evaluated were color, aroma/taste (flavor) texture and overall acceptability. A 9-point hedonic scale was used. The following instructions were given to the panelists:

a. Indicate gender (male or female), age bracket.
b. Testing for color: by visual inspection.
c. Testing for aroma and taste (flavor): by sniffing two times or more of a spoonful of each sample provided, and then chewing the spoonful for 1 to 5 min. Enough water was provided and the panelist was free to swallow or spit out the sample and rinse his/her mouth after each test.
d. Testing for texture: by touching and feeling some shreds of sample between the fingers.
e. Scoring: the panelists were given rating based on 9-hedonic scale: 9 – liked extremely; 8 – liked very much; 7 – liked moderately; 6 – liked slightly; 5 – neither liked nor disliked; 4 – disliked slightly; 3 – disliked moderately; 2 – disliked very much; and 1 – disliked extremely.
f. The panelists were to award a final score for overall acceptability.

The statistical analysis using a Completely Randomized Design (CRD) on a one-way classification model was used. Analysis of variance was done and the Friedman's Multiple Comparison test and Duncan Grouping Test at $P<0.05$ was used to compare the differences and similarities between Ugba samples (American society of Brewing Chemists, 1987). The organoleptic data (especially for controls best and least accepted samples) were compared to the compounds identified.

RESULTS AND DISCUSSION

Table 1 shows the characteristics and microorganisms identified during traditional method of preparation of Ugba (control). They were *S. saprophyticus, B. pumilus, B. subtilis and B. licheniformis*. Only bacteria isolates were isolated in this work, which were in agreement with the observation made by Mgbajunwa et al. (1998) and Obeta (1983). Out of the micro-organisms isolated by the workers, only *B. subtilis* and *S. saprophyticus* were isolated in this work.

The difference between my result and that of the former workers' could be due to the contamination of the wrapping leaves, which might have contributed since they usually do not sterilize the leaves.

Compounds identified in Ugba oil samples

Nineteen compounds were identified in the Ugba oil samples. Among the 19 compounds, eleven were naturally present in the unfermented sample (Table 2) and these include alcohols (butanol, phenyl alcohol and benzyl alcohol), esters (ethyl acetate, ethyl lactate, ethyl benzoate, ethyl heptadeconoate, ethyl stearate, ethyl oleate and ethlyl linoleate), and a phenol (ethyl phenol). During the 72 h fermentation of Ugba, more

Table 2. Compounds identified in extracted Ugba oil samples for unfermented sample and samples fermented for 3 days with different cultures.

Compounds	Unfermented sample	Control	Samples fermented with Bacteria cultures											Number of occurrence
			A	B	C	D	AC	AD	BD	BC	CD	BCD	ABCD	
Alcohols (7)														
Methanol		x					x		x				x	4
Ethanol		x	x						x	x	x	x	x	7
Buthanol	x		x							x				3
Benzyl alcohol	x													1
Phenyl alcohol	x									x				2
Furfuryl alcohol					x									1
Ethyl phenol	x		x	x	x	x	x	x			x	x		9
Esters (10)														
Ethyl acetate	x							x	x	x			x	5
Ethyl lactate	x								x	x				4
Ethyl benzoate	x	x		x	x	x			x					6
Ethyl heptadecanoate	x				x	x								3
Ethyl stearate	x						x			x	x			4
Ethyl oleate	x		x				x	x	x	x	x	x		8
Ethyl linoleate	x								x	x	x		x	5
Ethyl carbamate						x		x	x			x		4
Ethyl linoate								x						1
Ethyl octanoate		x	x	x	x					x	x			6
Carbonyls (2) (ketones)														
Methyl pentanone		x												1
Hexanone		x												1
Unidentified compounds		x						x	x					3

A – *Staphylococcus saprophyticus*; C – *Bacillus subtilis;* B – *Bacillus pumilu;* D – *Bacillus licheniformis.*

compounds were formed which included alcohols (methanol, ethanol and furfuryl alcohol), esters (ethyl carbamate, ethyl actonoate and ethyl linoate) and carbonyls–ketones (methyl pentanone and hexanone; found only in control sample). It was also observed that some compounds occurred in both unfermented and fermented samples, while some compounds in fermented Ugba samples (control and mixed starter samples of *Staphylococcus saprophyticus* and *B. subtilis; s. saprophyticus* and *B. Licheniformis*; and *B. licheniformis*) were unidentified (Table 2). The number of times these compounds occurred in various Ugba samples were also indicated in Table 2. The compound with the highest occurrence was ethyl phenol (9 times), followed by ethyl oleate (8 times) and ethanol (7 times). Some compounds occurred only once (benzyl alcohol in unfermented sample; ethyl linoate in *s. saprophyticus* and *B. licheniformis* sample; and furfuryl alcohol in *B. subtilis* sample.

Organoleptic properties of Ugba fermented with the starter cultures of bacteria isolates

Table 3 shows the sensory attributes (color, flavor, texture and overall acceptability) of control and experimental samples fermented for 72 h.

Color

Of the eleven samples studied, the color of five of them were neither liked nor disliked (score approximately 5.0). Specifically, they were those samples fermented with the following starter cultures: *B. pumilus; B. licheniformis*; mixed cultures of *B. pumilus and B. subtilis; S. saprophyticus* and *B. subtilis*; and control (traditionally prepared Ugba). Five samples fermented with starter cultures of *S. saprophyticus; B. subtilis, S. saprophyticus and B. licheniformis; B. pumilus, B. subtilis and B. licheniformis* were slightly liked (scores approximately 6.0). Only the sample fermented with the four-mixed culture *(S. saprophyticus, B. pumilus, B. subtilis and B. licheniformis* had color rated 'very much liked' (scored approximately 8.0). The sample fermented with *B. pumilus* and mixed culture *B. pumilus and B. subtilis* had the color rating significantly different (*P<0.05*) from the rest of the samples. Definitely, the color rating (score >5.5) of the control and other samples were significantly less acceptable than the color of sample fermented with the 4 mixed starters culture (score >7.6).

Flavor (aroma/taste)

With regard to flavor, only two of the samples *(B. pumilus and S. saprophyticus* and *B. subtilis*) were "neither liked nor disliked" (scored approximately 5.0). This rating ('neither liked nor disliked') was significantly different (at *P<0.05*) from the flavor which scored approximately 6.0 that is 'slightly liked') of other Ugba samples including the control. The sample fermented with mixed starter culture of *B. subtilis and B. licheniformis* had the highest score (6.4) which probably meant a better flavor than the rest of the samples.

Texture

Considering the texture of the samples, only two samples (*B. pumilus and B. subtilis; B. subtilis and B. licheniformis*) were "slightly liked" (scored approximately 6.0). The rest of the samples including the control were "neither liked nor disliked" (score <5.5).

Overall acceptability

The samples fermented with single starter cultures of *B. pumilus*, and *B. licheniformis* were "neither accepted nor rejected" (scored approximately 5.0) while the sample fermented with mixed starter culture of *B. subtilis and B. licheniformis* received "moderate acceptance" (scored approximately 7.0). The rest of the samples were only "slightly accepted" (scored approximately 6.0). Significant differences exist between overall acceptability of mixed starter culture of *B. subtilis and B. lichenifomis* sample and the acceptability of other samples (Table 3) by the sensory panel.

Considering the significant differences between the control sample and the other samples analyzed, the control (traditionally prepared Ugba) sample (score 5.9) was only better in flavor than sample fermented with single starter culture *B. licheniformis* (score 4.9). This was therefore matched by most of the other samples in all parameters evaluated with the exception of sample fermented with mixed starter culture *B. subtilis and B. licheniformis,* which had a higher and better score. Considering the findings, the sample fermented with mixed starter culture *B. subtilis and B. licheniformis* was 'most accepted' among the samples studied including control. Thus, fermentation of Ugba with mixed starter culture *B. subtilis* and *B. licheniformis* for 72 h could be recommended as a new technology for a better and reproducible product. The sample fermented with single starter culture *B. licheniformis* was the 'least accepted' sample having the least scores in flavor, texture (major quality attributes of Ugba) and overall acceptability.

Effect of compounds on organoleptic properties of Ugba samples

Flavor

Table 4 shows the mean levels (%) of compounds and the overall acceptability mean scores of some Ugba samples with emphasis on the control, least and best accepted samples. The high concentrations of ethyl carbamate (27.25%) and ethyl phenol (63.14%) might have contributed to the poor acceptability of the sample fermented by *B. licheniformis.* This is because ethyl carbamate and ethyl phenol had been described as having the aromas of burnt wood, antiseptic and horse stable (Gawel, 2004).

Thus, high level of ethyl phenol and ethyl carbamate (urethane) in a food could affect its flavor and acceptability. Probably, the absence of ethanol might have contributed to the poor acceptance of Ugba sample fermented with *B. licheniformis*, considering that ethanol

Table 3. Mean sensory scores of organoleptic attributed of Ugba samples fermented for 72 h.

	Ugba samples fermented with different starter culture bacteria isolates										
Parameters	***S. saprophyticus***	***B. subtilis***	***B. pumilus***	***B. licheniformis***	***S. saprophyticus* and *B. subtilis***	***S. saprophyticus* and *B. licheniformis***	***Control***	***B. pumilus, B. subtilis* and *B. licheniformis***	***B. pumilus* and *B. subtilis***	***B. subtilis* and *B. licheniformis***	***S. saprophyticus, B. pumilus, B. subtilis* and *B. licheniformis***	**LSD**
Color	5.7^{bc}	5.6^{bc}	4.8^{c}	5.3^{bc}	5.4^{bc}	6.4^{abc}	5.2^{bc}	6.2^{abc}	4.8^{ab}	6.4^{abc}	7.6^{a}	1.664
Aroma/taste (flavor)	6.0^{ab}	5.7^{ab}	5.5^{abc}	4.9^{c}	5.4^{bc}	6.2^{ab}	5.9^{ab}	6.2^{ab}	6.3^{ab}	6.4^{ab}	5.7^{ab}	0.848
Texture	5.0^{bc}	5.0^{bc}	5.1^{bc}	4.8^{c}	4.8^{c}	5.4^{ab}	5.3^{abc}	5.3^{abc}	5.5^{ab}	5.8^{a}	5.2^{bc}	0.4438
Overall Acceptance	5.6^{cd}	$5.9b^{cd}$	5.3^{cd}	5.0^{d}	5.5^{cd}	6.0^{bcd}	5.7^{cd}	6.3^{abc}	5.9^{bcd}	7.0^{a}	5.9^{cd}	0.845

could help to mellow down the effect of ethyl phenol in the sample (Gawel, 2004). The absence of ethanol and ethyl stearate, ethyl oleate, ethyl linoleate, ethyl heptadecanoate and ethyl octanoate (esters, known for their fruity aromas) (Wikipedia, 2007), might have contributed to the poor flavor of *B. licheniformis* fermented sample, while their presence might have contributed to the acceptable flavor of the sample fermented by *B. subtilis* and *B. licheniformis* (best). There were significant differences between the least acceptable sample and the best in levels of ethyl phenol, but there was no significant difference between the traditionally fermented sample (control) and the most acceptable sample (*B. subtilis* and *B. licheniformis*) in ethanol content (Table 4).

Color

The Ugba sample fermented with the 4-bacteria mixed culture of *S. saprophyticus, B. pumilus, B. subtilis* and *B. licheniformis* had an ethanol level of 89.91% and it did not seem that the compound influenced the color of the Ugba samples. This view was held because such compounds as ethanol which could have helped in preservation and better color were both in the control and the sample fermented by *B. subtilis* and *B. licheniformis* in appreciable levels (63.71 and 67.89%, respectively) yet their sensory scores differed.

Texture

It was observed that there was no indication that any of the organic compounds identified had any marked influence on the texture of the Ugba samples studied. It is difficult to explain why the sample fermented by *B. subtilis* and *B. licheniformis* gave the best textures even with the same treatment as other samples. The texture of Ugba is usually affected by degree of maturity of the African oil bean seeds, degree of cooking and days of fermentation. Softness being enhanced by cooking and fermentation period. Most times, long fermentation resulted in very soft (unacceptable) product (Ogueke and Aririatu, 2004).

Overall acceptability

The overall acceptability of the samples have tended more towards flavor and texture, hence the Ugba sample fermented by the mixed starter culture *B. subtilis* and *B. licheniformis* had the highest overall acceptability rating (score = 7.0, that is, moderately accepted). In addition, the proportion of the flavor compounds in this sample may have affected its flavor and overall acceptability (Table 4). Furthermore, Figure 2 showed the chromatograms of the unfermented Ugba, control and the best Ugba sample.

Conclusion

The microorganisms responsible for fermentation of Ugba were bacteria: *S. saprophyticus, B. pumilus, B. subtilis* and *B. licheniformis*. To study Ugba production by these bacteria, isolates from traditional Ugba preparation were used as starter cultures (in singles and in combinations). Sensory evaluation results showed that tasters in all parameters (colour, flavour, texture and overall acceptability) generally liked the Ugba sample fermented with a mixed culture of *B. subtilis* and *B. licheniformis*. This implied that the sample was 'overall best'. A total of 19 flavors compounds were identified in Ugba which included 7

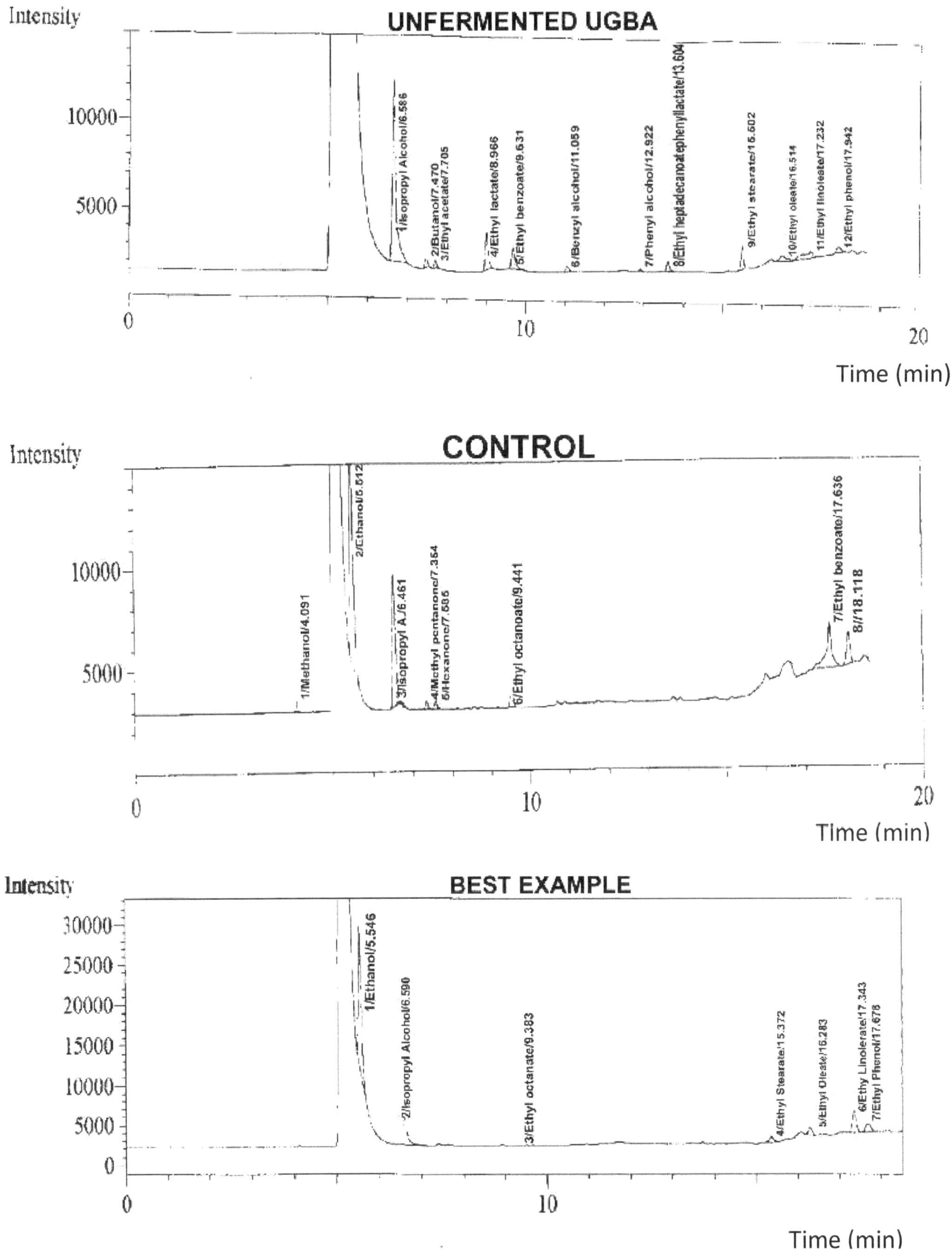

Figure 2. Chromatograms of flavor compounds for unfermented Ugba, control and best sample fermented for 72 h.

Table 4. Concentrations (%) of compounds that might affect the perceived flavours of some selected Ugba samples (control, best and least).

Overall acceptability/ compounds	■Control	Fermented Ugba samples									
		●CD	A	BC	BCD	ABCD	B	C	◆D	AD	AC
Acceptability	5.7^{bcd}	7.0^{a}	5.6^{bcd}	5.9^{bcd}	6.3^{abc}	5.9^{bcd}	5.3^{cd}	5.9^{bcd}	5.0^{d}	6.0^{bcd}	5.5^{bcd}
Ethanol	63.710^{d}	67.890^{d}	73.510^{b}	29.390^{f}	87.690^{a}	89.910^{a}	-	-	-	-	-
Methanol	1.38	-	-	-	-	-	-	-	-	-	-
Ethyl octanoate	4.720^{a}	2.690^{e}	1.610^{e}	7.750^{c}	-	-	32.600^{a}	28.680^{b}	-	-	-
Ethyl heptadecanoate	-	-	-	-	-	-	-	-	6.800	-	-
Ethyl cabamate	-	-	-	-	-	-	-	-	27.23	-	-
Ethyl phenol	-	6.940^{e}	21.910^{c}	-	-	-	64.420^{a}	16.820^{d}	63.140^{a}	40.700^{b}	39.640^{b}
Ethyl benzoate	16.180^{b}	-	18.400^{a}	-	-	-	3.00^{d}	5.280^{c}	2.830^{d}	-	-
Ethyl linoleate	-	14.140^{b}	-	24.430^{a}	-	9.200^{c}	-	-	-	-	-
Ethyl sterate	-	3.660^{c}	-	-	-	-	15.150^{b}	-	-	-	25.210^{a}
Ethyl oleate	-	4.700^{de}	3.020^{ef}	5.750^{cd}	0.840^{f}	-	-	-	-	13.880^{b}	16.340^{a}
Ethyl pentanon	1.57	-	-	-	-	-	-	-	-	-	-
Hexanone	1.65	-	-	-	-	-	-	-	-	-	-

Means with the same letters across the row are not significantly different at 95% probability level ($P \leq 0.05$). ●Best Ugba sample ◆ Least Ugba sample ■ Control. A - *Staphylococcus saprophyticus;* B - *Bacillus pumulis;* C - *Bacillus subtilis;* D - *Bacillus licheniformis.*

alcohols, 10 esters and 2 carbonyls: eleven of these compounds were found in unfermented sample, while 8 were mainly in fermented samples. The 'overall best' sample contained the following compounds: ethanol, ethyl stearate, ethyl oleate, ethyl linoleate, ethyl phenol and ethyl octanoate, and their percentage concentrations ranged from 2.69% for ethyl octanoate to 67.85% for ethanol. It was therefore concluded that these compounds produced during fermentation of African oil bean shreds influence the organoleptic properties of Ugba samples produced by starter cultures and traditional (normal) fermentation.

REFERENCES

American Society of Brewing Chemists (1987). Sensory Analysis. ASBC J. 45(3):102-105.

Beuchart LR (1984). Fermented Soybean Foods. Food Technol. 38:64.

Buchanan RE, Gibbson NE (1974). Berguey's Manual of Determinative Bacteriology. 8th Edition. Williams and Wikins Co. Baltimore, M.A., U.S.A.

Enujiugha VN (2003). Nutrient changes during the fermentation of African Oil Bean (*Pentaclethra macrophylla* Benth.) seeds. Pak. J. Nutr. 2(5):320-323.

Gawel R (2004). Brettanomyces character in wine. Australian society of wine Education National Convention. Hunter Valley, Australia. 4th-6th June.

Harrigan WF, Mc Cance ME (1976). Laboratory Methods in Food and Dairy Microbiology. Academic Press Inc. London.

Ihekoronye AI, Ngoddy PO (1985). Integrated Food Science and Technology for the Tropics. Macmillan Publishers Ltd, Hong Kong, pp. 172, 180-184, 190, 198.

Kolawole DO, Okonkwo BA (1985). Microbiology of the Fermentation of Oil Bean (*Pentaclethra macrophylla* – Benth.) seeds for "Ukpaka" a Nigerian condiment. NIFO J. 2(2,3), 3(1, 2, 3):149-151.

Kuboye MAO (1985). Traditional fermented foods and beverages of Nigeria. In: Development of indigenous fermented foods and technology in Africa. IFS Provisional Report. Proceedings of IFS/UN workshop, Doula, Cameroon 20:225-231.

Mbajunwa OK, Akingbala JO, Mulongoy K, Oguntimein G (1998). Starter culture evaluation for the production of Ugba rom African oil bean seed (*Pentaclethra macrophylla*). J. Sci. Food Agric. 77:127-132.

Njoku HO, Okemadu CP (1989). Biochemical changes during the natural fermentation of the African oil bean (*Pentaclethra macrophylla*) for the production of Ugba. J. Sci. Food Agric. 49:457-465.

Nunomura N, Sasaki M (1986). Soy sauce: In Legume Fermented Foods. Editors: Reddy, N.R. Pierson M.D. and Salunkhe, D.K. CRC Press Inc. USA, pp. 19, 23.

Obeta JAN (1983). A note on Microorganisms associated with the fermentation of seeds of the African oil bean tree (*Pentaclethra macrophylla*). J. Appl. Bacteriol. 54:433-435.

Ogueke CC, Aririatu LE (2004). Microbial and organoleptic changes associated with Ugba stored at ambient temperature. NIFOJ 22:133-140.

Okafor N, Uzuegbu JO (1987). Studies on the contribution of microorganisms to the organoleptic properties of Gari; A fermented food derived from cassava (*Manihot esculenta* Crantz). J. Food Agric. 2:99-105.

Reddy NR, Pierson MD, Salunkhe DK (1986). Legume-based fermented foods. CRC Press Inc, USA, p. 1.

Sathe SK, Reddy NR, Pierson MD (1986). Future of Legume Fermented Foods. In: Legume-based Fermented Foods. Editors: Reddy, N.R., Pierson, M.D. and Salunkhe, D.K. CRC Press Inc. USA, pp. 233-235.

Uzogara SG, Agu LN, Uzogara EO (1990). A review of traditional fermented foods, condiments and beverages in Nigeria: Their benefits and possible problems. Ecol. Food Nutr. 24:282-283.

Wang HL, Hesseltine CW (1981). Use of microbial cultures: Legume and cereal products. Food Technol. 35:79.

Wikipedia (2007). Aroma compound. In: Wikipedia Encyclopedia. Wikipedia Foundation, Inc. U.S.A. pp. 1-4. http://en.wikipedia.org/wiki/Aroma _compound. Oct.17, 2007.

Effects of germination time on the functional properties of maize flour and the degree of gelatinization of its cookies

Adedeji, O. E.[1], Oyinloye, O. D.[2] and Ocheme, O. B.[3]

[1]Department of Food Science and Technology, Federal University, Wukari, Taraba State, Nigeria.
[2]Department of Human Nutrition, University of Ibadan, Oyo State, Nigeria.
[3]Department of Food Science and Technology, Federal University of Technology, Minna, Niger State, Nigeria.

Flour was produced from germinated and ungerminated maize grains, and cookies were subsequently produced from the flours. Functional properties of the flours and degree of gelatinization of the cookies were determined. The pH decreased significantly ($p \leq 0.05$) as the germination time increased. pH values ranged between 5.67 and 6.56. Bulk density (loosed and packed) showed no significant difference ($p \leq 0.05$) between ungerminated sample and sample that germinated for 24 h, but there was a significant difference ($p \leq 0.05$) between ungerminated sample and samples that germinated for 48 and 72 h. The values ranged between 0.50-0.58 and 0.70-0.79 g/mL for loose and packed density, respectively. There was a significant difference ($p \leq 0.05$) among the samples in water absorption capacity and oil absorption capacity with the ungerminated sample having the least values of 0.94 and 1.03 ml/g for water absorption capacity and oil absorption capacity, respectively. There was a significant difference ($p \leq 0.05$) in swelling power of the samples with the ungerminated maize having the highest value of 19.81 mg/g. There was a significant difference ($p \leq 0.05$) in the forming capacity with the sample that germinated for 48 h having the least value. There was a significant difference ($p \leq 0.05$) in the forming capacity with the sample that germinated for 48 h having the least value. There was no significant difference ($p \leq 0.05$) between the ungerminated flour and the sample that germinated for 48 h and foaming stability time of 15, 30 and 60 s. Degree of gelatinization of the samples ranged between 74.50-93.10, 86.20-97.40 and 30.00-84.30 at baking temperature of 140, 160 and 180°C, respectively.

Key words: Germination, gelatinization, functional properties, maize flour.

INTRODUCTION

Maize (*Zea mays* L.), the American Indian word for corn literally means "that which sustains life". It is the third most important cereal grain of the world after wheat and rice providing nutrients for humans and animals and serving as a basic raw material for the production of starch, oil and protein, alcoholic beverages, food sweeteners and more recently, fuel (FAO, 1992). These three most important cereals have been reported to comprise at least 75% of the world's grain production and therefore, humanity has become dependent upon cereal grains for the majority of its food supply (Imtiaz et al., 2011).

Starch is the major constituent of cereal endosperms and an important structural component in many food products made from their flours (Sasaki, 2005). Corn starch is a valuable ingredient in the food industry, being widely used as a thickener, gelling agent, bulking agent and water retention agent (Singh et al., 2003). Corn starch granule has a polyhedral shape and diameter between 5 and 25 µm (Ribout, 2002). On the basis of amylose and amylopectin ratio, corn can be separated into normal, waxy and high amylose (Singh et al., 2005). In addition, sugary type corn with high sugar content also exists (Singh et al., 2005). Normal starch consists of about 75 wt% branched amylopectin and about 25 wt% amylose, that is linear or slightly branched.

The effective use, attributes and consequent acceptance of cereal flours by consumers are dependent on its functional properties and the degree of starch gelatinization (Miah et al., 2001; Brou et al., 2013), Mantzari (2010) reported that due to its complexity, corn starch exhibits certain unique properties (which are not encountered in other polysaccharides) which are correlated with its physicochemical and functional properties: temperature and enthalpy, gelatinization, pasting characteristics, enzymatic susceptibility, swelling and solubility (Rubi, 2009).

Several works have been done on germination as an alternative to genetic engineering in improving the nutriative values (such as amino acids, vitamins, minerals etc.), functional (temperature and enthalpy, gelatinization, pasting characteristics, swelling, and solubility) and chemical (pH, amylose, total starch etc.) properties of maize, particularly in the developing countries (Oluwamukomi et al., 2003; Obasi et al., 2009; Eneche, 2009; Gernah et al., 2011). Typically, chemical modification could be adopted to extend the range of specific physical properties available for different uses (Nur and Purwiyatno, 2010), however, the use of chemicals are not encouraged due to their side effects, besides, these chemicals are not readily available to vast population of the developing nations who rely mostly on these cereals. FAO/WHO (1985) stated the guidelines of an ideal weaning food to be nutrient dense, easily digestible, of suitable consistency and affordable.

This study was therefore undertaken to investigate the effects of germination time on some functional properties of maize starch and the degree of gelatinization of its cookies.

MATERIALS AND METHODS

Source of material

White maize grains (TZW, 2008 harvest) were purchased from Minna Central Market, Minna and were identified at the Department of Crop Production, Federal University of Technology, Minna, Niger State, Nigeria. These were utilized for research work between June, 2009 and January, 2010.

Cleaning

The maize grains were manually cleaned to remove husks, stone, cob, damaged and coloured seeds. These were achieved through winnowing, sieving and hand picking. Subsequently, the seeds were packaged in a 10 L plastic bucket, hermetically covered and stored in a refrigerator at 10±2°C from where samples were taken for processing and analyses.

Table 1. Cookie recipe.

Parameter	Measurement
Maize flour	100 g
Baking fat	50 g
Baking powder	5 g
Sugar	45 g
Egg (raw)	2
Powdered milk	30 g

Source: Asumugha and Uwalaka (2000) with little modification.

Germination

Germination was carried out according to the method described by Ariahu et al. (1999). The seeds were washed in 5% (w/v) sodium chloride solution to suppress mould growth and soaked in tap water in ratio of 1:3 (w/v) grain for 12 h at room temperature (32±2°C), the water drained at 4 h interval after which the seed were drained and divided into four equal portions and labeled A, B, C and D, spread separately on a clean jute bag, covered with damp cotton and were allowed to germinate for 0, 24, 48 and 72 h, respectively. Water was sprinkled at 12 h interval to facilitate the germination process. At the end of germination, root hairs were removed from the germinated seeds.

Production of germinated flour

The seeds were dried at 60°C in an oven to a moisture content of 10% and ground into flour using attrition mill (globe p44 Chima). Each flour sample was passed through a 0.5 mm mesh size sieve. They were packaged in an air tight polyethylene bags, stored in plastic containers with lids and then stored in cool dry place from where samples were taken for analyses.

Production of cookies

Cookies were prepared from the flour samples using the cream-in method as described by Asumugha and Uwalaka, (2000) with little modification. Table 1 shows the cookie recipe. Fat and sugar were mixed until fluffy. Whole eggs and powdered milk were added while mixing (HR-2815 Philips Model Mixer) and then mixed for a total of about 30 min. Flour and baking powder were mixed thoroughly and added to the cream mixture and were kneaded to form dough. The dough were rolled and cut into shapes of 5 cm diameter. Baking was carried out at 140, 160 and 180°C for 25 min. Cookies were cooled and stored till when needed.

Sample analysis

Functional properties

Loosed and packed bulk densities of the flour samples were determined using the method described by Wang and Kinsella (1976), water absorption capacity (WAC) and oil absorption capacity

Table 2. Functional properties of maize flour samples.

Parameter	Germination time			
	0 h	24 h	48 h	72 h
pH	6.56±0.01[a]	6.20±0.02[b]	6.28±0.04[b]	5.67±0.02[c]
LBD (g/mL)	0.58±0.03[a]	0.57±0.02[a]	0.55±0.00[b]	0.50±0.00[c]
PBD (g/mL)	0.79±0.00[a]	0.76±0.00[b]	0.75±0.00[b]	0.70±0.00[c]
WAC (ml/g)	0.94±0.00[d]	2.65±0.00[c]	2.73±0.03[b]	2.79±0.00[a]
OAC (mg/g)	1.03±0.00[c]	2.45±0.05[b]	2.52±0.01[b]	2.57±0.00[a]
EC (%)	47.22±0.00[c]	58.62±0.02[b]	60.00±0.00[b]	65.52±0.00[a]
SP (mg/g)	19.81±0.00[a]	19.10±0.02[b]	18.80±0.00[c]	15.00±0.00[d]
FC (%)	3.10±0.00[c]	2.50±0.02[b]	2.00±0.00[c]	2.50±0.00[b]

Values are means and standard deviations of triplicate scores. Values followed by different superscript in row are significantly different (p≤0.05) from one another. LBD, Loose bulk density; PBD, packed bulk density; WAC, water absorption capacity; OAC, oil absorption capacity; EC, emulsion capacity; ES, emulsion stability; SP, swelling power; GPC, gelation power capacity; FC, foaming capacity.

(OAC) were determined by methods described by Sasulki et al. (1996), emulsion capacity and stability were determined using the method described by Yasumatsu et al. (1992), foaming capacity was determined according to the method described by Narayana and Narsinga (1992), swelling power was determined by Akpada and Miachi (2001) method and pH was determined using AOAC (2000) method.

Determination of degree of starch gelatinization

Determination of degree of starch gelatinization of cookies was determined by the method described by Marshall et al. (1993). 2 g of the sample was macerated with 100 ml distilled water in a warming blender (HR-2815 Philips model). The suspension was centrifuged at 500 rpm for 10 min and duplicate aliquots (1 ml) were diluted with water to 10 ml and treated with 0.1 ml iodine solution. The absorbance of the sample were read at 600 nm with spectrophotometer (model 2903, perkin- Elmer co. Ltd.), against a reagent blank. A further suspension of the product (2 g) was prepared in 95 ml of distilled water (instead of 100 m distilled water) as described earlier. To this suspension, 5 ml of 10 M aqueous solution of potassium hydroxide was added and mixture was allowed to stand for 5 min with gentle agitation. The alkaline suspension was centrifuged and 1 ml of duplicate aliquots was treated with 1 ml of 0.5 M hydrochloric acid and diluted with water to iodine solution (0.1 ml) and their absorbance was measured as described earlier. The degree of starch gelatinization was calculated as:

$$\frac{A_1 \times 100\ (\%)}{A_2}$$

Where A_1 and A_2 are absorbance of the iodine complex prepared from the aqueous suspension before and after alkali solubilization, respectively.

Statistical analysis

The data obtained were statistically analyzed by subjecting them to analysis of variance (ANOVA) using the completely randomized design (CRD) with comparison made between the group means using the Duncan's new multiple range test to separate the mean (at 5% probability level) using SPSS (statistical package for social scientists), version 16.0, Windows 2006.

RESULTS

Table 2 shows the functional properties of maize flour samples. The pH decreased significantly (p≤0.05) as the germination time increased. Maize grains that germinated for 72 h had the lowest pH value of 5.67. Bulk density (loosed and packed) showed no significant difference (p≤0.05) between ungerminated sample and sample that germinated for 24 h, but there was a significant difference (p≤0.05) between ungerminated sample and samples that germinated for 48 and 72 h. There was a significant difference (p≤0.05) among the samples in water absorption capacity with the ungerminated sample having the least value.

Oil absorption capacity varied significantly (p≤0.05) among the samples with germination done for 72 h having the highest value. Emulsion capacity of the flour samples also differ significantly (p≤0.05) with germination for 72 h having the highest value. There was a significant difference (p≤0.05) in swelling power of the samples with the ungerminated maize having the highest value. There was a significant difference (p≤0.05) in the foaming capacity with the sample that germinated for 48 h having the least value.

Table 3 shows the foaming stability of the maize flour samples. There was no significant difference (p≤0.05) between the ungerminated flour and the sample that germinated for 48 h at foaming stability time of 15, 30 and 60 s both having higher foaming stability than samples that germinated for 24 and 72 h at the same foaming stability time.

Table 4 shows the effect of germination time on the degree of gelatinization of cookies produced from the maize flour samples. At 140°C, the degree of gelatinizetion of the ungerminated sample was higher than those of germinated samples, although it was not significantly different (p≤0.05) from the sample that germinated for 72 h. At 160°C, there was a significant difference (p≤0.05) among the samples with the ungerminated sample having

Table 3. Foaming stability of the maize flour samples.

Germination time (h)	Foaming stability time (seconds)			
	15	30	60	120
0	50.90±0.000[a]	41.00±0.000[a]	25.15±0.000[a]	15.00±0.000[a]
24	40.10±0.028[b]	35.00±0.028[b]	20.15±0.000[b]	10.00±0.000[b]
48	50.57±0.000[a]	42.00±0.000[a]	25.00±0.000[a]	9.50±0.000[b]
72	30.50±0.000[c]	20.10±0.028[c]	10.00±0.000[c]	10.00±0.000[b]

Values are means and standard deviations of triplicate scores. Values followed by different superscript in column are significantly different (p≤0.05) from one another.

Table 4. Effect of germination time on the degree of gelatinization of the cookies.

Germination time (h)	Temperature (°C)		
	140	160	180
0	93.10±0.000[a]	86.20±0.283[d]	30.00±0.000[d]
24	74.50±0.000[c]	91.90±0.000[c]	40.90±0.000[c]
48	90.50±0.283[b]	93.70±0.283[b]	65.40±0.283[b]
72	92.90±0.000[a]	97.40±0.283[a]	84.30±0.000[a]

Values are means and standard deviations of triplicate scores. Values followed by different superscript in column are significantly different (p≤0.05) from one another.

the least value and the sample that germinated for 72 h having the highest value.

DISCUSSION

The decrease observed in pH might have been as a result of secretion of enzymes resulting in the hydrolysis of complex organic molecules such as phytin and protein into simpler and more acidic compounds such as phosphate and amino acids, respectively. Evans et al. (2003) reported a marked increase in alpha amylase and other amylases during cereal germination. Egwim and Oloyede (2004) reported 72 h as optimum sprouting time for maximum amylase activity in maize. Results obtained for loosed and packed densities were in line with earlier work of Gernah et al. (2011). The reduction in bulk density observed might have been as a result of reduction in weight of the flour owing to the breakdown of complex denser compounds inherent in maize into simpler ones during germination (Gernah et al., 2011). Germination increased water absorption capacity of the samples, this contrasted the work of Imtiaz et al. (2011) but in line with the work of Gernah et al. (2011). The increase observed might have been as a result of the production of compounds having good water holding capacity such as soluble sugars. According to Okaka and Potter (1997), water holding capacity depends on the water bounding capacities of food components. Germination increased oil absorption capacity in line with earlier work of Imtial et al. (2011). Giani and Bekebain (1992) reported that germination of grains enhances the oil absorption capacity due to the entrapment of oil related to the non polar side chains of proteins. The increase observed in emulsion capacity could be due to an increase in the area of stabilized oil droplet at interface which is a function of the food components (Imtiaz et al., 2011). Germination decreased the swelling power of the samples probably as a result of disruption of hydrogen atoms inherent in maize by amylases and proteases into sugars and amino acid respectively (Okafor, 1987; Egwim and Ademonom, 2009). The decrease observed in foaming capacity might have been as a result of denaturation of protein molecules during milling and germination processes. Brou et al. (2013) reported that native protein provide higher foam capacity than denatured protein.

Brou et al. (2013) reported increasing foaming stability with increasing protein content while characterizing complementary food made from maize, millet, beans and soybeans. They further reported higher protein stability for native proteins. The increase in foaming stability observed for sample that germinated for 48 h might have been as a result of bioavailability of inherent proteins which were probably bound by antinutritional factors such as phytin in the sample. Singh and Raghuvanshi (2012) reported that antinutritional factors in cereals bind to both exogenous and endogenous proteins including enzymes of the digestive tract affecting utilization of proteins. The reduction in stability observed for sample that germinated for 72 h could have been due to denaturation of protein (Brou et al., 2013). The ungerminated sample had a higher stability than the germinated samples at 120 s. The result from the tables show that foaming stability of the samples decreased with time.

At 140°C, short germination period was not favourable probably due to the resistance of the amylose portion of the starch to the enthalpy produced at this temperature. The amylose content greatly influences the physicochemical properties of starch, such as gelatinization, retrogradation, and gelation (Parovuori, 1997; Czuchajowska, 1998; Fredriksson, 1998). Sasaki (2005) reported that starch with higher amylose content has more amorphous region and less crystal line, lowering gelatinization temperature. Germination increased the degree of gelatinizetion at 160°C probably due to the presence of adequate crystalline fractions in the starch molecules. The enthalpy of gelatinization reflects the loss of molecular order (Cooke and Gidley, 1992), and gelatinization temperature is considered a parameter of crystallite perfection (Tester and Morrison, 1990). At 180°C, the degree of gelatinizetion of the flour samples reduced remarkably probably due to disruption of the crystallite fractions of the samples.

Conclusion

The study indicated that germination has tremendous effects on the functional properties of maize flour samples considered, and on the degree of gelatinization of the cookies produced at various baking temperature. Germinated samples performed better in terms of water absorption capacity, oil absorption capacity and emulsion capacity. Sample that germinated for 48 h had the highest foaming capacity at time 15, 30 and 60 s. These are indications of improved nutritional value and functionality. Sample that germinated for 72 h performed best in water absorption capacity, oil absorption capacity and emulsion capacity. Germination resulted in cookies with better gelatinization at different baking temperature with sample that germinated for 72 h having the best result.

REFERENCES

Akpada MI, Miachi OE (2001). Proximate composition and selected functional properties of *Detarium microcarpum*. Plant Foods Hum. Nutr. 56:297-302.

AOAC (2000). Official methods of Analysis, Association of official Analytical chemist 17th Edition, Washington.

Ariahu CC, Ukpabi U, Mbajunwa KO (1999). Production of African Breadfruit (Treculia Africana) and Soyabean (*Glycine max*) based food formulations, 1: Effects of germination and fermentation on nutritional and organoleptic quality. Plant Foods Hum. Nutr. 54: 123-266

Asumugha VU, Uwalaka BC (2000). Chemical and organoleptic evaluation of snacks developed from cocoyam (*Colocasia esculenta, Xanthosoma mafafa*) and wheat (*Triticum spp*) composite flours. Niger. Agric. J. 31: 78-88.

Brou K, N'Da-Kouassi A, Kouadio J, Guehi T, N'Guessan K, Gnakri D (2013). Biochemical Characterization and Functional Properties of Weaning Food Made from Cereals (Millet, Maize) and Legumes (Beans, Soybeans). J. Food Chem. Nutr. 1 (1) : 22-32.

Cooke D, Gidley MJ (1992). Loss of crystalline and molecular order during starch gelatinization: origin of the enthalpic transition. Carbohydr. Res. 227:103–112.

Czuchajowska Z (1998) Structure and functionality of barley starches. Cereal Chem. 75: 747–754.

Egwim E, Adenomon M (2009). Predicting α-amylase yield and malt quality of some sprouting cereals using 2nd order polynomial model. Afr. J. Biochem. Res. 3(8):288-292

Egwim E, Oloyede B (2004). Comparison of αααα-amylase activity in some sprouting Nigerian cereals. Biokemistri 18(1):15-20

Eneche HE (2009). Rate of water absorption in maize grain during soaking. Proceedings of the AGM/Conference, Nigerian Institute of Food Science and Technology, Yola. pp. 41-42.

Evans DE, Van Weger B, Ma YF, Eghinton J (2003). The Impact of the Thermostability of alpha-amylase, beta-amylase and limit Dextrinase on Potential work Fermentability. J. Am. Soc. Brewing Chem. 61: 210-218.

FAO (1992). Maize in Human Nutrition. FAO Food and Nutrition Series, No. 25. ISBN92-5103013-8

FAO/WHO (1985). Energy and Protein Requirements. Report of the FAO/WHO/UNU Expert Consultation Technical Series No. 724, Geneva, Switzerland World Health Organisation.

Fredriksson H (1998). The influence of amylose and amylopectin characteristics on gelatinization and retrogradation properties of different starches. Carbohydr. Polym. 35: 119–134.

Gernah DI, Ariahu CC, Ingbian EK (2011). Effects of Malting and Lactic Fermentation on Some Chemical and Functional Properties of Maize (*Zea mays*). Am. J. Food Technol. 6:404-412

Giani SY, Bekebain DA (1992). Proximate composition and functional of raw and processed full fat fluted pumpkin (*Telfairia occidentalis*) seed flour. J. Sci. Food Agric. 3: 321-325

Imtiaz H, BurhanUddin M, Gulzar M (2011). Evaluation of weaning foods formulated from germinated wheat and mungbean from Bangladesh. Afr. J. Food Sci. 5(17): 897-903.

Mantzari G (2010). Effect of sorbitol addition on the physicochemical characteristics of starch–fatty acid systems. Carbohydr. Polym. 79(1):154-163

Marshall WE, Wadsworth JI, Verma LR, Velupillai L (1993). Determining the degree of gelatinisation in parboiled rice: comparison of a subjective and objective method. Cereal Chem. 70: 226–230.

Miah MAK., Haque A, Douglass MP, Clarke B (2001). Parboiling of rice. Part II: Effect of hot soaking time on the degree of starch gelatinization. Int. J. Food Sci. Technol. 37:539–545.

Narayana K, Narsinga Rao MS (1992). Functional properties of raw and heat processed winged bean (*psophocarpus tertragonolobus*) flour. J. Food Sci. 47: 534-538.

Nur A, Purwiyatno H (2010). Gelatinization properties of white maize starch from three varieties of corn subject to oxidized and acetylated-oxidized modification. Int. Food Res. J. 17: 961-968

Obasi N, Agim EL, Uchechukwu N (2009). Effects of different soaking periods on the chemical composition of pearl millet grains (*Pennisetum glancum*). Proceedings of the AGM/Conference, Nigerian Institute of Food Science and Technology, Yola. pp. 57-58.

Okafor N (1987). Processing of Nigerian Indigengenous Foods: A Chance of Innovation. Niger. Food J. 1: 32-34.

Okaka JC, Potter NN (1997). Functional properties of cowpea wheat flour blends in bread making. J. Food Sci. 42:828-833.

Oluwamukomi MO, Eleyinmi AF, Enujingha VN, Atofarati SO (2003). Nutritional, Physicochemical and Sensory Evaluation of Sorghum and Cowpea Based Weaning Formulations. Niger. Food J. 21:11-17.

Parovuori P (1997). Effects of enzymically modified amylopectin on the rheological properties of amylose-amylopectin mixed gels. Food Hydrocoll. 11:471–477.

Ribout M (2002). Using AFM to image the internal structure of starch granules. Carbohydr. Polym. 50: 123-132

Rubi G (2009). Blue Maize: Morphology and Starch Synthase Characterization of Starch Granule. Plant Foods Hum. Nutr. 64:18–24.

Sasaki T (2005). Effect of Wheat Starch Characteristics on the Gelatinization, Retrogradation, and Gelation Properties. JARQ 39 (4): 253 – 260.

Sasulki FW, Garrant MD, Slinkard AE (1996). Functional properties of fern legume flour. Inst. Food Sci. Technol. J. 9:66-69.

Singh N, Sandhu KS, Kaur M (2005). Physicochemical properties including granular morphology, amylose content, swelling and solubility, thermal and pasting properties of starches from normal,

waxy, high amylose and sugary corn. Prog. Food Biopolym. Res. 1: 44–54.
Singh N, Singh J, Kaur L, Sodhi NS, Gill BS (2003). Morphological, thermal and rheological properties of starches from different botanical sources. Food Chem. 81:219–231.
Singh P, Raghuvanshi RS (2012). Finger millet for food and nutritional security. Afr. J. Food Sci. 6(4): 77-84
Tester RF, Morrison WR (1990). Swelling and gelatinization of cereal starches. II. Waxy rice starches. Cereal Chem. 67:558–563.
Wang JC, Kinsella JE (1976). Functional properties of novel proteins Alfalfa leaf protein. J. Food Sci. 41:286-289.
Yasumatsu KI, Sawada K, Moritaka S, Misaki M, Tado J, Woda T (1992) . Whipping and emulsifying properties of sorghum product. J. Agric. Biol. Chem. 36: 719-725

Effect of paddy storage and processing parameters on quality of *Ofada* rice in the production of ready to eat flakes

Oludare Olumuyiwa Adekoyeni[1], Rahman Akinoso[2] and Akintunde Stephen Fagbemi[3]

[1]Food Science and Technology, Federal University Dutsinma, Katsina State, Nigeria.
[2]Food Technology, University of Ibadan, Oyo State, Nigeria.
[3]National Agency for Food and Drug Administration and Control (NAFDAC), Abuja, Nigeria.

Effect of paddy storage duration and processing parameters (soaking time, parboiling temperature and drying temperature) were optimized using response surface methodology for the production of ready to eat *Ofada* rice flakes which serves as alternative mean of harnessing its potential. There exist significance differences at $p<0.05$ for all the parameters except carbohydrate and metabolizable energy. The quadratic models were fit for prediction of effect of storage and processing parameters on quality of flakes produced from ofada paddy. The R^2 ranged from 0.9662 to 0.7318 which confirmed the fitness of the model. The predicted and validated values were related. The optimum storage duration and processing parameters for treatment of *Ofada* paddy in the production of ready to eat flake include storage of paddy for 9 months, soaking for 4 days and 17 h, parboiling at 106°C and drying at 30°C to yield optimum quality of *Ofada* rice flakes. The sensory assessment showed significant acceptability of colour, crispiness, aroma, taste and overall acceptability.

Key words: Ofada, processing parameters, response surface methodology, ready-to-eat flakes, optimization.

INTRODUCTION

Rice (Oryza sativa L) is enjoyed by many people as staple food especially varieties with distinctive aroma and flavor (Bryant and McChung, 2011). One of the popular indigenous rice varieties in Nigeria is *Ofada* rice. "*Ofada*" is a generic name used to describe all rice produced and processed in South-West, Nigeria. One of the early cultivated variety is OS6 and it is relished because of is aroma (NCRI and WARDA, 2007). Rice produced in Nigeria is consumed mostly in the form of boiled rice and as mashed porridge rolled into round balls both eaten with soup.

Nowadays, ready-to-eat foods are gaining much importance as they are convenient to use, easy to handle compared with ready-to-cook foods (Itasi et al., 2012). There is few or no utilisation of Nigeria rice flour in pastry production, however, production of breakfast cereals from maize, sorghum, millet is common. The use of rice for preparation of breakfast cereal such as noodles, flakes is

Table 1. Independent variables and levels.

Variables	Levels		
Storage duration (months)	1	5	9
Soaking time (days)	1	3	5
Parboiling temperature (°C)	80	100	120
Drying temperature (°C)	30	50	70

common in United States and Europe, therefore, utilization of Nigeria local rice for ready-to-eat flakes is another way of harnessing the potential of local rice. The aromatic characteristic, taste and immense nutritive value of *ofada* rice (OS6) compare to other local varieties are potentials that worth to be investigated in the production of high quality *ofada* ready-to-eat flakes.

According to Fred (2007), processing conditions of grain in flakes production could affect the quality of flakes. The major traditional unit operations in paddy processing include soaking, parboiling, and drying. The effects of storage history and ageing on rice were also documented (Zhou et al., 2001; Daniel et al., 1998). Therefore, the storage duration and processing conditions of paddy into rice flour for the production flake may have effect on quality of flake. Response surface methodology (RSM) is important in designing experiment, formulating, developing and analysing new and existing scientific studies cutting the cost, and measures several effects by objective test (Akinoso and Adeyanju, 2010; Montgomery, 2005). The main objective of the study was to optimize paddy storage duration and processing parameters (soaking time, parboiling temperature and drying temperature) of ofada paddy using response surface methodology for the production of quality ready-to-eat ofada rice flakes. The results of this study will make relevant technical data available for the present and prospective investors that would have relied on "trial and error" method in processing *ofada* rice and its utilisation for flake production. This will help to make the local rice sufficiently more competitive thereby increasing its demand.

MATERIALS AND METHODS

Ofada paddy (OS 6) was purchased at farm gate in Mokolokin-Ofada, a renowned area for *Ofada* rice production. The processing unit operations (soaking, parboiling, drying) used as treatment was adopted from the methods of processing of paddy adopted by the rice farmers in the area. D- Optimal response surface methodology was used for the design of the experiment. The independent and levels of variables is shown in Table 1. The rice stalk was threshed manually and cleaned to obtain paddy within 12 h, mixed thoroughly and stored in a dry cool place for processing at 1, 5 and 9 months as described in the experimental design.

Paddy (4kg) was soaked in cold water at ambient temperature (28±2°C) for typically 1, 3 and 5 day(s) to hydrate the kernels. The soaked paddy were parboiled at varied temperatures (80, 100, and 120°C) at constant pressure using digital autoclave for 15 min. The parboiled paddy were tempered for 30 min to cool and air dried in oven at 30, 50 and 70°C. The rice samples were milled (hulling and debranning) in grantex cono disc milling machine. The rice obtained were subsequently ground in a hammer mill, sieved (200 micron size) and analysed.

Production of *Ofada* rice flakes

The laboratory production of ready-to-eat flakes was carried out as described by Lu and Walker (1988). The milled rice was cleaned and ground into flour. The flour was sifted with sieve (60-mesh testing sieve) and the coarse residues were discarded. Flour (400 g) was mixed with 300 ml of water, 25 g of sugar, and 4 g of salt in a kitchen Kenwood mixer for 5 min. The dough was placed in a pasta extruder attachment and forced through a die with 5-mm hole. After the extrusion, it was cut 0.5 cm long pellets, and steamed using a pressure cooker for 15 min. After cooking, the pellets were tempered and pressed through a heavily spring roller. The resulting flakes were toasted on pans at 200°C for 20 min, cooled, and packaged in plastic.

Analysis of rice flakes

The analysis carried out on flakes include moisture, crude protein, crude fat, and ash contents, using AOAC,(2000); carbohydrates by difference and metabolizable energy by FAO (2002); phytic acid was carried out as described by Garcia-Estepa et al. (1999) and water absorption capacity as described by Walker et al. (1988).

RESULTS AND DISCUSSION

The result of chemical and functional qualities of ready-to-eat *ofada* rice flakes as affected by storage and processing parameters of *ofada* rice paddy is presented (Table 2).

Moisture composition of ready to eat *ofada* rice flakes

There were significant differences in the result of moisture contents obtained at $p < 0.05$. The minimum and maximum moisture content of the rice flakes were 3.10 and 6.89 respectively (Table 2). The least percentage moisture content was obtained from flake produced from *Ofada* paddy stored for 9 months and subjected to processing operations which involved 5 days of soaking, 80°C parboiling, and 30°C drying temperature while the maximum moisture in the flake was obtained from the paddy stored for 1 month, soaked for 1 day, parboiled at 120°C, and dried at 30°C temperature.

The R^2, adjusted R^2, and adequate precision are presented in Table 3 and coefficient of the model depicting effect of processing conditions of paddy and storage duration in the production of rice flour for the manufacturing of flakes is presented in Table 4. Positive coefficient of drying temperature showed that drying process of paddy has a major influence on moisture content of flake. The movement of water out of the flake during drying and into the flake during food processing (water absorption) has implications on quality (Fred, 2007).

Table 2. Result of chemical and functional composition of Ofada rice flakes.

Number	Soaking time (day)	Parboiling temp. (°C)	Drying Temp. (°C)	Storage Duration	Moisture (%)	Protein (%)	Ether extract (%)	Ash (%)	Carbohydrate (%)	Phytate (%)	Water absorption ratio	Metabolisable Energy (kcal/100 g)
1	5	120	70	1	5.9	8.5	0.89	2.93	81.78	0.8	2.37	368.46
2	1	80	70	1	5.6	8.81	1.49	2.4	81.7	1.51	3.43	374.81
3	1	80	30	5	6.25	9.73	0.3	1.28	82.44	1.47	4.34	382.63
4	5	80	30	1	5.6	8.77	0.49	1.57	83.57	0.69	2.65	373.03
5	5	80	30	9	5.65	12.84	0.49	1.59	79.43	1.15	4.29	383.53
6	3	80	50	5	5.15	10.26	0.52	0.95	83.12	1.46	4.77	389.33
7	5	120	30	5	5.4	9.64	0.4	0.2	84.36	0.39	4.51	391.11
8	1	80	70	1	5.69	8.9	0.97	2.43	82.01	1.36	3.41	371.67
9	5	120	70	9	5.75	12.21	0.51	1.68	79.85	0.51	4.41	383.09
10	3	120	50	5	5.85	10.5	0.37	1.13	82.15	0.62	4.17	384.95
11	3	120	30	9	3.7	13.1	0.66	1.6	80.94	0.92	4.55	392.28
12	5	100	50	5	5.25	10.19	0.31	1.03	83.22	0.64	4.69	387.72
13	1	120	70	5	5.1	9.76	0.24	1.03	83.87	1.01	4.88	388.19
14	3	100	50	9	3.73	13.01	0.76	1.72	80.76	1.09	4.74	392.02
15	1	120	50	9	5.6	12.81	0.32	1.52	79.75	1.14	4.32	383.37
16	5	120	70	1	5.24	8.62	0.86	2.94	82.34	0.72	2.38	370.89
17	1	100	50	5	5.3	9.96	0.46	1.12	83.16	1.36	4.59	387.84
18	3	100	50	5	4.05	6.99	0.46	2.63	86.87	0.89	2.95	378.99
19	5	80	30	1	5.01	8.81	0.56	1.65	83.97	0.71	2.67	375.42
20	1	120	30	1	6.89	9.36	0.63	1.42	85.79	1.42	3.22	385.49
21	1	100	30	9	3.33	12.76	0.44	1.74	81.73	1.53	4.52	392.42
22	5	80	70	5	6.85	9.94	0.43	1.01	81.77	1.42	4.66	381.73
23	1	80	70	9	3.8	12.93	0.52	2.05	80.7	1.67	4.21	389.46
24	5	80	30	9	3.1	12.61	0.38	1.65	82.26	1.22	3.92	393.55
25	1	120	30	1	6.82	9.28	0.67	1.41	81.82	1.25	3.24	369.66

Flake thickness influences the rate of diffusion of water during drying. However, Machado et al. (1998) reported that range of thickness did not correlate with water absorption.

Oeding (1996) has profound that higher steam temperature and great moisture during steaming resulted in increased water absorption, and also suggested that this was related to changes in pasting behaviour. The report justified the result of this experiment with rice parboiled at highest temperature measuring highest percentage of moisture while this decreased with reduction in parboiled temperature.

Protein composition of ready-to-eat *ofada* rice flakes

The results of the protein content of rice flakes ranged between 6.99 to 13.10 % (Table 2). Some of the values obtained in this study were comparable to protein content of flake made from wheat (USDA, 2013). There exist significant differences in the result for protein at $p < 0.05$ and values of R^2, adjusted R^2, and adequate precision are presented in Table 3. The closeness of R^2 and Adjusted R^2 to 1 and extent of greatness of adequate precision above 4 were indicators of the

Table 3. ANOVA of regression of chemical and functional qualities of Ofada rice flakes.

Parameters	p-value	R^2	Adjusted R^2	Adequate precision
Moisture (%)	0.0207	0.84	0.6161	7.088
Protein (%)	0.0019	0.9084	0.7802	7.58
Ether extract (%)	0.0029	0.8991	0.7579	9.795
Ash (%)	0.0267	0.8297	0.5914	7.966
Carbohydrate (%)	0.146	0.7318	0.3562	5.317
Phytate (%)	<0.0001	0.9662	0.919	16.673
Water absorption ratio	0.0229	0.8361	0.6066	5.835
Metabolisable energy (kcal/100 g)	0.0627	0.7882	0.4917	5.013

Table 4. Coefficient of the quadratic model for the effect of storage and processing parameters on ofada rice quality in production of ready to eat flakes.

Parameter	Moisture (%)	Protein (%)	Ether extract (%)	Ash (%)	CHO (%)	Phytate (%)	Water absorbtion	M.E (kcal/100 g)
Intercept	36.41	23.04	-0.76	-4.27	38.55	2.77	3.56	296.21
S	-2.31	-0.65	-0.02	0.42	2.9	-0.12	-0.45	3.72
P	-0.66	-0.37	0.03	0.16	1.01	-0.03	-0.03	1.44
D	0.21	0.19	0.03	-0.06	-0.39	0.01	0.04	-0.21
T	-0.27	-0.09	-0.24	-0.34	0.79	0.18	0.48	5.86
S^2	0.28	0.12	-0.03	-0.12	-0.3	$6.30X10^{-3}$	0.08	-0.27
P^2	$3.33x10^{-3}$	$2.02x10^{-3}$	$-1.44x10^{-4}$	$-9.17x10^{-4}$	$-4.89x10^{-3}$	$2.17x10^{-4}$	$2.19x10^{-4}$	$-5.92x10^{-3}$
D^2	$-2.30x10^{-3}$	$-1.45X10^{-3}$	$-1.55X10^{-5}$	$3.42X10^{-4}$	$3.68X10^{-3}$	$2.97X10^{-5}$	$-3.32X10^{-6}$	$5.65X10^{-3}$
T^2	-0.03	0.07	0.02	0.06	-0.11	$3.38X10^{-3}$	-0.05	-0.32
SP	$-2.63X10^{-3}$	$-1.63X10^{-3}$	$1.87X10^{-3}$	$2.62X10^{-3}$	$-3.99X10^{-3}$	$-2.11X10^{-3}$	$-1.75X10^{-3}$	$-1.44X10^{-3}$
SP	6.01	$2.12X10^{-3}$	$-1.55X10^{-4}$	$1.19X10^{-3}$	-0.01	$3.40X10^{-3}$	$7.94X10^{-4}$	-0.04
ST	0.04	$-6.07X10^{-3}$	$8.27X10^{-3}$	$-8.16X10^{-3}$	$-7.57X10^{-3}$	$-1.30X10^{-3}$	0.02	$5.19X10^{-3}$
PD	$-3.18X10^{-4}$	$-5.50X10^{-4}$	$-1.47X10^{-4}$	$4.73X10^{-4}$	$2.29X10^{-4}$	$-2.03X10^{-4}$	$-3.98X10^{-4}$	$-3.26X10^{-3}$
PT	$1.98X10^{-3}$	$-5.35X10^{-4}$	$2.42X10^{-4}$	$-1.50X10^{-3}$	$-3.28X10^{-3}$	$-1.43X10^{-3}$	$1.95X10^{-3}$	-0.01
DT	$1.81X10^{-3}$	$-9.5X10^{-4}$	$-1.06X10^{-3}$	$-2.08X10^{-3}$	$5.53X10^{-3}$	$-9.09X10^{-4}$	$-3.65X10^{-4}$	$3.95X10^{-3}$

S,soaking time; P, parboiling temp.; D,drying temp.; T, storage duration.

degree of fitness of the model. The coefficients of the model are presented in Table 4. Maximum protein content in rice flake was obtained from paddy stored for 9 months, and subjected to 3 days soaking, 100°C parboiling temperature, and 50°C drying temperature. It was revealed that ready-to-eat *Ofada* rice flakes produced from paddy stored for longer duration relatively shown improvement in protein content (Table 2).

The protein composition recorded in this experiment was higher than the result as reported by Nazni and Bhuvaneswari, (2011), and also higher than the nutritional protein composition of the popular 'Kelloggs' products.

Fat composition of ready to eat *ofada* rice flakes

The results of crude fat were significant at $p < 0.05$ and ranged from 0.24 to 1.49% with mean value of 0.57%. Generally, the crude fat contents were low (<1.5 %) with the maximum extract recorded from the flake produced from paddy stored for 1 month and treated under 1 day soaking, 80°C parboiling and 70°C drying temperatures respectively while the minimum value was from rice stored for 5 months, and processed by soaking for 1 day, parboiling at 120°C and drying temperature at 70°C. The coefficient of the model was presented (Table 4). The result obtained from this experiment is comparable to the general fat content recorded for most commercial rice flakes (Fatsecret, 2013). High fat content can easily cause rancidity of the package foods, however, maximum of 3% fat has been found in most commercial rice flakes. Gupta et al. (2012), recorded fat content range of 0.76 – 5.91 % in the production of rice flakes mix using dehydrated herbs.

Ash contents of ready to eat rice flakes

The ash content is the total mineral composition of rice

flakes and it ranged from 0.20 to 2.94 %. Maximum ash content was from rice flake produced from the paddy stored for 1 month followed by soaking of the rough rice for 5 days, parboiling at 120°C, and drying at 70°C while minimum value was obtained from rice stored for 5 months, soaked for 5 months, parboiled for 120°C and dried at 30°C drying temperature. It was observed that the effect of storage duration and drying temperature could have been responsible for the variation. The coefficient of the model revealed that soaking time and parboiling temperature treatment were major indicators of the ash contents of the flakes (Table 4). The milling operation might affect ash content of the flakes due to the presence of bran.

Carbohydrate contents of ready to eat *ofada* rice flakes

There was no significant difference (p>0.05) in the data obtained for carbohydrate. The carbohydrate contents ranged from 79.43 – 86.87% and "Lack of Fit test" is not significant which revealed that such model can still be used in prediction. The values of R^2, Adjusted R^2, and adequate precision of the model were presented in Table 3. The report has shown that storage duration of paddy and processing conditions of ofada rice does not have significant influence on the carbohydrate content of ready to eat rice flakes.

Water absorption of ready-to-eat *ofada* rice flakes

Water absorption is one of the major factors in determining the quality of flakes. The results obtained ranged from 2.37-4.88 and there were significant differences in the water absorption ratio values of the rice flakes (p<0.05). Observation of the data revealed that the rate of water absorption majorly varied with the storage duration of paddy before processing. However, the highest water absorption ratio was obtained in rice flakes processed from rice subjected to 1 day soaking, 120^{0}C parboiling temperature, 70^{0}C drying after storage for 5 months duration. However, values of R^2, Adjusted R^2, and adequate precision were presented (Table 3). The coefficient of the model showed a positive influence of drying temperature (D) and storage duration (T) on raw material processing (paddy) in manufacturing of flakes (Table 4).

Phytate composition of ready-to-eat *ofada* flakes

Rice contains some important anti-nutritional factors, most which are concentrated in the bran. All anti nutritional factors in rice except phytate are proteins and denatured by heat. The phytate levels from the experiment ranged from 0.39-1.67%. Variation in respect of phytate content was found to be significant (p < 0.05). The highest amount of phytate was recorded in the flake produced from rice stored for 9 months, processed by soaking for 1 day, 80°C parboiling temperature, and 70°C drying temperature while the least value of phytate was at 5 days soaking, 120°C parboiling, 30°C drying and 5 months storage duration of paddy. According to Noreen et al. (2009), soaking and boiling processes caused significant decrease in phytic acid in rice. This report may be related to the reason why soaking at longest day and highest degree of parboiling temperature resulted to the least value of phytate while the opposite of the processing conditions in terms of soaking and parboiling gave the highest phytate value in the rice flake. Humans have limited ability to absorb and hydrolyse phytate (Pawar and Ingle, 1988). Binding of minerals with phytic acid decrease bio availability of calcium, iron, phosphorus, zinc and other trace elements to human and other monogastric animals. This may lead to severe nutritional and consequently health problems in the consuming population (Thompson (1987).

Metabolization energy of ready to eat *ofada* rice flakes

The metabolizable energy in the flakes was between 368.46-393.55 kcal/100 g. There exist no significant difference (P > 0.05) in the metabolizable energy obtained from the rice flakes as affected by the processing condition and storage duration of paddy before processing. The R^2 and Adjusted R^2 were 0.7882 and 0.4917 respectively (Table 3).

Optimisation and validation of chemical quality of ready-to-eat *ofada* rice flakes

The optimisation of the ready to eat *ofada* rice flake was based on quality indices and level of desirability expected from high profile ready-to-eat flakes. These indices include maximum protein for good nutrition, minimum fat to prevent rancidity, minimum moisture to increase its shelf life and to prevent microbial activities, maximum ash represents total mineral composition which is also vital in nutrition because of their various functions, carbohydrate was not specified since the product have high carbohydrate value, minimum phytate (antinutritional factor usually found in rice), maximum metabolizable energy and maximum water absorption.

There were ten solutions suggested with desirability range of 0.745 to 0.783 with the first predicted solution having the highest desirability (Table 5). This involved storage of paddy for 9 months, soaking for 4 days and 17 h, parboiling at 106°C and drying at 30°C before milling to flour of desired particle size for the production of ready to

Table 5. Result of optimisation of quality of Ofada rice flakes.

S/N	Soaking Time (day)	Parboiling temp.	Drying Temp.	Storage Duration	Moisture (%)	Protein (%)	Ether extract (%)	Ash (%)	Phytate %	Water absorption ratio	Metabolisable Energy (kcal/100 g)	Desirebility
1	4.70	106.16	30.00	9.00	3.1	12.164	0.67	1.621	0.665	4.4035	393.192	0.7832
2	4.75	105.32	30.00	8.99	3.099	12.148	0.663	1.619	0.67	4.40266	393.191	0.7831
3	4.62	107.24	30.07	9.00	3.099	12.178	0.678	1.63	0.664	4.40116	393.166	0.7829
4	4.79	103.94	30.13	9.00	3.102	12.144	0.654	1.63	0.685	4.39364	393.112	0.7824
5	4.90	99.66	30.01	9.00	3.099	12.123	0.623	1.651	0.743	4.36695	392.962	0.7787
6	1.94	114.80	70.00	9.00	3.1	11.818	0.466	2.183	0.881	4.21874	388.686	0.7774
7	2.06	114.92	70.00	9.00	3.1	11.793	0.479	2.207	0.866	4.20355	388.584	0.7773
8	1.92	114.76	70.00	9.00	3.099	11.816	0.463	2.178	0.883	4.22206	388.71	0.7773
9	1.91	114.54	70.00	8.93	3.099	11.745	0.458	2.178	0.887	4.23492	388.817	0.7773
10	3.64	113.71	30.00	8.95	3.1	12.269	0.698	2.178	0.786	4.37523	392.861	0.7773

Table 6. Result of sensory evaluation of *ofada* rice flakes.

Ofada rice flakes	Colour	Taste	Aroma	Crispiness	Overall acceptability
First solution	6.7a	7.1a	6.9a	6a	7.30a
Sixth solution	6.3a	7.5a	7.2a	6.3a	6.4b
Tenth solution	7.8b	7.20a	6.80a	6.2a	6.6a

eat flakes to yield 3.10% moisture, 12.16% protein, 0.67% ether extract, 1.62% ash, 393.19 kcal/100 g, 0.67% phytate, and 4.40 water absorption ratio. However, the first three solutions as presented (Table 5) have similar desirability with high closeness of processing conditions and the same storage duration of paddy. Observations of the yields were also closely related.

The result of the validation of the best predicted solution showed yields of 4.08% moisture, 11.93% protein, 0.86% ether extract, 1.92% ash, 81.21%, 390.60 kcal/kg, 4.57 water absorption and 0.32% phytate. However, the results of the validation are closer to the result predicted by the response surface.

Sensory evaluation of ready-to-eat *ofada* rice flakes

The first, sixth and tenth predicted solutions were chosen for sensory evaluation based on variation of the levels of processing conditions. The ready to eat Ofada rice flakes manufactured were subjected to sensory analysis (colour, crispiness, taste, aroma, and overall acceptability) based on a nine point hedonic scale, where 1 is dislike extremely and 9 is like extremely. The result of the sensory analysis is presented in Table 6.

There exist no significant differences in crispiness, taste, and aroma, however, significant difference was shown in colour attribute with flakes produced from tenth predicted solution been rated better than the two. For the overall acceptability, the first solution was most accepted.

Conclusion

The effect of paddy storage duration and processing parameters (soaking time, parboiling temperature and drying temperature) were optimized for the production of ready to eat of ofada rice flakes. The coefficient of the model shown that soaking time of paddy has positive influence on ash, carbohydrate and metabolisable energy of rice flakes; parboiling temperature was positive on ether extract, ash, carbohydrate and metabolisable energy; drying temperature of paddy gave positive influence on moisture, protein, ether extract, phytate, metabolisable aenergy and water absorption; storage duration of ofada paddy also influence carbohydrate, phytate, water absorption, and metabolisable energy. The R^2 ranged from 0.9662 to 0.7318 which confirmed the fitness of the model. The optimum storage duration and processing parameters for treatment of ofada paddy for the production of ready to eat flake include storage of paddy for 9 months, soaking for 4 days and 17 h, parboiling at 106°C and drying at 30°C to yield optimum quality of ofada rice flakes. The sensory assessment showed good acceptability, colour, crispiness, aroma and taste.

Conflict of interests

The author(s) did not declare any conflict of interest.

REFERENCES

Akinoso R, Adeyanju JA (2010). Optimization of edible oil extraction from ofada rice bran using response surface methodology. J. Food Bioprocess Technol. 5(4):1372-1378.

AOAC (2000). Official method of analysis 17th edition, Horowitz edition intern, Maryland, USA. Vol. 1; 452-456.

Bryant RJ, McChung AM (2011). Volatile profiles of aromatic and non aromatic rice cultivars using SPME/GC-MS. Food Chem. 124: 501-513.

FAO (2002). Food Energy – Methods of analysis and conversion factors. Report of a technical workshop.

Fatsecret (2013). Rice flakes; nutritional facts. http://www.fatsecret.com/calories-nutrition retrieved on July 2014.

Fred G (2007). Role of heat treatment in the processing and quality of oat flakes. Academic dissertation submitted to the Department of Food Technology, University of Helsinki.

Gupta E, Sinha J, Dubey RP (2012). Utilization of dehydrated herbs in the formulation of value added snack "Rice Flakes Mix". J. Food Process S1-002. doi, 10(4172), 2157-7110.

Itasi HN, Rao BVRS, Jayndeep PA, Singh V (2012). Functional and antioxidant properties of ready-to-eat flakes from various cereals including sorhum and millets. 4(3):126-133.

Lu S, Walker CE (1988). Laboratory Preparation of Ready to Eat Breakfast Flakes from Grain Sorghum Flour. J. Am. Assoc. Cereal Chem. 65(4):377-379.

Machado MF, Oliveira FAR, Gekas V, Singh RP (1998). Kinetics of moisture uptake and soluble-solid loss by puffed breakfast cereals immersed in water. Int. J. Food Sci. Technol. 33:225-237.

Montgomery DC (2005). Design and analysis of experiments: Response surface method and designs, John Wiley and Sons Inc., New Jersey, USA.

National Cereal Research Institute (NCRI) and Africa Rice Centre (WARDA) (2007). Definition of Ofada Rice Qualities through Varietal Identification and Testing. PrOpCom publication Monograph Series #26, Abuja, Nigeria.

Nazni P, Bhuvaneswari J (2011). Optimization of mixture flakes and nuts to formulate ready to eat breakfast bar using response surface methodology. Int. J. Curr. Res. 3(3): 29-38.

Noreen, N., Shah, H., Anjum, F., Masood, T., and Faisal, S. (2009). Variation in mineral composition and phytic acid content in different rice varieties during home traditional cooking processes. Pak. J. Life Soc. Sci. 7(1):11-15.

Oeding J (1996). Einfluß der hydrothermischen Behandlung und des Flockierens auf die funktionellen und die physikalischen Eigenschaften von Haferflocke. Doctoral dissertation, University of Hohenheim.

Pawar VD, Inble UM (1988). Investigation on phytate protein mineral complexes in whey fractions of moth bean (Phaseolus aconitifolius jacq) flour. J. Food Sci. Technol. 25:190-195.

Thompson IU (1987). Reduction of PA in protein isolates by acylation technique. J. Am. Oil Chem. Soc. 64:712-717.

USDA (2013). Composition of foods, raw, processed, and prepared. USDA National Nutrient Database for Standard reference, release 26. Beltsville, Maryland 20705.

Walker RA, O'Brien ET, Pryer NK, Soboeiro M, Voter WA, Erickson HP, Salmon ED (1988) Dynamic instability of individual microtubules analysed by video light microscopy: rate constants and transition frequencies. J. Cell Biol 107:1437-1448.

Characterization and determination of the factors affecting anti-listerial bacteriocins from *Lactobacillus plantarum* and *Pediococcus pentosaceus* isolated from dairy milk products

Djadouni Fatima[1]* and Kihal Mebrouk[2]

[1]Faculty of Sciences, Mascara University, Algeria.
[2]Laboratory of Applied Microbiology, Department of Biology, Faculty of Sciences, Es-Senia University, Oran, Algeria.

***Lactobacillus plantarum* and *Pediococcus pentosaceus* strains isolated from dairy milk products, produced bacteriocins that displayed a wide spectrum of antimicrobial activity against *Listeria ivanovii* and food spoilage microorganisms. The two bacteriocins were thermally stable over a wide temperature range up to 100°C for 15 min and retained their activity at pH 2.0 to 6.0. Full bacteriocins activity were stable after three months of storage at 4 and -20°C, for 75 min of exposure to UV light, bacteriocin produced by *P. pentosaceus* was completely destroyed, but the second *L. plantarum* remained stable after the same time of exposure; treatment with proteolytic enzymes resulted in a remarkable stability of activity. Results obtained showed an increase in bacteriocins activity of the both strains against *L. ivanovii* on increasing the concentration of NaCl and KCl up to 5%, in presence of spices 5%, stability to acetone and hexane indicated that both surfaces were rather hydrophilic and bipolar. Further studies are required regarding suitable bioprocessing strategies for an efficient bacteriocin production process.**

Key words: *Lactobacillus plantarum*, *Pediococcus pentosaceus*, bacteriocin activity, temperature, pH, enzymes, solvents, salts, spices.

INTRODUCTION

Lactic acid bacteria (LAB) are wide spread in nature and predominate in microflora of milk and its products; many species are involved in the daily manufacturing of dairy products.

The LAB used in commercial starter cultures possesses numerous metabolic characteristics such as acidification activity, proteolytic activity, synthesis of bacteriocin, resistance to bacteriophage and production of exopolysaccharide are strain dependent.

All of these important activities contribute to the flavour, texture, and frequently the nutritional attributes of the products. Research has been focused on the role of starter and its required properties for the dairy industries (Crow et al., 1993; El-Soda et al., 1995; Limsowtin et al., 1996; Mayra.a-Makinen and Bigret, 1998; De Vuyst and Degeest, 1999).

Bacteriocins are bactericidal peptides or proteins produced by bacteria that inhibit the growth of closely related bacterial species (Klaenhammer, 1993). Currently, interest in bacteriocins is intense due to their inhibitory activity against food spoilage and foodborne pathogenic bacteria such as Listeria monocytogenes (Yamazaki et al., 2003).

So far, only nisin produced by *Lactococcus lactis*, is a commercial product and an approved food additive in most major food producing countries. *Lactobacillus plantarum* produced serial bacteriocins that were active against the pathogens and spoiling microorganisms in foods like plantaricin35d,bacteriocinsST28MSandST26MS(Todorov

*Corresponding author. E-mail: houarisalwa@yahoo.fr.

and Dicks, 2005).

Another bacteriocin that attracts research interest and Will likely be the next to be used in the food industry is pediocin (Ray, 1992; Turcotte et al., 2004), which is an antilisterial bacteriocin (Guyonnet et al., 2000; Simon et al., 2002) produced by several Pediococcus strains.

However, strains of *Pediococcus acidilactici* and *Pediococcus pentosaceus* have been reported to produce bacteriocins (Gonzales and Kunka, 1987; Biswas et al., 1991; Kim et al., 1992; Elegado et al., 1997; Gurira and Buys, 2005).

Some of these are designated as pediocin AcH by *P. acidilactici* H, E, F, and M (Bhunia et al., 1987; Ray et al., 1989; Kim et al., 1992), pediocin PA-1 by *P. acidilactici* PAC 1.0 (Gonzales and Kunka, 1987), pediocin JD by *P. acidilactici* SJ-1 (Schved et al., 1993), pediocin 5 by *P. acidilactici* UL5 (Huang et al., 1996), pediocin A by *P. pentosaceus* FBB-61 (Etchells et al., 1964; Flemming et al., 1975), pediocin N5p by *P. pentosaceus* (Strasser de Saad et al., 1995), and pediocin ST18 by *P. pentosaceus* (Todorov and Dicks, 2005).

Pediococcus damnosus has also been reported to produce a pediocin, designated as pediocin PD-1 (Green et al., 1997). It is now known for most studied pediocins that they are plasmid encoded (Ray, 1995; Le Marrec et al., 2000) and posttranslationally modified hydrophobic molecules (Green et al., 1997; Yin et al., 2003; Wu et al., 2004), which also share a similar N-terminal sequence (Henderson et al., 1992; Wu et al., 2004).

Pediocins form a group of bacteriocins belonging to the class IIa of bacteriocins, characterized as "antilisterial" (Papagianni, 2003). They inhibit several gram-positive spoilage and pathogenic bacteria.

The spectra of antimicrobial activity of pediocins produced by strains of *P. acidilactici* and *P. pentosaceus* have been found to be similar and this has been attributed to the phylogenetically close relation of the producer organisms (Collins et al., 1991).

Pediocin PD-1 by *P. Damnosus*, however, has been found to be inactive against other pediococci, a characteristic that makes it different from the known pediocins produced by *P. acidilactici* and *P. pentosaceus* strains (Green et al., 1997).

In this study, for the first time we have explored the isolation, identification, characterisation, and determination the factors affecting the activity of a novel antimicrobial peptide produced by *L. plantarum* and *P. pentosaceus* strains isolated from the dairy milk products.

MATERIALS AND METHODS

Isolation and selection of LAB strains

LAB strains were isolated from dairy milk products. Samples were plated directly on MRS and M17 agar (Merck, Germany) at 30°C for 2 to 3 days (pH 6.5) under aerobiosis and anaerobiosis conditions. They were routinely propagated and stored at -20°C supplemented with glycerol (20% v/v). Working cultures were sub-cultured twice (1% inoculum, 24 h, 30°C) prior to use.

Identification of the LAB isolates

The pure isolate selected as a potential bacteriocin - producer was identified on the basis of its cultural, morphological, physiological and biochemical characteristics. Selected LAB isolates were characterized by Gram stain, absence of spores and catalase test. Gram +ve, catalase and spores negative strains were maintained frozen until needed for antimicrobial activity testing (Cintas et al., 1998). Confirmation of the identification was based on the use of API 50 kits.

Bacterial strains

The indicator strain *Listeria ivanovii* used in this work was provided by the Laboratory of Bacteriology, Microbiology Department at the Faculty of Sciences, Es-Senia, Oran University. For the antimicrobial assay, the pathogenic culture (*L. ivanovii*) was grown in the nutrient agar media (NA) at pH 7.4 ± 0.2.

Determination of *in vitro* antimicrobial activity

The inhibitory activity of the selected LAB isolates against *L. ivanovii* was assayed by the agar spot test described by Fleming et al. (1975). The LAB isolates were spotted onto the surface of MRSm agar (with 2% glucose for *L. plantarum* and 4% maltose for *P. pentosaceus*) plates and incubated at 30°C for 24 h to allow colonies to develop.

L. ivanovii was inoculated into 10 ml of soft MRS agar (0.9% agar) and poured over the plate on which the LAB isolates were grown. After incubation at 30°C for 24 to 48 h under aerobiosis conditions, the plates were examined for the presence of inhibition zones. Inhibition was considered positive when the width of the clear zone around the colonies of the LAB isolates was 0.5 mm or larger.

Preparation of cell free supernatant (CFS)

To prepare CFS stock of *L. plantarum* and *P. pentosaceus*, an overnight pre-culture of each isolated strain was inoculated as 2% into flasks containing MRSm broth (volume is variable according to the experiment need). Flasks were incubated at 30°C for 12 h till the stationary phase was reached. The CFSs were separated by centrifugation at 8000 xg at room temperature for 20 min to remove viable bacterial cells and any insoluble particulate matter. The resulting supernatants were filter sterilized using 0.2 µm pore bacterial filters (Renner GMBH D-67125/Germany). The filtered CFSs was stored in the refrigerator for a maximum period of 2 weeks and periodically tested for the bacteriocin titer before being renewed. The crude bacteriocin preparation was relatively stable when prepared in this manner, with an insignificant decline in titer over 2 weeks of storage.

Physical and biochemical characterization of bacteriocins

The isolated crude bacteriocins were characterized with respect to the effect of physical and biochemical parameters on bacteriocin activity. In all tests, the filter sterilized CFSs subjected to the different parameters were tested for their antimicrobial activity against *L. ivanovii* using the ODM method at 600 nm (Naclerio et al., 1993). Untreated bacteriocin-containing CFSs of each producer strain, inoculated with the same indicator strains served as controls.

Heat resistance

The thermal stability of crude bacteriocin preparations was assessed by exposing the CFSs to different temperatures (Mota et al., 2004) ranging from 0C to 121°C (0, 30, 40, 50, 60, 70, 80, 90, 100 and 121°C, and 15 Lbs) for 15 min before being tested for their antimicrobial activity.

pH sensitivity

The effect of pH on the activity of bacteriocins was tested by adjusting cell-free supernatants from 2 to 12 (at increment of one pH unit) with sterile 1 N NaOH or 1N HCl (Albano et al., 2007). After 1 h of incubation at room temperature (25°C), the samples were tested for antimicrobial activity by the ODM.

Stability during storage

The method devised by Ogunbanwo et al. (2003a) was used to study the stability of bacteriocin preparations during different storage conditions. The crude bacteriocin was stored at –20 and +4°C for different interval of time (30, 45, 60 and 90 days). Samples were taken from the stored material to determine the bacteriocin activity as previously mentioned. Bacteriocin-containing CFSs of each producer strain that were not subjected to storage, and inoculated with the same indicator strains served as controls.

Effect of UV light

The effect of UV light was studied according to the method of Ogunbanwo et al. (2003a). Sterile Petri dishes containing aliquots of 10 ml crude bacteriocin preparations were exposed to the UV irradiation (Philips bulb, wavelength 340 nm, 220 to 240 V, 50 Hz) situated at a distance of 30 cm (Wanda and Bonita, 1991). Times of exposure to UV light ranged from 15 to 75 min. After each time interval, bacteriocin activity was estimated to UV light by ODM as previously stated together with unexposed bacteriocin-containing CFSs that served as the experimental controls.

Sensitivity to proteolytic and other enzymes

Selected enzymes were tested on CFSs (Bizani and Brandelli, 2002). Proteolytic enzymes including trypsin, pepsin, and papain were dissolved in 40 mM Tris-HCl (pH 8.2), 0.002 M HCl (pH 7), and 0.05 M sodium phosphate (pH 7.0) respectively to a final concentration of 0.1 mg/ml. Other enzymes such as lipase and α-amylase were dissolved in 0.1 M potassium phosphate (pH 6.0), and 0.1 M potassium phosphate (pH 7.0) respectively to a final concentration of 0.1 mg/ml. Equal aliquots of both filter sterilized CFSs of each test strain and each enzyme solution were mixed, incubated at 30°C for each enzyme for 2 h and heated in boiling water for 5 min to inactivate the enzymes. These sample mixtures and the controls (CFSs without enzyme treatment) were inoculated with the indicator strains as previously mentioned and tested for antimicrobial activity by the ODM.

Effect of organic solvents on bacteriocin activity

The sensitivity of bacteriocins to organic solvents such as acetone, chloroform, alcohol 90%, and hexane was investigated (Todorov et al., 2006). Freeze dried bacteriocin preparations from each producer strain were dissolved individually in each organic solvent at a final concentration of 10 mg/ml. The samples were incubated at 30°C for 1 h, after which the solvents were removed by evaporation at room temperature. The dried residue from the organic phase was re-suspended in sterile MRS broth (Ten et al., 1994) at a final concentration of 10 mg/ml and assayed together with the untreated controls (CFSs without solvent treatment) for antimicrobial activity as previously mentioned.

Effect of inorganic salts on bacteriocin activity

Different concentrations of inorganic salts such as NaCl and KCl (0, 0.5, 1, 3 and 5 % w/v) were examined for their inhibitory effect on bacteriocin preparations (Karaoğlu et al., 2003). Equal aliquots of both filters sterilized CFSs of each test strain and each salt solution were mixed, inoculated with the indicator strains as previously mentioned. The antimicrobial activity of bacteriocins was determined for sample mixtures and controls using ODM.

Effect of spices used as food additives on bacteriocin activity

The possibility of influencing the effectiveness of bacteriocin activity by the use of different spices (black pepper, red pepper, and garlic) that are used as food additives, for the purpose of food preservation, was tested according to the method of Verluyten et al. (2004). All spices were obtained locally and used without prior sterilization. The concentrations of spices were chosen to ascertain the maximum effects of the spices possible. The spices separately weighed as 0.01% w/v, were allowed to dissolve in 10 ml of sterile warm distilled water and mixed overnight by stirring. Spice solutions were centrifuged and filter-sterilized. The CFSs were filter sterilized, to which different volumes of the spice extracts were added independently to give a final concentration of 0.5, 1 and 3% (v/v) then inoculated with the indicator strains. Controls were prepared by growing the indicator strains into the spice solutions without adding the bacteriocin-containing CFSs. The bacteriocin activity was determined by ODM at 600 nm as previously mentioned.

RESULTS

Two isolates were chosen for further experiments and characterization of their antimicrobial activity. Their selection was based on the display of high bacteriocin activity and their potential of inhibiting the growth of *L. ivanovii* using the agar spot method. The average diameter of the inhibition zone ranged from 1 to 14 mm in size (Figure 1).

The effects of heat and pH were determined. Bacteriocins of the two strains of LAB (Figures 2 and 3) were considered to be extremely heat stable as antibacterial activity was not altered by heat treatment after 15 min at 121°C.

The pH stability was studied in the range of pH 2 to 12. It was observed that these bactériocins were active at pH values from 2 to 6.0 but reduced at higher pH levels. pH adjustment to 6.5 was chosen to eliminate the possible effect of organic acids.

The influence of UV light on bacteriocin activity was studied on both organisms. It was observed that bacteriocin produced by *P. pentosaceus* was completely destroyed after 75 min exposure to UV light (Figure 4). However, the bacteriocin produced by *L. plantarum* remained stable after the same period of exposure time.

The bacteriocin from *L. plantarum* (Figure 5) was inactivated to the tested enzymes and displayed a higher growth reduction potential against *L. ivanovii* compared to

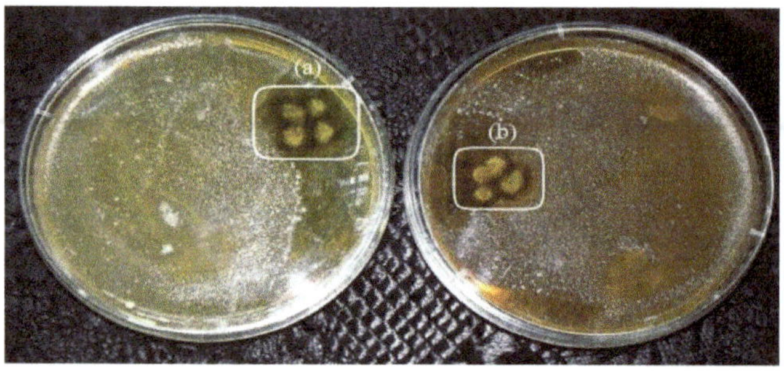

Figure 1. Anti-Listeria activity of bacteriocins from *L. pantarum* (a) and *P. pentosaceus* (b) determined by the agar spot on the MRSm agar plates at 30°C for 24 h.

that of *P. pentosaceus* which was sensitive to the rest of the selected enzymes (papain, pepsin, and lipase). However, since the activity of the filtrate was not completely inhibited, it is possible that the bacteriocin may also be bound to other molecules like a lipid or a carbohydrate moiety (Rakshita, 2011).

These data clearly show that the antimicrobial substance was of a proteinaceous nature. Storage of the active compounds at +4°C for three months and in a frozen state -20°C did not affect the antibacterial activity (Figure 6).

It was observed that the best organic solvents for activity of bacteriocin produced by *L. plantarum* hexane followed by acetone, where the activity was 50%, respectively. Concerning *P. pentosaceus*, the activity was stable when applying the tested organic solvents (hexane, acetone, and alcohol 90%) (Figure 7).

Different concentrations of NaCl and KCl (Figure 8) were selected to examine the effect of inorganic salts on the activity of the CFSs of the two LABs against the indicator strain used after 2 h of exposure. There was an increase in bacteriocin activity on the indicator strain with the increase in the concentrations of NaCl and KCl up to 5%.

The spices used in this experiment w/v (red pepper, black pepper, and garlic) in concentrations ranging from 0.5% affected the bacteriocins activity differently (Figure 9). The addition of 0.5% spices solution to both bacteriocins resulted in a significant activity against the indicator strains (up to 94% reduction of growth).

DISCUSSION

Bacteriocins could be interesting food conservatives on the industrial scale because of the thermostability of their relatively narrow spectrum of activity generally including Listeria and of their activity to weak concentrations on a broad range of value of pH. This study describes the partial characterization of a bacteriocins produced by *L. plantarum* and *P. pentosaceus* isolated from the dairy milk products. Choice was based on its broad antimicrobial spectrum against *L. ivanovii*. The ODM was preferred over the AWD assay in characterization experiments as aggregation, medium composition, and non- diffusible bacteriocins may influence the sensitivity of the AWD assay (Lewus et al., 1991).

In an attempt to characterize the physical and biochemical properties of the bacteriocins produced by the two strains presented in this study, different factors such as heat, pH, storage, UV light, enzymes, inorganic salts, organic solvents and spices were determined using *L. ivanovii* as indicator organism. Bacteriocins produced by *L. plantarum* and *P. pentosaceus* were heat stable for up to 15 min at 121°C. Storage for three months at -20 and 4°C did not affect bacteriocins activity. Full activity was retained even following incubation at 30°C for 2 h at pH values ranging from 2.0 to 6.0. These antimicrobial substances were sensitive to treatment with proteolytic enzymes, which was attributed to their proteinaceous nature. This is an interesting feature in view of its potential use as a food additive, in process like pasteurization, drying, tundalization, refrigeration and freezing.

Similar results reported by Suranjita et al. (2010) on the *L. lactis* W8 produced nisin concomitantly while fermenting milk to "dahi/curd", a traditional Indian fermented milk. Parada et al. (2007) indicated that some studies of characterization of bacteriocins show that these molecules can be active under certain ranges of temperature and pH. Sensibility to proteolytic enzymes evidences the proteinaceous characteristic of bacteriocins (De Martins et al., 2003). Complete inactivation or significant reduction in antimicrobial activity of the bacteriocins ST28MS and ST26MS produced by *L. plantarum* isolated from molasses was observed after treatment with proteinase K, pronase, pepsin and trypsin.

Ligocka et al. (2005) investigated that *Lactobacillus brevis* and *L. plantarum* isolated from compost green materials showed a high bactericidal activity (inhibition zone 1 to 4 mm) against *Listeria monocytogenes*, *Staphylococcus aureus* and *Escherichia coli*; but did not inhibit the growth of *Listeria innocua* strain. However, compared with the data given by Green et al. (1997) for pediocin PD-1 by *Pediococcus damnosus*, pediocin SA-1 appeared to be significantly more effective against *Listeria* spp. Purified pediocin SA-1 is heat stable for up to 60 min at 121°C. Storage for 4 weeks at -80, -20, +4 and 30°C did not affect bacteriocin activity. Full activity was retained even following incubation at 30°C for 1-week at pH values ranging from 3.0 to 12.0. No antimicrobial activity was detected after 30 min of incubation in the buffers of pH 2.0, 13.0 and 14.0. Purified pediocin SA-1 is resistant to treatment with trypsin, a-chymotrypsin, pepsin and papain, but not to proteinase K. (Anastasiadou et al., 2008).

The results reported by Diop et al. (2008) on bacteriocins produced by two strains of *L. lactis* subsp. *lactis* CWBI-B 1410 and CWBI-B1426 that were isolated from the fermented cereal and the seafood samples, indicated

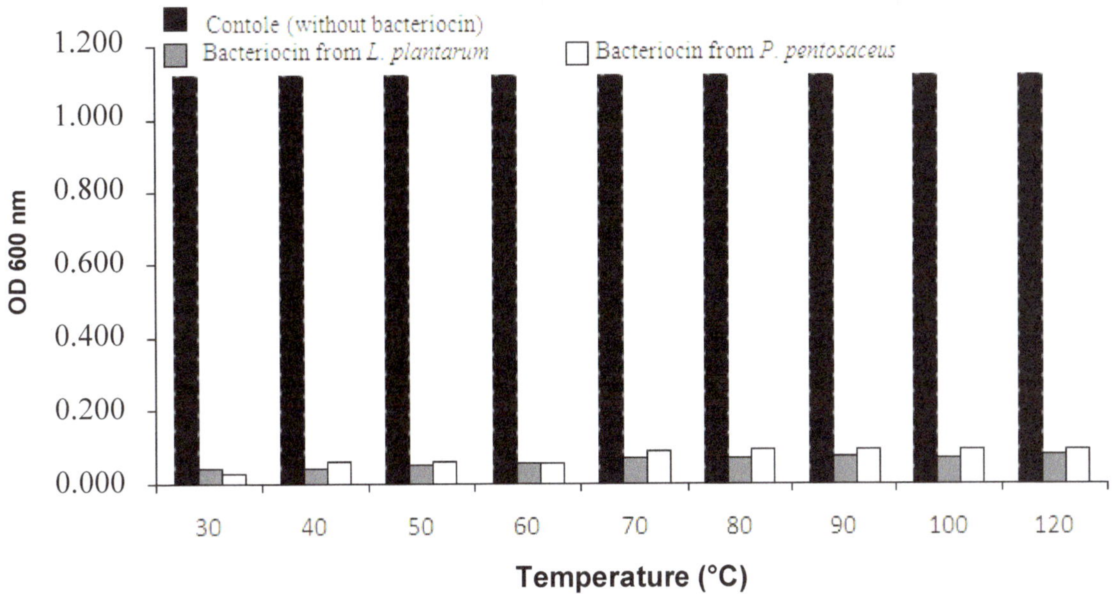

Figure 2. Effect of the heat treatment on antimicrobial activity of bacteriocins produced by *L. plantarum* and *P. pentosaceus.*

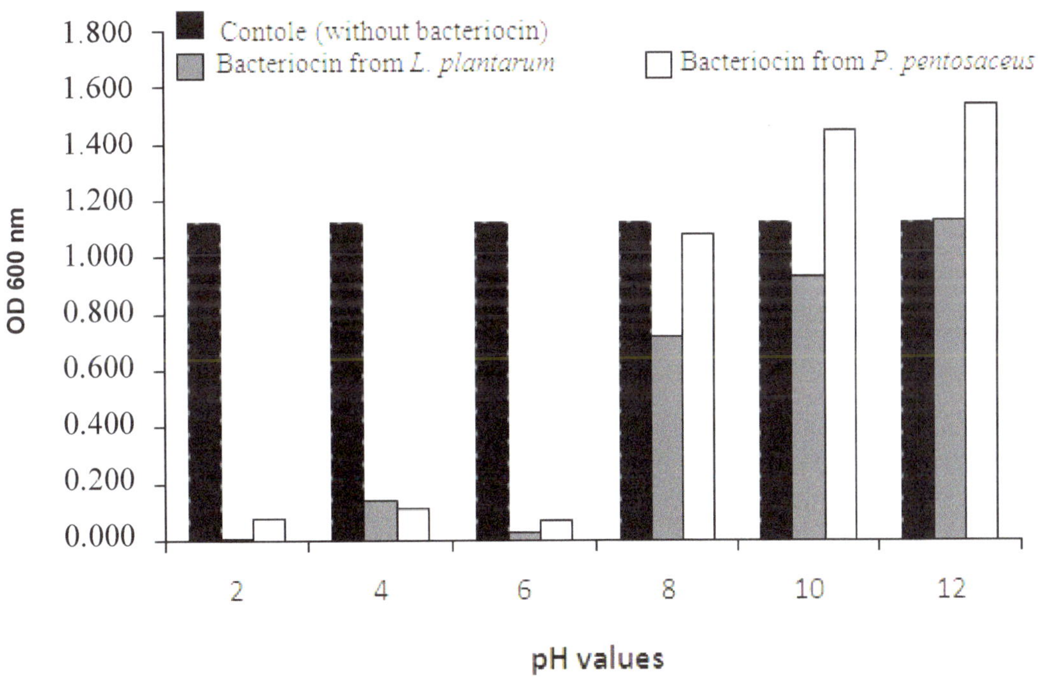

Figure 3. Effect of pH on the stability of inhibitors produced by *L. plantarum* and *P. pentosaceus.*

that they were digested by proteases and showed stability at 100°C for 30 min. Bromberg et al. (2005) showed that the bacteriocin-producing *L. lactis* subsp. *cremoris* CTC 204 was heat stable even at autoclaving temperature (121°C for 10 min) and was produced during refrigerated storage. Vinod et al. (2006) described that the bacteriocin produced by Lactobacillus CA44 was stable at up 100°C for 60 minutes but its activity declined compared to that at 68°C and was completely lost at 121°C for 15 min. Contradictory results were obtained by Valdes and Scherer (1994), where they found that the bacteriocin linocin M18 produced by Brevibacterium linens was heat sensitive. Joshil et al. (2006) reported that bacteriocin from *L. plantarum* CA44 (isolated from carrot fermentation) was

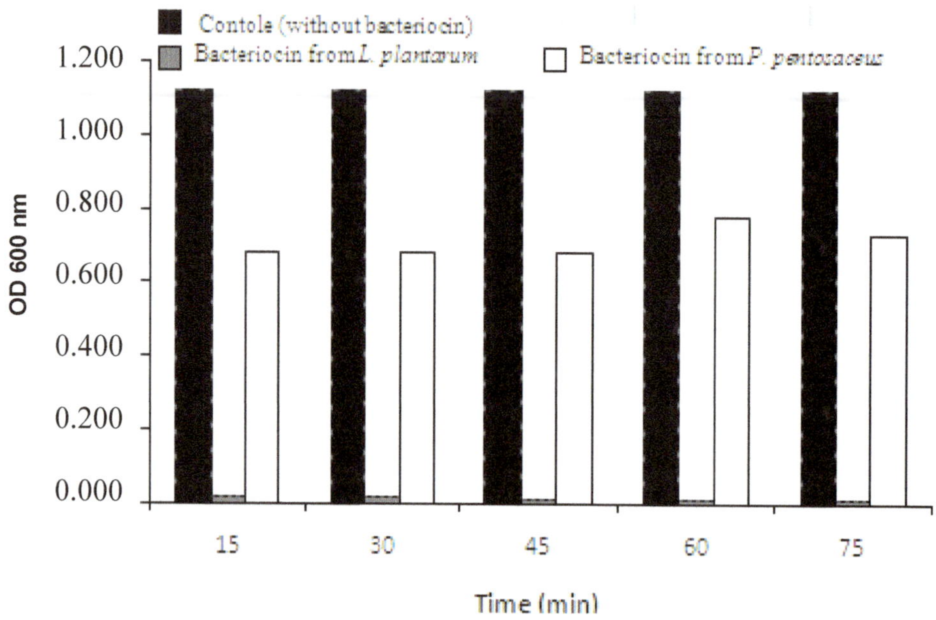

Figure 4. Effect of exposure to UV light on antimicrobial activity of bacteriocins produced by *L. plantarum* and *P. pentosaceus*.

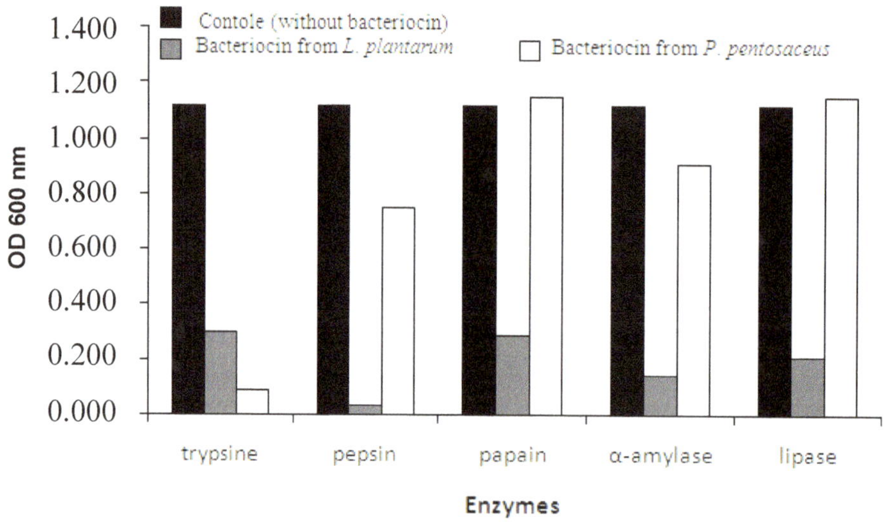

Figure 5. Study the effect of the proteolytic and non proteolytic enzymes (0.1 mg/ml) on the bacteriocins activity of the both strains *L. plantarum* and *P. pentosaceus* against *L. ivanovii* using the ODM method at 600 nm.

activated against *E. coli*, *S. aureus* and *Bacillus cereus*. It was stable up to 100°C but its activity declined at 68°C and was lost at 121°C. The maximum antimicrobial activity was retained with pH value of 4 to 5, but was adversely affected by the addition of papain.

Mkrtchyana et al. (2009) described that *Lactobacillus acidophilus* n.v. Er 317/402 strain Narine produces a small bacteriocin with a molecular weight of 1.1 kDa, designated acidocin LCHV. It was extremely heat stable (90 min at 130°C), was active over a wide pH range, and was found to be sensitive to proteolytic enzymes (trypsin, pepsin and proteinase K).

According to Navarro et al. (1999), *L. plantarum* J-51 bacteriocin remained stable at storage temperatures, such as 4, –20°C and room temperature, and under strong heating conditions (100°C for 60 min). Nevertheless, thermostability was lower after a previous storage at –20°C.

The influence of UV light on bacteriocin activity was studied on both organisms. It was observed that bacteriocin produced by *P. pentosaceus* was completely destroyed

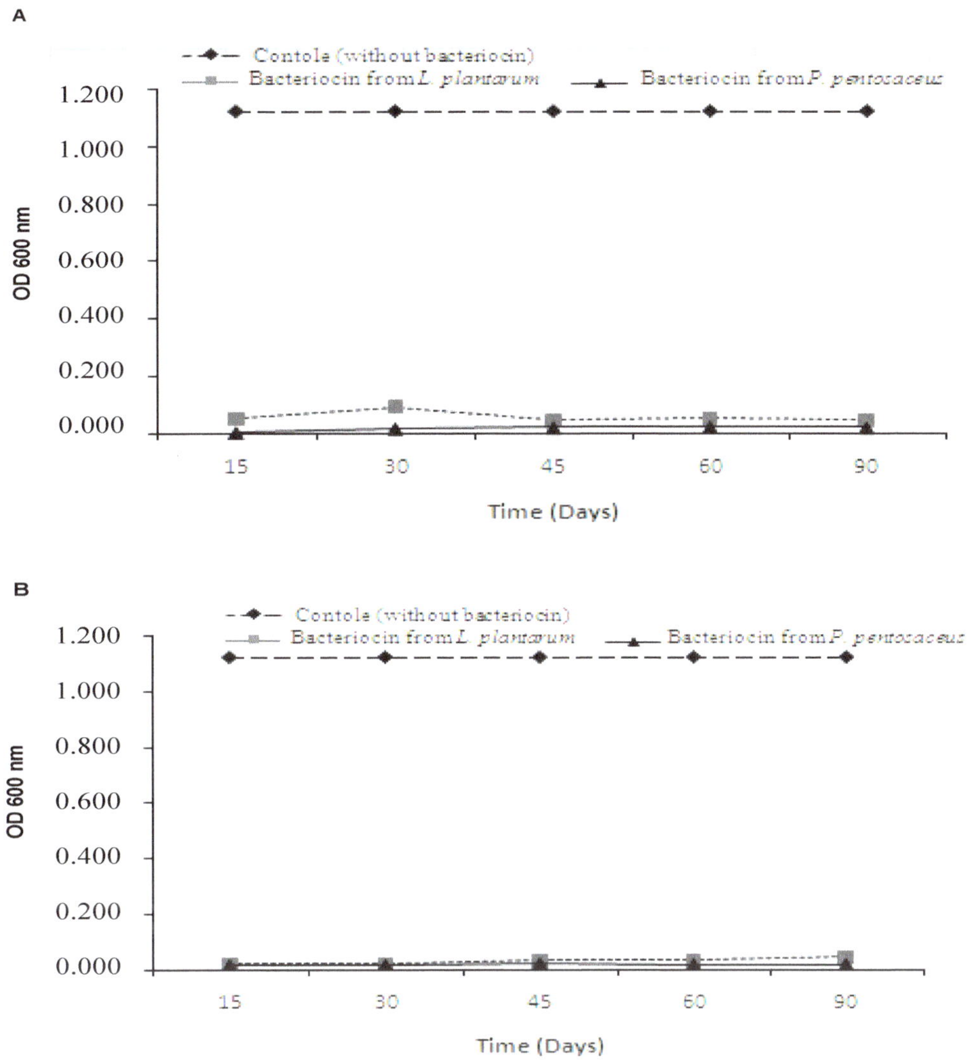

Figure 6. Effect of long time storage at +4°C (a) and -20°C (b) on the activity of bacteriocins produced by *L. plantarum* and *P. pentosaceus.*

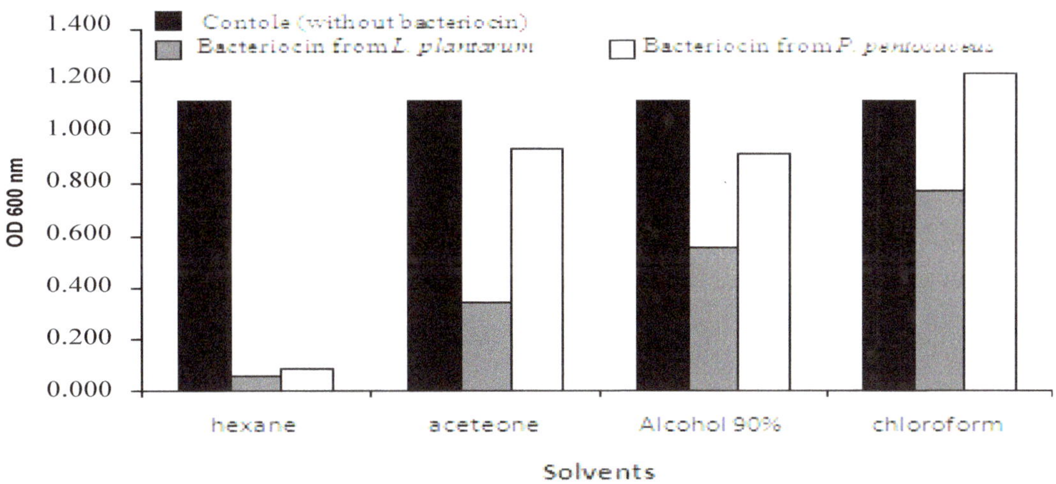

Figure 7. Sensibility of the bacteriocins of *L. plantarum* and *P. pentosaceus* strains to different solvents using *L. ivanovii* as indicator strain, using the ODM method at 600 nm.

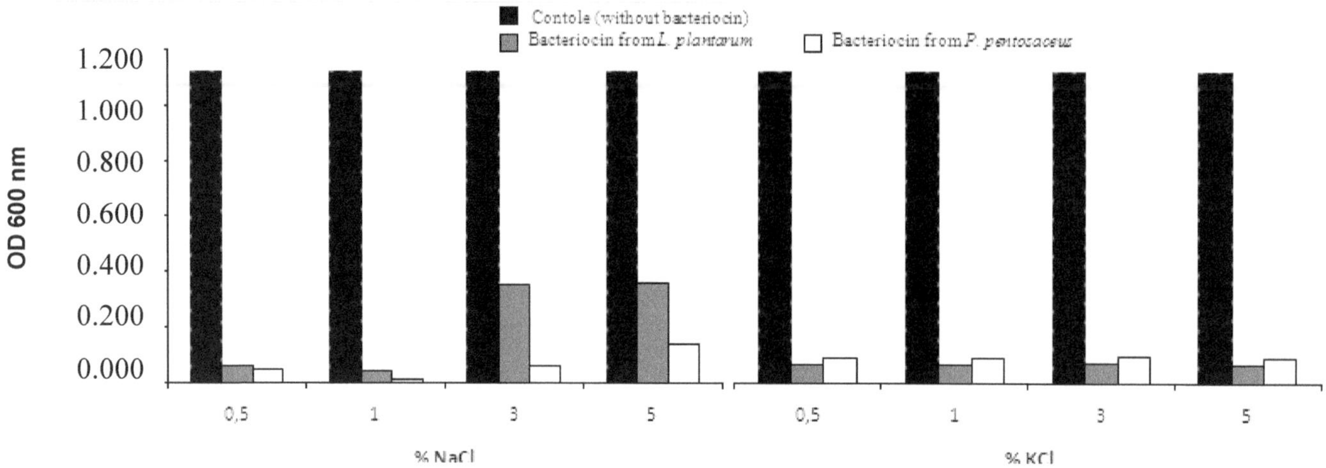

Figure 8. Study the effect of salts (NaCl and KCl) on antimicrobial activity of bacteriocins produced by *L. plantarum* and *P. pentosaceus.*

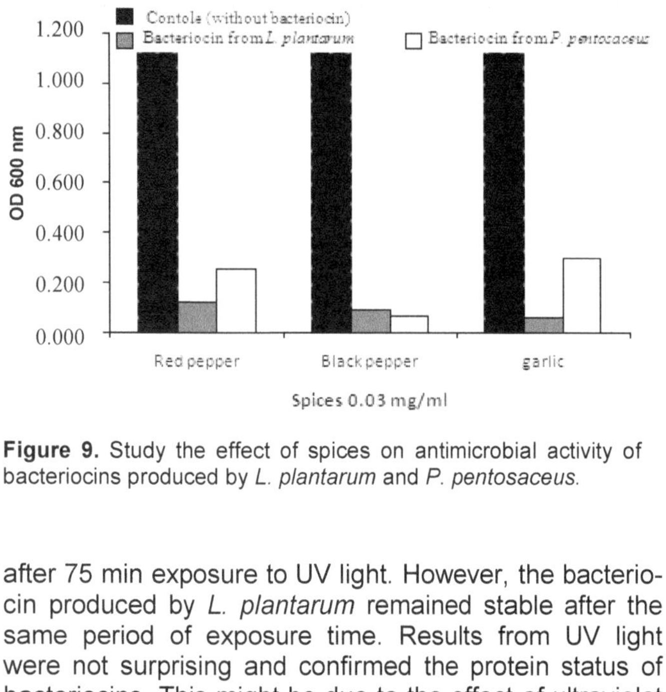

Figure 9. Study the effect of spices on antimicrobial activity of bacteriocins produced by *L. plantarum* and *P. pentosaceus.*

after 75 min exposure to UV light. However, the bacteriocin produced by *L. plantarum* remained stable after the same period of exposure time. Results from UV light were not surprising and confirmed the protein status of bacteriocins. This might be due to the effect of ultraviolet light on the protein nature causing modification and/or change in the ring structure of peptides, or affecting the protein function. The results are consistent with those described for bacteriocins produced by *L. plantarum* F1 and L. brevis OG1 after exposure to UV light from 0 to 5 min Ogunbanwo et al. (2003a)

The effect of some organic solvents on bacteriocin activity of the two LAB under test was also examined. Hexane, acetone, and alcohol 90% did not affect the activity of bacteriocin from P. pentosaceus. On the other hand, alcohol 90% and chloroform applied completely destroyed the bacteriocin produced from L. plantarum. These results suggest that acetone and hexane could be used in the extraction procedure to remove the bacteriocin from the aqueous phase and recover from the organic phase. Another explanation could prove the presence of lipid moiety in bacteriocin.

Some of results are in accordance to these stated for *P. pentosaceus*. Zeliha and Metin (2001) showed that the buchnericin LB produced by *Lactobacillus buchneri* LB completely lost its biological activity after treatment with organic solvents as formaldehyde, chloroform, acetone, 2-propanol, ethyl alcohol, hexane, iso-butanol and ethyl ether at a concentration of 10 mg/ml. Some of these results are similar to our results obtained for *L. plantarum*.

In order to show the effect of the addition of different ingredients on the bacteriocin efficacy, NaCl and KCl were used. There was an increase in bacteriocin activity of the both strains against L. ivanovii on increasing the concentration of NaCl and KCl up to 5%. These data suggest that the inorganic salts have a synergistic effect on bacteriocin efficacy when added with specific concentrations in foods.

Gänzle et al. (1999) reported that a synergistic effect has already been shown for curvacin A and sodium chloride, rendering the gram-negative pathogens *E. coli* and *Salmonella enterica* susceptible, while at low pH these pathogens also show increased sensitivity to the bacteriocins tested. Hagas et al. (2002) indicated that some common ingredients have a synergistic effect on bacteriocin efficacy or production, while some ingredients can reduce antimicrobial activity.

All of the spices tested in this study in concentrations of 5% did not affect bacteriocins activity differently. The addition of 5% solution spices to both bacteriocins resulted in a significant activity against the indicator strain (up to 94% reduction of growth). This indicates that small amounts of additives should be used in some foods to promote the activity of the bacteriocins applied, especially when more than one spice is used.

Extensive studies have been performed to determine the inhibitory properties of garlic, and many food-borne pathogens, both gram-positive and gram-negative bacteria, have been shown to be inhibited by it (Kumar and Berwal, 1998; Ross et al., 2001; Leuschner and Zamparini, 2002). Singh et al. (2001) indicated that bacteriocins or bacteriocinogenic starter cultures can be applied in various food products and may provide a synergistic effect using nisin and garlic extract. In both, such a positive interaction between nisin and garlic extract has been shown towards strains of *L. monocytogenes*.

Ettayebi et al. (2000) and Singh et al. (2001) described that the combination of spices together with bacteriocins that, albeit in a lower amount, can be produced *in situ* by the bacteriocinogenic starter culture may lead to a synergistic effect, rendering pathogens susceptible to *L. monocytogenes*. Such a synergistic inhibition has been shown between nisin on the one hand and garlic extract or thymol on the other.

Chumchalova´ et al. (2004) reported that the partial purification and characterization of heat-stable acidocin CH5, a bacteriocin produced by *L. acidophilus* CH5, showed that the active compound caused fast decrease in viable cell count of the indicator strain *Lactobacillus* delbrueckii subsp. *lactis* LTI30 which was in accordance to results obtained in this study.

ACKNOWLEDGEMENT

This research work was supported by Microbiology Research Laboratorie. Faculty of Sciences, Es-Senia University, Oran, Algeria.

REFERENCES

Albano H, Todorov SD, Van reenen CA, Hogg T, Dicks LM, Teixeira P (2007). Characterization of two bacteriocins produced by Pediococcus acidilactici isolated from Alheira, a fermented sausage traditionally produced in Portugal, J. Food, Microbiol. 116: 239-247.

Anastasiadou S, Papagianni M, Filiousis G, Ambrosiadis I, Koidis, PV (2008). Pediocin SA-1, an antimicrobial peptide from Pediococcus acidilactici NRRL B5627: Production conditions, purification and characterization. Food. Biotechnol. 99: 5384–5390.

Bhunia AK, Johnson MC, Ray B (1987). Direct detection of an antimicrobial peptide of Pediococcus acidilacti by SDS–PACE, J. Ind. Microbiol., 2: 312–319.

Biswas SR, Ray P, Johnson MC, Ray B (1991). Influence of growth conditions on the production of a bacteriocin pediocin AcH, by Pediococcus acidilactici H, Appl. Environ. Microbiol., 57: 1265–1267.

Bizani D, Brandelli A (2002). Characterization of a bacteriocin produced by a newly isolated Bacillus sp. Strain 8 A, J. Appl. Microbiol., 93: 512–519.

Bromberg R, Moreno I, Delboni, R, Cintra H, Oliveira P (2005). Characteristics of the bacteriocin produced by Lactococcus lactis subsp. cremoris CTC 204 and the effect of this compound on the mesophilic bacteria associated with raw beef, J. Microbiol. Biotechnol., 21: 351-358.

Chumchalova´ J, Stiles J, Josephsen J, Plockova M (2004). Characterization and purification of acidocin CH5, a bacteriocin produced by Lactobacillus acidophilus CH5, J. Appl. Microbiol., 96: 1082–1089.

Cintas LM, Casaus P, Holo H, Hernandez PE, Nes, IF, Harvastein LS (1998). Enterocins L50A and L50B, two novels bacteriocins from Enterococcus feacium L50, are related to staphylococcal hemolysins, J. Bacteriol., 180: 1988-1994.

Collins MD, Rodrigues UM, Ash C, Aguirre M, Farrow JAE, Martinez-Murcia A (1991). Phylogenetic analysis of the genus Lactobacillus and related lactic acid bacteria determined by reverse transcriptase sequencing of 16S rRNA. FEMS Microbiol. Lett., 77: 5–12.

Crow VL, Coolbear T, Holland R, Pritchard GG, Martley FG (1993). Starters as finishers: starter properties relevant to cheese ripening. Int. Dairy J., 3: 423-460.

De Vuyst L, Degeest B (1999). Heteropolysaccharides from lactic acid bacteria, FEMS. Microbiol. Rev., 23, 153-177.

Diop MB, DuBois-Dauphin R, Dortu C, Destain J, Tine E, Thonart E (2008). In vitro detection and characterization of bacteriocin-like inhibitory activity of Lactic Acid Bacteria (LAB) isolated from Senegalese local food products. Afri. J. Microbiol. Res. Sénégal, 2: 206-216.

Elegado FB, Kim WJ, Kwon DY (1997). Rapid purification, partial characterization, and antimicrobial spectrum of the bacteriocin, pediocin AcM, from Pediococcus acidilactici M, Int. J. Food Microbiol., 37: 1–11.

El-Soda M, Farkye N, Vuillemarad J, Simard R, Olson N, El Kouly W, Dako E, Medrano E, Gaber M, Li, L (1995). Autolysis of lactic acid bacteria. Impact on flavour development in cheese. In: Charalambous, G. (Ed.), Food Flavour: Generation Analysis and Process Influence. Elsevier Science B.V., Amsterdam, The Netherlands, pp. 2205–2223.

Etchells JL, Costilow RN, Anderson TE, Bell TA (1964). Pure culture fermentation of brined cucumbers, Appl. Environ. Microbiol., 12: 523–535.

Ettayebi K, El Yamani J, Rossi-Hassani BD (2000). Synergistic effects of nisin and thymol on antimicrobial activities in Listeria monocytogenes and Bacillus subtilis. FEMS Microbiol. Lett. 183: 191–195.

Flemming HP, Etchells JL, Costilow RN (1975). Microbial inhibition by an isolate of Pediococcus from cucumber brines, Appl. Microbiol., 30: 1040–1042.

Gänzle MG, Hertel C, Hammes WP (1999). Resistance of Escherichia coli and Salmonella against nisin and curvacin A, Int. J. Food Microbiol., 48: 37-50.

Gonzales CF, Kunka BS (1987). Plasmid-associated bacteriocin production and sucrose fermentation in Pediococcus acidilactici, Appl. Environ. Microbiol., 53: 2534–2538.

Green G, Dicks LMT, Bruggeman, G, Vandamme EJ, Chikindas, M.L (1997). Pediocin PD-1, a bactericidal antimicrobial peptide from Pediococcus damnosus NCFB 1832, J. Appl. Microbiol., 83: 127–132.

Gurira OZ, Buys EM (2005). Characterization and antimicrobial activity of Pediococcus species isolated from South African farm-style cheese, Food. Microbiol., 22: 159–168.

Guyonnet D, Fremaux C, Cenatiempo Y, Berjeaud JM (2000). Method for rapid purification of class IIa bacteriocins and comparison of their activities, Appl. Environ. Microbiol., 66: 1744–1748.

Hagas M, Garriga M, Pascual M, Aymerich MT, Monfort JM (2002). Enhancement of sakacin K activity against Listeria monocytogenes in fermented sausages with pepper or manganese as ingredients, Food Microbiol., 19: 519-528.

Henderson JT, Chopco AL, van Wassenaar PD (1992). Purification and primary structure of pediocin PA-1 produced by Pediococcus acidilactici PAC1 0. Arch. Biochem. Biophys., 295: 5–12.

Huang J, Lacroix C, Daba H, Simard RE (1996). Pediocin 5 production and plasmid stability during continuous free and immobilized cell cultures of Pediococcus acidilactici UL5 J. Appl. Bacteriol., 80: 635–644.

Joshi VK, Sharma S, Rana N, (2006). Production, purification, stability and efficacy of bacteriocin from isolates of natural Lactic acid fermentation of vegetables, Food. Technol. Biotechnol. India, 44 (3) 435-439.

Joshi VK, Sharma1 S, Neerja SR (2006). Production, purification, stability and efficacy of bacteriocin from isolates of natural lactic acid fermentation of vegetables, Food Technol. Biotechnol., 44 (3): 435–439.

Kim WJ, Ray B, Johnson MC (1992). Plasmid transfers by conjugation and electroporation in Pediococcus acidilactici, J. Appl. Bacteriol., 72: 201–207.

KlaenhammerTR (1993). Genetics of bacteriocins produced by lactic acid bacteria. FEMS, Microbiol. Rev., 12: 39-86.

Kumar M, Berwal JS (1998). Sensitivity of food pathogens to garlic (Allium sativum), J. Appl. Microbiol., 84: 213–215.

Leuschner RG, Zamparini J (2002). Effects of spices on growth and survival of Escherichia coli O157 and Salmonella enterica serovar Enteriditis in broth model systems and mayonnaise, Food. Cont., 13: 399–404.

Ligocka A, Paludzak Z (2005). Capability of Lactic Acid Bacteria to inhibit pathogens in sewage sludge subjected to biotechnological processes. Microbiol. Technol. Agric., 49: 23-27.

Ligocka A, Paludzak Z (2005). Capability of Lactic Acid Bacteria to inhibit pathogens in sewage sludge subjected to biotechnological processes, Microbiol. Technol. Agricol., 49: 23-27.

Limsowtin GKY, Powell IB, Parente E (1996). Types of starters. In: Cogan, T.M., Accolas, J.-P (Eds.), Dairy Starter Cultures. VCH Publishers, Inc., New York, pp. 101-129.

Mayra-Makinen A, Bigret M, (1998). Industrial use and production of lactic acid bacteria. In: Salminen, S., Wright, A. (Eds.), Lactic Acid Bacteria Micobiology and Functional Aspects. Marcel Dekker, Inc., New York, pp. 73-102.

Mkrtchyana H, Gibbonsb S, Heidelbergerb S, Zlohb M, Khalatbari Limakia H (2009). Purification, characterisation and identification of acidocin LCHV, an antimicrobial peptide produced by Lactobacillus acidophilus n.v. Er 317/402 strain Narine. Intl. J. Antimicrobial agents, 35: 255–260.

Motta AS., Cladera-Olivers F, Brandelli A (2004). Screening for antimicrobial activity among bacteria isolated from the Amazon basin. Braz. J. Microbiol., 35: 307-310.

Navarro L, Zarazaga M, Sa´enz, J, Ruiz-Larrea F, Torres C (1999). Bacteriocin production by lactic acid bacteria isolated from Rioja red wines, Agricult. Aliment. 3: 7135-7199.

Ogunbanwo ST, Sanni AI, Onilude AA (2003a). Characterization of bacteriocin produced by Lactobacillus plantarum F1 and Lactobacillus brevis OG1, Afr. J. Biotechnol., 2 (8): 219-227.

Papagianni M, Avramidis N, Filioussis G, Dasiou D, Ambrosiadis I (2006). Determination of bacteriocin activity with bioassays carried out on solid and liquid substrates: assessing the factor "indicator microorganism, Microb. Cell Fact. 5: 30.

Parada JL, Caron CR, Bianchi PA, Ricardo Soccol M, Ricardo Soccol C (2007). Bacteriocins from Lactic Acid Bacteria: Purification, Properties and use as Biopreservatives, Brazilian Archives. Biol. Technol., 50(3): .521-542.

Rakshita PK (2011). Antimicrobial activity of lactic acid bacteria and partial characterization of its bacteriocin. Master of Science in Microbiology. Mysore.

Ray B (1992). Pediocin(s) of Pediococcus acidilactici a food biopreservative. In: Ray, B., Daeschel, M. (Eds.), Food Biopreservatives of Microbial Origin. CRC Press, New York, USA, pp. 265–322.

Ross ZM, O'Gara EA, Hill DJ, Sleightholme HV, Maslin DJ (2001). Antimicrobial properties of garlic oil against human enteric bacteria: evaluation of methodologies and comparisons with garlic oil sulfides and garlic powder, Appl. Environ. Microbiol., 67: 475–480.

Schved F, Lalazar Y, Henis Y, Juven BJ (1993). Purification, partial characterization and plasmid-linkage of pediocin SJ-1, a bacteriocin produced by Pediococcus acidilactici, J. Appl. Bacteriol., 74: 67–77.

Simon L, Fremaux C, Cenatiempo Y, Berjeaud JM, Sakacin G (2002). A new type of antilisterial bacteriocin, Appl. Environ. Microbiol., 68: 6416–6420.

Singh B, Falahee MB, Adams MR (2001). Synergistic inhibition of Listeria monocytogenes by nisin and garlic extract. Food Microbiol., 18: 133–139.

Saad Stra¨sser de, AM, Pasteris, SE, Manca de Nadra MC (1995). Production and stability of pediocin N5p in grape juice medium, J. Appl. Bacteriol., 78: 473–476.

Suranjita Ma, Pran Krishna Cb, Swadesh RBa (2010). Potential production and preservation of dahi by Lactococcus lactis W8, LWT - Food a nisin-producing strain, Sci. Technol., 43: 337–342.

Ten Brink B, Minekus M, Vander Vossen JMBM, Leer RJ, Huis JHJ (1994). Antimicrobial activity of lactobacilli: preliminary characterization and optimization of production of acidocin B., a novel bacteriocin produced by Lactobacillus acidophilus M46, J. Appl. Bacteriol., 77: 140-148.

Todorov SD, Dicks LMT (2005). Pediocin ST18, an antilisterial bacteriocin produced by Pediococcus pentosaceus ST18 isolated from boza, a traditional cereal beverage from Bulgaria, Process Biochem., 40: 365–370.

Todorov SD, Danova ST, van Reenen CA, Meincken M, Dinkova G, Ivanova IV, Dicks LM (2006). Characterization of bacteriocin HV219, produced by Lactococcus lactis subsp. Lactis HV219 isolated from human vaginal secretions, J. Basic Microbiol., 46 (3): 226-238.

Turcotte C, Lacroix C, Kheadr E, Grignon L, Fliss I (2004). A rapid turbidometric microplate bioassay for accurate quantification of lactic acid bacteria bactriocins, Int. J. Food Microbiol., 90: 283– 293.

Verluyten V, Leroy F, de Vuyst L (2004). Effects of different spices used in production of fermented sausages on growth of and curvacin A production by Lactobacillus curvatus LTH 1174, Appl. Environ. Microbiol., 70 (8): 4807–4813.

Vinod KJ, Somesh S, Neerj R (2006). Production, purification, stability and efficacy of bacteriocin from isolates of natural lactic acid fermentation of vegetables, Food. Technol. Biotechnol., 44: 435-439.

Wanda J, Bonita A (1991). Partial purification and characterization of a bacteriocin produced by Propionibacterium thoenii, Appl. Environ. Microbiol. 57: 701-706.

Wu CW, Yin LJ, Jiang ST (2004). Purification and characterization of bacteriocin form Pediococcus pentosaceus ACCEL, J. Agric. Food Chem., 52: 1146–1154.

Yamazaki K, Inoue N, Montville TJ, Suzuki M, Kawai Y |(2003). Inhibition of Listeria monocytogenes in cold-smoked salmon by Carnobacterium piscicola CS526 isolated from frozen surimi, J. Food Protect., 66 (8): 1420–1425.

Yin LJ, Wu CW., Jiang ST (2003). Bacteriocins from Pediococcus pentosaceus L and S from pork meat, J. Agric. Food Chem., 51: 1071– 1076.

Comparisons of physiochemical, total phenol, flavanoid content and functional properties in six cultivars of aromatic rice in Bangladesh

Md. Asaduzzaman[1#], Md. Emdadul Haque[1#], Jiaur Rahman[1#], S. M. Kamrul Hasan[1], M. A. Ali[2] and Mst. Sorifa Akter[3] and Maruf Ahmed[1]

[1]Department of Food Processing and Preservation, Hajee Mohammad Danesh Science and Technology University, Dinajpur, Bangladesh.
[2]Department of Food Science and Nutrition, Hajee Mohammad Danesh Science and Technology University, Dinajpur, Bangladesh.
[3]Department of Food Science and Technology, State University of Bangladesh, Dhaka, Bangladesh.

Various aromatic rice varieties such as Kalizira, Begun Bichi, BRRI dhan-34, BRRI dhan-37, BRRI dhan-50, Philipine Katari were analyzed for physiochemical, total phenol, flavonoid contents and functional properties. All aromatic rice varieties had moisture contents (11.25 to 15.13%), protein (3.23 to 6.21%), fat (0.68 to 1.45%) and ash (0.88 to 1.46%). The maximum amount of amylose and starch content were obtained in BRRI dhan-37 and BRRI dhan-50 (23.01 and 72.606%, respectively). Total phenolic content was higher in BRRI-37 (474 mg/100 g); whereas, lower value was observed in BRRI dhan-34 (268.67 mg/100 g) variety. Both Philipine Katari and kalizira variety possessed highest level of flavonoid content among all rice varieties. Highest water absorption index value was found in BRRI dhan-37 and lowest in Begun Bichi variety. On the other hand, water soluble index value was varied by 1.32 to 2.12% in all aromatic rice varieties. Therefore, the study indicates that aromatic rice could be used as functional food ingredients as well as sources of natural phytochemicals.

Key words: Aromatic rice, physiochemical, functional properties, total phenol, flavonoid.

INTRODUCTION

Rice (*Oryza sativa* L.) is a most grown cereal crop in the developing countries. It is used as staple food for more than three billion people in Asia (Bhattacharjee et al., 2002). Rice also provides 50 to 80% of daily calorie in Asia (Khush, 2005). The global production of rice includes about more than 508,697,332 metric tons per year, where as, in Bangladesh, about 38,060,000 metric tons of rice was harvested in 2009 to 2010 (Food and Agriculture Organization, 2012). Although, the rice is a rich source of carbohydrate, it contains a moderate amount of protein and fat and also a source of B vitamins like thiamin, riboflavin and niacin (Fresco, 2005). Rice carbohydrate is mainly starch which is composed of amylose and amylopectin. Physicochemical properties and cooking charecterstics of rice depends on amylose content (Saikia et al., 2012). There are several types of rice around the world which are categorized on the basis of color like white rice, brown rice, red rice, black rice etc.

Rice can also be classified as aromatic and fine (non-aromatic) rice on the basis of aroma. Various aromatic rice is available in market such as Kataribhog, Radhunipagal, Chinigura, Badshabhog and Kalizira. Bangladesh Rice research Institute (BRRI) discovered BRRI dhan-34, BRRI dhan-37 and BRRI dhan-50 aromatic rice.

In Bangladesh northern region such as Rangpur, Dinajpur and Panchagarh district are famous for the production of aromatic rice. Generally, the price of the aromatic rice is three to four times greater than that of the non-aromatic or ordinary rice varieties in national and international markets. In Bangladesh, aromatic rice is cultivated during "Aman" season in November and December and its production was about 26,837 metric tons in 2009 (DAE, 2010). Aromatic rice is a good source of different phytochemical such as phenolic compounds, anthocyanin, flavonoids etc. These components are most likely involved in the reduction of degenerative human diseases due to their antioxidative and free radical scavenging properties (Basu et al., 2012). Usually, the use of synthetic antioxidants as Butylated Hydroxy-toluene (BHT) and butylated hydroxyanisole (BHA) in foods is discouraged due to their high levels of toxicity and carcinogenicity (Hirose et al., 1981). Therefore, natural antioxidants from plant extracts have attracted considerable attention due to their safety. In recent years, there has been a global trend toward the use of phyto-chemicals from natural resources such as vegetables, fruits, oilseeds and herbs, as antioxidants and functional ingredients. Besides, this research will give new information about bioactive compounds of aromatic rice in Bangladesh. Recently, farmers are encouraged to cultivate aromatic rice cultivars because of their potential to provide and promote human health by reducing the concentration of reactive oxygen species (ROS) and free radicals (Nam et al., 2006).

Based on the information as accumulated, a research was conducted to compare the physicochemical, func-tional and phytochemical properties of six verities of the aromatic rice grown in Bangladesh like Kalizira, Kataribhog, Bagun Bichi, BRRI dhan–34, BRRI dhan –37 and BRRI dhan –50.

MATERIALS AND METHODS

Three local and modern aromatic rice varieties namely Kalizira, Begun Bichi, Philipine Katari and BRRI dhan-34, BRRI dhan-37, BRRI dhan-50, respectively were purchased from a local market. The paddies were dried in the oven at 60°C for 1 h. Then, the paddies were husked manually and powder was made by using Blender machine (Jaipan CM/L- 7360065, Japan).

Proximate composition of rice flour

Moisture, crude protein, fat and ash content of flours were deter-mined by official methods (AOAC, 1998).

Physiochemical properties

Determination of starch

The starch content was determined by Lane and Eynon (1923) method with some modification. Rice powder (5 g) was taken in a beaker and 30 ml of water was added. It was transferred to the water bath and heated at 60°C for 25 min. Then, 100 ml of 95% ethanol was added and stirred for 15 min. It was filtered through Whatman filter paper no. 2 and residue was soaked in the 50% etha-nol solution for 1 h. After that, the residue was filtered and washed with 50% ethanol solution for 4 h. The residue was collected to a round bottom flask and 100 ml of water and 20 ml of HCl (6M) added. The flask was attached with the condenser and heated for 2.5 h. Then, it was allowed to cool and neutralized by adding NaOH solution (40%). After that, 10 ml of Fehling solution was taken into a conical flask and titrated against neutralized sample solution. When copper sulfate like color was observed, then, 3 drops of methylene blue indicator was added and continued titration.

The end point was indicated by brick- red color. The starch content was calculated by the following formula:

$$\% \text{ reducing sugar} = \frac{\text{Factor for Fehling's solution} \times \text{dilution}}{\text{Titre value} \times \text{weight of powder}} \times 100$$

$$\% \text{ starch} = \% \text{ reducing sugar} \times 0.9$$

Determination of amylose content

Amylose content was determined by Sompong et al. (2011) method with some modification. Rice powder (0.02 g) was taken into a volumetric flask. Then, 0.2 ml of ethanol (95%) was added. After that 1.8 ml of 1N NaOH was added and made 20 ml total volume by adding distilled water. It was kept 20 min at room temperature and boiled for 10 min at 45°C. It was filtered using Whatman filter paper no. 2. Then, 1 ml filtrate was transferred to a 50 ml tube and 0.2 ml of 1M acetic acid and 0.4 ml Lugol's solution were added and made total volume of 20 ml by adding distilled water. The mixture was mixed and kept for 20 min at room temperature. Then, the absorbance was taken at 620 nm using spectrophotometer (T80 U/VIS, United Kingdom). The amylose content was determined using potato amylose standard curve and was expressed as g/100 g extract.

Determination of phenol content

Determination of the phenolic content was done by Saikia et al. (2012) method with some modification. Rice powder (1 g) was extracted with 20 ml of 25% ethanol for 15 min and filtered through Whatman no. 2 filter paper. After that, 1 ml sample, 1 ml Folin reagent and 5 ml Na_2CO_3 were transferred to a volumetric flask and it was kept for 1 h at room temperature. Then the absorbance was taken at 765 nm using spectrophotometer (T80 U/VIS, United Kingdom). Total phenolics were calculated on the basis of standard curves of gallic acid, and expressed as mg/100 g.

Determination of flavonoid

Flavonoid content was done by Ali and Chang (2008) method with some modification. Rice powder of 10 mg was extracted with 10 ml of methanol. After that, 1 ml was taken into a test tube and 5 ml distilled water was added. Then, 0.1 ml of potassium acetate was added. After 5 min, 0.3 ml of $AlCl_3$ was added and solution was kept for 30 min at room temperature. The absorbance was taken at 415 nm using spectrophotometer (T80 U/VIS, United Kingdom). The flavonoid content was determined using Rutin standard curve and was expressed as mg/100 g.

Functional properties

Determination of water solubility index (WSI) and water absorption index (WAI)

WAI and WSI were determined following the method described by Anderson (1982). Each sample (0.83 g) was suspended in 10 ml of distilled water and stirred for 30 min. Subsequently, the dispersions

Table 1. Proximate compositions of the different varieties of aromatic rice.

Varieties	Parameters (g/100 g)			
	Moisture	Protein	Fat	Ash
Kalizira	11.55±0.01^{d}	5.18±0.015^{c}	0.937±0.01^{c}	0.923±0.00^{c}
Begun Bichi	12.39±0.02^{b}	5.74±0.035^{b}	0.680±0.01^{e}	1.070±0.01^{b}
BRRI dhan-34	11.48±0.01^{e}	5.44±0.548^{c}	1.331±0.05^{a}	1.210±0.01^{c}
BRRI dhan-37	15.13±0.01^{a}	5.69±0.075^{b}	0.782±0.02^{d}	1.050±0.02^{a}
BRRI dhan-50	11.25±0.01^{f}	6.21±0.015^{a}	0.873±0.05^{b}	1.460±0.05^{e}
Philipine Katari	12.11±0.02^{c}	3.23±0.005^{d}	1.450±0.02^{a}	0.887±.064^{e}

All values are express as mean ± SD.
Mean followed by different superscript letters in each column are significantly different (p<0.05).

were centrifuged at 4000 rpm for 30 min. The supernatants were poured into pre weighed petri dish and the residue was weighed after oven drying overnight at 70°C. WAI and WSI were calculated using following equations:

$$\text{WAI} = \frac{\text{Weight of sediment}}{\text{Weight of dry solids}}$$

$$\text{WSI} = \frac{\text{Weight of dissolved solids in supernatant}}{\text{Weight of dry solids}} \times 100$$

Statistical analysis

All measurements were carried out in triplicate for each of the sample. Data were statistically analyzed (SPSS for windows version 17.0) by one-way analysis of variance (ANOVA). Mean comparisons were performed using Duncan's multiple range tests for significant effect at P < 0.05.

RESULTS AND DISCUSSION

Proximate analysis

Proximate compositions of different varieties of aromatic rice are shown in Table 1. Moisture content of different aromatic rice varieties varied between 11.25/100 g to 15.13/100 g. These values were similar to those reported by Saikia et al. (2012) who found that moisture content in rice is 11.6 to 13.7/100 g and also by Sompong et al. (2011) found that moisture content in pigmented and non pigmented aromatic rice was 9.3 to 13.1/100 g. The highest value (15.13%) of moisture was found in BRRI dhan-37, while lowest value (11.25/100 g) was found in BRRI dhan-50. The difference in moisture value in different aromatic rice varieties might be due to the variation in moisture content in paddy after harvesting. Protein content of different aromatic rice varieties varied from 3.23 to 6.21/100 g that was lower than that found by Saikia et al. (2012) who reported protein value in pigmented and non pigmented aromatic rice 6.66 to 9.97/100 g and also by Sompong et al. (2011) who reported protein value in rice as 7.16 to 10.85/100 g. BRRI dhan-50 variety had highest protein value (6.21/100 g) compared with other aromatic rice varieties.

On the contrary, Philipine Katari variety had the lowest value of protein (3.23/100 g) than other varieties. The variation in protein value in different aromatic rice varieties could be due to the application of fertilizer (nitrogen content), growing conditions and time and also location of growing areas (Buresova et al., 2010). Fat values were appreciably different among the aromatic rice varieties and highest was found in Phillipine Katari (1.45/100 g), where lowest content (0.68/100 g) were studied in Begun Bichi. This observation was comparable to that found by Saikia et al. (2012) who reported fat value in pigmented and non pigmented aromatic rice 1.00 to 2.10/100 g.

The variation in fat value in different aromatic rice varieties may be due to oxidation of the fat because most of the fat content in rice grain is unsaturated and undergo oxidation easily by atmospheric oxygen. The ash content of different aromatic rice varieties varied between 0.88/100 to 1.46/100 g. This result was consistent to be found by Sompong et al. (2011) who reported ash content in rice as 0.82 to 1.74/100 g. Lowest ash value was found in Kalizira (0.88/100 g) variety while highest value (1.46/100 g) was studied in BRRI dhan-34 variety.

Physiochemical properties

Amylose and starch

The amylose content in rice is one of most important criteria of rice quality in terms of cooking and gelling properties (Abu-Kwarteng et al., 2003). BRRI (2004) reported that amylose content is a major factor that influences the eating quality of rice. In this study, the variation of amylose content among different aromatic rice varieties was 15.69 to 23.01/100 g (Table 2). This value was comparable to that found by Saikia et al. (2012) who reported amylase content in pigmented and non pigmented aromatic rice as 2.2 to 28.8/100 g. The highest amylose (23.01/100 g) content was found in BRRI dhan-37 variety and lowest value (14.23/100 g) in BRRI dhan-34. On the other hand, starch content of

Table 2. Physiochemical properties of different aromatic rice varieties.

Parameter (g/100 g)	Varieties					
	Kalizira	Begun Bichi	BRRI dhan-34	BRRI dhan-37	BRRI dhan-50	Philipine Katari
Amylose	15.69±0.30^{d}	17.80±0.48^{c}	19.70±0.36^{b}	23.01±0.43^{a}	14.23±0.14^{e}	16.34±0.62^{d}
Starch	68.07±5.63^{abc}	70.147±1.30^{ab}	63.323±1.02^{c}	65.457±1.05^{bc}	72.606±0.97^{a}	63.193±1.65^{c}

All values are express as mean ± SD.
Mean followed by different superscript letters in each raw are significantly different (p<0.05).

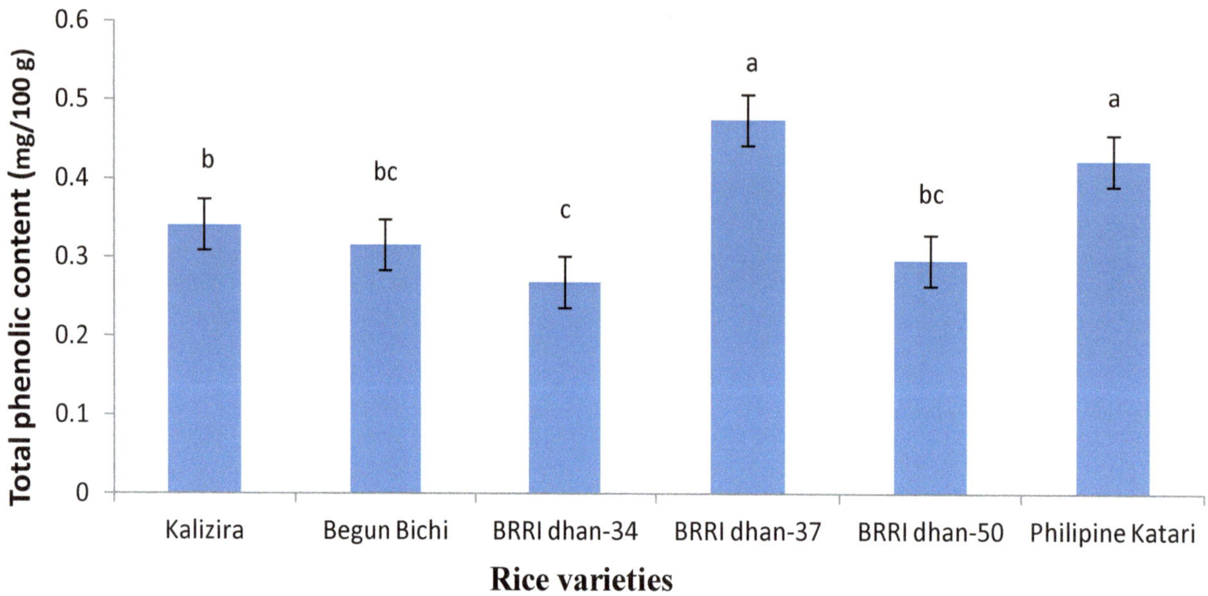

Figure 1. Total phenolic content among six aromatic rice varieties. Different superscript letters above the bars indicate statistically significant differences at p<0.05.

different aromatic rice varied between 63.193 to 72.60/100 g (Table 2). This value was comparable to that found by Saikia et al. (2012). The highest starch content (72.60/100 g) was found in BRRI dhan-50, while lowest value (63.32/100 g) was found in BRRI dhan-34. The variation in amylose and starch content in different aromatic rice varieties might be due to the application of fertilizer (nitrogen content), growing conditions, time and also location of the growing areas (Buresova et al., 2010).

Total phenolic contents

Total phenolic content (TPC) of the examined aromatic rice varieties are presented in Figure 1. Variations were found within the aromatic rice varieties, ranging from 295.34 to 474 mg /100 g GAE, with BRRI dhan-34 (268.67 mg/100 g) rice having the lowest value while the highest value came from BRRI dhan -37(474 mg/100 g). This result was similar to that obtained by Saikia et al. (2012), was in the range of 230 to 579 mg/100 g gallic acid equivalent (GAE) of aromatic pigmented and non-pigmented rice varieties. Another study showed that, total phenolic content of rice bran extract was 251 to 359 mg/100 g GAE by Igbal et al. (2005). The variation of total phenol content in different aromatic rice varieties may be affected by genotype, cultivation techniques and ripening process environmental variation (Mpofu et al., 2006). Also, Beta et al. (2011) reported that total phenolic content may vary because of moisture content. In this study, significant differences were found in moisture content among all aromatic rice varieties.

Flavonoids

Figure 2 shows flavonoid contents among six aromatic rice varieties were varied from 660 to 1280 mg /100 g Rutin equivalent (RE). This result was similar to flavonoid content in aromatic rice varieties reported by Saikia et al. (2012). BRRI dhan-37 rice had the lowest value (680 mg/100 g RE) while the highest value (1280 mg/100 g RE) came from philiphine katari rice. Even though, there were no significantly differences among all varieties. The variation of flavonoid content in different aromatic rice varieties may be affected by genotype and environmental variation.

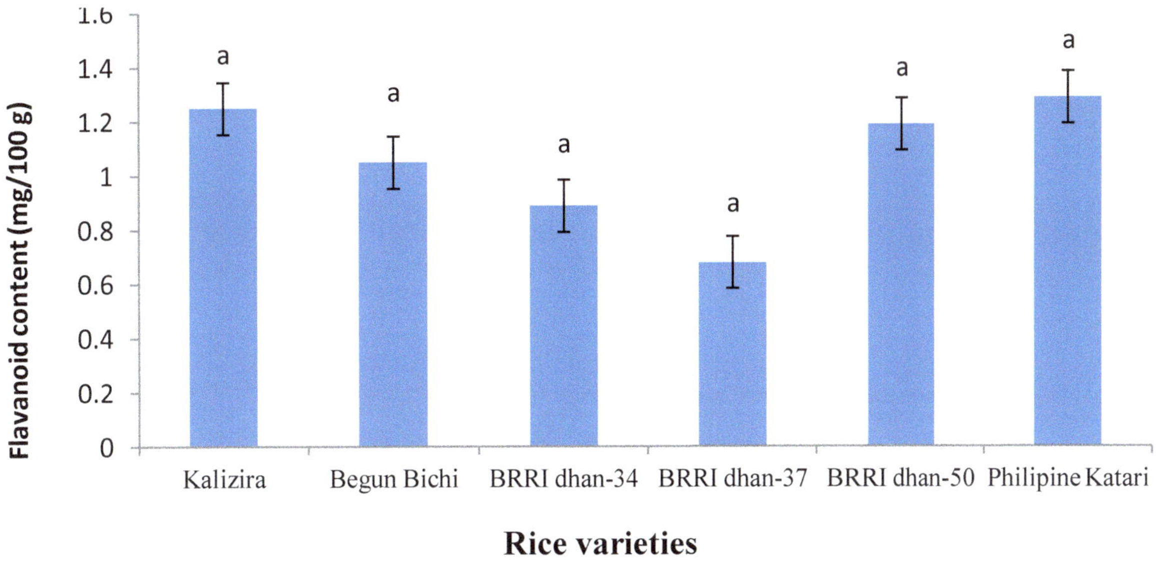

Figure 2. Flavanoid content among six aromatic rice varieties. Same superscript letters above the bars indicate statistically not significant at ($p<0.05$).

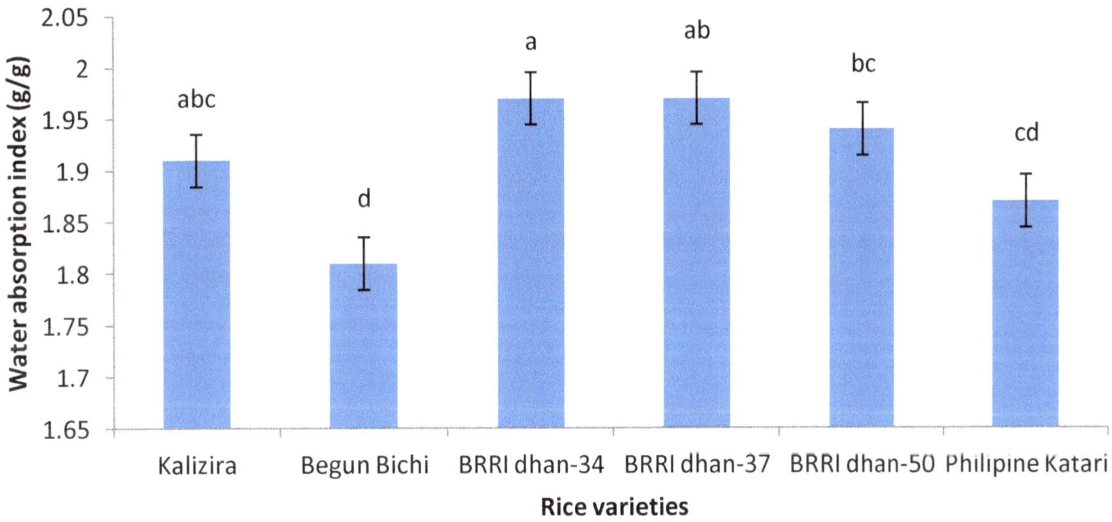

Figure 3. Water absorption index among six aromatic rice varieties. Different superscript letters above the bars indicate statistically significant differences at $p<0.05$.

Functional properties

The functional properties of different varieties of aromatic rice are represented in Figures 3 and 4. The study showed that the value of water absorption index (WAI) in different aromatic rice varied between 1.81 to 1.97 g/g (Figure 3). However, BRRI dhan -34 variety had highest WAI value while lowest value found in Begun Bichi variety. The variation in WAI might be due to the disparity in the amount of engagement of OH^- groups to form hydrogen and covalent bonds between starch chains and also depend on loss of starch crystalline structure (Gunarante and Hoover, 2002). The water soluble index (WSI) in different aromatic rice varieties was 1.32 to 2.12 g/g (Figure 4). Kalizira variety had highest WSI value while lowest value found in BRRI dhan-37. WSI value could be attributed to the semi crystalline structure and disruption of starch granules (Eliasson and Gudmundsson, 1996).

Conclusion

In this study, the physiochemical, phytochemicals content and functional properties of different aromatic rice varieties were investigated. The results show that various

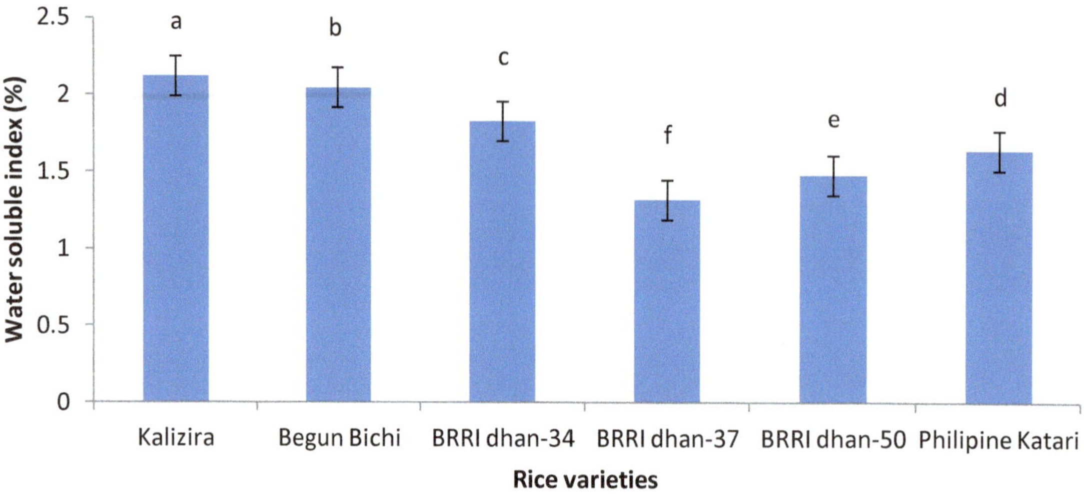

Figure 4. Water soluble index among six aromatic rice varieties. Different superscript letters above the bars indicate statistically significant differences at $p<0.05$.

aromatic rice varieties exhibit good physiochemical, functional and phytochemicals characteristics. Phytochemicals in aromatic rice grains have recently been ascribed to positive nutritional properties as for example prevention of cardiovascular diseases and cancer. On the other hand, non-aromatic rice grains have lack of phyto-chemicals. Therefore, the study reveals that aromatic rice could enhance the nutritional properties of functional food ingredients in terms of the present of various healths promoting phytochemicals such as phenolic and flavonoids contents.

ACKNOWLEDGEMENT

This study was supported by Ministry of Science and Technology Government of the Peoples Republic of Bangladesh (R & D Project Grant).

REFERENCES

Abu-Kwarteng E, Ellis WO, Oduro I, Manful JT (2003). Rice grain quality: a comparison of local varieties with new varieties under study in Ghana. Food Cont. 14: 507-514.

Ali MA, Chang YL (2008). Optimization of Phenolics and diatary fiber extraction from date seeds. Food Chem. 108: 977-985.

Anderson RA (1982). Absorption and solubility and amylograph characteristics of
roll-cooked small grain products. Cereal Chem. 59: 265–269.

Anonymous (2010). Department of Agricultural Extension (DAE) Dinajpur, Bangladesh.

AOAC (1998). Official Methods of Analysis of AOAC International (14th ed.). Washington, DC, USA: Association of Official Analytical Chemistry. Methods, 950.46, 938.08, 960.39, 955.04.

Basu S, Roychoudhury A, Sanyal S, Sengupta DN (2012). Carbohydrate and Antioxidative potential of the seed of three edible indica rice cultivars. Indian J. Biochem. 49: 115-123.

Beta T, Celli GB, Pereira-Netto AB (2011). Comparative analysis of total phenolic content, antioxidant activity, and flavanoids profile of fruits from two varities of Brazilian cherry (*Eugenia uniflora* L.) throughout the fruit development stages. Food Chem. 44: 2442-2451.

Bhattacharjee P, Singhal RS, Kulkarni PR (2002). Basmati rice: A review Int. J. of Food Biophys. 49: 115-123.

Buresova I, Sedlackova I, Famera O, Lipavsky J (2010). Effect of growing conditions on starch and protein content in triticale grain and amylose content in starch. Plant Soil Environs. *56*: 99–104.

Fresco L (2005). Rice is life. J. Food Comp. Anal. 18: 249- 253.

Eliasson AC, Gudmundsson M (1996). Starch: Physiochemical and functional aspect.In A.C. Eliasson (Ed.). Carbohydrates in Food. New York. pp. 431-503.

Food and Agriculture Organization (FAO) (2009-2010). International Exhibition to Promote rice Import & Export. 26-28 November 2012. Abu Dhabi National Exhibition Centre, UAE.

Gunaratne A, Hoover R (2002). Effect of heat-moisture treatment on the structure and physicochemical properties of tuber and root starches. Carbohyd. Polym. 4: 425-437.

Hirose M, Shibata, M, Hagiwara, A, Imaida, K (1981). Chronic toxicity of butylated hydroxy-toluene in Wistar rats. Food Cosmet. Toxicol.19:147-51.

Iqbal S, Bhanger MI, Anwar F (2005). Antioxidant properties and components of some commercially available varieties of rice bran in Pakistan. Food Chem. 93: 265- 272.

Khush GS (2005). What it will take to feed 5.0 billion rice consumers in 2030. Pl. Mol. Bio. 59: 1-6.

Lane JH, Eynon L (1923). Determination of reducing sugars by means of Fehling's solution with methylene blue as internal indicator. J. Soc. Chem. Indust. 42: 32–36.

Mpofu A, Sapirstein HD, Beta T (2006). Genotype and environmental variation in phenolic content, phenolic acid composition and antioxidant activity of hard spring wheat. J. Agric. Food Chem. 54: 1265–1270.

Nam SH, Choi SP, Kang MY, Koh JH, Kozukue N, Friedman M (2006). Antioxidative activities of bran extracts from twenty one pigmented rice cultivars. Food Chem. 94: 613–620.

Saikia S, Himjyoti D, Daizi S, Charu LM (2012). Quality characterization and estimination of phytochemical content capacity of aromatic pigmented and non pigmented rice varieties. Food Res. Int. 46: 334-340.

Sompong R, Ehn S, Martin L, Berghofer E (2011). Physicochemical and antioxidant properties of red and black rice varieties from Thailand, China and Sri Lanka, Food Chem. 124: 132-140.

Effect of pre-drying treatments on utilizability of sweet potato tubers for production of chips for confectionery flour

Oyebanji A. O.[1], Ajani A. O.[2], Adeize A.[1], Oyelakin M. O.[2], Ikotun I.[2], Agboola A. A.[3], Alimi, J. P.[2] and Awoite T.[2]

[1]Nigerian Stored Products Research Institute, Km 3 Asa-Dam Road, PMB 1489, Ilorin, Kwara State, Nigeria.
[2]Nigerian Stored Products Research Institute, Onireke-Dugbe, P.M.B. 5044, Dugbe, Ibadan, Oyo State, Nigeria.
[3]Nigerian Stored Products Research Institute, No, 1 Birikisu-Iyede St., Yabba, Lagos, Nigeria.

Yellow variety of sweet potato tubers lots were processed with different pre-drying treatments of peeled washed chips: Unblanched without Sodium Metabisulphite (UB-SMBS), blanched without sodium metabisulphite (B-SMBS), unblanched with sodium metabisulphite (UB+SMBS) and blanched with sodium metabisulphite (B+SMBS) for production of dried chips for confectionery flours. Low moisture contents in the range of 2.48 to 5.00% were achieved for the dried chips lots in 20 h of drying at about 55 to 60°C using NSPRI's Cabinet Crop Dryer without smoking. The dried chips had no visible mould growths in the duration of drying and mould loads were low (in the order of 10^2 cfu/g). However, different colours and textures were observed in dried chips and flours depending on pre-drying treatment. While UB-SMBS and UB+SMBS sweet potato chips' flours were suitable and acceptable for production of chin-chin, the B-SMBS and B+SMBS were unsuitable and unacceptable at 100% composition of the flour.

Key words: Sweet potato, blanching, sodium metabisulphite, dried chips, flour.

INTRODUCTION

Ipomoea batatas (L.) Lam, commonly called sweet potato is a tuberous root crop. Generally, sweet potato tuber is boiled, pounded or fried for eating in Nigeria, like is done for yam tubers. Kochlar (1981) reported that sweet potato has found uses in much the same way as Irish potatoes. Collins and Walter (1982) and Bouwkamp (1985) also reported that sweet potato can be processed for consumption by boiling or baking and can be preserved by canning or dehydration. CTA (2007) reported the boiling and mashing of sweet potato tubers with beans, the milling or pounding of the dried chips to make flour which can be mixed with either millet or cassava flour to make stiff porridge in Eastern Africa. Iwe and Onuh (1992) reported that sweet potato tubers are suitable for use as sweeteners in cereal based beverages (locally referred as "Kunu").

However, the processing of sweet potato for chips and flour is not common in Nigeria, perhaps due to lack of information on the production protocol and various characteristics of products for utilizations. Traditional drying of yam and cassava chips and cassava mash in Nigeria has been by open air sun-drying on flat surfaces, including bare ground, mats, polythene sheets and the notorious roadside drying of produce. Such dried crops have been plagued by contamination with dirt from blowing wind, animal and birds feeding action including contamination with faeces. The wetting and high moisture of crops during raining season, in continued and intermittent rains, especially in the more humid Southern Nigeria result in problems of delayed and inadequate drying of crops. Thus, compounding the problem of poor quality of traditional drying of crops, with high moisture retained in crops for too long under warm tropical ambient condition in Nigeria, the traditionally dried crops risked mouldiness, mycotoxin contamination and early insect infestation. Further, Akinola et al. (2006) reported that the drudgery, rudimentary and dirt contaminating nature of traditional food processing in Nigeria requires change for the better. Open air sun drying result in low value or out rightly rejection of products.

Sweet potato is an important crop in Nigeria, especially in Oyun and Offa areas of Kwara State, where the production of sweet potato is of social-cultural importance. However, the problem of soil fertility and more importantly the limited market for sweet potato tubers have been limiting production of sweet potato coupled with the increasing influence of cassava (NSPRI, 2002), even in Oyun/Offa area. Therefore, the need to give consideration to how to sustain the symbolic social-cultural heritage that sweet potato represents for the people the Oyun/Offa area is very important. Effort in this direction can be by actions such as creating a valuable market for harvested sweet potato tubers, especially through acceptable product development and protocol. Farmers would be encouraged to sustain and increase production of sweet potato tubers if the market for the harvest is guaranteed.

Production and processing of sweet potato can be a way for women and many youths in Nigeria to be gainfully employed. Also, opportunity exists for controlling post harvest deterioration of sweet potato tubers by processing into useful and acceptable products.

In Nigeria, quality drying of crops hold potential for production of primary products, preservation, and conservation, giving the relative ease of entry for women and the teaming unemployed youths than for involvement in the more complex industrial manufactures.

Concerning the poor drying problem of crops in Nigeria, Nigerian Stored Products Research Institute (NSPRI) has been intervening in promotion of its improved dryers, including Solar Tent, Solar Tray, Multipurpose dryer, Hybrid dryer (Oyebanji, 2003) and Electric Cabinet dryers for hygienic, clean and effective drying of produce. The opportunity for creating market for sweet potato tubers through the drying of the chips for confectionery flour needs to be investigated in Nigeria. In similar thinking, Adebayo (2006) reported the production of "High Quality Cassava Flour" (HQCF) for confectionery from dried freshly harvested cassava tubers in IITA. The HQCF was found suitable for use in confectioneries, at least in part mixing with wheat flour up to 20% for baking bread. Also, fermented dried cassava chips were made into fufu and lafun flours. Also, yam tubers are made to flour for amala meal and pando-yam flour in Nigeria.

The practice of peeling yam and cassava tubers and chipping into water to limit oxidative discolouration has been in Nigeria. As well, the use of sodium metabisulphite for retention of bright colour of products against discolouration and prevention of microbial spoilage controls during drying has been reported for tomato (Okanlawon et al., 2002). Efforts to process crops have in part been for production of utilizable products. Cassava tubers have been made into confectionery flour used at up to a maximum of 20% with 80% wheat flour for making bread. Processing of sweet potato tubers into adequately dried chips suitable for confectionery flour production would create market for sweet potato farmers and encourage increase in production. These would call for establishment of useful and acceptable processing protocol for drying sweet potato chips. It was therefore considered a possibility to utilise sweet potato tubers for production of confectionery flour for chin-chin, as snack in urban centres and for school children. Blanching, a form of per-boiling is being used in the production of yam flour (Elubo) for 'amala' meal and retention of bright colour and control of mouldiness on tomato during drying has been attributed to the pre-drying treatment of cut pieces of tomato with sodium metabisulphite by dipping (Okanlawon et al., 2002). Thus, the effects of blanching or not and combination with dip treatment of sweet potato in sodium metabisulphite or not as pre-drying treatments were investigated to determine the effects on quality characteristics of sweet potato chips' flours and the suitability for production of chin-chin at 100%.

MATERIALS AND METHODS

Sweet potato

Sweet potato tubers, *Ipomoea batatas* (L.) Lam., of the yellow variety were procured within 24 h of harvest from Ijagbo farm gate market for use in this study. The procured sweet potato tubers were quickly brought to the Community Processing Centre in Ijagbo, where the tubers were shared into four (4) lots for immediate in pre-drying treatments and drying of the chips for use in the production of confectionery flour which may be suitable for chin-chin production.

Sweet potato chips, pre-drying treatments and drying

Processors in Ijagbo were gathered at the Community Processing Centre with the assistance of extension officials of Kwara State Agricultural Development Project (KwADP) as link agency for the scheduled participatory research for development. The procured sweet potato tubers' lots were respectively peeled, chipped and washed in portable water manually by the participating processors for the pre-drying treatments for eventual appreciation of comparison of results. The pre-drying treatments were:

1. Unblanched without sodium metabisulphite (UB-SMBS)
2. Blanched without sodium metabisulphite (B-SMBS)
3. Unblanched with sodium metabisulphite (UB+SMBS)
4. Blanched with sodium metabisulphite (B+SMBS)

Blanching was by dipping manually chipped (2.5 x 20 x 30 mm) peeled washed sweet potato tuber into boiling water for 5 min and unblanched was the control and these were with or without sodium metabisulphite treatment at 0.5% in boiling water or water at ambient temperature, respectively. The respectively treated sweet potato chips were drained and air dried on raised platforms in sieve trays within 10 min for surface moisture to evaporate. Subsequently, the 4 differently treated sweet potato chips lots were respectively thin layer spread on net laid trays in NSPRI Multipurpose cabinet dryers to dry within 20 h at 55 to 60°C.

Analyses: Colour of chip, moisture content and mould load

Peeled sweet potato chips drained after washing were visually examined and colour was described to characterise changes that was due to treatment or drying. That is, dried chips from the respective pre-drying treatments were respectively visually examined for colour. Also, representative samples of the dried chips were analysed for moisture contents of milled portions by oven method of AOAC (2000) and mould load by serial dilution pour plate technique (Akani and Madumere, 2008; Oyebanji et al., 2011) on malt extract agar on 3 day old culture at 28°C.

Test milling into flour and chin-chin production

The dried chips were "milled" respectively using hammer mill. The "flours" from the 4 treatments were visually examined and described for flour forms. Samples were then taken from the 4 "flours", each at 100% (without wheat flour) and were respectively mixed with other ingredients or recipe and (2 ½ Cups of Sweet Potato "flour" + 1/5 Cup of Sugar (instead of 2/3) + ½ Teaspoon of salt + ¼ Teaspoon of Baking Powder + 2 eggs + 5 Tablespoons of Butter + 2 Tablespoons of Powered Milk) mixed with water for dough and flattened to 10 mm thickness for cutting into strips and pieces and fried into chin-chin in pan on gas burner with 4 cups of groundnut oil. Mixed pressed-rolled-cut pieces of 'dough' of respective flour lot was visually examined and described for the form of roll, dough and chin-chin. Pictures were taken to compliment descriptions of materials, methods and results. Further, a panel of 5 processors and chin-chin consumers was constituted to determine the acceptability of chin-chin made from the respective flour lots on the hedonic scale of 1 to 9 (Ihekoronye and Ngoddy, 1985; Munoz and King, 2007) for appearance, texture, flavour, taste and overall acceptability of chin-chin produced. Means of values were computed and data presented as a table to compare effects of the pre-drying treatments.

RESULTS AND DISCUSSION

Preparation of sweet potato chips for pre-drying treatments and drying

The manual peeling of sweet potato in this study by Ijagbo food processors at the workshop was consistent with the commonly observed practice for peeling similar root and tuber crops such as yam and cassava in Nigeria. However, the manual peeling of sweet potato observed was considered tedious, and so this operation would need need scale up for mechanization to remove the drudgery. The drudgery was not found in literature and was not anticipated. In any case, mechanical peeling machine development is still a challenge in Nigeria. The challenge has been attributed to the irregularity of root and tuber forms for ease of design and fabrication of effective peeling maching. Also, the chipping of the sweet potato called for the use of a chipping machine in the peculiar drudgery of manual chipping of sweet potato when compared with yam and cassava. Manual peeling of sweet potato in this study was at the rate of 50 kg/h/person, which meant approximately 400 kg/day of 8 h and labour charge of N400/day in Ijagbo. However, the activity was capable of contributing to alleviation of poverty in the associated employment of rural women and youths involved in such processing activities.

Effect of pre-drying treatments of sweet potato chips on colour of dried chips and flour quality

Table 1 summarizes findings which was characteristic of the sweet potato chips pre-drying treatments and drying effects. The yellow variety of sweet potato used for this study was characteristically cream white colour upon peeling. Following drying of the pre-treated sweet potato chips for 20 h in NSPRI's Multi Crop Cabinet Dryer (Plate 1), the UB-SMBS and UB+SMBS lots, had moisture contents of 3.85 and 2.48%, respectively and were brittle to touch, suggesting considerably low moisture contents as compared to the required safe moisture content of about 10% required for flour mills in Nigeria. While subsequent practice may not require this long drying duration, this result showed that the chips dried very well in terms of the extent of moisture reduction much below the safe level, and so can have extended shelf life against mouldiness in hermetic containers and against insect infestation in the relative cleanliness of the drying environment as compared to traditional open air drying. Similarly, B-SMBS and B+SMBS treated dried chip lots had low moisture contents of 4.98 and 5.00%, respectively (Table 1), but were glassed and flexed, which showed change of form from the expected.

The clean dried products with low moisture levels obtained in the respectively treated and subsequently dried sweet potato chips were indications of the effectiveness of the improved crop dryer used. The low mould load in sweet potato lot dried in NSPRI's Multipurpose dryer contrasted with the high load associated with traditional road side dried fufu made from cassava. Moulds detected in samples were still indications of contamination in the drying environment and so continued to call for improvement in sanitation and hygiene of the processing equipment and personnel. The contaminants on crops will become important in the spoilage of the dried products if holding moisture condition becomes too high in a situation of poor storage, under warm tropical condition. Dried sweet potato chips' appearance varied depending on the

NSPRI's Multipurpose dryer

Sweet potato chips dried in NSPRI's Multipurpose dryer at Ijagbo

Plate 1. Sweet potato chips in NSPRI's Multi Crop Cabinet Dryer at Ijagbo.

pre-drying treatment and expectedly from the fresh peeled sweet potato chips. While fresh peeled sweet potato used in this study was cream white, the dried chips with unblanched pre-drying treatments were grayish white while the blanched pre-drying treatments changed to yellow irrespective of whether with SMBS or not. Result for the blanched sweet potato chips showed additional colourising effect of blanching different from that due to the drying of the chips. Also, the blanched lots (B-SMBS and B+SMBS lots) were glassed. It was not surprising that the dried blanched lots could not be well milled; they came out as broken pieces of flakes and granules upon milling. This was because the associated starch became gelatinized in the high blanching temperature. On the other hand, the dried unblanched sweet potato chips' lots milled into flour easily.

Effect of pre-drying treatments of sweet potato chips on chips' flour quality for dough rolling, cutting and frying into chin-chin

Dough roll and cutting tests were positive for UB-SMBS, UB+SMBS and B-SMBS flour, but was negative for B+SMBS. That is, the B+SMBS flour lot did not roll and could not be cut into pieces for chin-chin. The mixture of B+SMBS in confectionery ingredients scattered as granules upon frying in vegetable oil. Thus, result showed that adverse effect of sodium metabisulphite treatment on

Table 1. Characterizations of sweet potato chips before and after drying for different pre-drying treatments: UB-SMBS, B-SMBS, UB+SMBS and B+SMBS and effect on milling quality and chin-chin production.

Parameter	Pre-drying treatment			
	UB-SMBS	B-SMBS	UB+SMBS	B+SMBS
Fresh peeled chips' colour	Cream White	Cream White	Cream White	Cream White
Multipupose dryer dried sweet potato chips	Greyish White	Yellowish	Greyish White	Yellowish
Moisture content (%) of 20 hr dried chips	3.85	4.99	2.48	5.00
Mould load (cfu/g) of the dried chips	3×10^2	2×10^2	3×10^2	3×10^2
Milling quality test of dried chips to flour	Positive	Negative	Positive	Negative
Rolling of ingredient mixed 'flour' for duff	Positive	Positive	Positive	Negative
Cutting test of 'rolled' duff into pieces	Positive	Positive	Positive	Negative
Frying of pieces in vegetable oil for Chin-Chin	Positive	Negative	Positive	Negative

Table 2. Appearance, flavour, texture, taste and overall acceptability of chin-chin made from different treated chips' flours.

Parameter	Acceptability (Scale 1 – 9)			
	UB-SMBS	B-SMBS	UB+SMBS	B+SMBS
Appearance of the chin-chin	8.66^a	2.78^c	7.00^b	1.78^d
Flavour of chin-chin	7.11^a	4.89^c	6.22^b	3.11^d
Texture of chin-chin	7.89^a	6.89^b	7.89^a	2.00^c
Taste of chin-chin	8.33^a	6.00^c	7.22^b	2.00^d
Overall acceptability of chin-chin	8.44^a	2.76^c	6.89^b	1.11^d

Values along rows with different letters are significantly different at p=0.05. The high overall acceptability of UB-SMBS (8.44) and UB+SMBS (6.89) samples in production of chin-chin contrasted with the low values for B-SMBS (2.76) and B+SMBS (1.11).

rolling and cutting of dough was influenced by the blanching. Table 2 shows the acceptability of chin-chin produced from the different treated chips' flours in terms of appearance, flavour, texture, taste and overall acceptability on the hedonic scale of 1-9. Further, Plate 2 showed that frying of the un-blanched lots (UB-SMBS and UB+SMBS) in vegetable oil gave acceptable chin-chin, but blanched lots without metabisulphite (B-SMBS) when rolled and cut, scattered during frying while the B+SMBS could not be rolled, cut or fried for chin-chin. That is, frying test confirmed that sweet potato chips need not be blanched or treated with sodium metabisulphite, if the flour is intended for chin-chin production at 100% sweet potato flour. However, the chin-chin made from UB-SMBS and UB+SMBS flours were considered acceptable to chin-chin consumers but attempts from B-SMBS and B+SMBS were rejected by the panel of 5 processors and chin-chin consumers. Thus, dried chips or flour should be packed in hermetic containers tightly filled and closely packaged to store in a cool atmosphere to control insect infestation during storage without using insecticides. The hygienic drying in the improved dryer should enhance storability of the dried chips against insect infestation. The sweet potato flour at 100% when needed will be ready for use in making chin-chin. Part mixture of sweet potato flour with wheat flour was not considered in this study, but part mixture of high quality cassava flour with wheat for production of bread has been promoted in Nigeria to create market for cassava farmers and processors.

Conclusion

There is no difference in the effect of pre-drying treatment (addition or not of sodium metabisulphite) of un-blanched sweet potato chips on colour of dried chips' flour quality for dough roll, cutting or frying for chin-chin. Thus, it becomes economical to do without the application of sodium metabisulphite in the production of sweet potato chips for 100% sweet potato flour for chin-chin production. Further, any risk of misuse or abuse of use of sodium metabisulphite would be avoided, with the indication that there was no need for sodium metabisulphite dip treatment of sweet potato chips. That is, the recommended protocol for production of sweet potato flour for 100% use in chin-chin production is: Procure freshly harvested sweet potato tubers (within 48 h of harvest), wash, peel, chip, wash, drain, air dry (in sieve under net cover on a platform for about 10 min to remove surface moisture) and then properly dry chips (in improved dryer for effective moisture reduction and against contamination

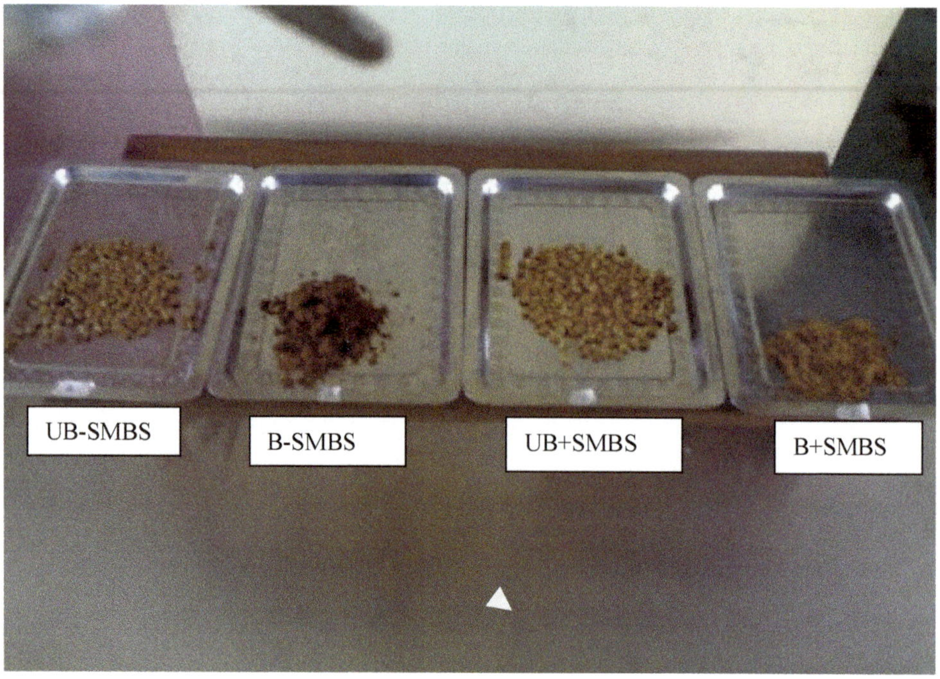

Fried sweet potato pieces from flours of different pre-drying treatments of chips

Plate 2. Showing sweet potato chin-chin fried from UB-SMBS, B-SMBS, UB+SMBS and B+SMBS, respectively. NB: A is UB-SMBS, B is B-SMBS, C is UB+SMBS and D is B+SMBS.

to safe moisture level) and allow chips to cool, then mill into flour. The sweet potato processing venture would enable employment of rural women and youths and the sweet potato flour would be marketable, creating income for investors in processing and confectioneries as well as market for sweet potato farmers.

ACKNOWLEDGEMENTS

We acknowledge the Root and Tuber Expansion Programme's funding of this project and the collaboration of Kwara State Agricultural Development Programme as our linkage agency to Ijagbo food processing group (for participatory product development and demonstrations of use of NSPRI's Multi Crop Cabinet Dryer for hygienic drying of Sweet Potato Chips in Ijagbo Community to promote the adoption of recommendations).

REFERENCES

Adebayo BA (2006). How to make High Quality Cassava Flour (HQCF): IITA Research to nourish Africa: ISBN: 978-131-281-5, 1-14.

Akani NP, Madumere B (2008) Mycoflora of partially processed periwinkles (*Tympanotonus fascatus*) from local markets in Port Harcourt, Nigeria. Nig. J. Mycol. 1(1)138-142.

Association of Official Analytical Chemists (2000). Official Methods of Analysis of Association of Analytical Chemists. Arilington 2000 17th Edition.

Bouwkamp JC (1985) Sweet potato products: A natural resource for the tropics. CRC Press Inc. Florida 139-258.

Collins WW, Walter WM J (1982). Potential for increasing value of sweet potatoes. In:Sweet potato Proc. First Int. Symp.V llareal RL, Briggs TD (ED) AVRDC. Shanhua. pp. 355-63.

CTA (2007). Making sweet potato chips and flour: In CTA Practical Guide Series No. 6:1-6 ISSN 1873-8192.

Ihekoronye AI, Ngoddy PO (1985). Integrated food science and technology for the tropics. MacMillan Ed. Ltd. 175-187.

Iwe MO, Onuh JO (1992). Functional and sensory properties of soybeans and sweet potato flour mixtures Lebensin wiss-U Technology 25:569-573.

Kochlar SL (1981). Tropical Crops: A text book of economic botany MacMillian Publishers Ltd. pp. 230-232.

Munoz AM, King SC (2007). International consumer product testing across cultures and countries: ASTM International MNL 55.

Okanlawon SO, Ibrahim MH, Oyebanji AO (2002). Effect of predrying treatment on storage of dried tomato. Trop. Sci. 42:40-41.

Oyebanji AO (2003). Increasing cassava production in Nigeria: Post harvest handling challenges and prospects. Raw Materials Update 4(1): 31-33.

Oyebanji AO, Ibrahim MH, Okanlawon SO, Awagu EF (2011). Moisture content, mouldiness, insect infestation and acceptability of market samples of dried 'tatase' pepper and tomato in Kano, Nigeria. J. Stored Products Post harvest Res. 2(10)200-207.

NSPRI (2002). Utilization of sweet potato. In proceedings of Nigerian Stored Products Research Institute (NSPRI) workshop in participatory technology development methods. Community analysis report on postharvest practices on cassava and sweet potato in Ijagho, Oyun LGA of Kwara state on 6th-7th Sept. 2002. Organized by Capacity Building for Decentralized Development (CCBD). Edited by Ayoola J, Udo-Ekong CR. pp. 11-12.

14

Effect of substituting sugar with date palm pulp meal on the physicochemical, organoleptic and storage properties of bread

J. E. Obiegbuna[1], P. I. Akubor[2], C. N. Ishiwu[1] and J Ndife[3]

[1]Department of Food Science and Technology, Nnamdi Azikiwe University, Awka, Anambra State, Nigeria.
[2]Department of Food Science and Technology, University of Nigeria, Nsukka, Enugu State, Nigeria
[3]Department of Food Science and Technology, Kaduna Polytechnic, Kaduna State, Nigeria.

The effect of substituting granulated sugar with date palm fruit pulp meal (DPFPM) in bread production was evaluated. Date palm fruit pulp was oven dried (46°C, 8 h) and milled. Granulated sugar was substituted at 0, 25, 50, 75 and 100% with DPFPM in bread recipe and bread produced using straight dough bulk fermentation method. Proofing ability of the dough, oven spring, bread loaf weight, volume and specific volume were evaluated. Proximate composition, incidence of mold growth on bread loaves and sensory properties of the loaves were also determined. On dry weight basis, crude fat, crude fiber, crude protein and ash contents significantly ($p < 0.05$) increased from 6.25 to 7.72, 1.52 to 4.45, 16.85 to 21.26, and 2.26 to 3.84%, respectively, whereas carbohydrate decreased from 73.12 to 62.73% with increasing substitution of sugar with DPFPM. The decrease in proofing ability of the dough (2.20 to 2.08) was insignificant ($p > 0.05$) while the decreases in oven spring (2.30 to 1.87 cm), loaf volume (848.3 to 762.3 cm^3) and specific volume (3.15 to 2.64 cm^3/g) were significant ($p < 0.05$). Loaf weight increased ($p < 0.05$) from 265.5 to 288.8 g. Visually observable colonies of mold on bread loaves during storage initially decreased but later increased with increasing DPFPM. Substituting sugar with DPFPM had no adverse overall effect on the acceptability of bread loaves; it however increased the nutritional value but decreased the physical quality of the bread loaves.

Key words: Bread, granulated sugar, date palm fruit pulp meal, substitution, physicochemical properties.

INTRODUCTION

Bread is one of the commonest staple food eaten world over. It is composed of dough from wheat, yeast, water and other ingredients, which has been fermented and subsequently baked. When the dough is made, the yeast starts to work on the fermentable sugars, transforming them into alcohol and carbon dioxide, thus increasing the dough volume. Sugar, though used as ingredient in bread, is not considered essential or indispensable in bread making. This is because flour contains a small amount of natural sugar and some starch is converted to sugar during fermentation process. The natural sugar content of the flour, however, is very limited in amount and for most forms of bread, sugary agent must be added to meet the usual yeast requirements throughout the fermentation process and also supply the necessary sweetness and other functions performed by sugar in bread baking

(Wihlfaht, 2007). Apart from providing substrate to yeast for CO_2 production and imparting sweetness to baked goods, other functions of sugar include the production of golden brown color of the crust, improvement of texture of the crumb, retention of moisture in the crumb and adding to the nutritional value of bread.

Many sugar substitutes are available for use in baking especially in a situation where low-calorie alternative may affect positively a medical condition such as diabetes where sugar consumption is severely limited. Some natural alternatives to white sugar for bakery products namely raw honey, maple syrup, molasses, corn syrup, stervia, xylitol, agava nectar, brown rice syrup, evaporated cane juice, black strap molasses, date sugar and organic sugar have been listed by Anon (2010), and Khan (2010). Khan (2010) reported that one cup of date sugar is equivalent to one cup of granulated or brown sugar. NHB and AIB (1990) and Ebiringa and Echebiri (2004) used honey as sugar substitute in bread production, Akubor and Yusuf (2007) used date palm syrup in cake production.

Date palm (*Phoenix dactylifera* L.) fruit, locally called 'debino' in Hausa language, is from the family of Arecaceae (Al-daihan and Bhat, 2012). It is cultivated for its edible sweet fruit. The fruit is a drupe in which an outer fleshy part (exocarp and mesocarp) referred to as pulp (Besbes et al., 2010) and pericarp (Duke 1981) surround a shell (the pit or stone) of hard endocarp with a seed inside. The fruit is available in different forms namely whole pitted and unpitted, dehydrated pieces, diced, extruded date pieces and macerated fruit (Berreveld, 1993; Glasner et al., 2002). Barreveld (1993) reported that a date palm fruit of 20% moisture content provides 3,000 kcal/ kg of date flesh.

Ahmed et al. (1995) and Vandercook et al. (1980) documented that date fruits are high energy food sources with 72 to 80% sugar content at maturity. At khalal stage, nearly all 80 to 85% of the sugar is sucrose. As ripening progresses, the sucrose is hydrolyzed into reduced sugars such as glucose and fructose. Al-Shahib and Marshal (2003) stated that the fruit of date palm contains a high percentage of carbohydrate (total sugar, 44 to 88%), fat (0.2 to 0.5%), 13 salts and minerals, protein (2.3 to 5.6%), vitamins and a high percentage of dietary fiber (6.4 to 11.5%). Ahmed et al. (1995) and Vandercook et al. (1980) reported that date fruit are good sources of iron and potassium, a fair source of calcium, chlorine, copper, magnesium and sulphur, and a minor source of 16 amino acids and vitamins A, B_1 and B_2.

Many products useful to humans are produced from date palm. The primary product is the fruit which is eaten fresh, dried or various processed forms (Glasner et al., 2002; Kader 1992). Dry or soft dates are eaten out of hand or may be seeded and stuffed, or chopped and used in great variety of ways on cereal pudding, bread, cakes, cookies, ice cream or candy bars (Barreveld 1993; Glasner et al., 2002); and can be made into juice, vinegar, wine, beer, sugar, syrup, honey, pickle, paste, dip and food flavor (Barreveld 1993; Glasner et al., 2002).

Date fruit pulp has been used more as a sugar source than as a fruit (Barreveld 1993). Date sugar is manufactured by processing dried dates into powder form and is great for sprinkling on breakfast cereal and baked goods (Khan 2010). In this study, date pulp meal was used as replacement for granulated sugar in bread production with the aim of evaluating the physicochemical, organoleptic and storage properties of the bread.

MATERIALS AND METHODS

Raw material procurement

The ripe date palm fruit used in this work was purchased from a hawker in Kaduna, Nigeria. Wheat flour (Golden Penny), granulated sugar (Dangote, Nigeria), margarine (Blue Band), yeast (Angel), iodized salt (Dangote), ascorbic acid, Edlen Dough Conditioner (EDC-95) and powdered milk (Cowbell) were purchased from Ega Market in Idah, Kogi State, Nigeria.

Processing of date palm fruit into meal

The seeds of the date palm fruits were removed and discarded. The pulp (pericarp) was oven dried at 45°C for 8 h (VMR Scientific, Model 1370, Oregon, USA) and subsequently milled using hand milling machine.

Production of bread

The bread was produced using the recipe shown in Table 1. Date palm fruit pulp meal (DPFPM) was used as replacement for granulated sugar at the following ratios: 80:0, 60:20, 40:40, 20:60 and 0:80 (sugar: DPFPM), which culminated to 0, 25, 50, 75 and 100% substitution levels of sugar. The dry ingredients were weighed or measured in the required quantities and mixed. Then, 500 ml of water was added to form the dough and mixed thoroughly. The dough was then kneaded to smooth consistency. During the kneading, margarine mixed with vegetable oil was added. When kneading was completed, the resulting dough was cut in weights of 350 g, molded into oblong shape and put into baking pans already greased with vegetable oil. The dough was proofed at room temperature (32 ± 2°C) for 1 h 40 min. It was then transferred to baking oven (Animex, Enugu, Nigeria) preheated to 230°C and baked for 40 min, removed from the oven and left to cool to room temperature (32 ± 2°C) before packaging in polyethene films.

Methods of analyses

Proximate analyses

Moisture (air oven method), crude fat (Soxhlet extraction method), protein (microKjeldhal method, Nx6.25), ash and crude fiber of granulated sugar, DPFPM and bread loaf samples were determined as described by AOAC (1990) methods. Carbohydrate content was estimated by difference {100- (%moisture + % Crude fat + % crude protein + % ash + % Crude fiber)}.

Table 1. Recipe for bread production.

Ingredient	Quantity
Wheat flour	1000 g
Yeast	80 g
Salt	50 g
Sugar/ date pulp meal	80 g
Margarine	30 g
Improver (ascorbic acid)	2 g
Edlen dough conditioner (EDC-95)	2 ml
Powdered milk	8 g
Water	500 ml

Table 2. Proximate composition of granulated sugar (table sugar) and pitted date palm fruit pulp meal (DPFPM)

Parameter	Granulated sugar	DPFPM
Moisture (%)	3.11	6.56
Fat (%)	0	1.88
Crude fiber (%)	0	3.66
Crude protein (%)	0	5.51
Ash (%)	1.01	2.95
Carbohydrate (%)	95.88	79.44

Proofing ability

This was carried out by measuring the height of the dough in the pan before and after proofing. The difference in height was the proofing ability (Rakkar 2007).

Oven spring

This was determined from the difference in the height of proofed dough before and after baking (Idowu et al., 1996).

Loaf weight, volume and specific volume determinations

The weight of bread loaf was measured using an analytical weighing balance. The volume of the loaf was determined by rape seed displacement method (AACC, 2000) with slight modification. Sesame seed was used in place of rape seed. The seeds were layered in the box of known dimension. Bread loaf was inserted and the seed poured to fill the remaining space till the seeds were running over. A straight edge rule was used to level the seed. The volume of the seeds was measured using a measuring cylinder and the volume of bread loaf calculated as the volume of the rectangular box less the volume of sesame seed. Loaf specific volume was calculated as the ratio of loaf volume to its weight.

Fermentation/ baking loss determination

This was determined as the difference in weight of the dough before proofing and the average weight of bread loaves.

Assessment of incidence of mold growth on bread loaf surfaces

This was carried out by counting the number of mold colonies visually observed on bread loaf surfaces during storage. The bread loaves were stored (unwrapped and wrapped in transparent polyethylene) on clean laboratory benches at room temperature (32 ± 2°C).

Sensory evaluation of bread loaves

A twenty-man panel consisting of equal numbers of male and female of 15 years and above evaluated the bread samples. The samples were served simultaneously in white plastic plates in a sensory evaluation laboratory illuminated by white fluorescent tube. The panelists evaluated the preference for the samples on the following characteristics: crumb color, crust color, flavor, texture and overall acceptability. The evaluation was carried out on a seven point Hedonic scale with 1= disliked very much and 7 = liked very much. Tap water was provided for rinsing of mouth to remove after taste.

Statistical analysis

Data were subjected to analysis of variance (ANOVA) and significant means discriminated by Tukey's LSD test (Gomez and Gomez 1984). Significance was accepted at $p < 0.05$.

RESULTS AND DISCUSSION

Proximate composition of granulated sugar, DPFPM and bread loaves

The proximate composition of granulated sugar and date palm fruit pulp meal (DPFPM) are presented in Table 2. The table reveals that apart from carbohydrate, ash (1.01%) was the only solid identified in the granulated sugar. This was unlike DPFPM which contained fat (1.88%), crude fiber (3.66%), crude protein (5.51%) and ash (2.95%) in addition to carbohydrate (79.44%). As expected, the granulated sugar is in a purer form than DPFPM.

The proximate composition of bread loaves, shown in Table 3, indicates that crude protein, fat, ash and crude fiber contents significantly ($p < 0.05$) increased with increasing replacement of granulated sugar with DPFPM whereas the carbohydrate content significantly ($p < 0.05$) decreased. The significance was recorded at 50% replacement with DPFPM for carbohydrate, 75% replacement for crude protein, ash and crude fiber; and at 100% for crude fat. Significant ($P < 0.05$) differences also existed between full (100%) replacement of granulated sugar with DPFPM and 25% replacement of granulated sugar in protein, ash, crude fiber and carbohydrate contents.

Proofing ability and oven spring

The proofing ability and oven spring (Table 4) decreased

Table 3. Proximate composition (dry weight basis) of bread as affected by sugar substitution with date palm fruit pulp meal (DPFPM).

Parameter	Percentage (%) Sugar Substitution with DPFPM (w/w)					
	0	25	50	75	100	LSD
Moisture* (%)	26.12	25.98	25.76	26.36	26.44	-
Fat (%)	6.25^{a}	6.68^{ab}	7.01^{ab}	7.38^{ab}	7.72^{b}	1.30
Crude fiber (%)	1.52^{a}	2.30^{ab}	2.98^{abc}	3.76^{bc}	4.45^{c}	1.48
Crude protein (%)	16.85^{a}	17.85^{ab}	19.05^{abc}	20.21^{bc}	21.26^{c}	2.41
Ash (%)	2.26^{a}	2.65^{ab}	3.14^{abc}	3.42b^{c}	3.84^{c}	1.02
Carbohydrate (%)	73.12^{a}	70.52^{ab}	67.82^{bc}	65.23^{c}	62.73c	3.75

Values are means of duplicate determinations. Means with the same superscript within a row were not significantly ($p > 0.05$) different. LSD, Least significant different.

Table 4. Physical properties of dough and bread as affected by sugar substitution with date palm fruit pulp meal (DPFPM).

Parameter	Percentage (%) Sugar substitution with DPFPM (w/w)					
	0	25	50	75	100	LSD
Proofing ability (cm)	2.20^{a}	2.18^{a}	2.14^{a}	2.10^{a}	2.08^{a}	-
Oven spring (cm)	2.30^{a}	2.22^{ab}	2.16^{ab}	2.02^{bc}	1.93^{c}	0.21
Weight of bread loaf (g)	269.5^{d}	272.4^{cd}	275.5^{c}	279.9^{b}	288.8^{a}	4.12
Volume of bread loaf (cm^3)	848.3^{a}	817.6^{b}	795.6^{c}	771.6^{d}	762.3^{e}	5.75
Specific volume (cm^3/g)	3.15^{a}	3.00^{ab}	2.89^{ab}	2.76^{ab}	2.64^{b}	0.43
Fermentation/ oven loss (%)	23.00^{a}	22.17^{a}	21.29^{ab}	20.03^{b}	17.49^{c}	1.92

Values are means of quadruplicates determinations. Means with the same superscript within a row were not significantly ($p > 0.05$) different. LSD, least significant different.

from 2.20 to 2.08 cm and 2.30 to 1.87 cm, respectively, as granulated sugar was replaced with DPFPM. The decrease in proofing ability was insignificant ($p > 0.05$), whereas that of oven spring was significant ($p < 0.05$). Proofing is an expression for the step in creating yeast breads during which the yeast makes the bread to rise (Rakkar, 2007); and it is measured by dough height which depends to some extent on the volume of gas (CO_2) production during fermentation (proofing), as well as protein matrix with its level of gluten (Badifu et al., 2005). During proofing, the yeast metabolizes simple sugar derived by hydrolysis of wheat starch and sugar optionally added to dough, producing CO_2 that was trapped by gluten matrix of the dough. The consequence is the rising (leavening) of the dough as a result of the stretching of the gluten matrix. Simple sugar is more readily provided by the hydrolysis of granulated sugar which is mainly sucrose than the wheat flour starch as the latter is more complex and may be more difficult to hydrolyze by yeast enzymes. This sucrose was also likely more abundant for yeast activity in granulated sugar than in DPFPM as evidenced in the higher carbohydrate contents of granulated sugar (Table 2). The decrease in proofing ability of the dough as the level of DPFPM increased may be as a result of the decrease in the amount of readily available fermentable sugar.

Oven spring is the ability of the dough to rise during baking. It is an important baking quality that measures the dough strength against heat related changes (Idowu et al., 2010). It is an indication of the proper strength of gas cell walls that entraps the gas and their elasticity by which they expand without collapsing due to increased yeast activity during initial phase of baking (Rakkar 2007). It is also a common knowledge that one of the main reasons for oven spring (rising in the hot oven) is the formation of bubbles in the dough and the formation of steam expanding the bubbles. Hence, increased yeast activity during initial phase of baking and/ or the formation of bubbles in the dough and the formation of steam expanding the bubbles may have caused the oven spring. Extra protein and fiber introduced by increasing amount of DPFPM may have bound some water making it unavailable for bubble and steam formation thereby reducing the consequent expansion.

Table 5. Mean sensory scores of bread as affected by replacement of sugar with date palm fruit pulp meal (DPFPM).

Parameter	%Sugar substitution with DPFPM					
	0	25	50	75	100	LSD
Crust color	6.3^{a}	5.7^{ab}	5.4^{ab}	5.5^{ab}	5.1^{b}	1.15
Crumb color	2.5^{b}	5.6^{a}	5.3^{a}	5.3^{a}	5.3^{a}	1.39
Flavor	5.2^{a}	4.1^{a}	3.8^{a}	4.5^{a}	4.4^{a}	1.53
Texture	5.6^{a}	4.8^{a}	5.0^{a}	5.0^{a}	5.4^{a}	0.96
General acceptability	4.7^{a}	3.7^{a}	4.0^{a}	4.2^{a}	4.2^{a}	1.32

Means with the same superscript within a row were not significantly ($p > 0.05$) different. LSD, least significant different.

Weight, volume, specific volume and processing loss of bread loaves

The weight of bread loaves significantly ($p < 0.05$) increased from 283.3 to 299.5 g with increasing replacement of granulated sugar with DPFPM. Shittu et al. (2007) reported the basic determinant of loaf weight to be the quantity of dough baked and the amount of moisture and carbon dioxide diffused out of the loaf during baking. Badifu et al. (2005) also attributed the increase in loaf weight to the relative high moisture content, the nature of carbohydrate and high bulk density of flour. In this study, this could be a result of increase in crude fiber, protein, fat and ash that could not be easily fermented by yeast and which have probably absorbed and immobilized water, making it not easily vaporized during baking.

The volume and specific volume of bread loaves significantly ($p < 0.05$) decreased from 848.3 to 762.3 cm^3 and 3.15 to 2.64 cm^3/g, respectively, as granulated sugar was replaced with DPFPM (Table 4). The decrease in volume could be a cumulative effect of lower proofing ability and oven spring which are functions of gas evolution. Shittu et al. (2007) attributed the variation in loaf volume of the bread samples produced from the same formulation, proofing time and dough size to different rate of gas evolution. The sugar available for yeast activity and consequently for evolution of gas could be lower in granulated sugar/ DPFPM blends due to the contents of protein and crude fiber (Table 2). Specific volume is the volume per unit weight of loaf. It is the integral of weight and volume of the loaf related to the rising power of the loaf during baking (Ayo et al., 2008) and has been generally adopted in the literature as a more reliable measure of loaf size (Shittu et al., 2007). The reduced specific volume with increasing DPFPM signified that the decrease in volume of loaves was not proportional to increase in weight. The rate at which the volume decreased outweighed the rate at which the weight increased. The specific volume of 2.64 to 3.15 cm^3/g for bread samples in this study is low. According to Lin et al. (2009), China Grain Products Research and Development Institute (CGPRDI) in 1983 documented that the specific volume of standard bread should be 6 cm^3/g and should not be less than 3.5 cm^3/g. Specific loaf volume is a function of hydration. The low specific loaf volume in this study could be as a result of minimal hydration of 50%, having used 500 ml of water for 1000 g of flour to form the dough. Ocheme et al. (2010) used hydration of 56% (56 ml of water for 100 g of wheat flour) and obtained specific loaf volume of 3.86 cm^3/g.

Table 4 shows that processing loss decreased significantly ($p < 0.05$) with increasing replacement of granulated sugar with DPFPM. The loss, deduced as the difference of the dough weight before proofing and the weight of baked loaf, could be a cumulative effect of fermentation loss during proofing and water evaporated during baking. Rakkar (2007) made a similar deduction of the weight of baked bread from the initial weight of the dough and attributed the difference to the moisture loss alone. It is likely that the fiber and protein introduced into the dough by DPFPM held some moisture, preventing them from evaporation.

Sensory properties of bread loaves

The mean sensory scores on preference for bread loaves are shown in Table 5. Whereas the preferences for crust and crumb colors were significantly ($p < 0.05$) affected by replacement of granulated sugar with DPFPM, those of texture, flavor and overall acceptability of the loaves was not. Only the preferences for crust and crumb color of loaves with 100% granulated sugar were different ($p < 0.05$) from those of 100% DPFPM. The preferences for the color of both the crust and the crumb of the blends were not different ($p > 0.05$). However, whereas the preference for crumb color increased with increasing replacement of granulated sugar with DPFPM, that of crust color decreased. The non-difference in preference

Table 6. Incidence of mold growth as affected by replacement of sugar with date palm fruit pulp meal (DPFPM).

Storage period (day)	**Number of observable mold colonies on bread loaves**				
	Percentage (%) Sugar substitution with DPFPM				
	0	**25**	**50**	**75**	**100**
Wrapped bread loaf samples					
1.	0	0	0	0	0
2.	0	0	0	0	0
3.	0	0	0	0	0
4.	15	12	0	0	0
5.	35	21	12	17	29
6.	64	48	46	60	72
Unwrapped bread loaf samples					
1.	0	0	0	0	0
2.	0	0	0	0	0
3.	0	0	0	0	0
4.	1	0	0	0	0
5.	8	2	2	3	6
6.	15	4	3	5	10

for flavor of bread may be expected if the sweetness of the same mass of date sugar and granulated or brown sugar is equivalent (Khan, 2010) unless contributions of phytochemical found in date meal, as impure substance, could come into play. The inverse preference for color of the bread crust and crumb may have affected its possible effect on the overall acceptability.

Incidence of mold growth on bread loaf surfaces

The number of mold colonies visually observed on bread loaves during six days of storage is presented in Table 6. Wrapped bread samples baked with 100% granulated sugar and 25% (20 g) replacement with DPFPM had 15 and 12 colonies, respectively, at the fourth day of storage whereas unwrapped bread sample baked with only granulated sugar had one visually observed colony. On the fifth and sixth days, highest numbers were observed on the loaves baked with granulated sugar alone for both wrapped and unwrapped loaves. The number of colonies were, however, higher on the wrapped bread loaves. The visually observed colonies seemed to increase with increasing replacement of granulated sugar with DPFPM. The initial delay of mold growth on bread loaves baked with DPFPM and its blends with granulated sugar could have suggested the possession of antifungal properties by DPFPM. However, the increasing number of mold colonies with increasing amount of DPFPM has proved otherwise. Mold growth is a function of water activity of a product or relative humidity of an environment. As bread stales, water is emitted as a result of retrogradation of gelatinized starch molecules. This water migrates to the surface of loaf where it is vaporized. The vapor saturated the microenvironment of wrapped loaves and increased the relative humidity which now favors the growth of mold. The vapor is not held back as much for unwrapped loaves leading to lesser number of colonies. Non-sugar components of DPFPM such as protein and ash may have enriched the bread loaves, making it more favorable substrate for mold growth. Higher moisture content of bread with higher DPFPM may have also favored the mold growth.

Conclusion

The use of date palm fruit pulp meal as sugary agent to replace granulated sugar in bread production improved the nutritional value of bread. The overall organoleptic quality was not affected even at 100% replacement. The physical property of the bread, judged majorly by the loaf specific volume that relates loaf weight to its volume, decreased with increasing replacement of granulated sugar. As the decrease was not significant at 75% replacement with date fruit pulp meal but was at 100% replacement, replacement of granulated sugar with a proportion of the date fruit pulp should be encouraged. Using high proportion may delay the incident of mold growth on the loaves but may encourage greater growth at prolonged storage. Antimycotic may be incorporated in the bread to suppress this activity.

REFERENCES

AACC (2000). Approved methods of American Association of Cereal Chemists, 10th Ed. AACC International, St Paul, MN.

Ahmed AI, Ahmed AWK, Robinson RK (1995). Chemical composition of date varieties as influenced by the stage of ripening. Food Chem. 54:305-309

Akubor PI, Yusuf D (2007). Effect of incorporation of date palm syrup on the sensory and physical properties of wheat flour biscuit. J. Appl. Sci. Environ. 1:25-28.

Al-daihan S, Bhat RS (2012). Antibacterial activities of extracts of leaf, fruit, seed and bark of *Phoenix dactylifera*. Afr. J. Biotechnol. 11(42):10021-10025

Al-Shahib W, Marshall RJ (2003). The fruit of the date palm: Its possible use as the best food for the future? Int. J. Food Sci. Nutri. 54(4):247 - 259.

Anon (2010). Natural sugar substitute: 10 healthier alternatives to refined sugar. American Diabetes. www.americandiabetes.com/livingdiabetes Accessed May 2012.

AOAC (1990). Official methods of analyses. Association of Analytical Chemists, 15th edn. Arlington, Virginia, USA 2: 143 - 202.

Ayo JA, Nkama I, Haruna US, Bitrus Yelmi, Onajaife F (2008). Effect of dough improver on the physical and sensory quality of acha (Digitaria exilis) flour bread. Nig. Food J. 26(1):102-110.

Badifu GIO, Chima CE, Ajayi YI, Ogori AF (2005). Influence of mango mesocarp flour supplement to micronutrient, physical and organoleptic qualities of wheat-based bread. Nig. Food J. 23: 59 - 68.

Barreveld WH (1993). Date palm products. Food and Agriculture Organization of the United nations. Agriculture Service Bulletin № 101, United Nations, Rome, Italy.

Besbes S, Ghorbel R, Sala RB, Masmoudi M, Jedidi F, Attia H, Blecker C (2010). Date fiber concentrate: Chemical composition, functional properties and effect on quality characteristics of beef burgers. J. Food Drug Analys. 18(1):8 - 14.

Duke JA (1981). Handbook of Legumes of World Economic Importance. Plenium Press, New York. p 33.

Ebiringa DC, Echebiri SI (2004). Effect of substituting honey for sucrose in bread making. Nig, Food J. 22:189 - 194.

Glasner B, Botes A, Zaid A, Emmens J (2002). Date harvesting, packinghouse management and marketing aspects. In: Zaid A (ed.) Date palm cultivation. Food and Agriculture Organization Plant Production and Protection paper № 156. Food and Agriculture Organization of the United Nations, Rome, Italy, p. 177 - 208

Gomez KA, Gomez AA (1984). Statistical Procedure for Agricultural Research, 2nd edn. John Wiley and Sons, New York. pp. 25-28, 188-190.

Idowu MA, Ojo MR, Shttu TA (2010). Quality evaluation of some commercial baker's yeast in Nigeria. Nig. Food J. 28 (1): 164 - 172

Idowu MA, Oni A, Amusa BM (1996). Bread and biscuit making potentials of some Nigerian cocoyam cultivars. Nig. Food J. 14:1-12.

Kader AA (1992). Postharvest biology and technology: An Overview. In: Kader AA (ed.) Postharvest technology of horticultural crops (2nd edn), Publication 311, University of California Division of Agriculture and Natural Resources, Berkeley, CA. p. 15-20.

Khan F (2010). Natural Ingredient You Can Use to Replace Sugar. Demand Media Inc. http://www.livestrong.com/ Accessed June 2012.

Lin L, Liu H, Yu Y, Lin S, Mau J (2009). Quality and antioxidant property of buckwheat enhanced wheat bread. Food Chem. 37:461-467.

NHB and AIB (1990). Honey, its Utilization in Bakery Products. National Honey Board, Colorad. pp. 1 - 5.

Ocheme OB, Oloyede OO, Mahmud AH (2010). Production and evaluation of bread using blends of wheat flour and fermented plantain flour. Nig. Food J. 28 (2):284 - 293.

Rakkar PS (2007). Development of a Gluten-Free Commercial Bread. A Thesis Submitted in Partial fulfillment of the Requirement for the Degree of Master of Applied Science (Food Science). Auckland University of Technology, Auckland. pp. 22-23.

Shittu TA, Raji AO, Sanni LO (2007). Bread from composite cassava-wheat flour: I. Effect of baking time and temperature on some physical properties of bread loaf. Food Research International 40:280-290.

Vandercook CE, Hasegawa S, Maier VP (1980). Dates. In: Nagy S, Shaw PE (eds), Tropical and Subtropical Fruits: Composition, Properties and Uses. AVI Publishing Company, Westport, CT. p. 506 - 541.

Wihlfaht JE (2007). A Treatise on Baking: Essential factors Influencing Bread Quality. ACH Foods Inc. http://abrfaq.info/treatise/200 Accessed June 2012.

15

Studies on foodborne bacteria in commercially hawked ready-to-eat fish in Jos and its environments

[1,2]Chukwu, O. O. C., [2]Chukwuedo, A. A., [2]Otalike, P., [3]Chukwu, I. D., [2]Echeonwu, N. O. G., [2]Bitrus, J. G., [2]Akubo, S.I.

[1]Departments of Molecular Biology National Veterinary Research Institute, Vom, Nigeria.
[2]Medical Microbiology, Federal College of Veterinary and Medical Laboratory Technology, National Veterinary Research Institute, Vom, Nigeria.
[3]Central Diagnostic Laboratory, National Veterinary Research Institute, Vom, Nigeria

Studies on foodborne bacteria in hawked fish (*Trachurus capensis*) were carried out with 200 flesh and 24 surface swab samples collected at various selling points in Jos and surrounding suburbs. The samples were separately cultured in bacteriological broth and agar media. The isolates were identified and their viable counts determined. The analysis gave the following bacteria species: *Escherichia coli*, 26 (16.8%); *Bacillus* species, 25 (16%); *Shigella* species, 22 (14.2%); *Staphylococcus aureus*, 19 (12.6%); *Proteus* species, 12 (7.7%); *Klebsiella* species, 8 (5.2%); *Neisseria catarrhalis*, 6 (3.9%); *Staphylococcus epidermidis*, 4 (2.6%); *Streptococcus faecalis*, 4 (2.6%); Enterobacter species, 4 (2.6%); *Pseudomonas* species, 2 (1.2%); *Lactobacillus* species, 2 (1.2%) and *Citrobacter* species, 1 (0.6%). These results show the bacteria load species distribution in the hawked fish. The result is quite informative with respect to public health hazard and calls for urgent improvement in sanitation maintenance of good hygiene by food producers, handlers and vendors in the study area as a possible control measure through the processing chain.

Key words: Foodborne bacteria, Fish, *Trachurus capensis,* bacterial load, Public Health.

INTRODUCTION

Ready-to-eat foods can be described as the status of food being ready for immediate consumption at the point of sale. These could be raw, undercooked or cooked, hot or chilled and can be consumed without further heat treatment (Tsang, 2002). Different terms have been used to describe such ready-to-eat foods. These include convenient, ready or instant and fast foods, such as pastries, meat pie, sausage, rolls, burger, salad or coleslaw, fried meat, fried chicken, milk and milk products and fish (Caserani et al., 1974; Patience et al., 2002).

Fish has been an important source of protein for humans throughout recorded history. They are harvested either from wild fisheries or farmed in much the same way as livestock and poultry. They are also exploited by recreational fishers and fish-keeper, and are exhibited in public aquaria (Tidwell and Allan, 2001).

In other to meet the fish food demand, there is the possibility of contamination during the processing, preparation and packaging of the fish, it may encounter bacteria which may make it unsuitable for consumption. Torok et al. (1997) stated that outbreaks of food borne diseases are caused by foods that are contaminated intrinsically or extrinsically during harvesting, processing or preparation. The bacteria may come from the fish contact surfaces, the surrounding environment as well as fish handlers or vendors. Sapers et al. (2005) and Peariso (2005) stated that, during food processing, contamination is possible from infected food handlers and that poor hygiene is a major factor.

Serious consequences relating to national productivity and development can arise from lack of hygiene and sanitation in such outlets. There have been several reports on the health risks associated with the consumption of processed seafood, ranging from allergic reactions, stomach and intestinal cancerous growths, a general degeneration of peripheral cellular tissues to gradual breakdown of the digestive and excretive systems (Edema et al., 2005).

Phyllis (2007) illustrates how critical factors come together to produce tragic and large preventable results. Modern microbes often team up with old practices, short sighted decisions, or current consumer trends to produce an outbreak. Recipes that do not include an adequate final cooking step have become increasingly popular with consumers and can be a significant source of food borne illness. According to Higgins (2007), anyone who works in food safety sooner or later discovers that one of the most valuable tools for prevention is simply reading about and understanding how past outbreaks have occurred. This study is aimed to evaluate the bacterial load and distribution of bacterial pathogens in processed ready-to-eat *Trachurus capensis* at some of their various sale outlet points in Jos metropolis and its environs.

MATERIALS AND METHODS

Sample collection

Ready-to-eat fried fish (*T. capensis*) samples were purchased from the different selling spots located in the Jos and its environs. These include the Jos, Bukuru, Kaduna-Vom (K-Vom) and Vom Christian (Vwang) areas and screened for pathogenic bacteria. Outlets such as, hawkers, kiosks and restaurants were utilized. A total of 200 fish (*Trachurus capensis*) were purchased from all the selling spots; 20 each were bought from the kiosks and hawkers, while 10 each from the restaurants in each of the location. These were collected from April 2010 to May 2010.

The samples were aseptically collected separately in sterile paper foils and labeled accordingly. They were immediately transported to the bacteriology laboratory, Federal College of Veterinary and Medical Laboratory Technology (FCVMLT) Vom for prompt analysis. The sterile swab sticks were dipped into sterile nutrient broth and selenite fluid respectively which were then used to swab the contact surfaces (10 x 10 cm^2) utilized by the vendors for the preparation and hawking of the fishes such as cutting boards or working tables, showcases and hawking trays. They were also conveyed alongside the fish samples immediately to the college bacteriology laboratory for prompt analysis.

Enumeration of bacteria load

The outer layer of the fish samples were aseptically swabbed with sterile swab sticks. The swab stick samples were held into Bijou bottles containing 5 ml each of sterile nutrient broth. They were cut aseptically into the bottles using sterile scissors. Bacterial load of each of the fish sample was enumerated by carrying out bacterial colony count using the Miles and Mizras method of bacteria colony count (Ochei and Kolhatkar, 2008).

Culture

The outer layers of the fish samples were aseptically swabbed with sterile swab sticks. The swabbed samples held into the respective sterile Bijou bottles containing the nutrient broths and selenite fluids separately were cut using sterile scissors. They were incubated at 37°C overnight.

The swabbed sample from the contact surfaces used by the fish vendors were also cut aseptically into the respective sterile bijou bottles containing the nutrient broth and selenite fluid respectively. These were similarly incubated at 37°C overnight.

All the organisms observed to grow both in the nutrient broth and selenite fluid culture were aseptically sub-cultured onto blood agar and MacConkey agar. The blood agar plates were incubated both aerobically and anaerobically while the MacConkey agar plates were incubated aerobically at 37°C overnight. The organisms in the selenite fluid were aseptically sub-cultured using a sterile wire loop unto *Salmonella Shigella* agar and were incubated aerobically at 37°C overnight.

All the organisms observed to grow in any of the culture plates were picked and sub-cultured unto nutrient agar slopes and stored in the refrigerator for further test for identification and characterization.

Identification and characterization of bacteria isolates

The cultural characteristics of the isolates were examined based on colonial morphology (macroscopically and microscopically), physiologically and biochemically according to Barrow et al. (1993).

RESULTS

A total of 224 samples (200 fish samples, *T. capensis* and 24 swab samples) were analyzed. All the 224 samples (100%) were found positive that is, all samples yielded growths of at least one bacterium species. The highest total coliform counts of 5.7 x 10^{11} cfu/ml was obtained in fish samples at Jos area followed by 5.6 x 10^{11} cfu/ml at Bukuru area, 3.7 x 10^{11} cfu/2 ml at K-Vom area and 3.0 x 10^{11} cfu/ml at Vwang. Table 1 shows the distribution of swab samples collected from the different areas, selling points and their numbers. Table 2 shows bacteria isolates from fish samples, their distribution and occurrence rates. Table 3 shows the bacteria isolates from fish at different bacteria species total and mean distributions at the various locations. The histogram illustrating comparatively the percentage occurrence of bacterial isolates is shown in Figure 1.

The bacteria isolates included; *Salmonella* species, *Bacillus* species, *Proteus* species, *Citrobacter* species, *Shigella* species, *Enterobacter* species, *E. coli, Klebsiella* species, *Lactobacillus* species, *Neisseria catarrhalis, Pseudomonas* species, *S. aureus, S. epidermidis and S. faecalis.*

DISCUSSION

The presence of a good number of aerobic bacterial isola-

Table 1. Total coliform count and their distribution at the various selling points

Location		Total coliform count (cfu/cm^2)	Number of sample
Jos	Hawkers	3.1 x 10^{11}	20
	Kiosks	2.5 x 10^{11}	20
	Restaurants	5.1 x 10^{9}	10
	Total	**5.7 x 10^{11}**	**50**
Bukuru	Hawkers	3.3 x 10^{11}	20
	Kiosks	2.3 x 10^{11}	20
	Restaurants	4.0 x 10^{9}	10
	Total	**5.6 x 10^{11}**	**50**
Vwang	Hawkers	2.6 x 10^{11}	20
	Kiosks	3.2 x 10^{10}	20
	Restaurants	4.8 x 10^{9}	10
	Total	**3.0 x 10^{11}**	**50**
K/Vom	Hawkers	3.4 x 10^{11}	20
	Kiosks	2.6 x 10^{10}	20
	Restaurants	3.8 x 10^{9}	10
	Total	**3.7 x 10^{11}**	**50**

tes give an overall picture of the level of contamination of fried fish (*T. capensis*) and the sanitary conditions of its preparation and handling. The results obtained in this study are highly suggestive that these products are prepared and sold under conditions that would permit the survival and multiplication of various organisms. The presence of these isolates which included some pathogenic bacteria can pose serious public health problems to the unsuspecting consumers.

According to Doyle and Evans (1999), food borne diseases are diseases resulting from ingestion of bacteria, toxins and cells produced by microorganisms present in food. During food processing, contamination is possible from infected food handlers as poor hygiene is a factor (Sapers et al., 2005; Peariso, 2005). *E. coli* and *S. aureus* are normal flora in human and animals, their presence in foods are indications of excessive human handling (Adamolekun and Adamolekun, 1992).

In Nigeria, a number of ready-to-eat foods have been reported to have high incidence of pathogenic bacteria (Adesiyun 1995; Chukwu et al., 2006; Chukwu et al., 2009; Okonko et al., 2009). *E. coli*, *Klebsiella*, *S. aureus*, *Bacillus* species, *Enterobacter* and *Pseudomonas* has been isolated from foods implicated in illness (Oluwafemi and Simisaye, 2005; Okonko et al., 2009). *S. aureus* can cause food poisoning even with very small amount (100 to 200 µg) of its heat stable enterotoxin (Evenson et al., 1988; Jay, 2000). *E. coli* were found to be the most prevalent bacteria in all the fish samples analyzed followed by *Bacillus* species. *Staphylococcus* species are normal flora of human body and is found on the skin cavities, mucous membranes and also in the air. Based on these reports, it is not surprising that the fish samples screened, tested positive to *S. aureus* isolates. Handling and aerial microbial load could therefore be the major sources of contamination in hawked foods. The reservoir of *Salmonella* in the environment is the intestinal tract of animals such as chickens, turkey and pigs; man too can act as a carrier of *Salmonellae*. Both the animal and human sources play an important role in cross contamination of cooked foods.

Bacteria colony count of the fried fish samples were also carried out with the highest total coliform counts of 3.3 x 10^{11} being obtained from hawkers followed by 2.5 x 10^{11} from kiosks which shows exposure to contamination. This study has revealed high contamination of the ready-to-eat fish from the highest dilution calls for proper reheating before consumption. Bacteria like *Bacillus* species are readily present in air and dust. Contamination of ready-to-eat fried fish may also result from other sources like the utensils used in the preparation, water used in the washing of the fishes, working tables, showcases and trays used in processing and carrying of the fishes. The isolation of *Neisseria catarrhalis* in the fish may also indicate some unhygienic practices of the food handlers or the consumers because this bacterium is a normal flora in nasal or respiratory tract, its presence therefore would indicate sneezing or coughing over the products at one stage or another.

Of all the isolates, *E. coli* had the highest percentage occurrence of 26 (17%) followed by *Bacillus* species, *25 (16%); Shigella* species, 22 (14%); *Salmonella* species, 20 (12.5%); *S. aureus*, 19 (12.3%); *Proteus* species, 12 (7.7%); *Klebsiella*, 8 (5.2%), *Neisseria catarrhalis*, 6 (3.9%); *Enterobacter* species, 4 (2.5%); *S. epidermidis* 4 (2.5%), *Streptococcus faecalis*, 4 (2.5%); *Lactobacillus* species 2 (1.2%); *Pseudomonas* species, 2(1.2%) and *Citrobacter* species 1 (0.6%).

Total bacteria isolates from the kiosks and hawked products are 38.71 and 40.00%, respectively higher than that from the restaurant with 21.29%. This could be attributed to a number of reasons; Hawkers and kiosks vendors usually do not cover the fishes during and after preparation. Some buyers often have the habit of inspecting the fish products with their bare hands while talking or sneezing before finally choosing the one they want thereby contaminating them. Food poisoning/ illnesses are entirely preventable by practicing good sanitation and food handling techniques (Betty and Richard, 1994). If the monitoring of the bacteriological quality of this fish food can be improved during preparation, handling and sampling, the bacterial load of the fish products (*T. capensis*) can be eliminated.

In conclusion, all the fish samples analyzed appears to be contaminated with bacteria. This culminates a potential public health danger if allowed unaddressed or unabated. The Public Health Officers in Jos and its environs

Table 2. Bacteria Isolate from Fish Samples, their distribution and occurrence rates

Location	Number collected	Number positive	% Positive	Bacteria	Number of Isolate	% Occurrence
Jos	56	54	98.2	*Escherichia coli*	8	14.8
				Streptococcus faecalis	4	7.4
				Proteus spp.	5	9.2
				Staphylococcus aureus	8	14.8
				Shigella spp.	10	18.5
				Pseudomonas spp.	2	3.7
				Bacillus spp.	10	18.5
				Salmonella spp.	6	11.1
				Neisseria catarrhalis	1	1.9
Bukuru	56	49	87.5	*Neisseria catarrhalis*	5	10.2
				Enterobacter spp.	4	8.2
				Bacillus spp.	6	12.2
				Salmonella spp.	8	16.3
				Escherichia coli	10	20.4
				Staphylococcus aureus	5	10.2
				Staphylococcu epidermidis	4	8.2
				Klebsiella spp.	3	6.1
				Proteus spp.	4	8.2
Vwang	56	26	46.4	*Citrobacter* spp.	1	3.9
				Escherichia coli	3	11.5
				Lactobacillus spp.	2	7.7
				Klebsiella spp.	3	11.5
				Shigella spp.	6	23.1
				Bacillus spp.	5	19.2
				Staphylococcus aureus	6	23.1
K/Vom	56	26	46.4	*Escherichia coli*	5	19.2
				Bacillus spp.	4	15.4
				Salmonella spp.	6	23.1
				Klebsiella spp.	2	7.7
				Proteus spp.	3	1.5
				Shigella spp.	6	23.1
Total	**224**	**155**	**69.2**		**155**	

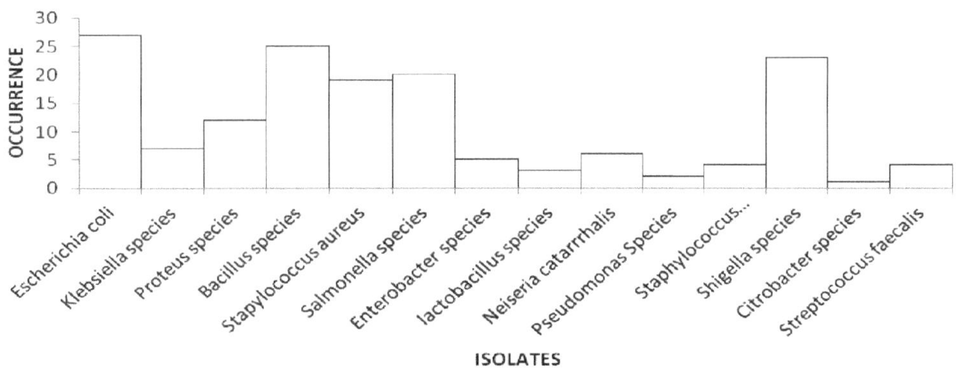

Figure 1. Histogram showing the percentage occurrence of bacteria isolates.

Table 3. Bacteria isolates from fish and their mean distribution

Bacterial Isolate	Kiosk	Hawker	Restaurant	Total	Mean (X)
Escherichia coli	9	11	6	26	8.67
Klebsiella species	3	3	2	8	2.67
Proteus species	6	5	1	12	4.00
Bacillus species	9	12	4	25	8.33
Staphylococcus aureus	7	7	5	19	6.33
Salmonella species	7	8	5	20	6.67
Enterobacter species	2	2	0	4	1.33
Lactobacillus species	0	2	0	2	0.67
Neisseria catarrhalis	4	2	0	6	1.33
Pseudomonas species	1	1	0	2	0.67
Stapylococcu epidermidis	2	0	2	4	1.33
Shigella species	10	6	6	22	7.33
Citrobacter species	0	1	0	1	0.33
Streptococcus faecelis	0	2	2	4	1.33
Total	**60**	**62**	**33**	**155**	**51.66**

should screen food - handlers periodically, so that carriers or sufferers of communicable diseases can be disallowed from handling foods meant for other people and should be treated as well. Public health officers should inspect the premises meant for preparation and selling of foods periodically. The public health unit should organize basic training on personal hygiene for food handlers. This will go a long way in improving the hygienic standards of their food products.

ACKNOWLEDGEMENT

The authors are grateful to the staff of the Departments of Molecular Biology and Bacteriology of Federal College of Veterinary and Medical Laboratory Technology, Vom for their assistance in sampling. Also, to the Staff of Central Diagnostic Laboratory, National Veterinary Research Institute, Vom for their assistance in sample diagnosis.

REFERENCES

Adamolekun WE, Adamolekun B (1992). Bacteria associated with food processing. Nig. Med. Pract. 24:43-45.

Adesiyun AA (1995). Bacteriologic quality of some Trinidadian ready to consume foods and drinks and possible health risks to consumers. J. Food Protect. 58(3):651-655.

Barrow IG, Steal KJ, Feltham AK (1993). Cowan and Steal manual for Identification of Medical bacteria, 3rd Edition Kambridge University Press Kambridge New York.

Betty CH, Richard, JB (1994). Bacterial causes of food poisoning in food poisoning and hygiene. 8th edition, Edward annod. Publishing Limited London. pp. 250-257.

Caserani V, Kinston R. (1974). Practical Cookery, 4th edition Edward Arnold Publishers London. pp. 1-10.

Chukwu OOC, Ogbonna, CIC, Chukwu DI, Olabode OA, Onwuliri FC and Nwankiti, OO (2006). *Listeria monocytogenes* in Nigerian processed meat and ready-to-eat diary product. Nig. J. Microbiol. 20 (2):900-904.

Doyle MP, Evans PD (1999). Food borne pathogens of recent concern. Ann. Revised Nutr. 6:25-41

Chukwu OOC, Olabode OA, Chukwuedo AA, Umoh GE, Esiekpe KM (2009). Bacteriological Evaluation of Pre-cut Fruits sold in Kano Metropolis, Kano State, Nigeria. East Afr. J. Public Health 6(1):51-4.

Edema MO, Omemu AM, Bankole MO (2005). Microbiological safety and quality of ready to eat foods in Nigeria. In: the book of abstract of the 29th Annual conference and general meeting (Abeokuta 2005) on microbes as agents of sustainable development organized by Nigerian society for Microbiology (NSM), University of Agriculture.

Evenson ML, Hinds MW, Berstein RS, Bergdoll MS (1988). Estimation of human dose of *Staphylococcus enterotoxin* A from a large outbreak of *Staphylococcal* food poisoning involving chocolate milk. Int J. Food microbiol 7:311-316.

Higgins C (2007). Food safety: old habits, new perspectives. In: Book and Media. J. Emerg. Infect. Dis. 13(6):960-961.

Jay JM (2000). Modern Food Microbiology 65th ed. Gaithersberg, Maryland. Aspen Publishers.

Ochei J, Kolhatkar A (2008). Medical Laboratory Science Theory and Practice Department of Microbiology College of Medicine Sultan Qaboos University Muscat 8th edition Tata McGraw-Hill Publishing Company Limited New Delhi.

Okonko IO, Donbraye E, Babatunde SOI (2009). Microbiological Quality of Sea food Processor and water used in two different sea processing plant in Nigeria EJEAFCHE 8(8):621-629.

Oluwafemi F, Simisaye MT (2005). Extent of microbial contamination of sausages sold in two Nigeria cities. In: The Book of Abstract of the 29th Annual Conference and General Meeting (Abeokuta 2005) on Microbes As Agents of Sustainable Development, Organized by Nigeria Society for Microbiology (NSM), University of Agriculture, Abeokuta, from 6-10th Nov. pp. 28.

Patience Mensa, Dorothy Yeboah-Manu, Kwaku Owusu-Darko, Anthony Ablordey (2002). Bull. World Health Org. 80:546-554.

Peariso D (2005). Preventing Foreign Material contaminations in Foods. Malden, MA: Blackwell Publishing Limited.

Phyllis E (2007). Food safety: old habits, new perspectives. ASM Press. Herndon, Virginia, USA. pp. 414.

Sapers GM, Gorny JR, Yousef AE (2005). Microbiology of Fruits and Vegetables. Boca Raton, FL: CRC Press.

Tidwell JH, Allan GL (2001). "Fish as food; Aquaculture's Contribution Ecological and Economic Impact and Contributions of Fish Farming and Capture Fisheries".

Torok TK, Tanze RV, Wise RP, Livengood JR, Sokolow R, Manvans S. (1997). A large community outbreak of Salmonellosis caused by International Contamination of a restaurant salads/bars. J. Am. Med. Assoc. 278(8):389-395.

Tsang D (2002). Microbiological guidelines for ready-to-eat food. Road and Environmental Hygiene Department Hong Kong. pp 115-116.

16

Studies on different drying, canning and value addition techniques for mushrooms (*Calocybe Indica*)

Gopinath Lakshmipathy, Arunkumar Jayakumar, Meera abhilash and Shantha Prema Raj

Department of Industrial Biotechnology, Dr. MGR Educational and Research University, Maduravoyal, Chennai - 600095, India.

Drying characteristics and the quality of dried mushrooms were analyzed. *Calocybe indica* slices (5 mm) were dried in cross flow dryer, open sun dryer, solar dryer, through flow dryer, vibro dryer, vacuum dryer in the respective drying chambers at 65, 36, 55, 60, 55 and 45°C. Mushrooms were dried from about 84 to 3% moisture content (w.b) in about 8 h. Solar drier was cost effective and efficient in drying in the overall analysis. Potassium meta bisulphite (0.25 g/l) and 3% H_2O_2 and was proposed for washing mushrooms which supports storage up to six days with minimal contamination. Mushroom jam was made with different sources of pectin; however, jam made with natural pectin (guava) showed highest sensory score. Mushroom squash Formulation - III contains flavor enhancers such as amla and lemon extract.

Key words: *Calocybe indica,* drying, driers, mushroom, post harvest, value addition, jam, canning and squash.

INTRODUCTION

Generally, for long-term storage of mushrooms, canning and drying processes along with some value addition technology are employed. The quality of the preserved product is rarely comparable with that of fresh mushrooms, and these processes are not always suitable for all types of mushrooms.

Fresh mushrooms are highly perishable with short shelf life under ambient environment, temperature and humidity and their commercialization becomes difficult. Among the various techniques employed for preservation of mushrooms, drying seems to be an effective approach to extend shelf life and ensure distribution. Various drying techniques are employed to dry different food products. Each technique has its own advantages and limitations (Deka, 2000). Studies comparing traditional sun drying and other drying techniques show that the use of solar dryer leads to a considerable decline of the drying time and to a significant development of the product quality in terms of color, texture and taste.

Canning is the most common process used for preserving mushrooms. Preservation of mushrooms by canning has turn out to be considerably more specialized in recent years. Jam production was adopted as a preservation method to overcome the losses that occur in the mushroom cultivation. Standard formulations can be developed according to their end use, consumer fondness, market demand, food laws, buyer's specifications and economic utilization (Deka and Sethi, 2001). Mushroom squash is not quite popular as it does not have a greater variety of beverages. Preparation of nutritive ready-to-serve (RTS) beverages is considered to be a suitable and economic substitute for utilization of mushroom squash (Bala et al., 2009). Lack of technology of the processing of mushroom and inadequate information on the chemical properties increases the degree of losses.

The aim of this work was to investigate the effect of different washing, drying and canning methods on quality of mushroom and to develop formulations for some value added mushroom products such as jam and squash.

EXPERIMENTALS

Determination of chemical for washing mushroom

Mushrooms were washed to increase the shelf life period with the chemicals such as citric acid, potassium meta bisulphite (KMS), hydrogen peroxide (H_2O_2) and chlorine. Parameters noted were color change, texture, storage time and taste.

The KMS was prepared at a range of 0.25 g/l, 1 g of citric acid was diluted with distilled water. Similarly H_2O_2 (3%) and chlorine of 20 ppm was used for washing. Washed mushrooms were observed for their shelf life for a period of six days. Total microbial count was measured using plate count method described in the FAO Manual (FAO, 1995).

Effect of different dryers on dehydration of mushrooms

Mushrooms were processed at optimal temperatures. Fully matured milky mushrooms were cut into 5 mm thick slices. No pre-treatment/blanching was done and the mushroom slices were dried from an initial moisture content of approximately 90% (w.b.) to the final moisture content of about 10% (d.b.) in cross flow dryer (65±1°C), open sun dryer (36±3°C), solar dryer (55±2°C), through flow dryer (60±1°C), vibro dryer (55±1°C) and vacuum dryer (45±1°C).

Estimation of dehydration time of mushrooms using various dryers

Mushroom sample was divided into six equal parts and were dried properly in dehydrators at temperatures indicated above. A sample from each dryer was removed every 30 min. The moisture content was measured by weighing the mushroom samples then placing them in a hot air oven at 80°C and reweighing the samples after they have completely dried.

Estimation of total drying time

The total drying time to reduce the moisture content of mushroom from approximately 86% (d.b.) to < 6% (d.b.) was estimated. Depending on the operating conditions of the dryers, the drying time varied from 4 h 45 min to 18 h.

Enumeration of microbial population from the dried mushroom by total plate count method

Microbial analyses were done by serially diluting the samples and the dilutions were pour plated at a dilution of 10^{-6}.

Estimation of microbial and sensory quality of canned mushrooms

Batch I, Fresh milky mushroom was selected and washed with clean water. The washed mushrooms were filled into cans, covered with various concentrations of brine solution containing citric acid and NaCl and pre heated at 85°C for 5-7 min, and then the cans were sealed immediately and sterilized at 116°C for 35 min, then cooled and stored at ambient temperature.

Batch II, mushroom was placed in cans containing various concentrations of sodium chloride and calcium ions. Similar canning procedure was performed. The microbial quality was analyzed with plate count technique.

Preparation of mushroom jam

The washed mushrooms were pulped with mortar and pestle; and the crude mixture pulp was sieved through an 80 nylon mesh.

Pulps from apple and banana were processed into fruit jams according to the FAO guideline (FAO, 1997). Along with that natural pectin (guava), synthetic pectin, sea weed were tested as different gelling agent. Guava pectin was extracted by boiling the guava pieces (one part with two parts of water) for 30 min. Mushroom pulp was 33-45 parts, along with mixture of fruits such as apple and banana were in 50% of the total mushroom for the preparation of jam. Total soluble solids (TSS) and total titratable acidity (TTA) were determined in the pulp before jam preparation to establish any additional requirement to meet the recommended acid to sugar ratio. After analyzing sugar ratio citric acid and sodium chloride were added to the mixture. The mixture was processed to the desired TSS of 69% and checked by using a hand refractometer (Hand refractometer 58 to 90° Brix). Finally, the jam was mixed with class II preservative sodium benzoate (100 ppm). Later, the hot jam was filled into sterilized glass bottles and left to cool and aluminum foiled. Then they were stored at room temperature (27 to 32°C). Storage stability of the prepared jams was assessed by following their changes in total soluble solids using refractometer, titratable acidity using AOAC standard method by titrating 10 ml of the pulps and juices against 0.1 M NaOH standard solution using phenolphthalein indicator (AOAC, 2000) and total microbial plate microbiological analysis was performed according to the FAO Manual (FAO, 1995).

Preparation of mushroom squash

Calocybe indica was washed with clean running water to remove external dust particles and to reduce the microbial load on the surface of the mushroom. Mushroom was crushed in screw type juice extractor machine for extraction of juice. Amla (Indian gooseberry) juice and lemon juice were added along with the extract. The juice was filtered through muslin cloth. A calculated quantity of potassium meta-bi-sulphite (750 ppm) was dissolved in a part of water according to the treatments and well mixed in the blended juice with the help of stirrer. Sugar syrup made with the addition of citric acid, sugar syrup and water was added to the extracts at different proportion. Squash was made by mixing the extract of *C.indica*, along with extracts of amla, lemon and sugar syrup in three different formulations. The squashes were diluted with purified RO water until it reached 52° brix in refractometer. The squashes were adjusted to have an acidity level of 0.16. The bottles containing squash were stored at room temperature (28 ± 8°C) for further storage studies and analyzed at 60 days intervals for six months.

Sensory analysis

The sensory quality was analyzed during the storage period such as taste, colour, texture, appearance and overall acceptance.

Table 1. Effects of wash on mushrooms by viable plate count method.

Surface washing agents	Period of storage - log CFU.ml^{-1}			
	0 Days	2 Days	4 Days	6 Days
Control-water	5.9	6.3	8.4	10.1
KMS – 0.25 g/L	4.3	4.7	5.6	7.84
1 g citric acid + 5 g NaCl	4.7	5.17	6.2	8.68
3% H_2O_2	3.9	4.29	5.14	7.19
20 ppm Chlorine	4.5	4.9	5.89	8.24

Table 2. Effects of dehydration by different dryers.

Dryer	Temperature (°C)	Moisture % (db)	Drying time (min)	Plate count log (afterdrying) CFU.ml^{-1}
Cross flow	65±1	5.64	285	3.14
Open Sun	36±3	11.5	1080	7.78
Solar	55±2	4.52	350	3.53
Through flow	60±1	4.31	325	3.72
Vibro	55±1	4.18	375	4.24
Vacuum	45±1	3.74	455	5.65

These qualities were estimated by a five member trained panel having age of 25 to 35 years. The panel consists of three females and two males who were trained to be familiar with sensory properties of fruit squash and jam. The sensory testing method was an acceptance test in which the sensory parameters were scored on a descriptive hedonial scale of 1-6. The sensory parameters investigated included the following: (i) taste, (ii) colour, (iii) texture, (iv) odour and (v) overall acceptance. Descriptions of each score were as follows: 5– like very much, 4– like slightly, 3– neither like nor dislike 2– dislike slightly and 1– dislike very much. Sensory tests were replicated thrice.

RESULTS AND DISCUSSION

Effects of mushroom washing on the shelf life of mushrooms

The mushrooms were washed with various surface sterilizing agents such as 1 g citric acid + 5 g NaCl, KMS - 0.25 g/L, 3% H_2O_2 and 20 ppm chlorine. Washed mushrooms were observed for their shelf life for 6 days period. Microbial loads were significantly reduced to 4.3 log CFU.ml^{-1} in mushrooms treated with potassium meta bisulphate (KMS) (Table 1). Overall, the microbiological results show that the H_2O_2 wash was more efficient in reducing bacteria counts. Previously conducted studies proposed 3% H_2O_2 as efficient in reducing bacterial counts (Ndabikunze et al., 2011). Our research results recommend KMS and H_2O_2 for washing mushrooms, after washing, the water was drain off completely which supports of storage up to six days with minimal contamination.

Effects of dehydration by different dryers on *C. indica*

The total drying time to reduce the moisture content of mushroom from approximately 84 (w.w) to about 3% (d.w.) is summarized in Table 2. Depending on the operating conditions, the drying time varied from 4 h 45 min to 18 h. It was observed that the total drying time decreased upon increasing the temperature for a given drying air velocity and batch size. The drying time also decreased upon increasing the drying air velocity for a given temperature and batch size. For a given drying air velocity and temperature, the total drying time was more for larger batch size, which was expected.

Percentage moisture was proportional to the amount of time spent in the dehydrator. As the amount of time in the dehydrator was increased, the percentage moisture decreased. The results show that mushrooms dried in solar dryer and vacuum dryer showed moisture content (4.52 and 3.74) and drying time (3.53 and 5.65), respectively.

Effect of temperature on microbiological growth in mushroom

The microbial quality of dried mushrooms is shown in Table 2. Open dried mushrooms had a significant higher number (7.78 log CFU.ml^{-1}) of microorganism than all other dehydrated mushrooms and took 18 h to get dried. Increased moisture content of the open dried mushroom was higher when compared to other dryers, which could

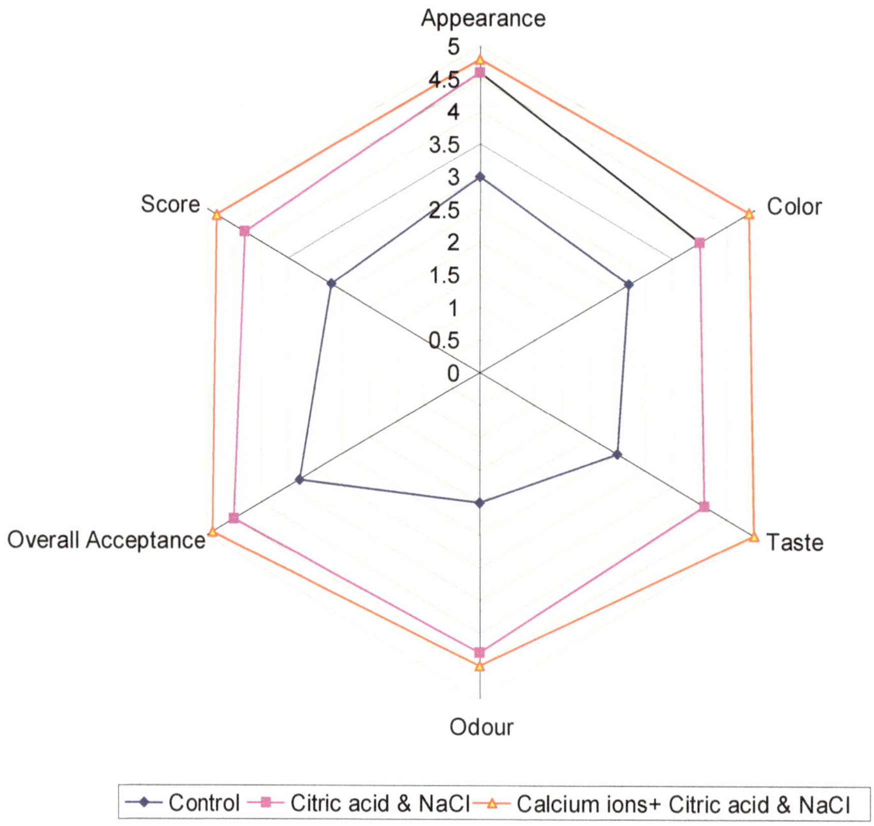

Figure 1. Sensory evaluation of canned mushrooms.

have influenced the microorganism on the dried mushrooms. Least colonies were observed in cross flow driers and solar dryer, 3.14 log and 3.53 log $CFU.ml^{-1}$ respectively. The moisture content of mushroom reached 89.41 to 6.14% (wb) in 8 h of drying in the solar tunnel drier while it took 8 h of drying to bring down the moisture content in a similar sample to 15.0% in traditional sun drying method (Tiisekwa et al., 2002).

Effect of canning on microbial and sensory quality of mushroom

There were no colonies found in the mushrooms canned in brine containing citric acid and NaCl and exhausted, and similarly another batch of mushroom placed in cans containing sodium chloride and calcium ions were sampled frequently which showed no colonies on 4th, 8th and 12th storage months.

Microbiological analysis of glass containers, after 63-91 days, showed that canned recipes were in a condition of commercial sterility (Belitz and Grosch, 1999). Sensory analysis of canned mushroom results showed that the texture, color and taste were retained when the mushrooms were sterilized and canned with calcium ions with sodium chloride (Figure 1).

Effect of storage time on total acidity in mushroom jam

There were no variations in acidity observed during storage period in the present study. Similarly, previous study results pertaining to the titratable acidity revealed that acidity increased during storage. The mean value of acidity recorded were 0.6629 for storage period of 4th, 8th and storage months. The total soluble solids of mushroom jam were 68% during storage at the 12th month. For jam to be kept for a long time, it must have a TSS value of at least 65% (Martinez-carrera et al., 1998; Sapers et al., 1999), pH (3.0 to 3.5) and pectin (0.5%), which was the case with all formulated jams.

Effect of storage time on total microbiological growth in mushroom jam

The microbial quality of jams at different time of storage

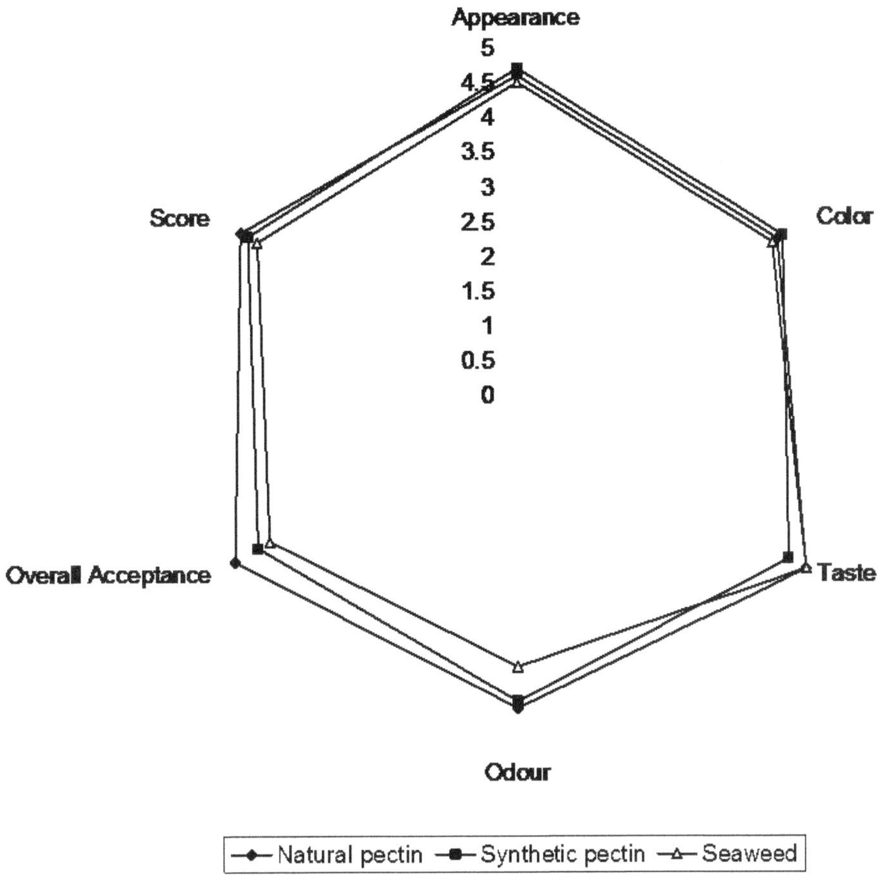

Figure 2. Sensory evaluation of mushrooms jam.

was not changed during its storage. In similar studies, Jam had 2 log CFU.ml^{-1} of microorganism initially. However, there was no significant microbial growth and no coliform growth in jam throughout the storage time. The results shows that jams formulated with additional pectin could be stored for six months at room temperature without spoilage (Deka and Sethi, 2001).

Sensory evaluation of mushroom jams

The sensory evaluation of the formulated jams is presented in Figure 2. In general, addition of natural pectin and synthetic pectin had significant effect on the taste of the jams produced. Jam prepared from sea weed alone had poor odor and a pleasant odor was produced by jam made of natural pectin. The appearance of jam was ranked significantly high among all other jams formulated with synthetic pectin. Fruit jams formulated with natural pectin was highly accepted and could be used as a substitute of the expensive commercial pectin in the preparation of indigenous fruit jams. Similar studies revealed that due to the high content of commercial pectin, the taste was rich (Saka et al., 2007; Manoj et al., 2009).

Enumeration of microbial population from mushroom squash

The data showed negligible count of microbial load (3 Log CFU.ml^{-1}), on mushroom squash. There was negligible growth of moulds and yeasts in lime – aonla and mango- pineapple spiced RTS beverages, which further reduced during storage due to inhibitory effect and antioxidative properties of spices (Bhardwaj and Mukherjee, 2011). Other studies reported that no bacterial growth was observed in the spiced mixed fruit juice, RTS beverages (Kirk and Sawyer, 1991).

Physico-chemical characteristics of squash

Effects on total soluble solids

The TSS (51%) observed in the mushroom squash did

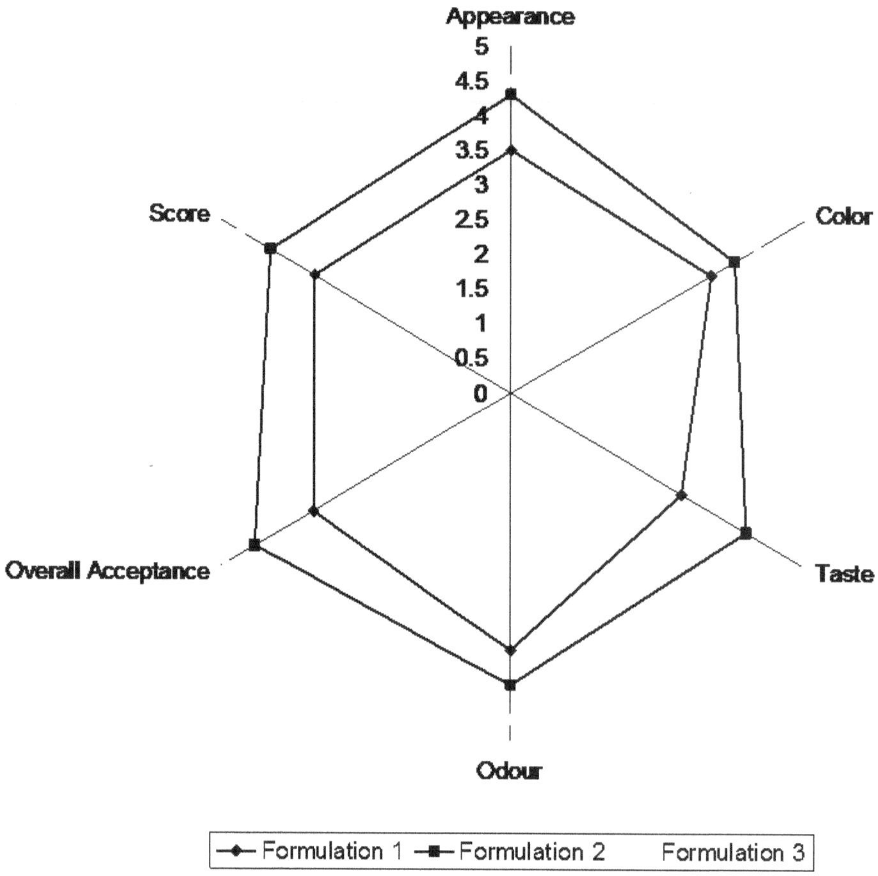

Figure 3. Sensory evaluation of mushrooms squash.

not change throughout its storage period in the current study. The TSS increased with gradual passage of storage time, which might be due to hydrolysis of polysaccharides into monosaccharide and oligosaccharides (Bala et al., 2009). Similar results were also reported in juice blends and found an increasing trend in total soluble solids during storage at ambient and low temperature in lime - aonla and mango-pineapple spiced RTS beverages (Bhardwaj and Mukherjee, 2011; Kirk and Sawyer, 1991).

Effects on acidity

There was a no significant decrease in titratable acidity content during storage. Previous studies on squashes observed that maximum acidity, 0.530% was recorded in the kinnow juice blended with ginger juice and Aonla juice. It was found that the acidity of the RTS beverage prepared from lime-aonla, mango-pineapple, guava-mango blends decreased with addition of spices with advancement of storage period up to six months under different storage conditions (Bhardwaj and Mukherjee, 2011).

Sensory evaluation of mushroom squash

In the present study, three formulations were evaluated for flavour, colour and organoleptic taste scores of squashes during the storage period (Figure 3). The colour, flavor and organoleptic taste of the squashes were found to be superior as for the formulation III when compared with squashes prepared from other formulations. In a similar study, the blend of kinnow juice (87%) + pomegranate juice (10%) + ginger juice (3%) recorded higher score for colour (7.27), flavour (7.43) and organoleptic taste (7.71) as compared to other blends at the end of storage (Bala et al., 2009).

Conclusions

Based on the study conducted, it clearly shows that washing will enhance the fresh mushroom life by removing the microbial load. Potassium meta bisulphite

(0.25 mg/L) and H_2O_2 (3%) showed the maximum removal of microorganisms results with 4.3 and 3.9 log CFU.ml^{-1}, respectively, at an acceptable concentration. On the basis of the above results it can be concluded that drying rate and quality was maximum for solar drier with moisture of 4.52% and 3.53 log CFU.ml^{-1}. When microbial analysis was done for canning there were no colonies observed for any of the sampling during its storage period of one year, similarly, the texture, taste and odor was maximum when the mushrooms were canned with calcium ions and sodium chloride. These results are in accordance with reports that total soluble solids and acidity did not change on storage. The overall acceptability of jam scored high (4.68) with the use of natural pectin. From the results of the present study, it may be concluded that formulation III is possible to satisfy consumer taste and preferences. These juice blends can be stored effectively for a period of six months at room temperature. It is suggested to extend the findings for future application or the main contribution in its applications.

ACKNOWLEDGEMENT

The authors are grateful to Dr. MGR Educational and Research Institute University, Maduravoyal, Chennai, India for supporting this research. Special thanks goes to Dr. P. Sathiya moorthi and P. Ravikumar, Vinzi Sperri, Directorate of Agri Business Development, Tamil Nadu Agricultural University, Coimbatore, Tamilnadu, India.

REFERENCES

Bala BK, Morshed MA, Rahman MF (2009). International Solar Food Processing Conference.

Belitz D, Grosch W (1999). Food chemistry, second edition. Springer-Verag. Berlin German. p 987.

Bhardwaj RL, Mukherjee S (2011). Effects of fruit juice blending ratios on kinnow juice preservation at ambient storage condition. Afr. J. Food Sci. 5(5):281-286.

Deka BC (2000). Ph.D. Thesis, IARI, New Delhi.

Deka BC, Sethi V (2001). Ind. Fd. Packer. pp. 42, 58.

Kirk RS, Sawyer R (1991). Pearson's composition and analysis of foods 9th Edition John Wily and sons Inc. New York. pp. 570.

Manoj Kulshreshtha, Anupama Singh, Deepti and Vipul 2009. Effect of drying conditions on mushroom quality. J. Eng. Sci. Technol. 4(1):90 - 98.

Martinez-carrera D, Sobal M, Aguilar A, Navarro M, Bonilla M, Larque-saavedra A (1998). Canning technology as an alternative for management and conservation of wild edible mushrooms in Mexico. Micol. Neotrop. Appl. 11:35-51.

Ndabikunze BK, Masambu BN, Tiisekwa BPM, Issa-Zacharia A (2011). The production of jam from indigenous fruits using baobab (*Adansonia digitata* L.) powder as a substitute for commercial pectin Afr. J. Food Sci. 5(3):168-175.

Saka J, Isabel R, Akinnifesi F, Victoria NV, Mhango J (2007). Physicochemical and organoleptic characteristics of Uapaca kirkiana, *Strychnos cocculoides*, *Adansonia digitata* and *Mangiferia indica* fruit products. Int. J. Food Sci. Technol. 42:836-841.

Sapers GM, Miller RL, Choi SW, Cooke PH (1999). Structure and composition of mushrooms as affected by hydrogen peroxide wash. J. Food Sci. 64:889-892.

Tiisekwa BPM, Ndabikunze BK, Samson G, Juma M (2002). Proceedings of the Region Agroforestry Conference. Warmbaths, South Africa. p. 68.

17

Chemical evaluation of processed maize (*Zea mays*), indeginous cowpea spp. (*Vigna unguiculata*) and melon seed (*Citrullus vulgaris*) and organoleptic attributes of their blends

Madukwe, Edith U., Okoye, Vivian, Ayogu, Rufina N. B. and Obizoba, Ikemefuna C.

Department of Home Science, Nutrition and Dietetics, University of Nigeria, Nsukka, Nigeria.

The thrust of this work was to evaluate the chemical composition of processed maize, cowpea and melon; and assess the sensory attributes of their blends. The proximate, iron, zinc, tannins and phytate composition of the processed samples were assessed using standard assay methods. Blends were formulated using 14 g protein (average protein requirement for infants 0 to 3 years) basis and subjected to sensory assay. A nine-point hedonic scale was used in assessing sensory attributes. Maize gruel served as the control. The protein, iron and zinc contents of the samples ranged from 2.19 to 35.00%, 3.63 to 5.5 mg/100 g and 0.01 to 0.12 mg/100 g, respectively. Their Phytate: Zinc Molar Ratios (PZMR) were less than 10. Sensory result of the blends was positive. Differences observed within the gruels were not significant (p>0.05). Enriching the maize gruel with either cowpea and/or melon reduced the bulk while meeting the nutritional needs of infants. The use of indigenous legumes (pulses and oil seeds) as complements to the cereal gruel given to infants should be encouraged.

Key words: Complementary diet, gruel, chemical composition, enrichment, organoleptic attributes.

INTRODUCTION

Infant feeding practices in Nigeria is still inappropriate. This has a lot of influence on child survival in developing countries. The traditional complementary infant foods are basically carbohydrate gruels from cereals and root crops. These are deficient in protein (both in quantity and quality). They are bulky and of low nutrient density (Onofiok and Nnanyelugo, 1998). There has been an increased awareness on enrichment of traditional complementary foods of infants and meals of the under fives. Various cereal/legume mixes have been popularized to improve the nutritive value of complementary foods. Soybean and groundnut have received much attention (Nnam, 2002; Opara et al., 2012) with neglect of other locally available and cheap alternatives. Most Nigerian indigenous legumes /oil seeds and nuts which are cheap sources of plant protein are under exploited and are rarely used for infant foods. These crops such as the very tasty local specie of cowpea (*Vigna unguiculata*) called "*oraludi*" by Eastern Nigerians and melon seed (*Citrullus vulgaris*) could be explored as alternatives and nutritionally adequate complementary food.

MATERIALS AND METHODS

Yellow maize (*Zea mays*), dehulled melon (*C. vulgaris*) seeds and a local variety of cowpea (*V. unguiculata*; local name: *oraludi*) used for this work were purchased from Nsukka main market in Enugu State, Nigeria.

Table 1. Ratio of composite blend formulation for sensory evaluation.

Code	Composite blends	Ratio	Gram
AK	Fermented maize (48 h)	100:0:0	319.6:0:0
ND	Fermented maize: boiled cowpea roasted for 15 min	70:30:0	223.75:7.75:0
EO	Fermented maize, boiled cowpea roasted for 20 min	70:30:0	223.75:7.25:0
IS	Fermented maize: boiled cowpea roasted for 15 min: roasted melon seed paste	70:20:10	223.75:5.15:2
RH	Fermented maize: boiled cowpea roasted for 20 min: roasted melon seed paste	70:20:10	223.75:4.83:2
TJ	Fermented maize: roasted melon seed paste	70:30:0	223.75:6:0

AK = Fermented maize (48 h) paste; ND = Fermented maize, boiled cowpea roasted for 15 min (70:30); EO = Fermented maize, boiled cowpea roasted for 20 min (70:30); IS = Fermented maize, boiled cowpea roasted for 15 min, roasted melon seed paste (70:20:10); RH = Fermented maize, boiled cowpea roasted for 20 min, roasted melon seed paste (70:20:10); TJ = Fermented maize: roasted melon seed paste (70:30).

Sample preparation

Six kilograms of maize grains were picked clean, washed and soaked in water in the ratio of 1: 3 w/v for 48 h. Thereafter, it was washed and wet-milled with water into slurry which was sieved using muslin cloth to remove husk. The filtrate was poured into a cotton bag and squeezed to remove excess water. Twenty grams of the sample were used for chemical analysis while the remaining was packed in a polyethylene bag and stored in a freezer (at -20°C) until needed. Two hundred grams of dehusked melon seeds were washed with clean water and par boiled for 10 min such that the water dried up after cooking (to prevent loss of nutrients in the cooking water). The seeds were dried in a hot air oven (Model No 320, Gallenkamp, England) at 80°C for 30 min, roasted for 15 min and milled into fine paste using electric grinder. Twenty grams of the sample were taken for chemical analysis while the remaining portion was stored in polyethylene bag and refrigerated at +4°C until needed. Five hundred grams of cowpea grains were steeped in clean water at room temperature, dehulled by attrition; boiled till soft (but not mushy). Thereafter, it was sun dried for 6 h. The sun dried sample was divided into 2 equal portions. One portion was roasted for 15 min and the other for 20 min. They were separately hammer milled (Model ED-5 Thomas Wiley, England) into fine flour (70 mm mesh screen) and stored in labelled sealed polyethylene bags at room temperature until used.

Formulation of composites

The crude protein of each sample was estimated by the micro-Kjedahl method (Pearson, 1976). The composites were formulated on 14 g protein basis in the ratio of 100.0.0, 70.30.0 and 70:20:10 of fermented maize, boiled roasted cowpea and parboiled roasted melon seeds. Six composites blends were formulated as follows:

i) Fermented maize (48 h) 100:0
ii) Fermented maize, boiled cowpea roasted for 15 min 70:30:0
iii) Fermented maize, boiled cowpea roasted for 20 min 70:30:0
iv) Fermented maize, boiled cowpea roasted for 15 min, roasted melon seed paste 70:20:10
v) Fermented maize, boiled cowpea roasted for 20 min, roasted melon seed paste 70:20:10
vi) Fermented maize, roasted melon seed paste 70:30

Table 1 shows the ratios of the composite blends and the quantity involved.

Chemical analysis

Association of Analytical Chemists (AOAC, 1990) methods were used for chemical analyses. Crude protein was determined by micro- Kjeldahl method; crude fibre by acid hydrolysis; fat by soxhlet extraction method; ash by dry ashing method while carbohydrate was determined by difference. Minerals were evaluated using atomic absorption spectrophotometric method; phytate by Harland and Oberleas (1986) and tannins by Price and Butler (1977).

Preparation of gruel

Composites (blends) were formulated using 14 g protein (average protein requirement for infants 0 to 3 years). For each of the products, the following recipe was used. For the 5 samples, 224.0 g of fermented maize paste (base) was dissolved in 50 ml of water (at room temperature) to form a slurry. Nine hundred millilitres of boiling water was added to the slurry while stirring until it gels (base). To each of the base, appropriate quantity of composites and 5 g of granulated sugar were added and stirred until well distributed. The samples were allowed to cool to 40°C (serving temperature) and separately kept in thermos flask to maintain the serving temperature (at 40°C) for sensory evaluation.

Sensory evaluation

Thirty students were randomly selected by balloting from third and final year students of the Department of Home Science, Nutrition and Dietetics, University of Nigeria, Nsukka. It was based on their previous participation in similar works. A nine-point hedonic scale (Williams, 1981) was used to test for flavour, colour, texture and general acceptability of the gruels. The degree to which the product was liked was expressed as: like extremely (9 points), like very much (8 points), like moderately (7 points), like slightly (6 points), neither like nor dislike (5 points), dislike slightly (4 points), dislike moderately (3 points), dislike very much (2 points) and dislike extremely (1 point). Like extremely to like slightly constituted good; while dislike slightly to dislike extremely constituted poor. Neither like nor dislike showed that the product was neither good nor bad. The sensory assessment was carried out in the Food Research Laboratory of Department of Home Science, Nutrition and Dietetics, University of Nigeria, Nsukka. The laboratory was adequately lit and free from distraction. The judges were arranged in such a way that they could not see the grading of each other. The samples were presented in plain coloured bowls and each judge was provided with a teaspoon and a glass of water to rinse his/her mouth after each testing. The testing was done around 11.00 am and the samples were presented at 40°C (serving temperature) in portions of 250 ml.

Table 2. Proximate composition of fermented maize, boiled cowpea and roasted melon seed paste (%/100 g).

Composite	Moisture (%)	Protein (%)	Ash (%)	Crude fibre (%)	Fat (%)	CHO (%)
$AK_{(48)b}$	46.00±0.05	2.19±0.02	1.50±0.05	0.10±0.02	1.60±0.05	48.61±0.01
BCR_{1a}	8.10±0.03	27.13±0.14	3.15±0.18	2.70±0.05	1.6±0.04	57.32±0.05
BCR_{2a}	8.04±0.04	28.88±0.09	3.00±0.16	2.75±0.01	2.35±0.02	54.94±0.32
PER_a	2.28±0.2	35.00±0.01	3.60±0.01	4.85±0.05	39.95±0.04	14.32±0.04

Mean ± SD of 2 determinations; a = Analysis on a dry weight basis; b = Analysis on a wet weight basis; AK = Fermented maize (48 h) paste, CHO = Carbohydrate; BCR_1 = Boiled cowpea, roasted 15 min (flour); BCR_2 = Boiled cowpea, roasted 20 min (flour); PER = 10 min parboiled melon, roasted 15 min (paste).

Table 3. Iron and zinc content of fermented maize, boiled cowpea and roasted melon seed (mg/100 g dry weight basis).

Mineral	AK_{48}	BCR_1	BCR_2	PER
Iron (mg/100 g)	3.63±0.04	4.33±0.14	4.50±0.04	5.53±0.15
Zinc (mg/100 g)	0.07±0.05	0.01±0.01	0.01±0.02	0.12±0.08

Mean ± SD of 2 determinations; AK = Fermented maize (48 h) paste; BCR_1 = Boiled cowpea, roasted 15 min (flour); BCR_2 = Boiled cowpea, roasted 20 min (flour); PER = 10 min parboiled melon, roasted 15 min (paste).

Table 4. Value of tannins, phytate and phytate zinc molar ratio (PZMR) of fermented maize, boiled cowpea and roasted melon seed paste (mg/100 g dry weight basis).

Variables	AK	BCR_1	BCR_2	PER
Tannin (mg/100 g)	5.60	14.47	14.85	8.81
Phytate (mg/100 g)	0.20	0.56	0.52	0.12
PZMR	0.28	5.05	4.68	0.63

Mean ± SD of 2 determinations; AK = Fermented maize (48 h) paste; BCR_1 = Boiled cowpea, roasted 15 min (flour); BCR_2 = Boiled cowpea, roasted 20 min (flour); PER = 10 min parboiled melon, roasted 15 min (paste); PZMR = Phytate zinc molar ratio.

Statistical analysis

Means, analysis of variance (ANOVA) and Duncan's new multiple range test (DNMRT) were the statistical tools employed. Significance was accepted at $P \leq 0.05$.

RESULTS

Table 2 shows the proximate composition of fermented maize, boiled cowpea and roasted melon seed paste (%/100 g). It showed that PER had the least moisture value of 2% as against BCR_1 and BCR_2 that recorded 8% each. In terms of protein, crude fibre and fat values, PER showed appreciably higher values than BCR_1 and BCR_2. In ash content, PER, BCR_1 and BCR_2 shared similar values but BCR_1 and BCR_2 recorded lower values as against PER. Table 3 reveals the iron and zinc content of the samples. The values for iron ranged from 3.63 to 5.53 mg/100 g. PER had the highest value (5.53 mg) while BCR_1 and BCR_2 shared similar values (4.33 and 4.5 mg/100 g). The range for zinc was 0.01 to 0.12 mg/100 g. PER still had the highest value (0.12 mg/100 g) while BCR_1 and BCR_2 both had 0.01 mg/100 g. AK_{48} showed higher values (0.07 mg/100 g) than BCR_1 and BCR_2. Table 4 presents the tannins and phytate in the samples and their phytate zinc molar ratios. The tannins values ranged from 5.6 to 14.85 mg/100 g. BCR_1 and BCR_2 (14.47 and 14.85 mg/100 g, respectively) showed higher values than the others with AK_{48} having the least value (5.6 mg/100 g). The phytate values ranged from 0.12 to 0.56 mg/100g. BCR_1 and BCR_2 showed higher values (0.56 and 0.52 mg/100 g, respectively) than both PER and AK (0.12 and 0.2 mg/100 g). However, the PZMR of all the samples was less than 10. The range was 0.28 and 5.05. Table 5 shows the sensory scores of the respective samples. In terms of flavour, the score ranged from 6.00 to 7.07. RH had the highest score (7.07) while AK had the least (6.00). However, the flavour scores of the samples showed no significant difference ($p>0.05$). The scores for colour attribute ranged from 6.07 to 7.40. Though, IS had the highest score, the difference in the colour scores was not significant ($p>0.05$). The scores on texture varied. It ranged from 6.93 to 7.37. ND had the highest score (7.37), followed by IS (7.30) and AK had the least (6.93). However, there was no significant difference ($p>0.05$) in the scores. The general acceptability scores ranged from 6.29 to 7.29. RH was the most generally accepted (7.29). It was followed by ND (6.96). Nevertheless, the general acceptability scores of the samples revealed no significance difference ($p>0.05$).

Table 6 shows that 213.6 g maize paste (AK) would be required to meet 1/3 of the 14 g protein requirement of infants 6 to 12 months of age while only approximately 154 g of enriched maize paste would be required to meet

Table 5. Sensory evaluation of traditional fermented maize gruel enriched with either boiled cowpea flour and/or roasted melon seed paste (as consumed).

Composite	Flavour	Colour	Texture	General acceptability
AK	$6.00 \pm 0.34_a$	$7.20 \pm 0.23_a$	$6.93 \pm 0.32_a$	$6.29 \pm 0.4_a$
ND	$6.970 \pm 0.26_a$	$7.33 \pm 0.24_a$	$7.37 \pm 0.2_a$	$6.96 \pm 0.24_a$
EO	$6.57 \pm 0.17_a$	$7.23 \pm 0.24_a$	$7.11 \pm 0.18_a$	$6.54 \pm 0.23_a$
IS	$6.83 \pm 0.28_a$	$7.40 \pm 0.22_a$	$7.30 \pm 0.27_a$	$6.86 \pm 0.30_a$
RH	$7.07 \pm 0.2_a$	$6.93 \pm 0.29_a$	$7.23 \pm 0.24_a$	$7.29 \pm 0.26_a$
TJ	$6.27 \pm 0.37_a$	$6.07 \pm 0.37_a$	$7.00 \pm 0.32_a$	$6.78 \pm 0.31_a$

Mean ± SEM of 30 respondents; AK = Fermented maize (48 h) paste; ND = Fermented maize, boiled cowpea roasted for 15 min (70:30); EO = Fermented maize, boiled cowpea roasted for 20 min (70:30); IS = Fermented maize, boiled cowpea roasted for 15 min, roasted melon seed paste (70:20:10); RH = Fermented maize, boiled cowpea roasted for 20 min, roasted melon seed paste (70:20:10); TJ = Fermented maize: roasted melon seed paste (70:30).

Table 6. Portion (uncooked size) that meets the nutrient requirement of infant (6 to 12 months).

Composite	1/3 daily protein requirement (g)	Medium 10% absorption		Low 5% absorption		Energy requirement for infants 9 to 12 months (%)
		% Fe req•	% Zn req	% FE req	% Zn req	
AK	213.10	70.0	2.5	36.80	1.35	49.50
ND	154.27	49.53	1.73	25.94	0.93	35.99
EO	154.17	49.58	1.73	25.94	0.93	35.99
IS	154.02	49.65	1.73	25.94	0.93	35.99
RH	154.22	49.68	1.73	25.94	0.93	35.99
TJ	153.16	49.89	1.73	25.94	0.93	35.99

Mean ± SEM of 30 respondents; AK = Fermented maize (48 h) paste; ND = Fermented maize, boiled cowpea roasted for 15 min (70:30), EO = Fermented maize, boiled cowpea roasted for 20 min (70:30); IS = Fermented maize, boiled cowpea roasted for 15 min, roasted melon seed paste (70:20:10); RH = Fermented maize, boiled cowpea roasted for 20 min, roasted melon seed paste (70:20:10); TJ = Fermented maize: roasted melon seed paste (70:30); • = requirement.

the same requirement. Under low (5%) absorption, AK would provide 36.8% of iron and 1.35% of zinc. Under medium (10%) absorption, it would furnish 70% iron and 2.5% zinc. For all the other samples, 154 g of each would under medium (10%) absorption provide about 50% of daily iron requirement and 1.73% of zinc. Under low (5%) absorption, 154 g of each of the test samples would provide about 26% of infant daily iron need and 0.93% of zinc. All the samples (154 g) including the control (213 g) would meet one-third energy requirement of infant 9 to 12 months of age.

DISCUSSION

The lower values of protein, ash and crude fibre for fermented maize alone shows that infants should not be fed on maize gruel alone. Enrichment and supplementation should be used to make up for what is missing in this readily available traditional infant food. The higher values seen in enriched maize gruels lend credence to this. The high values of iron in the enriched composites suggests that infants fed on the samples require rich vitamin C sources to enhance iron absorption since non haem iron is not readily available biologically (Yip, 2001). With the addition of vitamin C, these composites can be relied upon to provide the recommended nutrient intake of the infants for iron since at about 6 months of life, iron stores become depleted and the infant needs exogenous iron to cope with the challenges of rapid growth and increase in body size (Dallman; 1980; Yip, 2001). It was also observed that boiled and roasted cowpea is a good source of protein and ash (minerals). However, roasted melon had superior nutritional value. This was probably due to its low moisture content. The lower the moisture values of food, the higher the nutrients per 100 g (Okeke and Obizoba, 1986). Oil seeds are good sources of protein, minerals and fat soluble vitamins (Ene-Obong, 2001; Pamplona-Rogers, 2004; FAO, 2012). The provitamin A in oil seeds have been reported to enhance iron and zinc status. Dietary vitamin A either as preformed or provitamin A increases the bioavailability of inorganic iron by counteracting its inhibition by phytic acid (Solomons, 2001b). This implies that zinc bioavailability will also be improved since its absorption is also impaired by phytic acid.

The high values of protein, crude fibre and fat for boiled cowpea roasted for 20 min shows that roasting for a longer time not only eliminates moisture but also concentrates the nutrients in the food. This is essential in infant feeding because not only that bulk is reduced, nutritionally, the infant fares better. Good zinc status has been associated with enhanced immunity (Dibley, 2001; Solomons, 2001a). These composites have been shown to provide zinc of medium bioavailability since the phytate zinc molar ratios were a little above 5. This implies a lower risk of diarrhoea and other infections of infancy. The high general acceptability scores of the composites were expected. This is a result of the organoleptic attributes of flavour, colour and texture of the composites. This observation was similar to previous observations (Onoja and Obizoba, 2009). The stomach of the infant can take only a small quantity of food at a time. It is important that at each feed, the infant receives adequate quantity of nutrients to meet with its increased body demands. The composites are ideal for this. Only 154 g could provide the required protein and iron for the infants. However, in preparing the feeds, much water should be avoided to prevent creating bulk and reducing the nutrient value.

Conclusion

Enrichment of cereals with local variety of cowpea (*oraludi*) and melon seed has good nutritional implications. This should be incorporated into infant feeding to fight the problem of malnutrition in infants and increase their chances of survival.

REFERENCES

AOAC (1990). Official methods of analysis (15th Ed.). Association of Official Analytical Chemists: Arlington, V.A., USA.

Dallman PR, Siemes MA, Stekel A (1980). Iron deficiency in infancy and childhood. Am. J. Clin. Nutr. 33: 86 – 118.

Dibley MJ (2001). Zinc. In BA Bowman, RM Russel (Eds.), Present knowledge in nutrition (8thEd.). International Life Sciences Institute: Washington DC.

Ene – Obong HN (2001). Eating right. University of Calabar Press: Calabar.

FAO (2012). West African Food composition Table. Stadlmayr B, Charrondiere UR, Enujiugha V, Romaric N, Bayili G, Fagbohoun E, Samb E, Addy P, Barikmo I,Ouattara F, Oshaug A, Akinyele I, Annor GA, Bomfeh K, Ene-Obong H, Smith IF, Thiam I, Burlingame B. (Eds.). FAO: Rome.

Harland BF, Oberleas D (1986). Anion exchange method for determination of phytate in foods: Collaborative study. J. Assoc. Off. Anal. Chem. 69(4): 667 – 669.

Nnam NM (2002). Evaluation of complementary foods based on maize, groundnut, pawpaw and mango flour blends. Nig. J. Nutr. Sci. 23: 1 – 4.

Okeke EC, Obizoba IC (1986). The nutritive value of all vegetable protein diets based on legumes, cereals and tubers in weanling rats. Plant Food Hum. Nutr. 36: 213 – 222.

Onofiok NO, Nnanyelugo DO (1998). Weaning foods in West Africa: Nutritional problems and possible solutions. Food Nutr. Bull. 19(1): 27 – 33.

Onoja US, Obizoba IC (2009). Nutrient composition and organoleptic attributes of gruel based on fermented cereal, legume, tuber and root flour. J. Trop. Agric. Food. Environ. Ext. 8(3): 162 – 168.

Opara B, Uchechukwu IN, Omodamiro RM, Paul N (2012). Production and sensory evaluation of food blends from maize-plantain-soybean as infant complementary food. J. Stor. Prod. Posthar. Res. 3(6): 75 – 79.

Pamplona–Roger GD (2004). Healthy foods. Homles – Calleja Translations: Madrid.

Pearson D (1976). The chemical analysis of foods. Churchill Livingstone, Edinburgh.

Price ML, Butler LG (1977). Rapid visual estimation and spectrophotometric determination of tannin content of sorghum grain. J. Agric. Food Chem. 25: 1268 – 1273.

Solomons NW (2001a). Dietary sources of zinc and factors affecting its bioavailability. Food Nutr. Bull. 22(2): 138 – 154.

Solomons NW (2001b). Vitamin A and carotenoids. In: BA Bowman, RM Russel, (Ed.), Present knowledge in nutrition (8th Ed.). International Life Sciences Institute Press: Washington DC. pp. 127-145.

Williams AA (1981). Scoring methods in the scoring analysis of foods and beverages. Food Technol. 17: 163 – 175.

Yip R (2001). Iron. In: BA Bowman, RM Russell (Eds.), Present knowledge in nutrition (8th Ed.). International Life Sciences Institute Press: Washington DC. pp. 311-328.

18

Effect of heat-moisture-treated cassava starch and amaranth malt on the quality of sorghum-cassava-amaranth bread

Calvin Onyango[1], Eunice A. Mewa[2], Anne W. Mutahi[2] and Michael W. Okoth[2]

[1]Food Technology Division, Kenya Industrial Research and Development Institute, Kenya.
[2]Department of Food Technology and Nutrition, University of Nairobi, Kenya.

Gluten-free batter and bread was prepared from modified cassava starch, sorghum and amaranth flour. Heat-moisture treatment (80°C, 18, 24 or 30% moisture content, and 8, 16 or 24 h incubation) of cassava starch increased its crystallinity, onset pasting temperature and water absorption index; and decreased its swelling power, water solubility index and peak viscosity. Heat-moisture treated cassava starch was made into batter with sorghum and amaranth in the ratio of 50:40:10. The other ingredients, weighed on flour-weight-basis, were water (75%), sugar (6.7%), fat (2%), salt (1.7%), and amaranth malt at 0, 1 or 2.5%. The interaction effect (that is, starch modification x amaranth malt) significantly ($P \leq 0.05$) affected batter consistency. Breads made from heat-moisture treated starch were softer than those containing native starch. The interaction effect (that is, starch modification x amaranth malt) was significant ($P \leq 0.05$) only for the texture profile analysis property of cohesiveness. Further investigation of the effect of heat-moisture treatment showed that the interaction effect (that is, moisture content x incubation time) significantly ($P \leq 0.05$) affected only the Texture Profile Analysis properties of hardness and chewiness.

Key words: Amaranth, cassava, gluten-free bread, heat-moisture treatment, malt, sorghum.

INTRODUCTION

As a continent, Africa is the largest producer of sorghum, with sorghum representing around 70% of the cereals produced in West Africa, 30% in East Africa, and 10% in Southern Africa (Dendy, 1995). More than 252 million metric tons of cassava was produced worldwide in 2011, of which Africa accounted for 56% (FAO, 2011). The prevalence of these crops in Africa is because they are well adapted to hot semi-arid regions, and their ability to resist drought makes them especially valuable foods in these areas which suffers from chronic food shortages. Thus, nutritional intervention through development of processed products, such as bread, from these crops can reach out to millions of people in the region. Viewed from a health perspective, the lack of gluten in these crops makes them safe for people suffering from coeliac disease (Schober, 2009; Schoenlechner et al., 2008).

The manufacture of gluten-free bread is technologically challenging due to the lack of gluten in the gluten-free flours. Consequently, starch is the most important structure-forming ingredient in gluten-free breads (Onyango et al., 2011a; Onyango et al., 2011b; Schober, 2009). Addition of pure starch to gluten-free sorghum bread causes gelatinization to occur more readily and completely, resulting in an early increase in batter viscosity, and facilitating the development of a cohesive and consistent crumb network that traps gas bubbles and prevents loss of carbon dioxide and crust collapse (Schober et al., 2007; Taylor et al., 2006).

It also dilutes the endosperm and bran particles of sorghum which disturb the homogeneity of the starch gel and obstruct uniform gas cell formation in the batter. The amount of starch and its botanical origin (Onyango et al., 2011a) affects the crumb texture of gluten-free bread. Starch modification can also influence crumb quality. Onyango et al. (2011b) found that incorporation of pre-gelatinised starch in sorghum-based gluten-free bread increases crumb adhesiveness: an undesirable crumb property. It is acknowledged that processed foods have better functional properties when appropriately modified starches are incorporated in them (Miyazaki et al., 2006). In this study, we hypothesized that other starch modification techniques, such as heat-moisture treatment (HMT), can improve the quality of gluten-free bread.

Starch modification can be physical, chemical or enzymatic (Miyazaki et al., 2006), but in view of the increasing demand by consumers for additive-free foods, there is growing interest for physically modified starch because it alters starch functionality without introducing foreign substances (Jacobs and Delcour, 1998). Heat-moisture treatment is a form of physical modification of starch, without gelatinization or damage to granular integrity with respect to size, shape or birefringence, through controlled application of heat and moisture (Jacobs and Delcour, 1998). It refers to the exposure of starch to high temperatures, commonly above the gelatinization temperature, at restricted moisture content of about 18 to 27% (Collado and Corke, 1999). Heat-moisture treatment causes molecular reordering of starch polymers leading to changes in its physicochemical properties. For instance, swelling of starch granules decreases (Adebowale and Lawal, 2003; Gunaratne and Hoover, 2002); and onset pasting temperature increases (Gunaratne and Hoover, 2002) whereas the peak viscosity decreases (Singh et al., 2005; Adebowale and Lawal, 2003). The first aim of this work was to determine the functional effect of HMT cassava starch on the rheological properties and crumb texture of sorghum-cassava-amaranth batter and bread, respectively. Amaranth was added to the composite flour mixture because of its good nutritional properties. The seeds are rich in protein and have higher levels of essential amino acids, and better protein digestibility and bioavailability than cereals, such as sorghum (Schoenlechner et al., 2008).

During germination (malting) of grains, hydrolytic enzymes progressively degrade the starch and the protein in the seed endosperm into dextrin, glucose, and amino acids. In gluten-free bread making, malt addition decreeses batter viscosities at both proofing temperature and during heating leading to a decrease in bread density and formation of more open crumb thus improved bread volume (Makinen et al., 2013). However an excessive malt addition will result in excessive liquefaction and dextrinization, yielding bread with a wet sticky crumb (Khalil et al., 2000). In an attempt to further improve the textural properties of bread, this study also focused on evaluating the suitability for use of malted amaranth flour in gluten-free baking.

MATERIALS AND METHODS

Modification of cassava starch by heat-moisture treatment

Cassava starch [11% moisture content (MC), 87.6% starch] was purchased from Universal Starch Public Company Ltd., Bangkok, Thailand. Its MC was adjusted to 18, 24 or 30% w/w using distilled water. The samples were transferred to sealed containers and equilibrated at 4°C for 12 h. Thereafter, they were incubated at 80°C for 8, 16 or 24 h then dried in a cabinet drier at 4°C to about 10% MC.

Physical characteristics heat-moisture treated cassava starch

Damaged starch was determined using the Megazyme Starch Damage Assay kit, K-SDAM 05/2008, (Megazyme International Ireland Ltd., Co. Wicklow, Ireland) and expressed on dry weight basis. Water solubility index (WSI) was determined by adding 1 g into 20 ml distilled water in test tubes and heated in a water bath at 60°C for 30 min. At the end of heating, the mixture was centrifuged at 1,200 x *g* for 20 min and 10 ml of the supernatant decanted and dried to constant weight at 110°C. Water solubility index was expressed as the percent of dried solid weight based on weight of the sample.

Water absorption index (WAI) was determined by weighing 5 g into pre-weighed 50 ml centrifuge tubes. Distilled water (25 ml) was added to each tube, the cap secured and the tubes vigorously hand-shaken for 5 s to suspend the flour. The suspension was allowed to solvate and swell for 20 min with intermittent shaking at 5, 10, 15 and 20 min. The samples were centrifuged at 1,000 x *g* for 15 min after which the supernatant was decanted and pellet drained by touching with an adsorbent-paper towel. Water absorption index was determined as the difference between the weight of the drained tube and the weight of the original empty tube/weight of dry sample.

Swelling power was determined by weighing 0.1 g into pre-weighed test tubes containing 10 ml distilled water. The tubes were heated in a water-bath at 60°C for 30 min with intermittent hand-shaking for 5 s after 5, 15 and 25 min. The tubes were centrifuged at 1,000 x *g* for 15 min and the supernatant decanted and the weight of the starch paste determined. Swelling power was calculated as weight of starch paste divided by the weight of dry starch.

A viscograph-E (Brabender GmbH and Co. KG, Duisburg, Germany) was used to evaluate the pasting profile of the starch. Samples (ca. 27 g) were transferred into a canister and water (ca. 433 ml) was added. They were then heated from 30 to 90°C at 1.5°C/min and held at 95°C for 10 min. The onset pasting temperature (°C) and peak viscosity in Brabender units (BU) were determined.

Starch crystallinity was determined using fourier transform infrared spectroscopy (FT-IR). The FT-IR spectra were obtained on a Nicolet iS10 FT-IR spectrometer (Thermo Fisher Scientific Inc., Massachusetts, USA). About 0.8 mg sample was mixed with potassium bromide (80 mg), homogenized and compressed with a manual press. For each sample, 16 scans were recorded at 25±1°C in the wave number interval of 400-4000 cm^{-1} with a spectral resolution of 4 cm^{-1} and co-added. A background spectrum was recorded and subtracted from the sample spectrum. Spectra were collected and processed using OMNIC 8.2.388 software (Thermo Fisher Scientific Inc., Massachusetts, USA). The spectral region 1200-800 cm^{-1}, which is sensitive to the structural state of starch (van Soest et al., 1995) was truncated and baseline corrected by subtracting a two point linear function touching 1200 and 880 cm^{-1} and the resulting spectra normalised to unit area. Voigt function peaks were fitted at 1047 and 1022 cm^{-1} prior to calculating the ratios of peak heights at 1047/1022.

Preparation and characterization of native and malted amaranth flours

Amaranth (*Amaranthus cruentus*) flour and grains were purchased from Allgrain Company Ltd., Nairobi, Kenya. The grains were soaked in excess water for 6 h after which they were spread thinly on a wire tray and allowed to drain for 6 h at 25°C. Hereupon they were covered with a fine cotton cloth and left to germinate for 30 h. The germinated grains were placed on drying racks and sun-dried to 10% moisture content. The dried malt (together with external roots and shoots) was finely milled to pass through a 0.8 mm sieve.

Moisture content, protein (N x 5.85), fiber, ash, and lipid contents of native and malted amaranth flours were determined according to AACC Method 44-15A, 46-11A, 32-10, 08-01, and 32-25 (AACC, 1995), respectively. Total carbohydrate was determined as the difference between 100 and the sum of values for moisture, lipid, protein, fiber and ash. Starch content was determined using the megazyme total starch assay kit (Megazyme International Ireland Ltd., Co. Wicklow, Ireland). Total soluble sugars were determined by the phenol-sulphuric acid method (Dubois et al., 1956). Reducing sugars were determined using the Nelson-Somogyi alkaline copper reduction method (Somogyi, 1952). Diastatic power (that is, joint α- and β-amylase activity) was determined according to the procedure of American Society of Brewing Chemists (ASBC, 1958). Free amino nitrogen (FAN) was determined by adding 1 g sample to 40 ml 5% trichloroacetic acid at 30°C and extraction carried out for 1 h at 30°C. The tubes were swirled at 15 min inter-vals to suspend the contents. The extract (10 ml) was centrifuged at 4,500 x *g* for 10 min and 1 ml of clear supernatant diluted to 25 ml with distilled water. The samples were subjected to ninhydrin assay according to AOAC Method 10.180 (AOAC, 1980).

Effect of heat-moisture treated cassava starch and amaranth on rheology of gluten-free batter

Composite flour was prepared from modified cassava starch, sorghum (gadam variety; its composition, on dry-weight basis, was: protein, 13.2%; fiber, 2.0%; ash, 1.6%; lipid, 3.6%; and carbohydrates, 79.6%) and amaranth flour at a ratio of 50:40:10. The other ingredients, weighed on flour-weight-basis (fwb), were water (75%), sugar (6.7%), baker's fat (2%), salt (1.7%) and amaranth malt at 0, 1 or 2.5%. The dry ingredients were manually mixed in a wide bowl before adding to the mixing bowl containing water and baker's fat. The ingredients were mixed using a Kenwood KM264 kitchen mixer (Kenwood Limited, Hampshire, England) for 3 min to obtain homogenous batter. The batters were incubated at 30°C for 1 h before measuring their consistencies using an HDP/FE forward extrusion cell of a TA.XT.*plus* Texture Analyzer (Stable Microsystems, Surrey, United Kingdom) equipped with a 50 kg load cell (Onyango et al., 2011a).

Effect of heat-moisture treated cassava starch and amaranth on crumb texture of gluten-free bread

Batter was prepared as described above with the addition of instant active dry yeast (3% fwb). Initially, we investigated the effect of HMT cassava starch and amaranth malt (0, 1 and 2.5% fwb) on the crumb quality of gluten-free bread. Thereafter, we investigated the effect of only HMT cassava starch (that is, incubation time and moisture content) on crumb quality of gluten-free bread and therefore did not add amaranth malt to the formulation. The batters (400 g) were weighed into baking pans and proofed for 15 min at 33°C and 85% relative humidity. The batters were covered with a lid and baked at 210°C for 30 min. The loaves were unmold and cooled for 2 h, packed in moisture-permeable polythene bags, closed with a twist tie and stored for 22 h at 25°C. The loaves were sliced into 10 mm thick slices using a bread slicer (MacAdams Baking Systems, Cape Town, South Africa). Two slices were taken from the centre after 24 h. A ring (30 mm diameter) was punched out from each slice. Texture profile analysis was determined using a TA.XT.*plus* texture analyzer (Stable Microsystems, Surrey, United Kingdom), equipped with a 50 kg load cell (Onyango et al., 2011a).

Experimental design

Experiments on cassava starch modification, batter consistency and crumb texture were designed as multiple factorial designs with three replicates. The data was analyzed using MINITAB 16 software (Minitab Inc, Pennsylvania, USA). Analysis of variance was carried out and the P-value used to determine significance of interaction between variables at $\alpha \leq 0.05$, $\alpha \leq 0.01$ or $\alpha \leq 0.001$. Amaranth flour and malt characterization experiments were done in triplicate. The data was analysed using *t*-test to evaluate the differences in means between native and malted flours. The P-value was used to determine the level of significance at $\alpha \leq 0.05$.

RESULTS AND DISCUSSION

Physical properties of heat-moisture treated cassava starch

The amount of damaged starch in native cassava starch was 0.34%. Damaged starch contents in HMT cassava starch ranged from 0.11% (80°C, 18% MC, 24 h) to 4.27% (80°C, 30% MC, 16 h). Generally, the mean damaged starch content increased with increasing MC and incubation time. Adjustment of MC to 18, 24 and 30% gave HMT starches with 0.11 to 0.24, 0.54 to 0.66 and 2.78 to 4.27% damaged starch, respectively. The interaction effect of % MC x incubation time significantly affected ($P \leq 0.001$) the amount of damaged starch in HMT starch (Table 1). Damaged starch refers to small particles of starch broken away from the main starch granules. Ideally starch damage does not occur during HMT (Jacobs and Delcour, 1998), however, Kawabata et al. (1994) found that in industrially produced HMT starch, the polarized cross of starch granules becomes slightly unclear and the centric hila of the granules are directly damaged due to non-homogeneity of starch. We also speculated that some starch may have been damaged during the post-HMT operations such as drying and milling). Damaged starches hydrate more easily during dough preparation. The level of starch damage therefore improves the water absorption and dough mixing properties of flour and is of technological significance. Damaged starch has much greater water retention capacity; however, too much starch damage leads to sticky dough, strong proofing, and undesirable red crust colour (Bettge et al., 1995).

Water solubility index, WAI and swelling power of native cassava starch were 1.24%, 1.87 g and 9.76 g/g, respectively. Table 1 shows the statistical summary of the effect of HMT of cassava starch on WSI, WAI and swelling power. The interaction effect was significant for WSI ($P \leq 0.001$), WAI ($P \leq 0.001$) and swelling power ($P \leq 0.01$). Water solubility index of HMT starch ranged from

Table 1. F-values of physical properties of heat-moisture treated cassava starch.

Variable	DS (%)	WSI (%)	WAI (g)	SP (g/g)	Onset pasting temperature (°C)	Peak viscosity (BU)
MC (%)	3012.82***	17.09***	968.75***	6.00**	4543.38***	1378.13***
Time (h)	63.11***	13.58***	128.91***	5.10**	529.31***	1125.19***
MC (%) x time (h)	59.47***	6.83***	60.91***	2.92**	175.86***	236.66***
r^2	0.99	0.83	0.99	0.65	0.99	0.99

**Significant at P ≤ 0.01; ***significant at P ≤ 0.001. DS, Damaged starch; WSI, water solubility index; WAI: water absorption index; SP, swelling power; MC, moisture content.

0.33% (80°C, 30% MC, 16 h) to 1.33% (80°C, 18% MC, 8 h). Water absorption index ranged from 1.93 g (80°C, 18% MC, 16 h) to 2.37 g (80°C, 30% MC, 24 h). Starch whose MC were adjusted to 30% exhibited higher WAI (2.12 to 2.37 g) than those with 18 or 24% MC (1.93 to 2.02 g), irrespective of incubation time. Swelling power ranged from 4.60 g/g (80°C, 30% MC, 24 h) to 8.13 g/g, (80°C, 18% MC, 8 h). These swelling power values represented the extremes of the experimental design, whereby the treatment with the lowest MC and shortest incubation time had the highest swelling power, and that with highest MC and longest incubation time had the lowest swelling power. The decrease in swelling power and WSI is attributed to changes in the packing arrangement of starch crystallites and interactions between or among starch components in the amorphous region of the granule (Hoover and Vasanthan, 1994). The properties of the swollen granules and the soluble materials leached out from the granules cooperatively control viscosity parameters during pasting which in turn affects bread density and crumb properties. On the other hand, the increase in WAI in HMT starch implies that hydrophilic character is increased by starch modification. Damaged starch may also be responsible for the increase in WAI in HMT starch. Damaged starch granules absorb about four times as much water as intact starch granules (Stauffer, 2007) and thus an increase in damaged starch increases WAI.

The onset pasting temperature and peak viscosity of native cassava starch was 66°C and 576 BU, respectively. Heat-moisture treatment of cassava starch increased the onset pasting temperature (66.90-75.15°C) and decreased the peak viscosity (175 to 452 BU). The interaction effect significantly (P ≤ 0.001) affected the onset pasting temperature and peak viscosity (Table 1). These findings are similar to other studies in which starches from other botanical origins were investigated (Watcharatewinkul et al., 2009; Olayinka et al., 2008). The increase in onset pasting temperature and decrease in peak viscosity reflect molecular reorganization as a result of HMT. As shown in the FT-IR data, HMT enhances perfection of starch crystallites. This is possibly initiated by incipient swelling and the resulting mobility of amorphous α-glucans which facilitate ordering of double helices (that is, increased inter- and intramolecular hydrogen bonding) (Lawal, 2005). Consequently, HMT starch requires more heat before structural disintegration and paste formation occurs. Furthermore, the enhancement of crystallinity after HMT limits starch swelling and structural disintegration during heating in the batter which causes difficulty in gelatinizing (Miyazaki et al., 2006), both of which contribute significantly to batter viscosity and bread density.

Fourier-transform infrared spectroscopy was used to probe the molecular order of cassava starch. Band peaks at 1047 cm^{-1} relate to molecular order and crystallinity of starch polymers whereas peaks at 1022 cm^{-1} correspond to the disordered or amorphous phase (van Soest et al., 1995). The ratios of the heights of the bands at 1047 and 1022 cm^{-1} express the amount of ordered starch to amorphous starch (van Soest et al., 1995). The height ratio for native starch at 1047/1022 was 0.631. Heat-moisture treatment of starch increased the degree of crystallinity to mean ratios of 0.653 to 0.682 (Table 2). Crystallinity was highest in starch treated at 80°C, 30% MC and 16 h and lowest in starch treated at 80°C, 18% MC and 8 h. Incubation time had no significant effect (P > 0.05) on starch crystallinity, whereas increasing MC increased starch crystallinity (P ≤ 0.05). Exposure of starch to high temperatures and restricted MC elevates the glass transition temperature, the trigger for polymeric reorganization within granules (Tester and Debon, 2000). Thus, HMT promotes interaction of polymer chains by initially disrupting the crystalline structure and dissociating the double helical structure in the amorphous region followed by rearrangement of the disrupted crystals (Gunaratne and Hoover, 2002).

Composition of amaranth flour and malt

Table 3 shows the proximate composition, on dry-weight basis, of amaranth flour and malt. The protein (N x 5.85), lipid and ash contents were close to values that have been previously reported for the same species (Menegassi et al., 2011). However, Menegassi et al. (2011) reported higher starch (69%) and fiber contents (13%) for this species. Protein, FAN and diastatic power were higher (P ≤ 0.05) in malted flours, whereas ash and starch contents were higher (P ≤ 0.05) in native flours (Table 3). The biochemical and physiological changes that take place during malting lead to reduction in the levels of protein, fiber, lipid, ash and carbohydrates (Elmaki et al., 1999). These

Table 2. Effect of heat-moisture treatment on cassava starch crystallinity.

Moisture content (%)	Peaks fitted at 1047/1022			Mean
	8 h	16 h	24 h	
18	0.609	0.639	0.655	0.634 ± 0.02^{a}
24	0.669	0.655	0.672	0.665 ± 0.01^{ab}
30	0.680	0.684	0.682	0.682 ± 0.00^{c}
Mean	0.653 ± 0.04^{x}	0.659 ± 0.02^{x}	0.670 ± 0.01^{x}	

Values followed by the same superscript letter (a-c) in the same column and x in the same row are not significantly different at $P \leq 0.05$.

Table 3. Composition of amaranth flour and malt.

Amaranth	Protein*	Fibre	Ash	Lipid	CHO	Starch	TSS	RS	DP	FAN
Flour	13.37 ± 0.41^{a}	5.53 ± 1.92^{a}	2.78 ± 0.05^{b}	6.93 ± 0.70^{a}	71.39^{a}	53.14 ± 0.68^{b}	59.88 ± 10.42^{a}	0.06 ± 12.22^{a}	8.93 ± 2.81^{a}	9.07 ± 8.69^{a}
Malt	15.98 ± 0.29^{b}	4.96 ± 0.51^{a}	2.37 ± 0.04^{a}	6.65 ± 0.18^{a}	70.04^{a}	49.12 ± 0.15^{a}	69.19 ± 11.74^{a}	0.09 ± 30.36^{a}	60.64 ± 19.03^{b}	83.17 ± 19.74^{b}

All values are given as g/100 g dry-matter-basis, except DP as (dry-matter-basis) and FAN as mg/l. *N x 5.85. CHO, Carbohydrate; TSS, total soluble carbohydrates; RS, reducing sugars; DP, diastatic power; FAN, free amino nitrogen. Values followed by the same superscript letter in the same column are not significantly different at $P \leq 0.05$.

changes were not evident in our study probably because malt was prepared from grains that came from a different batch than that used to prepare native flour. Nonetheless, since malt quality is defined primarily in terms of diastatic power and FAN (Dewar et al., 1997), the different values ($P \leq 0.05$) of these parameters were reliable indicators of different flour qualities.

Effect of heat-moisture treated cassava starch and amaranth on rheology of gluten-free batter

Batter was prepared from native or HMT cassava starch, sorghum and amaranth flour in the ratio of 50:40:10. The other ingredients were water (75%), sugar (6.7%), baker's fat (2%), salt (1.7%), and amaranth malt at 0, 1 or 2.5% fwb. Extrusion force of batter containing native starch and 0, 1 or 2.5% fwb amaranth malt was 110.71, 180.21 and 186.53 N, respectively. Extrusion forces of batters containing HMT starches and different levels of amaranth malt ranged between 127.56 and 283.64 N. All batters prepared from cassava starch exposed to HMT at 30% MC gave extrusion forces that exceeded the load cell (that is, > 490 N) of the texture analyser and thus were not included in the analysis. The interaction effect significantly ($P \leq 0.01$) affected the consistency of the batter (Table 4). Batter consistency tended to increase with increasing MC and incubation time of starch modification and amaranth malt content. During HMT, increase in gel firmness has been attributed to the increased cross-linking between starch chains in the particular amylose portion. This allows for the formation of more junction zones in the continuous phase of the gel, resulting in increased gel hardness (Hoover and Manuel, 1996).

Effect of heat-moisture treated cassava starch and amaranth on crumb texture of gluten-free bread

Texture profile analysis properties of hardness, chewiness, cohesiveness and resilience of bread prepared from HMT starch, sorghum, amaranth flour and 0 to 2.5% fwb amaranth malt ranged between 22.71 to 36.11 N, 11.73 to 17.97 N, 0.51 to 0.58 and 0.30-0.36, respectively. None of the independent variables affected crumb springiness ($P > 0.05$). The interaction effect was significant only for the texture profile analysis property of cohesiveness ($P \leq 0.05$, Table 5). Crumb cohesiveness tended to decrease with increasing amaranth malt content at all HMT combinations (data not shown). In order to evaluate the influence of only starch modification on crumb texture, we also baked the bread without amaranth malt. The inter-

Table 4. F-values of extrusion force of sorghum-cassava-amaranth batter.

Variable	F-value
Starch modification	20.23***
Amaranth malt (%)	189.14***
Starch modification x amaranth malt (%)	4.83**
r^2	0.97

**Significant at P ≤ 0.01; ***significant at P ≤ 0.001.

Table 5. F-values of texture profile analysis parameters of sorghum-cassava-amaranth bread modified with amaranth malt.

Variable	Hardness (N)	Chewiness (N)	Cohesiveness[a]	Resilience[a]
Starch modification	7.23***	3.82**	3.14**	2.64*
Amaranth malt (%)	3.30*	Ns	27.92***	26.96***
Starch modification x amaranth malt (%)	ns	Ns	1.98*	ns
r^2	0.78	0.66	0.81	0.82

[a]Dimensionless terms. *Significant at P ≤ 0.05; **significant at P ≤ 0.01; ***significant at P ≤ 0.001; ns not significant at P ≤ 0.05.

Table 6. F-values of texture profile analysis parameters of sorghum-cassava-amaranth bread without amaranth malt addition.

Variable	Hardness (N)	Chewiness (N)
Moisture content (%)	8.95**	ns
Time (h)	21.00***	4.28*
Moisture content (%) x time (h)	6.71**	3.08*
r^2	0.91	0.74

*Significant at P ≤ 0.05; **significant at P ≤ 0.01; ***significant at P ≤ 0.001; ns not significant at P ≤ 0.05.

interaction effect significantly affected the Texture Profile Analysis properties of hardness (P ≤ 0.01) and chewiness (P≤0.05, Table 6). Springiness, cohesiveness and resilience were not affected (P > 0.05) by starch modification.

Texture profile analysis is a useful technique for objecttive assessment of food texture because it measures several parameters (that is, hardness, springiness, cohesiveness, resilience, chewiness, gumminess, fracturability and adhesiveness) from a single test (Bourne, 2002). However, not all parameters may be simultaneously useful. In bread, for example, crumb hardness (or firmness) is the recommended indicator of crumb quality (AACC Method 74-09, 1995) because of its close association with consumer perception of freshness. Hardness of bread increases with storage. This is attributed to the ageing phenomenon or staling which results in an increase in starch crystallinity of the crumb. In this study, hardness was affected (P ≤ 0.01) by HMT (Table 6) and it tended to decrease with increasing MC and incubation time of starch modification (data not shown). Cohesiveness and resilience showed high levels of significance (P ≤ 0.001) when amaranth malt was added to the batter (Table 5) but differences were insignificant (P > 0.05) in treatments without amaranth malt. Crumb cohesiveness and resilience of gluten-free bread is caused by the gel network of starch after HMT. This network could be affected by malt addition due to the high amylase activity. Chewiness is classified as a secondary textural parameter because it is the product of the primary textural parameters of hardness, springiness and cohesiveness (Bourne, 2002). It represents the energy required to masticate a solid food product to a state ready for swallowing. Chewiness was not affected (P > 0.05) by amaranth malt (Table 5) but it was affected (P ≤ 0.05) by HMT (Table 6). Similarly to hardness, chewiness tended to decrease with increasing moisture content and incubation time of starch modification. The decrease in bread hardness and chewiness could be due to formation of a more open crumb.

Conclusion

Heat-moisture treatment caused molecular reordering of cassava starch polymers leading to changes in its physicochemical properties. These changes affected the rheological properties and crumb texture of gluten-free sorghum-cassava-amaranth batter and bread, respectively. Batter consistency tended to increase with increasing moisture content and incubation time of starch modification and increasing amaranth malt content. The most sensitive texture profile analysis parameters were crumb hardness and chewiness, which tended to decrease with increasing moisture content and incubation time of starch modification. Addition of amaranth malt significantly affected crumb cohesiveness and resilience which tended to decrease with increasing level of malt.

REFERENCES

AACC International. (1995). Approved Methods of the American Association of Cereal Chemists, 9th Ed. Methods 44-15A, 46-11A, 32-10, 08-01, 32-25, 74-09. Minnesota.

Adebowale KO, Lawal OS (2003). Microstructure, physicochemical properties and retrogradation behaviour of Mucuna bean (*Mucuna pruriens*) starch on heat moisture treatments. Food Hydrocoll. 17: 265-272.

AOAC. (1980). Official Methods of Analysis. 14th 331 ed. Method 10.180. Washington.

ASBC (1958). Methods of Analysis of the American Society of Brewing Chemists. 6th. ed. Minnesota.

Bettge AD, Morris CF, Greenblat GA (1995). Accessing genotypic softness in single wheat kernels using starch granule-associated friabilin as a biochemical marker. Euphytica 86: 65-72.

Bourne MC (2002). Food Texture and Viscosity. (2nd ed.). London: Academic Press, Chapter 4).

Collado LS, Corke H (1999). Heat-moisture treatment effects on sweetpotato starches differing in amylose content. Food Chem. 65: 339-346.

Dendy DAV (1995). Sorghum and Millets: Chemistry and Technology. St. Paul, MN: Am. Assoc. Cereal Chem. pp.11-26.

Dewar J, Taylor JRN, Berjak P (1997). Determination of improved steeping conditions for sorghum malting. J. Cereal Sci. 26: 129-136.

Dubois M, Gilles KS, Hamilton JK, Rebers PA, Smith F (1956). Colorimetric method for determination of sugars and related substances. Anal. Chem. 28: 350-356.

Elmaki HB, Babiker E, El Tinay AH (1999). Changes in chemical composition, grain malting, starch and tannin contents and protein digestibility during germination of sorghum cultivars. Food Chem. 64: 331-336.

FAO (2011). FAOSTAT database. http://faostat.fao.org/. Rome: Food and Agricultural Organization of the United Nations. Rome. Accessed on 18 April 2013.

Gunaratne A, Hoover R (2002). Effect of heat-moisture treatment on the structure and physicochemical properties of tuber and root starches. Carbohydr. Polym. 49: 425-437.

Hoover R, Manuel H (1996). The effect of heat-moisture treatment on the structure and physico-chemical properties of normal maize, waxy maize, dull waxy maize and amylo-maize V-starches. J. Cereal Sci. 23: 153-162.

Hoover R, Vasanthan T (1994). Effect of heat-moisture treatment on the structure and physicochemical properties of cereal, legume, and tuber starches. Carbohydr. Res. 252: 33-53.

Jacobs H, Delcour JA (1998). Hydrothermal modifications of granular starch, with retention of the granular structure: A review. J. Agric. Food Chem. 46: 2895-2905.

Kawabata A, Takase N, Miyoshi E, Sawayama S, Kimura T, Kudo K (1994). Microscopic observation and X-ray diffractometry of heat-moisture treated starch granules. Starch/Stärke. 46: 463-469.

Khalil AH, Mansour EH, Dawoud FM (2000). Influence of malt on rheological and baking properties of wheat-cassava composite flours. LWT – Food Sci. Technol. 33: 159-164.

Lawal OS (2005). Studies on the hydrothermal modifications of new cocoyam (*Xanthosoma sagittifolium*) starch. Int. J. Biol. Macromol. 37: 268-277.

Makinen OE, Zannini E, Arendt EK (2013). Germination of oat and quinoa and evaluation of the malts as gluten-free baking ingredients. Plant Foods Hum. Nutr. 68: 90-95.

Menegassi B, Pilosof AMR, Areas JAG (2011). Comparison of the properties of native and extruded amaranth (*Amaranthus cruentus* L. – BRS Alegria) flour. LWT – Food Sci. Technol. 44: 1915-1921.

Miyazaki M, Hung PV, Maeda T, Morita, N (2006). Recent advances in application of modified starches for bread making. Trends Food Sci. Technol. 17: 591-599.

Olayinka OO, Adebowale KO, Olu-Owolabi BI (2008). Effect of heat-moisture treatment on physicochemical properties of white sorghum starch. Food Hydrocoll. *22:* 225-230.

Onyango C, Mutungi C, Unbehend G, Lindhauer MG (2011a). Modification of gluten free sorghum batter and bread using maize, potato, cassava or rice starch. LWT – Food Sci. Technol. 44: 681-686.

Onyango C, Mutungi C, Unbehend G, Lindhauer MG (2011b). Rheological and textural properties of sorghum-based formulations modified with variable amounts of native or pregelatinized cassava starch. LWT - Food Sci. Technol. 44: 687-693.

Schober TJ (2009). Manufacture of gluten-free specialty breads and confectionery products. In: E. Gallagher (Ed.). Gluten-free Food Science and Technology. London: John Wiley and Sons Ltd. pp. 130-180.

Schober TJ, Bean SR, Boyle DL (2007). Gluten-free sorghum bread improved by sourdough fermentation: biochemical, rheological and microstructural background. J. Agric. Food Chem. 55: 5137-5146.

Schoenlechner R, Siebenhandl S, Berghofer E (2008). Pseudocereals. In: E. K. Arendt, & F. D. Bello. (Eds.). Gluten-free Cereal Products and Beverages. London: Academic Press. pp. 149-190.

Singh S, Raina, CS, Bawa AS, Saxena DC (2005). Effect of heat-moisture treatment and acid modification on rheological, textural, and differential scanning calorimetry characteristics of sweet potato starch. J. Food Sci. 70: E373-E378.

Somogyi M (1952). Notes on sugar determination. J. Biol. Chem. 195:19-23.

Stauffer CE (2007). Principles of dough formation. In: S. P. Cauvain, 387 & L. S. Young (Eds.). Technology of breadmaking. New York: Springer Science+Business Media, LLC. pp. 299-332

Taylor JNR, Shober TJ, Bean SR (2006). Novel food and non-food uses for sorghum and millets. J. Cereal Sci. 44: 252-271.

Tester RF, Debon JJJ (2000). Annealing of starch – a review. Int. J. Biol. Macromol. 27: 1-12.

van Soest JJG, Tournois H, de Wit D, Vliegenthart JFG (1995). Short-range structure in (partially) crystalline potato starch determined with attenuated total reflectance Fourier-transform IR spectroscopy. Carbohydr. Res. 279: 201-214.

Watcharatewinkul Y, Puttanlek C, Rungsardthong V, Uttapap D (2009). Pasting properties of a heat-moisture treated canna starch in relation to its structural characteristics. Carbohydr. Polym. 75: 505-511.

Assessment of some aspects of phytonutrients of cashew apple juice of domestic origin in Nigeria

Daramola, B.

Department of Food Technology Federal Polytechnic, Ado-Ekiti, Ekiti State, Nigeria.

Phytochemical screening, assessment of some biochemical characteristics and antioxidants indices of juice obtained from red and yellow varieties of cashew apple of domestic origin in Western Nigeria was were done with view to gain insight to its phytonutrients and health maintenance potentials. Assessment revealed the presence of phytochemical of the therapeutic importance principally: total phenolics, flavonoids, tannins and anthraquinones. Quantitative assessment showed that ascorbic acid (mg/mL of juice) ranged from 0.24 to 4.00. Total sugar (mg glucose/mL of juice) ranged from 3.50 to 111.88; titratable acidity (mg citric acid equivalent/mL of juice) ranged from 5.89 to 15.71 and total phenolic content (mg ferulic acid equivalent/mL of juice) ranged from 1.20 to 1.68. Results of reducing power and hydroxyl scavenging power suggest antioxidative endowment.

Key words: Cashew apple juice, domestic origin, phytochemical screening, antioxidant potentials.

INTRODUCTION

Fruits and vegetables are important for good health and certainly good for all age categories because the class of foods forms an important portion of a healthy diet (Zhang et al., 2010). Regarding phytonutrients, both fruits and vegetables contain many hundreds of compounds with potential antioxidant activity, including the antioxidant vitamins C and E. There are also carotenoids, chlorophyll and a wide variety of phytochemicals such as a simple phenolics compounds, flavonoids glycosides and in some foods, complex polymeric tannins (eprocyanidins and gallotannins) (Pellegrini et al., 2007). As a corollary to this, there exist a plethora of literature expressing epidemiological evidence suggesting positive association between consumption of a diet high in fruits and vegetables and decreased of lifestyle diseases (cancer, cardio vascular hypertension, etc) (Agte and Tarwadi, 2005).

Of all the tropical fruits trees in Western Nigeria, cashew tree is the most widely domesticated food plant, ahead of mango and pawpaw trees. The cashew (*Anacardium occidentale* L) is native to tropical America. Originating in Brazil, the plant has become naturalised in many Tropical countries such as Vietnam, India and the African countries such as Nigeria, Tanzania, Ivory Coast, Mozambique and Benin. Like any other fruit juices, cashew apple juice is notably characterised by sugar, organic acids and vitamin C (Akinwale, 2000). Further studies have indicated antitumor, antimicrobial (Cavalcante et al., 2005) and antioxidant activity (Alves et al., 2010) among others. However, the source of the samples is either agricultural research station or commercial farm. Treatments offered in agricultural experimental stations or commercial farms for the sake of yield improvement or disease resistance modifies the naturality of the juice nutrients obtained from such fruits and therefore, its biological value. Like any other biological system, other factors that could cause variation in nutritional and biochemical profile of cashew apple juice are: genetic, edaphic, application of nutrients, climatic, processing method, geographical and pest control. Therefore, it is incorrect to assume the nutrient profile or biochemical compositions of cashew apple juice from structural sources such as experimental stations or commercial farms are same with juice from domestic origin. Added to this, it is inappropriate to rely on imported data regarding cashew apple juice for local applications. Besides, only miniscule amount of cashew apple of domestic origin is consumes as relish. Leaving a larger portion to waste inform ground drop apple due to lack of processing technology. The quantity of the wasted cashew apple could be adequate for cottage juice production or more. Therefore, there is need for quality data on the juice of apple of domestic origin. Such information could be useful for juice quality standardization and nutrient-health promotion campaign.

Hence, the objective of this study was the evaluation of the nutritional and biochemical characteristics of cashew apple juice of domestic origin in Western Nigeria.

MATERIALS AND METHODS

Fully matured yellow and red varieties of cashew apples obtained from cashew tree on the premises of Federal Polytechnic, Ado-Ekiti, Nigeria in April 2011 were used for the expression of juice while the two apples varieties were ripe, and unripe sample was selected for diagnostic study. Juice was extracted by two methods tagged: squeezed and 'blended'. Juice obtained by squeezed method was by manual pressing of apple to yield juice while the other method involve extraction of juice using juice extractor which blended the apple to yield higher obtained using muslin cloth for filtration prior to preservation and clarification treatments.

Each group was pasteurized at a temperature of 65°C for 30 min after the addition of gelatine at 50 mg/100 mL (Nagaraja et al., 2011). Sodium benzoate (0.7 g/L) was added for preservation, cooled at room temperature (28 ± 2°C) and used for subbasement analysis. The juice samples were allowed to stand for overnight in a refrigerator. The clarified juice was filtered using muslin cloth. The quantity of the juice recovered was expressed as percentage of the cashew apple used. Some physical characteristics of the juice was subjectively assessed and described in terms of clarity, cloudiness and colour as adjudged by the eye.

Phytochemical screening

Phytochemical investigations on cashew apple juice/liquor were accomplished as described by Trease and Evans (2002). Briefly, Meyers reagent was used for alkaloids, Molish Test for glycosides, Iron (III) chloride Test, for phenolic , reducing compounds and tannins, Mg-HCl reagent for flavonoids, Born Trager test for anthraquinones, Fehling solution for reducing sugars, Sulphuric acid test for carotenoids and frothing method was for detection of saponins.

Refractive index determination

Refractive index of samples was measured using Abbey refractometer (ABBE 325, ZUZI) and the value recorded without modification.

pH determination

The pH measurement was accomplished by dispensing 50 mL of sample into 100-mL beaker and read directly on a pH meter (omega H.HPx digital pH meter) which has been previously standardized with buffer 4 and 9 solutions. Value was recorded when equilibrium pH was attained.

Titratable acidity determination

Titratable acid was determined by titrating 100 mL of the sample against 0.1 M NaOH using phenolphthalein as indicator of end point, and results were expressed in citric acid equivalent/100 mL of sample.

Total sugar determination

Total sugar in samples was determined colourimetrically using phenol-sulphuric acid method as described by Dubois et al. (1956). Briefly, sample was treated with phenol sulphuric acid and absorbance measured using spectrophotometer at wavelength of 490 nm. Total sugar was calculated using glucose standard curve.

Ascorbic acid determination

Ascorbic acid content was determined by the indophenols method (Ruck, 1969). Ascorbic acid in juice samples was extracted with 0.4 % oxalic acid and the extract titrated with standardized sodium 2,6 dichlorophenol indophenols.

Evaluation of total phenolic content

Total phenolic content was evaluated according to the method described by Taga et al. (1984). Briefly, A 100 µL of Folin-Ciocalteau reagent (2N wrt acid Fluka Chemic AG-Ch-9470 BUCHS) was added to each sample (20 µL) and well mixed after addition of 1.58 mL of water. After 30 s, 300 µL of 2% sodium carbonate solution was added and the sample tubes were left at room temperature for 2 h. The absorbance (A) of the developed blue colour was measured at 750 nm using Unicam Helios and UV/VIS/Spectrophotometer. A plot of A_{750nm} against corresponding concentration was used to calculate phenolic content (mg/g ferulic acid equivalent).

Determination of reducing power

Reducing power of each sample was determined in accordance with the method of Oyaizu (1986). Simply, each sample (1 mg/ mL) in methanol (2.5 mL) was mixed with sodium phosphate buffer (pH 6.6). The buffered sample was mixed with conditioning reagents (1% K_3-Fe-CN_6, 10% TCA, 0.1 % $FeCl_3$) centrifuged, diluted using distilled water and absorbance was measured at 700 nm.

Hydroxyl radical scavenging activity

The hydroxyl radical scavenging activity was determined according to the method of Klein et al. (1991) with little modification. 1 ml of iron-EDTA solution (0.13% ferrous ammonium sulphate and 0.26% EDTA), 0.5 mL of EDTA (0.018 %), and 1 mL of DMSO (0.85% v/v in 0.1 M phosphate buffer, pH 7.4) were added to 2 mL of juice sample, and the reaction was initiated by adding 0.5 mL of 0.22% ascorbic acid. The test tubes were capped tightly and heated on a water bath at 80°C for 15 min. The reaction was terminated by the addition of 1 mL ice cold TCA (17.5% w/v). 3 mL of Nash reagent (75.0 g of ammonium acetate, 3 mL of glacial acetic acid and 2 mL of acetyl acetone were mixed and raised to 1 L with distilled water)

Table 1. Phytochemical characteristics of cashew apple juice .

Sample	PC	RC	Glycosides	Flavonoid	Alkaloid	Saponin	Carotenoid	Anthraquinone	Tannin	RS
CJ-YAB	+	+	-	-	+	-	-	-	+	++
CJ-YAS	++++	+++	+	+	+	-	++	++	++++	++
CJ-RAB	+++	+++	++++	+++	+	-	+	+	++++	++
CJ-RAS	++	++	+	++	+	-	-	+++	++	++
CJ-UYAS	+++	+++	+++	+	+	-	-	++++	++	+
CJ-URAS	++	+++	++	++	+	-	-	++	++	+

was added to all of the test tubes and left at room temperature for 15 min for colour development. The intensity of the yellow colour formed was measured spectrophotometrically at 412 nm against reagent blank. The percentage hydroxyl radical scavenging was calculated as follows: % hydroxyl radical scavenging activity = 1-(difference in absorbance of sample/difference in absorbance of blank) x 100.

Statistical analysis

Statistical comparison of some biochemical characteristics of cashew apple juice samples was accomplished using student s t-test. Difference at p=0.10 was considered to be significant.

RESULTS AND DISCUSSION

Phytochemical screening of cashew apple juice

The result of the phytochemical screening of the cashew apple juice as accomplished in this study is presented in Table 1. All the samples showed positive for phenolics and its sub-class principally, flavonoids, and tannins. This is in agreement with the results of Sivagurunathan et al. (2010). In addition, NNMDA (2006) reported presence of tannins as well as traces of alkaloids in cashew apple fruit. However, no saponins were detected in this study.

Other phytochemicals detected were anthraquinones, glycosides and similar compounds (sugars) at varied quantities. One of the modes of therapeutics activity of phytochemicals is by reduction therefore screened. All the samples showed vivid reducing property. The result of phytochemical screening revealed that cashew apple juice whether obtained from squeezed or blended was endowed with some bioactive components that could exhibit antioxidative activities. Results of reducing power, vitamin C and phenolic content of the juice suggest it can be a source of antioxidant food drink. This is because natural antioxidants can be phenolic compounds (tocopherols, flavonoids, phenolic acids), and nitrogen compounds (alkaloids, chlorophyllics, amino acids, amines, peptides). Others are carotenoids and ascorbic acids (Velioglu et al., 1998) with concomitant biological antioxidant functions in the nature of antimutagenicity, anti-aging, anticarcinogenicity (Cook and Samman, 1996). Pellegrinic et al. (2007) stated that besides minerals and vitamins, also fruits contain other phytochemicals some of which include simple phenolic compounds, organic acids, flavonoids glycosides and complex polymeric tannins. These phytochemicals are characterized with antioxidant potentials and associated health benefits. This assertion informed the phytochemicals screening of juice obtain from cashew apple of domestic origin with view to ascertain its phytochemical endowment.

Selected physical and physicochemical characteristics of the cashew apple juice

Two varieties of cashew apple fruit namely the red and yellow were used in this study. Besides, an additional sample of unripe yellow variety was used. The juice was obtained from the fruit samples by squeeze and blend processes. Selected physical and physicochemical characteristics of the cashew apple juice obtained from the expression processes are presented in Table 2.

All the juice samples obtained by squeeze were of higher clarity in comparison to samples obtained from blends preparations. This is reasonable because blends will facilitate comminution of fibres into the mash and eventually in the juice resulting to cloudy juice. The colour of the obtained juice is a reflection of the colour of the skin of the variety of cashew fruit used. However, colour intensity was influence by the method (squeezed or blended) of juice expression. The juice yield (%) (Table 2) is high for expression obtained using blending in comparison to manual squeeze. This observation is logical as more liquor is liberated by blending from structural and non-structural cells of the fruit apple. However, such yield requires further processing (centrifugation or sieving) to obtain juice of comparable clarity with that obtained using squeeze. The refractive index of all the samples ranged from 1.351 to 1.354. This value could be useful for quality control assessment during preparation of cashew apple juice.

The pH of the juice samples ranged from 3.61 to 3.82, this is similar with an earlier report by Talasila et al. (2011). This classifies cashew apple juice as acid food same class as that of orange (Ihekoronye and Ngoddy, 1985).

Table 2. Physical and Physicochemical characteristics of cashew apple juice.

Sample	Parameters				
	pH	R.I.	Juice yield (%)	Colour	Clarity
CJ-YAB	3.72	1.354	62.58	Light yellow	-
CJ-YAS	3.75	1.354	40.48	Clear	+
CJ-RAB	3.82	1.352	59.79	Pink	-
CJ-RAS	3.70	1.352	43.23	Light pink	+
CJ-UYAS	3.65	1.351	35.23	Milk white	+
CJ-URAS	3.61	1.351	32.40	Milk white	+

CJ-YAB, Cashew juice-yellow apple blended; CJ-YAS, cashew juice-yellow apple squeezed; CJ-RAB, cashew juice-red apple blended; CJ-RAS, cashew juice- red apple squeezed; CJ-UYAS, cashew juice-unripe yellow apple squeezed; CJ-URAS, cashew juice-unripe red apple squeezed.

Table 3. Selected biochemical characteristics of cashew apple juice.

Sample	Biochemical characteristic					
	Ascorbic acid	Total sugar	Titratable acidity	Total phenolic content	Reducing power	Hydroxyl radical scavenging index
CJ-YAB	0.255	7.500	5.891	1.200	100	36.58
CJ-YAS	4.000	95.000	6.480	1.350	80	55.57
CJ-RAB	2.580	96.887	7.534	1.350	90	44.18
CJ-RAS	3.000	111.880	10.799	1.650	100	39.44
CJ-UYAS	0.240	4.000	15.714	1.680	110	78.60
CJ-URAS	0.242	3.500	15.000	1.640	110	76.70

CJ-YAB, Cashew juice-yellow apple blended; CJ-YAS, cashew juice-yellow apple squeezed; CJ-RAB, cashew juice-red apple blended; CJ-RAS, cashew juice- red apple squeezed; CJ-UYAS, cashew juice-unripe yellow apple squeezed; CJ-URAS, cashew juice-unripe red apple squeezed. Ascorbic acid = mg ascorbic acid equivalent/ml of juice; Total sugar = mg glucose equivalent/mL of juice; Titratable acidity = mg citric acid equivalent/mL of juice; Total phenolic content = mg ferulic acid equivalent/mL of juice; Reducing power = mg ferulic acid reducing power equivalent/100mL; hydroxyl radical scavenging = %.

Selected biochemical characteristics of the cashew apple juice

The results of assessed selected biochemical characteristics of the cashew apple juice are shown in Table 3. The ascorbic acid content (mg/mL) ranged from 0.240 to 4.000. The result is similar to earlier report by Nagaraja et al. (2011). However, a cursory look at the values singly revealed highest value for the juice obtained from squeeze processes most especially CJ-YAS, while the least value was obtained from juice from unripe fruit. The trend of result can be explained on the basis that the other samples had lower values probably due to oxidative enzymic action resulting from destruction of cellular structure of fruits as a result of blending used for expelling juice from the fruit.

The total sugar content (mg glucose equivalent/mL) ranged from 3.500 to 111.880; were similar to values reported by Nagaraja et al. (2011). Examination of the values with respect to the method of juice expression showed that blend preparation gave liquor with lower total sugar. This suggests that blending offered the release of enzyme which probably caused glycosylation thereby causing reduction in the number of carbonyl group for sugar reaction. Juice obtained from CJ-UYAS were characterized with lowest amount of sugar; this is reasonable since the fruit was unripe.

Titratable acidity (mg citric acid equivalent/mL) of the sample ranged from 5.891 to 15.714. Although, the literature at my disposal did not give report on acidity of cashew apple juice but the fact that the sample with the lowest pH has the highest value of titratable acidity proved the plausibility of the trend of results obtained in this study. In addition, the preparation from unripe fruit has the highest 15.000 to 15.714 titratable acidity values in comparison to low 5.891 to 10.799 values obtained for ripe samples irrespective of mode of juice expression. More importantly, the trend of result is justified by the fact that fruit acidity decrease with ripeness. Total phenolic content (mg ferulic acid equivalent/mL) of the samples ranged from 1.20 to 1.68 (Table 4). The trend of result shows that higher value of total phenolics for juice was obtained from unripe fruits and juice obtained by squeeze method from red variety. Generally, the juice samples were characterized by tangible amount of total phenolic content.

Table 4. Statistical comparison of some biochemical characteristics of cashew apple juice.

Parameters	Apple variety		Students T test value
	Yellow	Red	
Ascorbic acid	0.24 – 3.40	0.242 – 0.300	0.968
Total sugar	0.004 – 0.095	0.0035 – 0.11188	1.5693
Titratable acidity	5.891 – 15.714	7.534 – 15.000	1.204
Total phenolic content	0.0012 – 0.00168	0.00135 – 0.00164	4.1675*
Reducing power	0.0008 – 0.001	0.0009 – 0.0011	0.794
Hydroxyl radical scavenging	36.58 – 78.60	39.44 – 76.70	0.5048

Value 0.01, 0.05, 0.10 = 31.598, 4.303, 2.92, n = 3.

Two antioxidant indices namely relative reducing power and hydroxyl radical scavenging index were employed to gain insight to health maintenance potentials of the juice. To my knowledge, this has not been reported in literature. The relative reducing power (mg ferulic acid reducing power equivalent/100 mL) of the samples ranged from 80 to 110. All the samples were characterized with good relative reducing power suggesting a high health maintaining potentials. The hydroxyl radical scavenging index (%) of the samples varied from 36.58 to 78.60. The high variance within the hydroxyl radical scavenging index could be explained in terms of many factors in nature of oxidation and reduction that influence hydroxyl radical scavengation which manifest through the biochemical changes in ascorbic acid, titratable acidity and total phenolic compounds. The high value of hydroxyl radical scavenging index for the juice obtained from unripe fruit was probably due to synergistic activity of ascorbic acid, titratable acidity and total phenolic content of the samples.

Statistical analysis of the assessed phytonutrients and antioxidant indices of the juice samples were accomplished using pair comparison students t-test with respect to yellow and red variety, regardless of method of juice expression. Analysis showed that no significant difference was detected in all the parameters evaluated except in total phenolic content in which significant difference exist at p=0.10. The consequence of this result is that either of the fruit variety could be used to prepare juice with no significant difference in the biochemical properties assessed in this study.

Conclusion

Phytochemical and nutritional assessment showed that juiced obtained from two varieties of cashew apple of domestic origin in Western Nigeria is endowed with phytochemical and nutritional constituents that could play a role in health maintenance. As suggested by assessed antioxidant tests. The findings may be useful for quality control and nutrient-health promotion campaign.

REFERENCES

Agte V, Tarwadi K (2005). Antioxidant and micronutrient quality of fruit and root vegetables from the Indian subcontinent and their comparative performance with green leafy vegetables and fruits. J. Sci. Food Agric. 85:1469-1476.

Akinwale TO (2000). Cashew apple juice: its use in fortifying the nutritional quality of some tropical fruits. Eur. Food Res. Technol. 211:205-207.

Alves ER, Brito ES, Tabernero M, Perez-Jimenez J, Ruffino MSM, Saura-Calixto F (2010). Acerola and cashew apple as source of antioxidants and dietary fibre. Int. J. Food Sci. Technol. 45:2227-2233.

Cavalcante AAM, Rubensam G, Erdtmann B, Brendel M, Henriques JAP (2005). Cashew (*Anacardium occidentale)* apple juice lowers mutagenicity of aflatoxin B1 in *S. Typhimurium* TA 102. Genet. Mol. Biol. 28:328-333.

Cook NC, Samman S (1996). Flavonoids: Chemistry, metabolism carprotective effects and dietary sources. Nutr. Biochem. 7:66-76.

Dubois M, Gillies KA, Hamilton JK, Roberts PA, Smith P (1956). Colorimetric method for determination of sugar and related substances. Anal. Chem. 28:350-356.

Ihekoronye AI, Ngoddy PO (1985). Integrated Food Science and Technology for the Tropics. Macmillan, London.

Pellegrini N, Barbara C, Salvatore S, Brenna OV, Galaverna G, Rio DD, Bianchi M, Bennett RN, Brighenti F (2007). Evaluation of antioxidant capacity of some fruit and vegetable foods: efficiency of extraction of a sequence of solvents. J. Sci. Food Agric. 87:103-111.

Klein SM, Cohen G, Cederbaum AI (1991). Production of formaldehyde during metabolism of dimethyl sulphoxide by hydroxyl radical generating system. Biochemistry 20:6006-6012.

Nagaraja KV, Bhuvaneshwari, Swamy KRM (2011). Biochemical characterization of cashew (*Anacardium occidentale* L) apple juice and promace in India. http://www2.bioversity international.org/publiscations/pgrnewsletter. Plant Genet. Resour. 149:9-13.

NNMDA (2006). Nigeria Natural Medium Development Agency: Medicial plants of Nigeria. South-West Zone Vol. 1. Federal Ministry of Science and Technology, VI Lagos, Nigeria.

Oyaizu M (1986). Studies on extracts of browning reaction: Antioxidative activities of extracts of browning reactions prepared from glucosamine. Japanese J. Nutr. 44:307-315.

Ruck JA (1969). Chemical methods for analysis of fruit and vegetable products. SP 50, Summerland Research Station, Department of Agriculture, Canada.

Sivagurunathan P, Sivasankan S, Muthukkaruppan SM (2010). Characteristization of cashew apple (*Anacardium occidentale* L) fruits collected from Ariyalur District. J. Biosci. Res. 1:101-107.

Taga MS, Miller EE, Pratt DE (1984). Chia seeds as a source of natural lipid antioxidants. J. Am. Oil Chem. Soc. 61:928-932.

Talasila U, Vechalapu RR, Shaik KB (2011). Preservation and shelf life extension of cashew apple juice. Internet J. Food Saf. 13:275-280.

Trease GE, Evans WC (2002). Pharmacognosy. 15th Edn Harcourt Publishers, Edinburgh U.K.

Velioglu YS, Mazza G, Gao L, Domah BD (1998). Antioxidant activity and total phenolics in selected fruits, vegetables and grain products. J. Agric. Food Chem. 6:4113-4117.

Zhang XM, Karangwa E, Khizar H, Rao L, Nahimiyimana DS, Foh MBK, Li L, Xia SQ (2010). Optimization of processing parameters for clarification of blended carrot-orange juice and improvement of its carotene content. Adv. J. Food Sci. Technol. 2:268-278.

Extending the use of an underutilised tuber I: Physicochemical and pasting properties of cocoyam (*Xanthosoma sagittifolium)* flour and its suitability for making biscuits

EJOH Shirley Isibhakhomen[1]*, OBATOLU Veronica Adeoti[2], OLANIPEKUN Oyeyoyin Taiwo[2] and FARINDE Elizabeth Oluremi[2]

[1]Department of Human Nutrition, University of Ibadan, Ibadan, Oyo State, Nigeria.
[2]Institute of Agricultural Research and Training (IAR&T), Obafemi Awolowo University. P.M.B. 5029, Moor Plantation, Ibadan, Oyo State. Nigeria.

Cocoyam is one of the important root food crops especially among the low income earners in Nigeria. However, the utilisation options of cocoyam are mainly limited to direct consumption as whole boiled or pounded tuber only, thus making it an underutilised crop. Physicochemical and pasting properties of cocoyam flour, its suitability and acceptability in biscuit making were determined. Red and white cocoyam varieties were processed into flour by sun-drying method. The physicochemical and pasting properties were compared with that of commercial wheat flour (CWF). Cocoyam flour and wheat flour blend in the ratio 10:90, 20:80, 50:50 and 100% were used to prepare biscuits and sensory evaluation was conducted. Blanched cocoyam flours had significantly higher ($p<0.05$) starch and amylose content than the unblanched flours: 21.36 and 21.87% starch; 44.35 and 44.20% amylose, blanched white cocoyam flour (Bl-WCF) and blanched red cocoyam flour (Bl-RCF), respectively; 21.12 and 15.56% starch and 39.36 and 43.39% amylose, unblanched white cocoyam flour (U-WCF) and unblanched red cocoyam flour (U-RCF), respectively. Peak viscosity was significantly higher ($p<0.05$) in blanched flours (262.79RVU and 267.58RVU, Bl-WCF and Bl-RCF, respectively) than in unblanched flours and CWF (161.79RVU, 200.58RVU and 145.54RVU, U-WCF, U-RCF and CWF, respectively). Sensory evaluation showed that biscuits made from two flour blends had high acceptability especially at 10% substitution level.

Key words: Cocoyam flour, physicochemical properties, pasting properties, biscuits, sensory evaluation

INTRODUCTION

Cocoyam (*Xanthosoma sagittifolium)* belongs to the family Araceae. It is one of the six most important root and tuber crops world-wide (Jennings, 1987; Onwueme and Charles, 1994). The corm, cormel and leaves are important source of carbohydrates for human nutrition, animal feed (Ndoumou et al., 1995; Nyochembeng and Garton, 1998) and of cash income for farmers in Asia, Africa, and Latin America (Onwueme and Charles, 1994). It is grown by farmers with small holdings mainly for its edible tubers and is used as subsistence staple in many parts of Nigeria in particular and generally in the arid sub-tropics in Africa. The average production figure for

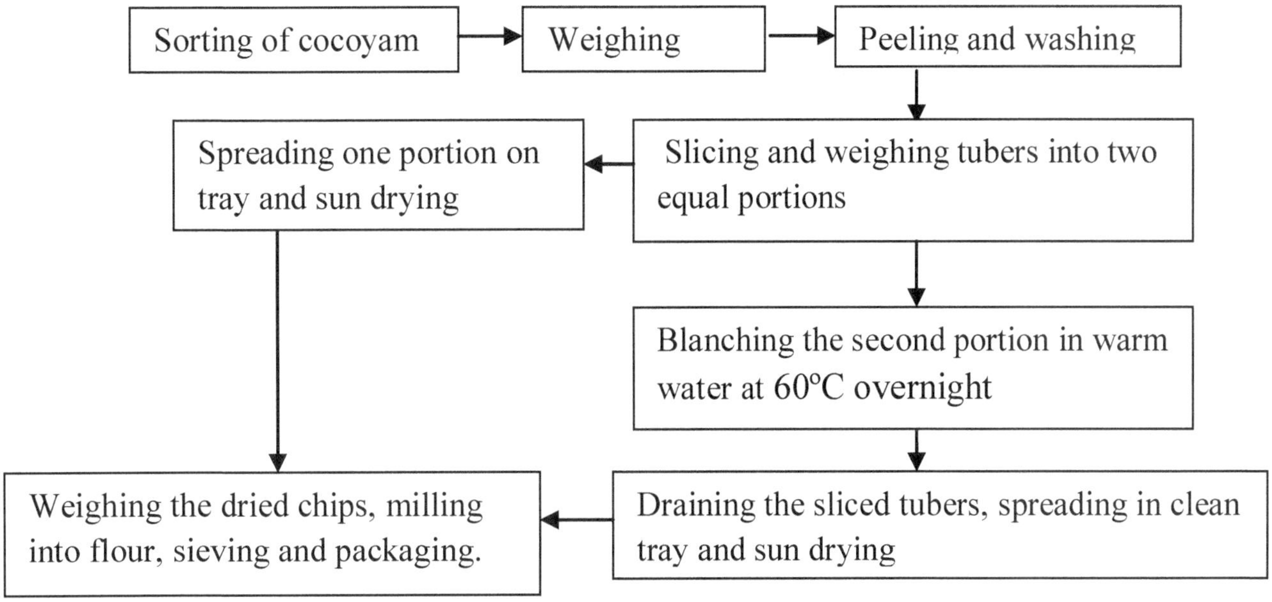

Figure 1. Flow chart for the processing of cocoyam tuber into flour.

Nigeria is 5,068,000 metric tonnes which accounts for about 37% of total world output of cocoyam (FAO, 2007).

In Nigeria, cocoyam is one of the important root food crops especially among the low income earners. However, the use of cocoyam is limited to direct consumption through boiling of the tuber, frying in oil or pounding into *fufu* (a dumpling similar to pounded yam) and eaten with soup and it is also used in thickening soups especially in eastern Nigeria. Cocoyam, despite its uses, is regarded as an underutilized tuber, and an insufficiently studied crop (Watanabe, 2002). This is partly because it suffers very stiff competition from yam and cassava, which is preferred for consumption. The palatability problems of cocoyam: bitter and astringent taste and scratchiness in the mouth and throat, as reported by Sefa-Dedeh and Agyir-Sackey (2004) and Iwuoha and Kalu (1995), has also contributed to its being less accepted and thus categorized as an underutilized tuber crop.

Unlike cocoyam, the extended uses of other tuber crops such as cassava and sweet potato have been extensively investigated especially in flour production. These have been used to make composite flour with wheat flour for a variety of pastry products, which have market value and have contributed to reduced importation of wheat flour (Sanful and Darko, 2010). However cocoyam remains an underexploited tuber crop in this regard. The utilization and market options of cocoyam are very limited due to limited processing and utilisation options. Low quality cocoyam flour and dried chips are the cocoyam products which are traditionally processed for small scale and household use. Hence, cocoyam flour is not yet a commercially traded product in Nigeria. Little adaptive research to substitute wheat flour with cocoyam flour in pastry making has been done. It could be assumed that a partial wheat flour substitution with cocoyam flour like other root tubers and grain flour in snack processing will extend the uses of cocoyam tuber. If adopted, it will generate income for cocoyam farmers and processors, as well as provide varieties of nutritious snack items that will be available and acceptable to consumers.

The aim of the study was to determine the physicochemical and pasting properties of cocoyam flour from two processing techniques, and assess the suitability of cocoyam and wheat flour blend quality for preparing biscuits.

MATERIALS AND METHODS

Collection and preparation of test samples

White and red varieties of cocoyam tubers used in this study were purchased from *Omi* market in Oyo State, Nigeria.

Flour Production

Two types of cocoyam flour were prepared from peeled red and white cocoyam tubers. Each of the varieties were cut into chips, and divided into two equal halves. They were given two treatments: one half was sundried directly before dry milling, while the other half was blanched before sun drying and milling into flour, for white and red cocoyam tubers respectively (Figure 1). The two treatments yielded four different flours: 1) Red cocoyam flour (sundried only); 2) White cocoyam flour (sundried only); 3) Red cocoyam flour (soaked in warm water and sundried) and 4) White cocoyam flour (soaked in warm water and sundried).

Table 1. Flour cocoyam flour/ wheat flour substitution levels (10, 30, 50, 100%).

Sample code	White cocoyam flour (%)	Red cocoyam flour (%)	Wheat flour (%)
A	-	-	100
B	10	-	90
C	20	-	80
D	50	-	50
E	100	-	
F	-	10	90
G	-	20	80
H	-	50	50
I	-	100	-

Laboratory analysis

The cocoyam flour were assessed for their total starch content, amylose and amylopectin content as well as their pasting properties using Rapid visco analyser model RVA Newport Scientific Super Model 3 with a Computer IBM compatible, capable of running RVA control software. The following were recorded: peak viscosity, trough, breakdown, final viscosity, setback from trough, peak time and pasting temperature. Commercial wheat flour was also analysed for the same parameters as control. The physicochemical and pasting properties of the samples were carried out in duplicates.

Experimental design and product development (using the unblanched red and white cocoyam flour)

Ten (10) biscuits constituted the experimental conditions (manipulations). The manipulations (that is substitution levels with wheat flour) were: sample A was produced with 100% all purpose wheat flour (Dangote brand) (250 g); B produced with 90% wheat flour and 10% white cocoyam flour (175 g/25 g); C produced with 70% wheat flour and 30% white cocoyam flour (125 g/7g); D produced with 50% wheat flour and 50% white cocoyam flour (125 g/125 g); 100% white cocoyam flour (250g); E produced with 90% wheat flour and 10% red cocoyam flour (175 g/25 g); F produced with 70% wheat flour and 30% red cocoyam flour (125 g/75 g); G produced with 50% wheat flour and 50% red cocoyam flour (125 g/125 g); I produced with 100% red cocoyam flour (250 g); J produced with commercially sold biscuit (coastal biscuit). The quantities of all other ingredients namely: margarine (125 g), sugar (55 g), salt (2.5 g), baking powder (5 g), grated nutmeg (1.5 g), vanilla essence (5 ml) and water (60 - 65 ml), remained constant.

The biscuits were produced according to the recipe described by Sanni et al. (2006) for cassava cookies with slight modifications. All dry ingredients were first weighed into a clean dry plastic bowl and thoroughly mixed. Margarine was added to the content of the bowl and mixed; finally water was added to achieve non sticky soft dough. The dough was placed on a floured wooden work table, rolled out and cut into round shapes with a biscuit cutter. The surface was pricked with a fork to prevent the dough from rising. The biscuits were baked in a hot oven set at 175°C (350°F) for 30 - 35 minutes.

Sensory evaluation: participants

Eighty five (85) undergraduate students of the University of Ibadan participated as panelist (untrained). Each participant evaluated all 10 biscuit samples (A to J) to rate the sensory attributes of colour, appearance, taste, flavor, texture, and overall acceptability using a 5 - point hedonic scale: 5, like extremely; 4, like slightly; 3, neither like nor dislike; 2, dislike slightly; 1, dislike extremely. Groups of 10 student panelist were used in every instant for the sensory evaluation exercise. Necessary precautions were taken to prevent carryover of flavor during the test by ensuring that panelists rinsed their mouth properly with clean drinking water after evaluating each biscuit.

Statistical analysis

Statistical Package for the Social Sciences (SPSS version 17.0) was used for analysis. One way ANOVA was used to compare the means of the physicochemical and pasting properties of the wheat flour and cocoyam flours. Duncan multiple range test was used for mean separation. Repeated measures one-way ANOVA design (that is within subjects design, whereby each participant evaluated all ten biscuit samples) was used to compare the means for each sensory attribute of the biscuits. The biscuits were the independent variables (IV), while the ratings based on the 5 - point hedonic scale was the dependent variable (DV).

RESULTS

The results of the physicochemical characteristics (sugar, starch, amylose and amylopectin content, swelling capacity and solubility) of cocoyam flours and commercial wheat flour are presented in Table 2. The sugar content of all the samples analysed ranged from 0.205 to 0.852%, unblanched white cocoyam flour (U-WCF) and unblanched red cocoyam flour (U-RCF), respectively. The values obtained for commercial wheat flour (CWF) and U-RCF were not significantly different ($p<0.05$). The starch content of the cocoyam flour samples ranged from 15.563% (U-RCF), to 21.874% blanched red cocoyam flour (Bl-RCF). The starch content of the blanched flours were significantly different ($p<0.05$) from starch content of the unblanched flours. The percentage amylose content of the flour samples showed that CWF, had the least

Table 2. Physicochemical characteristics of cocoyam flour and commercial wheat flour.

Treatment	%Sugar	%Starch	%Amylose	%Amylopectin	Swelling power	%Solubility
CWF	0.79±0.05[a]	21.28±0.15[b]	31.23±0.07	68.77±0.07[d]	9.19±0.74[a]	10.11±1.21[c]
U-WCF	0.21±0.04[d]	21.12±0.08[b]	39.36±0.14	60.64±0.14[c]	10.97±0.28[a]	14.44±0.01[ab]
U-RCF	0.85±0.01[a]	15.57±0.35	43.39±0.14	56.61±0.14[b]	11.36±0.18[a]	16.14±1.08[ab]
Bl-WCF	0.33±0.04[c]	21.37±0.32[ab]	44.35±0.07	55.65±0.07[a]	10.63±2.21[a]	14.01±0.29[b]
Bl-RCF	0.58±0.01[b]	21.88±0.04[a]	44.20±0.14	55.80±0.14[a]	11.74±0.03[a]	14.45±0.13[a]

CWF, Commercial wheat flour; U-WCF, unblanched white cocoyam flour ; U-RCF, unblanched red cocoyam flour; Bl-WCF, blanched white cocoyam flour; Bl-RCF, blanched red cocoyam flour. Means with the same letter in the same column are not significantly different p<0.05.

Table 3. Pasting Properties of cocoyam flour.

Treatment	Peak Viscosity	Trough	Breakdown	Final viscosity	Set back	Peak time	Pasting temp (°C)
CWF	145.54±5.60[d]	84.54±8.78[c]	61.00±3.18[ab]	170.13±7.49[d]	85.59±1.29[c]	5.93±0.18[a]	61.95±0.42[a]
U-WCF	161.80±4.77[c]	155.88±3.71[b]	5.92±1.06[d]	243.88±4.42[c]	88.00±0.71[c]	5.77±0.04[a]	61.73±0.04[a]
U-RCF	200.59±5.54[b]	165.38±5.83[b]	35.21±0.30[c]	270.38±1.70[b]	105.00±4.13[b]	5.04±0.12[b]	61.93±0.39[a]
Bl-WCF	262.79±2.42[a]	197.34±0.64[a]	58.92±3.06[b]	312.75±4.48[a]	108.88±5.13[b]	4.97±0.07[b]	61.78±0.04[a]
Bl-RCF	267.58±5.30[a]	203.88±2.24[a]	70.25±7.54[a]	316.17±2.83[a]	118.84±0.59[a]	4.78±0.12[b]	61.80±0.00[a]

Means with the same letter in the same column are not significantly different p<0.05.

Table 4. Gender and age of participants.

Parameter	Frequency	Percent (%)
Gender		
Male	22	29.3
Female	53	70.7
Age		
<20yrs	41	54.7
20-25yrs	27	36.0
26-30yrs	4	5.3
>30yrs	2	2.7

amylose content (31.23%), followed by U-WCF, which contained 39.36%. Amylose content of Bl-WCF and Bl-RCF, were not significantly different at (p<0.05), but the amylose content of CWF, U-WCF and U-RCF were significantly different at (p<0.05). The swelling power of all the samples were not significantly different (p<0.05).

Table 3 shows the pasting properties of cocoyam flours and commercial wheat flour. The peak viscosity (PV) and final viscosity (FV) of all the samples were significantly different (p<0.05) except for Bl-WCF and Bl-RCF which were not significantly different (p>0.05). Set back between CWF and U-WCF were not significantly different; U-RCF and Bl-WCF, were not significantly different (p>0.05). Peak time for CWF and U-WCF were not significantly different; U-RCF, Bl-WCF and Bl-RCF, were not significantly different (p>0.05).

There were significant differences in breakdown viscosities of blanched and unblanched cocoyam flours (p<0.05). The peak time which is a measure of the cooking time was not significantly different between CWF and U-WCF (p>0.05); U-RCF, Bl-WCF and Bl-RCF, were also not significantly different (p>0.05).

CWF had the highest value (5.93±0.18) for peak time while Bl-RCF had the least value (4.78±0.12). Pasting temperature for all flour samples analysed, were not significantly different from each other (p>0.05); the values ranged from 61.73±0.04°C (Bl-WCF) to 62.95±0.42°C (CWF).

Table 4 shows the gender and age distribution of the participants (untrained panelists): 29.3% were male, and 70.7% were female. About 54.7% of the participants were less than 20 years of age; 36% were between 20 -25 years; 5.3% were 26 - 30 year and 2.7% were above 30 years.

Table 5 shows the mean ratings of the sensory attributes of the biscuit products by 85 untrained panelists. Repeated measures ANOVA was conducted to assess whether there were differences between the average ratings of the sensory attributes of the 10 biscuit products: colour, appearance, flavor , taste, texture and overall acceptability.

The results in Table 5 show that there was significant difference in the ratings of the sensory attributes of the 10 biscuits by the participants: biscuit crust colour; F(7.37,

Table 5. Sensory evaluation of biscuits products from wheat/cocoyam flour blend.

Sample code	Colour	Appearance	Flavor	Taste	Texture	Overall acceptability
A	4.36±0.80	4.08±1.09	4.39±0.85	4.44±0.76	4.36±0.86	4.58±0.61
B	3.75±1.06	3.42±1.16	3.81±0.97	3.86±1.07	4.02±1.06	3.85±0.94
C	3.66±1.04	3.51±1.12	3.56±1.05	3.78±1.08	3.96±0.95	3.73±1.05
D	3.19±1.29	3.16±1.23	3.39±1.17	3.08±1.26	3.26±1.10	3.14±1.05
E	2.68±1.42	2.71±1.33	2.68±1.05	2.01±1.06	3.25±1.14	2.42±1.07
F	3.98±1.02	3.64±1.11	3.85±0.93	3.91±1.01	4.20±0.90	4.00±0.82
G	3.54±1.14	3.34±1.18	3.58±1.10	3.40±1.24	3.72±1.03	3.56±0.97
H	3.59±1.26	3.56±1.28	3.35±1.21	3.13±1.18	3.31±1.25	3.20±1.12
I	3.49±1.49	3.65±1.47	3.26±1.26	2.67±1.37	3.46±1.27	3.07±1.40
J	4.39±0.96	4.35±1.05	3.80±1.14	4.09±1.05	4.65±0.63	4.29±0.84

A, 100%wheat flour biscuit; B, 90%wheat flour/10%white cocoyam flour biscuit; C, 70%wheat flour/30%white cocoyam flour biscuit; D, 50% wheat flour/ 50% white cocoyam flour biscuit; E, 100% white cocoyam flour biscuit; F, 90% wheat flour/ 10% red cocoyam flour biscuit; G, 70% wheat flour/ 30% red cocoyam flour biscuit; H, 50% wheat flour/50% red cocoyam flour biscuit; I, 100% red cocoyam flour biscuit; J, commercially sold biscuit (coastal biscuit).

618.86) = 19.84, p<0.001, eta^2 = .19; appearance; F (7.35, 617.14) = 14.73, p<.001, eta^2=.15; flavor (aroma); F (8.47, 711.44)=19.37, p<0.001, eta^2= .19; taste; F (9, 756) = 41.71, p<0.001, eta^2 = .33; texture; F (8.29, 687.68) = 24.46, p<0.001, eta^2=.23; overall acceptability; F (8.15, 676.33) = 40.29, p<0.001, eta^2 = .33. Apart from biscuit taste (because Mauchly's test of sphericity was not significant, hence sphericity was assumed), all the other sensory attributes were done with Huynh-Feldt correction (because Mauchly's test of sphericity was significant and epilsons was >.75).

The means and standard deviations for the 10 biscuit products were listed in order of highest to least percentage wheat flour content and white cocoyam flour content (Table 1) and highest percentage wheat flour content and red cocoyam flour content (Table 1). Product J is a commercially sold biscuit (coastal biscuit) which served as control.

An examination of these means suggests that participants rated the biscuit products having more wheat flour content better than the biscuit products with higher cocoyam flour content. The polynomial contrasts are indicated in support of this: there was significant quadratic trend for biscuit crust colour and biscuit texture (p<0.001), and there was significant linear and quadratic trend for biscuit appearance, taste, and overall acceptance (p<0.05, p<0.001, respectively). For biscuit flavor, the linear, quadratic and cubic trends were significant (p<0.001, p<0.001, p<0.05, respectively).

Non parametric test

A Friedman test was conducted to assess if there were differences among the mean ranks of the biscuit sensory attribute ratings; biscuit crust colour: χ^2 (9, N = 85) = 127.59, p<0.001; biscuit appearance: χ^2 (9, N = 85) = 115.52, p<0.001; biscuit flavor (aroma): χ^2 (9, N=85)= 146.37, p<0.001; biscuit taste: χ^2 (9, N = 85) = 247.65, p<0.001; biscuit texture: χ^2 (9, N = 84) =193.57, p<0.001; biscuit overall acceptance: χ^2 (9, N = 84) = 258.13, p<0.001.

The results indicate that there were differences among the 10 mean ranks of the sensory attributes and orthogonal contrasts were performed using Wilcoxon tests with Bonferroni correction to ascertain the pair of biscuit products that differed significantly.

In all cases, the significant contrasts show that biscuits that had higher %age of wheat flour content were rated more highly in their sensory attributes. Hence as wheat flour content decreased, the rating decreased. Sensory attributes of colour and appearance for biscuits A and J were not significantly different (p>0.05). Biscuits B, F and J were not significantly different in flavor and taste ratings (p>0.05).

Finally, mixed ANOVA was conducted to determine if there were differences in gender ratings of the sensory attributes of the biscuit products (Table 6). The main effect of the individual sensory attributes of crust colour, appearance, flavor, taste, texture and overall acceptance of the biscuit were significant (p<0.001).

However, gender effect was not significant in all cases (p>0.05), and the interaction between the individual sensory attribute ratings and gender was also not significant (p>0.05).

This implies that although different ratings were given by both males and females, their ratings did not quite differ from one another.

DISCUSSION

The need to improve the value chain of African traditional foods especially those that are underutilised, like coco-

Table 6. Sensory evaluation of biscuit products by gender.

Sample code	Colour		Appearance		Flavor		Taste		Texture		Overall Acceptability	
	M	F	M	F	M	F	M	F	M	F	M	F
A	4.29±0.60	4.40±0.88	3.82±1.09	4.21±1.08	4.36±0.83	4.40±0.86	4.29±0.90	4.51±0.68	4.11±0.96	4.48±0.79	4.39±0.63	4.68±0.58
B	3.86±1.04	3.70±1.07	3.32±1.19	3.47±1.15	3.86±1.00	3.79±0.95	3.89±1.07	3.84±1.08	3.68±1.31	4.20±0.88	3.79±0.88	3.95±0.84
C	3.93±0.94	3.53±1.07	5.00±7.71	3.47±1.17	3.39±1.10	3.65±1.03	3.68±1.16	3.82±1.05	3.89±0.92	4.00±0.97	3.64±1.03	3.78±1.06
D	3.46±1.26	3.05±1.30	3.39±1.26	3.05±1.22	3.32±1.12	3.42±1.19	3.11±1.37	3.07±1.21	3.18±0.98	3.32±1.16	3.04±1.00	3.20±1.09
E	2.71±1.41	2.67±1.43	2.93±1.21	2.60±1.39	2.82±1.06	2.61±1.05	2.36±1.25	1.84±0.92	3.43±1.10	3.20±1.13	2.54±1.00	2.36±1.10
F	4.11±0.88	3.91±1.09	3.75±1.00	3.58±1.16	3.82±1.02	3.86±0.90	3.93±1.12	3.89±0.96	4.11±0.99	4.25±0.86	4.11±0.79	3.95±0.84
G	3.71±1.15	3.46±1.14	3.50±1.17	3.26±1.19	3.57±0.96	3.58±1.18	3.86±1.01	3.18±1.28	3.61±0.99	3.77±1.06	3.89±0.79	3.39±1.02
H	3.75±1.14	3.51±1.31	3.57±1.26	3.56±1.30	3.43±1.00	3.32±1.31	3.21±1.67	3.09±1.20	3.43±1.14	3.29±1.29	3.43±0.96	3.09±1.18
I	3.32±1.31	3.58±1.58	3.50±1.43	3.72±1.50	3.43±1.35	3.18±1.23	2.71±1.51	2.65±1.32	3.29±1.36	3.55±1.23	3.07±1.49	3.07±1.37
J	4.43±0.92	4.37±0.99	4.46±0.88	4.30±1.13	4.11±0.88	3.65±1.23	4.25±0.97	4.02±1.09	4.86±0.36	4.54±0.71	4.68±0.55	4.09±0.90

yam, is timely in recognition that local foods are the basis for the African sustainable diets. Hence the need to explore various options for extending the food uses of many of our traditional food crops.

In the light of the above, the physicochemical and pasting properties of cocoyam flour processed with different methods was determined in order to ascertain its suitability and acceptability in biscuit making. The cocoyam flour properties were compared with wheat flour properties.

Amylose content is one of the factors that influences the clarity of starch pastes, as lower amylose starches are easily dispersed, increasing transmittance and clarity (Mweta et al., 2008). Unblanched white cocoyam flour (U-WCF) had amylose content of 39%, which was closest to that of CWF (31%) (Table 2). This implies that it is more likely to be desirable for use in confectionery food industry.

Higher swelling capacity indicates higher viscosity; BI-RCF had the highest swelling capacity, which also corresponded to its high peak viscosity (Table 3). Although at *p<0.05* the swelling capacity of the flour samples reported in this study were not significantly different (Table 2), but Jangchud et al., 2003, observed that blanching increased the swelling capacity of sweet potato flours. Swelling capacity refers to the expansion accompanying spontaneous uptake of solvent while, solubility index (SI) is the amount of water soluble solids per unit weight of the sample (Omueti et al., 2009). Swelling capacity is regarded as quality criterion in some good formulations such as bakery products (Osungbaro et al., 2010). The solubility of the blanched and unblanched flour samples were significantly different at *p<0.05,* compared to CWF. These solubility values for cocoyam flours in this study fell within the range 10.00 - 26.67 %, reported by Ojinnaka et al. (2009), for cocoyam starch.

Pasting properties

Pasting properties of flours are used to determine their suitability and desirability of baked products in the food industry. These parameters include viscocity, gelatinization time, temperature and retrogradtion tendencies. In this study, BI-RCF had the highest peak viscocity (PV), 267.580 RVU, while CWF had the least PV, 145. 450RVU; followed by U-CWF, 161.795RVU. Peak viscosity (PV) indicates the water-binding capacity of the starch or mixture; it is the ability of the starch to swell before it physically breaks down (Odedeji and Adeleke, 2010). In this study, the high PV of the blanched cocoyam flour corresponded with their high amylose content, and starch content while the unblanched cocoyam flours, which had lower amylose content and starch content, had lower PVs (Tables 2 and 3).

Peak viscosity is often correlated with the final product quality. It also provides an indication of the viscous load likely to be encountered during mixing (Adebowale et al., 2008). This suggests that the blanched cocoyam flours will form a thicker viscous gel on cooking, more than the unblanched cocoyam flours and CWF. Hence the blanched cocoyam flours may be more suitable for products which require high gel strength (Odedeji and Adeleke, 2010). The PV and final viscosities (FV) of the blanched flours were significantly higher than the values obtained for the unblanched flours and the CWF. However, the reverse was reported for potato flours by Jangchud et al. (2003), the peak and final viscosities of blanched flours were reported to be lower than those of unblanched flours.

FV is the most commonly used parameter to define the quality of a particular starch-based sample, as it indicates the ability of the material to form a viscous paste or gel after cooking and cooling as well as the effect of texture modifiers on the physicochemical resistance of the paste to shear force during stirring (Odedeji and Adeleke, 2010). The value, 167.46±8.2RVU, obtained for final viscosity of cocoyam flour (unfermented) by Oke and Bolarinwa (2012), differed slightly from the values obtained in this study for unblanched red and white cocoyam flour (270.38±1.78RVU and 243.88±4.42RVU), respectively.

The trough, which can also be referred to as shear thinning, holding strength, or hot paste viscosity, is the minimum viscosity value in the constant temperature phase of the RVA profile and measures the ability of paste to withstand breakdown during cooling (Ikegwu et al., 2010; Adebowale et al., 2008). There were significant differences between the troughs of blanched and unblanched cocoyam flour. The blanched cocoyam flours had higher holding strength (trough) than the unblanched flours. This means that the blanched cocoyam flours were likely to withstand breakdown during cooling than the unblanched flours.

Breakdown in viscosity to a holding strength (trough) usually occurs when a sample is subjected to a period of constant high temperature and mechanical shear stress. The ability of a mixture to withstand heating and shear stress that is usually encountered during processing is an important factor for many processes especially those requiring stable paste and low retrogradation/syneresis (Adebowale et al., 2008). In this study, the breakdown values obtained especially for the unblanched red and white cocoyam flours, 35.21±0.30 and 5.92±1.06 RVU, respectively, were much lower than the values obtained by Oke and Bolarinwa (2012).

However, the breakdown viscosity for these authors for unfermented and fermented cocoyam flours were closer to the values obtained for blanched red and white cocoyam flours, respectively. Unfermented cocoyam flour, 24 h cocoyam flour and 48 h cocoyam flour, breakdown viscosities were 62.84±5.1, 58.83±1.9 and 66.96±5.3 RVU, respectively (Oke and Bolarinwa, 2012); in the present study, breakdown viscosities for the blanched red and white cocoyam flour were 70.25±7.54 and 58.92±3.06 RVU, respectively. Furthermore the latter values were closer to the breakdown viscosity of the CWF obtained in this study, which was 61.00±3.18RVU. The implication is that Bl-WCF, Bl-RCF and CWF have a better chance of withstanding heating and shear stress during processing.

The "setback" value is related to the amylose content, retrogradation and reordering of starch molecules (Chanapamokkhot and Thongngam, 2007; Jangchud et al., 2003). High set back is associated with syneresis, or weeping, during freeze/thaw cycles (Perten Instruments, 2010). CWF had the lowest set back of 85.585RVU, while Bl-RCF had the highest set back value; 118.835RVU. The set back of U-WCF, 88.000RVU was not significantly different from that of CWF. This suggests that both CWF and U-WCF will have the lowest rate of retrogradation and syneresis.

According to Omueti et al. (2009), low set back is an indication that the starch has a low tendency to retrograde or undergo syneresis. However, the high set back values for cocoyam flour in this study were contrary to those of Oti and Akobundun (2007) results, where they obtained 0.00 setback values for their cocoyam-soybean-crayfish flour blends.

The peak time is a measure of the cooking time (Adebowale et al., 2008). Table 3 shows that the blanched cocoyam flours had the lower peak times than the unblanched cocoyam flours and CWF (5.930): Bl-RCF, 4.775 min and Bl-WCF, 4.97min. This could be attributed to the fact that partial gelatinization had occurred during blanching of the cocoyam chips before processing it into flour. The peak time for unblanched cocoyam flours (5.77±0.04 and 5.04±0.12), respectively was close to the peak time obtained for unfermented cocoyam flour (5.68±0.1 min) by Oke and Bolarinwa (2012).

Pasting temperature is the temperature at the onset of gelatinization (Adebowale et al., 2008). In this study, the pasting temperature ranged from 61.73±0.04 to 61.95±0.42°C; these values for blanched and unblanched cocoyam flours and the CWF were not significantly different. Values for pasting temperature for unfermented and fermented cocoyam flours obtained by Oke and Bolarinwa (2012) were close to the pasting temperature values obtained for unblanched and blanched cocoyam flours in the present study.

Sensory evaluation

Generally, for all the sensory attributes, colour, appearance, taste, flavor, texture and overall acceptance, there was significant statistical difference in the rating within each attribute. This implies that the differences observed were actually due to the manipulations i.e. the substitution levels of wheat and cocoyam flour affected the preference of the panelists.

Colour

In terms of biscuit crust colour, the 100% wheat flour biscuit and the commercial biscuit were the most pre-

Plate 1. Samples of baked wheat/ cocoyam flour biscuits.

preferred. This was closely followed by the 10% red cocoyam flour and white cocoyam flour substituted biscuits, respectively. This preference was similar between the male female participants. The preference for the biscuit crust colour reduced considerably as the quantity of cocoyam flour in the biscuit increased; the biscuits became very dark brown in colour (Plate 1).

The results in Table 5 shows that higher ratings were given to the biscuit products with lower quantities of cocoyam flour substitution. This implies that the crust colour of these biscuit products A (100% wheat flour biscuit), J (commercial biscuit), F (10% red cocoyam flour biscuit), B (10% white cocoyam flour biscuit), and C, (30% white cocoyam flour) were most preferred because they had the highest rating.

Between biscuits A and J, the mean rating was not statistically significant. The mean comparisons between the cocoyam flour substituted biscuits were not significant. When the ratings were examined by gender, there was no statistical difference between the preference for the colour of biscuit crust between male and female participants.

However, the crust colour was given different ratings by both gender in this order of preference: female and male participants' ratings, respectively: A (4.40±0.88), J (4.37±0.99,), F (3.91±1.09, 4.11±0.88), B (3.70±1.07), C (3.53±1.07); male participants' rating: A (4.29±0.60), J (4.43±0.92), C (3.93±0.94) and B (3.86±1.04).

Although the means (Table 5) shows that males rated the colour of the biscuit crust of biscuit C higher than biscuit B and vice versa for the female, these were not statistically significant. Biscuits D (50% white cocoyam flour), E (100% white cocoyam flour), and I (100% red cocoyam flour) were the least preferred.

This could be attributed to the fact that as the cocoyam flour content in the biscuit products increased, the crust colour became darker (Plate 1) thereby making them less desirable.

Appearance

The rating for the biscuit appearance shows that biscuits A (100% wheat flour biscuit), J (commercial biscuit), and C (30% white cocoyam flour biscuit) were the most preferred.

The statistical difference between the mean ratings was also significant. When further statistical analysis was done to separate the means, only biscuits, B and C, F and G, G and H, H and I, were not statistically significant. This time around, none of the biscuits substituted with red cocoyam flour had a favourable rating.

When the observation was disaggregated by gender, though there was no statistical difference observed, mean ratings by male and female were however different: biscuit J (4.46±0.88, 4.30±1.13), A (3.82±1.09, 4.21±1.08), C (5.00±7.71, 3.47±1.17). Hence, the order of preference for male participants is biscuits J, C and A; while the order of preference for female participants is biscuits J, A, and C.

Flavor

The sensory ratings showed that the most preferred products were the 100% wheat flour biscuit, 10% white cocoyam flour, 10% red cocoyam flour substituted biscuits and commercial biscuits. Between male and female participants respectively, the preference for these biscuits were also similar.

Taste

The ratings for taste of the biscuits by the participants showed that 100% wheat flour biscuit, commercial biscuit, and 10% white and red cocoyam flour substituted biscuits, respectively, were the most preferred products. The rating between male and female participants were similar however, an observation of the mean ratings in Table 4 shows that female participants also had a high mean rating for the taste of 30% white cocoyam flour substituted biscuits in addition to the former.

Texture

The mean ratings shows that products from 100% wheat flour, commercial biscuits, 10% red and white cocoyam flour substituted biscuits, respectively and 30% white cocoyam substituted biscuits were the most preferred by the participants. Preference for products between the male and female participants, respectively, showed similar trends. However, the male participants had a low mean rating for the texture of the 10% white cocoyam flour substituted biscuits.

Overall acceptability

The mean rating followed the trend for all other sensory attributes: 100% wheat flour biscuit, commercial biscuit, 10% red and white cocoyam flour substituted biscuits, were the most preferred products overall. This trend was observed to also be similar between male and female participants, respectively. Furthermore, from the observed results, taste of the biscuits had the strongest significant positive correlation with the overall acceptance of the various biscuit products.

The findings of this study are in line with previous similar studies, whereby wheat flour was partially substituted with flours from other sources especially from tubers (Aniedu, 2006; Sanful and Darko, 2010; Adeleke and Odedeji, 2010), in order to extend their uses. It was observed that at substitution levels with white cocoyam flour beyond 30%, preference decreased while for red cocoyam flour, the preference decreased significantly when substitution level exceeded 10%.

Studies by Ojinnaka et al. (2009) showed that the use of cocoyam starch in cookies production at 5 and 10% levels respectively, had desirable sensory attributes and were acceptable by panelists in their study. Adeleke and Odedeji (2010), equally reported that blending sweet potato flour (which is a tuber), with wheat flour up to 20% substitution level, produced samples which can be used for production of confectioneries with improved functional properties. Plantain and wheat flour blend of 20 and 40% partial substitution have also been reported to be suitable and acceptable in making biscuits (Mepba et al., 2007). Sanful and Darko (2010) observed in their study that cassava and cocoyam flour could be used to substitute wheat flour for up to 30% in the production of rock cakes, which are acceptable by consumers.

Conclusion

This study shows that the physicochemical and pasting properties of the unblanched red and white cocoyam flours were similar to those of the commercial wheat flour. Also, partial substitution of wheat flour with cocoyam flour is possible and biscuits made from this combination with up to 10% substitution for both red and white cocoyam flours were acceptable. Hence, the use of cocoyam flour partially substituted with wheat flour should be further explored in pastry making. This will extend as well as increase the utilisation options for this underutilised tuber beyond its current basic use. This diversification in the value chain for cocoyam can also become an increased source of income for cocoyam farmers and processors in Nigeria.

REFERENCES

Adeleke RO, Odedeji JO (2010). Functional properties of wheat and sweet potato flour blends. Pak. J. Nutr. 9:535-538.

Aniedu C (2006). Development of secondary food forms of cocoyams (*Colocasia* spp). J. Agric. For. Soc. Sci. 4(2): 126-130.

Chanapamokkhot H, Thongngam M (2007). The chemical and physico-chemical properties of sorghum starch and flour. Kasetsart J. (Nat. Sci.) 41:343-349

FAO (2007). FAOSTAT Statistics Division of the Food and Agriculture Organization, http://faostat.fao.org.

Ikegwu OJ, Okechukwu PE, Ekumankana EO (2010). Physico-chemical and pasting characteristics of flour and starch from achi *Brachystegia eurycoma* seed. J. Food Technol. 8(2):58-66.

Iwuoha CI, Kalu FA (1995). Calcium oxalate and physicochemical properties of cocoyam (Colocasia esculenta and Xanthosoma sagittifolium) tuber flours as affected by processing. Food Chem. 54:61-66.

Jangchud K, Phimolsiripol Y, Haruthaithanasan V (2003). Physicochemical properties of sweet potato flour and starch as affected by blanching and processing. Starch 55:258-264

Jennings DL (1987). Starch crops. In: CRC Handbook of plant Science in Agriculture. Volume II. (Eds. B. R. Christie). CRC Press, Inc. Boca Raton, Florida, USA. pp. 137-143.

Mepba HD, Eboh L, Nwaojigwa SU (2007). Chemical composition, functional and baking properties of wheat-plantain composite flours. Afr. J. Food Agric. Nutr. Dev. 7(1):1-22.

Mweta DE, Labuschagne MT, Koen E, Benesi IRM, Saka JDK (2008). Some properties of starches from cocoyam (Colocasia esculenta) and cassava (Manihot esculenta Crantz.) grown in Malawi. Afr. J. Food Sci. 2: 102-111.

Ndoumou DO, Tsala GN, Kanmegne G, Balangé AP (1995). *In vitro* induction of multiple shoots, plant generation and tuberization from shoot tips of cocoyam. C. R. Acd. Sci. Paris, Sciences de la vie/ life Sciences 318:773-778.

Odedeji JO, Adeleke RO (2010). Pasting characteristics of wheat and

sweet potato flour blends. Pak. J. Nutr. 9 (6): 555-557.
Ojinnaka MC, Akobundu ENT, Iwe MO (2009). Cocoyam starch modification effects on functional, sensory and cookies qualities. Pak. J. of Nutr. 8(5):558-567.
Omueti O, Otegbayo B, Jaiyeola O, Afolabi O (2009). Functional properties of complementary diets developed from soybean (*Glycine max.*), groundnut (*Arachis hypogea*) and crayfish (*Macrobrachium Spp*). EJEAFChe. 8(8):563-573.
Onwueme IC, Charles WB (1994). Cultivation of cocoyam. In: Tropical root and tuber crops. Production, perspectives and future prospects. FAO Plant Production and Protection Paper 126, Rome. pp. 139-161.
Osungbaro TO, Jimoh D, Osundeyi E (2010). Functional and pasting properties of composite Cassava-Sorgum flour meals. Agric. Biol. J. N. Am. 1(4):715-720
Oti E, Akobundu ENT (2007). Physical, functional and amylograph pasting properties of cocoyam-soybean-crayfish flour blends. Niger. Food. J. 25(1):161-170.
Sanful RE, Darko S (2010). Production of Cocoyam, Cassava and Wheat flour composite rock cake. Pak. J. Nutr. 9(8): 810-814.
Sanni LB, Maziya-Dixon AO, Onabolu BE, Arowosafe AE, Okoruwa RU, Okechukwu AGO, Dixon ADI, Waziri IP, Ezedinma C, Ssemakula G, Lemchi J, Akoroda M, Ogbe F, Tarawali G, Okoro E, Geteloma C (2006). Cassava recipes for household food security. IITA Integrated Cassava Project, Ibadan, Nigeria.
Watanabe KZ (2002). Challenges in biotechnology for abiotic stress tolerance on root and tubers. JIRCAS Working Reports 75-83.

Effect of wheat flour substitution, maize variety and fermentation time on the characteristics of *Akara*, a deep oil fried dough product

Kameni A.[1] , Kouebou C.[2], Aboubakar D. A. K.[2] and C. The[1]

[1]IRAD Nkolbisson, PO Box 2067, Yaounde, Cameroon.
[2]Food Technology Unit, IRAD Garoua, Box 415, Cameroon.

An investigation into the possibility of using local cereal resources (maize) to develop composite flour suitable for the production *of Akara* (a deep fried fermented dough) was undertaken. The experimental design was a 4 x 5 x 5 factorial experiment. Factor one was the maize variety (CMS 8806, CMS 8501, CMS 9015 and CMS 8704), factor two the substitution level (0, 10, 20, 30 and 40%) of wheat for maize flour and factor three was the fermentation times (0, 30, 60, 90 and 120 min). Dough nuts were produced from composite flour from wheat and maize flour. For the production of maize fine flour, wet milling of de-hulled, soaked grain was made followed by drying and sieving. Composite flours (32.4%) and other ingredients (13.5% sugar, 0.5% instant bakers' yeast and 0.3% iodised salt) were mixed with water, whipped for 10 min and cut into samples of 40 g which were manually shaped into balls. They were allowed to ferment and rise prior to deep frying. Flour samples were analyzed for their physico-chemical and functional properties. A mixed panel of women operating in dough nut production and some regular consumers was trained for sensory evaluation. The sensory attributes considered were colour, taste, texture and overall acceptability on a five point scale. Results showed that maize cultivars did not affect the quality of fine flours from wet milling. In composite flours, proteins, lipids and ash contents were significantly reduced as the level of maize flour increased. An inverse trend was observed with water absorption capacity. Dough swelling capacity and *Akara* density decreased with increase in maize substitution levels independently of the fermentation time and maize variety. Similarly, sensory scores decreased with increases in the substitution level independently of the maize cultivar used for composite flour preparation. However, *Akara* from composite flours had overall acceptability scores between 4.2 and 2.8 on a five-point scale. *Akara* prepared from CMS 8806 and CMS 8704 all of yellow color showed the highest sensory scores. From the results obtained, it could be suggested that composite flour from wheat and maize with up to 40% substitution level is acceptable for local commercial production of *Akara* with preference given to yellow maize.

Key words: Maize, flour, doughnut, characteristics.

INTRODUCTION

Traditional cereal foods play an important role in the diet of the people of Africa, particularly in cereal producing zones. Flour from various cereals is one of the main raw material used in the production of popular food products with high acceptability, good storage characteristics and affordable cost. One of such food product is *Akara*, a deep oil fried fermented dough, produced by women, using simple processing methods. It is a breakfast food largely consummed by children and adults in West and Central Africa, especially in Cameroon (Cerdan et al., 2004). Initially, dough for *Akara* was made from 100% imported wheat flour exposing production to price fluctuation due to importation and transportation costs. In an attempt to minimize production costs in rural areas, village women have developed composite flours by incorporating various amounts of rice or maize flour into wheat flour for *Akara* production. This has led to a lot of variability in taste, texture, size and appearance in the *Akara* from one producer to another and from one village to another. Studies conducted to develop composite flour suitable for bread production suggested that bread of acceptable characteristics could be obtained from composite flour with up to 40% substitution (Tiekoura, 1994; Keregero and Mtebe, 1994; Iwuoha et al., 1997). Bread with 40% or less sorghum composite flour had scores of relative acceptance above 70% as compared to the standard 100% wheat bread. From above 40% sorghum composite flour, bread produced received low acceptability scores of 6.9 on a nine-point scale (Carson et al., 2000). With maize flour, incorporation rate of up to 15% was shown to be acceptable suggesting that the type of cereal used affects the optimum acceptable level of flour incorporated. However, these results have had little application among bakers because the relative benefit from maize composite flour does not justify the extract expenses, time and production adjustments needed. Doughnut producers on the other hand are voluntarily adding various amounts of cereal flours in their dough to obtain besides *Akara*, different types of fried dough products *Cen-cen*, *Taara-pott-en, Wardi* and *Wayna* (Lopez and Muchnik, 1997).

The characterization and standardisation of the use of composite flour for doughnuts would strengthen and encourage its production. Based on availability, maize flour is probably the cheapest possible ingredient to be considered. However, the physico-chemical composition of maize grain from various cultivars varies considerably (FAO, 1993; IRA-NCRE, 1990) and this may affect the functionality of resulting maize flour. The method by which flour is produced (Watson, 1984; Munck, 1995) should equally be considered in developing composite flours. This study was designed to evaluate the effects of wheat flour substitution rate on composite flours from maize cultivars on the properties of *Akara*.

MATERIALS AND METHODS

Maize grain and wheat flour sources

Maize (*Zea mays* L.) grains used were obtained from the collection of Cameroon Maize Selection (CMS) of the maize breeding program of the Institute of Agricultural Research for Development (IRAD), Cameroon. Four leading cultivars were considered based on their high adoption by farmers. They included CMS 8501 and CMS 8704 both suitable for high rain fall or humid climate with long maturing cycle, high yielding potentials (5 to 7 ton/ha) and with white and yellow colors, respectively. The other two cultivars suitable to arid climate were CMS 8806 and CMS 9015 with yellow and white colors, respectively, with short maturing cycles and moderate yields of 4 to 5 ton/ha. All these varieties were grown under the same environmental conditions. Commercial baker's wheat flour (12% protein) purchased from the local market was used as a control.

Pre-treatment of maize grains and flour production

Flours from maize were produced in a wet milling process (Ndjouenkeu et al., 1989). De-hulled grains free of bran particles were soaked in clean tap water (1:3 w/v) for 4 h after which they were removed, partially sun dried on a mat for 2 h and milled in a hammer mill. The flour obtained was sun dried for 6 h and fractioned through a 400 µm sieve. Fine fraction of particle diameters less than 400 µ were used to formulate composite flours by thoroughly mixing in a baker mixer.

Experimental design

Composite flours in ratios of 90:10, 80:20, 70:30 and 60:40 w/w for wheat and maize flours, respectively, were reconstituted and used in the production of *Akara*. Whole wheat flour was used as control. A 4 x 5 x 5 factorial experiment was used with four maize varieties (CMS 8806, CMS 8501, CMS 9015 and CMS 8704), five substitution levels (0, 10, 20, 30 and 40% maize) and five fermentation times (0, 30, 60, 90 and 120 min).

Dough samples and doughnut production

Composite flours (32.4%) and other ingredients (13.5% sugar, 0.5% instant bakers' yeast and 0.3% iodised salt) were used for dough preparation. The ingredients were introduced in potable water (53.3 to 65.3%) according to the specific water absorption capacity (WAC) of composite flour, mixed for 3 to 5 min after which appropriate mass of flour was added and kneaded for 2 to 3 min. The dough was whipped for 10 min using a fork mixer (Bonnet, France) and cut into samples of 40 g which were manually shaped into balls. The balls were divided into two groups from which one group was fermented and tested for swelling capacity after 0, 30, 60, 90 and 120 min. The other group was fried in deep oil for 5 min after 90 min of incubation at room temperature (25 to 28°C).

Physico-chemical analysis

The dehulling and milling rates of maize grains were calculated as the ratio of the weight of dehulled grains and milled flour (400 µm) to that of whole and dehullled grains, respectively. Samples of 1000 grains were weighed, and whole wheat and composite flours analysed for moisture, proteins, lipids and ash contents as described by official methods of analysis (AOAC, 1999). Moisture content was determined as weight loss of 10 g sample after drying for 24 h at 102°C. Lipid content was calculated by weight loss after 16 h hexane extraction in a soxhlet apparatus. Protein content was determined by Kjeldahl digestion technique followed by spectrophotometric determination of resulting nitrogen. The ashing was done in a muffle furnace at 550°C for 24 h. Total carbohydrate was obtained by calculation (Egan et al., 1981).

Functional property analysis of flour and dough samples

Water absorption capacity was determined by the method of Phillips (Phillips et al., 1988). A sample of 1 g of the composite orwheat flour was mixed in 10 ml of distilled water and mechanically mixed for 30 min in a KS 10 agitator. The mixture was centrifuged

Table 1. Physico-chemical properties of maize and wheat flours (means ± sd)*.

Property	Flour sources				
	Wheat	CMS 8806	CMS 9015	CMS 8501	CMS 8704
Dehulling yield (%)		72.1±1.5[c]	70.8±2.0[a]	71.3±1.1[b]	72.7±1.5[c]
Flour yield (%)		57.3±1.2[b]	63.3±2.1[c]	57.1±1.1[b]	49.2±1.9[a]
1000 Grain weight (g)		188.4±0.9[a]	287.6±2.2[c]	305.8±3.1[d]	271.7±0.4[b]
Carbohydrates (%MS)	70.4±0.1[a]	78.5±0.3[b]	79.1±0.1[bc]	78.6±0.3[b]	78.4±0.2[b]
Moisture (%MS)	12.9±0.1[a]	13.9±0.0[c]	13.5± 0.4[b]	14.1± 0.1[c]	13.1±0.0[b]
Proteins (%MS)	12.0±0.1[b]	4.1±0.3[a]	4.6±1.0[a]	4.1±0.1[a]	4.7±0.4[a]
Lipids (%MS)	3.8±0.1[c]	2.5±0.2[a]	2.4±0.0[a]	2.7±0.1[a]	3.4±0.1[b]
Ash (%MS)	0.8±0.0[c]	0.7±0.0[b]	0.5±0.0[a]	0.3±0.0[a]	0.4±0.1[a]

Means in the same line with different letters are significantly ($p<0.05$) different.

at 4500 rpm for 30 min on a desktop centrifuge (Bioblock Scientific MLWT.62.1). The resulting sediment (M_2) was weighed and then dried at 105°C for 24 h and the dry weight (M_1) was determined. The WAC was then calculated as follows:

$$WAC (\%) = 100 \times (M_2-M_1)/M_1$$

Dough swelling capacity (DSC) was determined from the relative swelling (Delhaye et al., 1984) of 40 g dough sample gently filled to mark in a 10 cm graduated cylinder. After incubation at 30°C for the desire fermentation time, the DSC was estimated as the ratio of the height of each dough sample to that of wheat multiplied by 100.

Physical and sensory analysis

Physical parameters (weight and diameter) of fried doughnut were measured. Sensory evaluation was conducted using the quantitative descriptive analysis (Powers, 1984) by a 12 member mixed panel. Panelists were selected based on their past habit of consuming the conventional wheat *Akara*. They underwent training sessions on how to evaluate the organoleptic characteristics of *Makala* samples. Sensory attributes considred were flavor, color, taste, firmness and overall acceptability. The intensity of each attribute was scored on levels of likeness ranging from 1 (extreme dislike) to 5 (like extremely). Coded doughnut samples were presented to the panelists in disposable plates without any additional ingredient. Water was provided for rinsing the mouth between samples.

Statistical analysis

All measurements were carried out in triplicate and data obtained were treated with the statistical package SPSS (1993). The analysis of variance (ANOVA) was used to test the effects of the factors on the properties measured and the Duncan's multiple range test was used to separate treatment means whenever there were significant differences.

RESULTS

Flour physico-chemical characteristics

The physico-chemical composition of maize and wheat flours is presented in Table 1. The maize cultivars used were characterised by specific distinct 1000 grain weight, dehulling and milling rates. However, their fine flour fractions exhibited similar composition as far as protein, carbohydrate and moisture contents are concerned. The incorporation of various levels of these maize flours into wheat flour significantly affected the chemical properties of resulting composite flours (Figures 1 and 2).

Protein, lipids and ash contents significantly ($p<0.05$) decreased as the level of maize flour substitution increased independently of the cultivars used. The high lipid content of the grains of maize cultivar CMS 8704 led to composite flour with highest fat contents. Similar results where observed in *bro* (maize composite bread) made with blend of wheat and maize flours in the ratio of 60:40, respectively (Tiekoura, 1994).

Water absorption capacity

The maize cultivars used affected WAC of composite flours (Figure 3) with the yellow cultivars CMS 8806 showing the highest ($p<0.05$) values of WAC followed by CMS 8704. Wheat flour samples showed significantly lower ($p<0.05$) WAC values despite their higher protein content. WAC of composite flour increased significantly ($P<0.05$) with increase in the the quantities of wheat flour substituted. These results contradict past reports (Mbofung et al., 2002) indicating that most composite flours had lower WAC as compared to wheat flours. The quality of maize flour used may explain these differences in WAC. In the present case, the flour fraction used was made of very fine particles (<400 µ) obtained from wet milling. Similar results obtained earlier (Enwere, 1998) confirmed the fact that starch quality in terms of particle size equally plays an important role in the absorption of water.

Dough swelling capacity during fermentation

The maize cultivars used did not affect the swelling capacity of dough from composite flour which was 92 to 94%

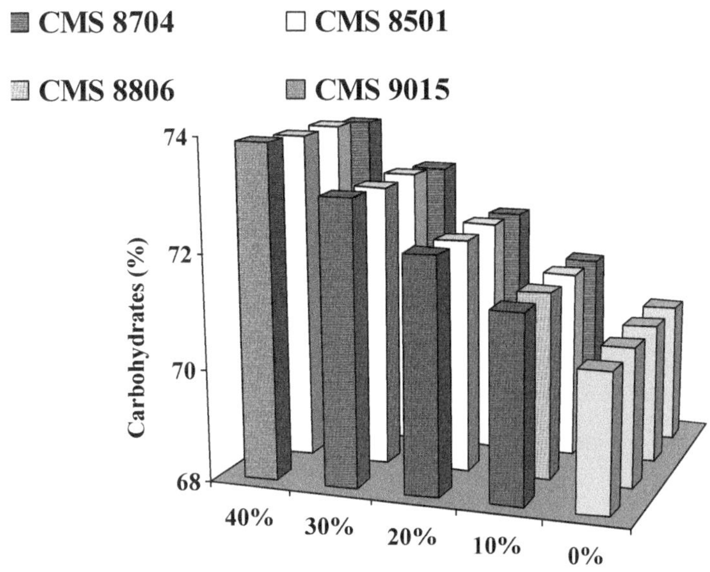

Figure 1. Variation in carbohydrate contents of composite flours as affected by the levels of substitution and maize varieties.

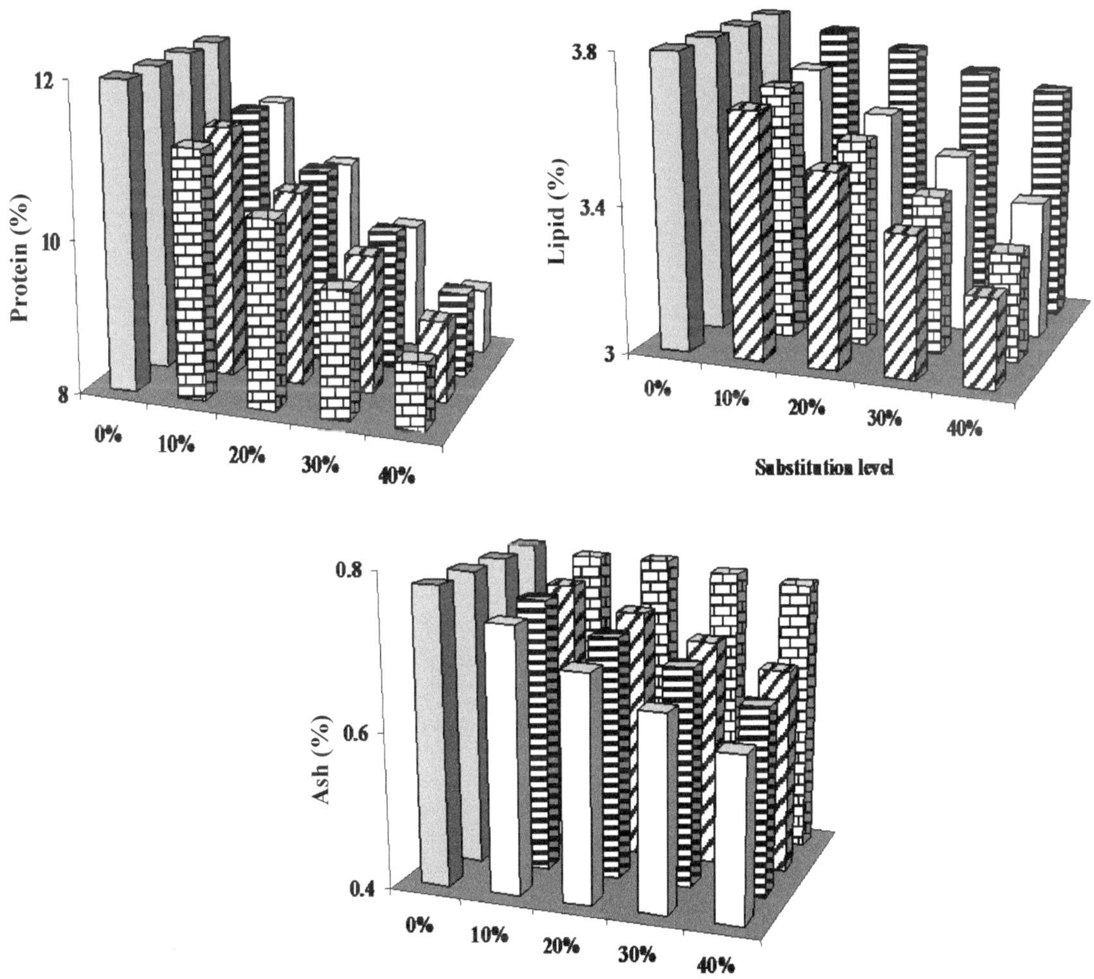

Figure 2. Variation in protein, lipid and ash contents of composite flours as affected by the level of substitution and maize varieties.

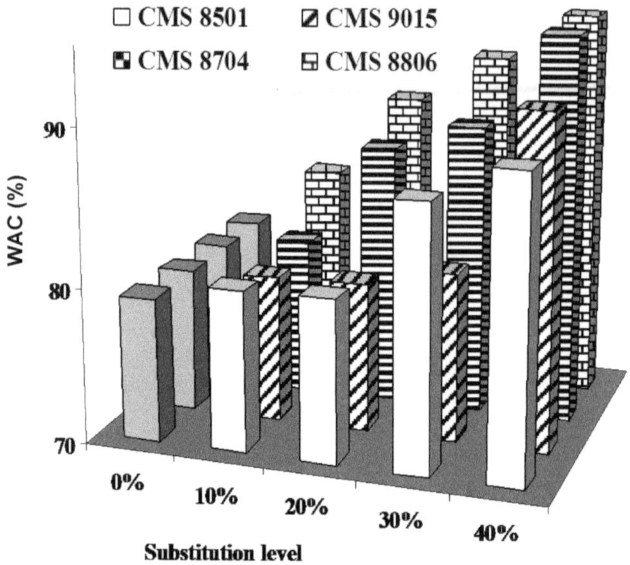

Figure 3. Water absorption capacity of wheat and composite flours as affected by the levels of substitution and maize varieties.

Table 2. Relative rising and weight losses of the dough as function of production parameters.

Dough parameter		Rising of the dough	Weight losses (g)
	0	1.00[a]	0.44[c]
	10	0.94[b]	0.44[c]
Substitution rate	20	0.91[c]	0.46[b]
	30	0.90[c]	0.48[a]
	40	0.88[e]	0.49[a]
	0	1.00[a]	0.0[e]
	30	0.98[b]	0.17[d]
Fermentation time in minutes	60	0.86[e]	0.43[c]
	90	0.88[d]	0.70[b]
	120	0.91[c]	1.02[a]

Means in the same column with different letters are significantly ($p<0.05$) different.

as compared to the control samples from wheat. The incorporation of various levels of maize flour into wheat flour and the fermentation time significantly influenced the rising of fermented dough. Dough swelling capacity (DSC) decreased with decreases in incubation time and increases in maize flour incorporation rates (Table 2). Dough rose to reach a peak after 90 min of fermentation followed by a decline. This was an indicator that dough fermentation time should not exceed 90 min, beyond which cracks appear in the dough ball, and thus facilitate the escape of all the gas trapped, resulting in a drop in its volume. For substitution levels ranging from 10 to 30%, the differences in terms of rising height between the control and the dough from composite flour did not exceed 10%. This was considered as an acceptable margin and therefore 30% substitution level was the possible highest level that would not hamper dough production. The lower gluten content of the composite flours as a result of higher substitution rate could explain the poor rising attributes of these flours. Dough fermentation is known to be affected by flour composition (Pyler, 1979; Delhaye et al., 1984) and this should be taken into account when designing dough products from composite flours. The dillution of the gluten network has affected the volume of gas produced and its retention, resulting in lower dough rising at high substitutions (He and Hoseney, 1991). An inverse trend was obtained in sorghum composite dough with the addition of vital wheat gluten to reinforce the gluten network and gas retention, resulting in higher volume (Cheong and Sun, 1998; Carson et al., 2000).

Table 3. Physical characteristics of fried doughnuts as a function of maize variety and substitution rates of wheat for maize flour.

Dough parameter		Weight of doughnuts prior to frying (g)	Diameters of doughnuts (cm)	Oil absorbed per 100 g (g)	Weight losses after frying (g)
Maize varieties	CMS 8806	39.51[a]	5.11[c]	7.46[e]	10.50[a]
	CMS 9015	39.48[a]	5.10[c]	7.17[f]	9.87[b]
	CMS 8501	39.48[a]	5.01[d]	7.46[e]	10.55[a]
	CMS 8704	39.49[a]	4.96[d]	8.23[c]	9.88[b]
Substitution levels (%)	0	39.31[b]	5.53[a]	6.12[f]	7.88[f]
	10	39.52[a]	5.39[b]	6.54[d]	9.39[d]
	20	39.53[a]	5.17[c]	7.65[d]	10.06[c]
	30	39.54[a]	4.92[d]	8.52[b]	11.28[b]
	40	39.56[a]	4.21[e]	9.07[a]	12.38[e]

Means in the same column with different letters are significantly ($p<0.05$) different.

Table 4. Sensory scores of *Akara* samples containing 0 to 40% maize flour.

Wheat flour substitution level	Origin of flours used for substitution	Sensory attributes of corresponding doughnut (*Makala*)				
		Colour	Texture	Taste	Flavour	Overall acceptance
0% Control	100% wheat	4.41	4.00	4.22	4.11	4.33
10%	CMS 8806	4.00	4.00	4.11	3.96	4.22
	CMS 8501	3.87	4.00	4.00	3.96	4.00
	CMS 8704	3.89	3.88	4.00	4.00	4.11
	CMS 9015	3.89	3.80	4.00	3.96	4.05
20%	CMS 8806	4.00	3.90	4.11	3.96	4.20
	CMS 8501	3.87	3.00	4.11	3.86	4.00
	CMS 8704	3.98	3.86	4.10	4.10	4.00
	CMS 9015	3.89	3.66	4.05	3.94	4.02
30%	CMS 8806	3.98	3.87	4.00	4.02	4.10
	CMS 8501	3.20	2.86	3.66	3.78	3.02
	CMS 8704	3.86	3.85	4.01	4.00	3.96
	CMS 9015	3.90	3,10	3.83	3.61	3.12
40%	CMS 8806	3.64	3.87	3.24	3.10	3.11
	CMS 8501	3.11	2.18	2.46	2.60	2.80
	CMS 8704	3.42	3.58	2.86	2.92	2.98
	CMS 9015	3.06	2.05	2.81	2.45	2.86

Akara physical and sensory attributes

The physical properties of *Akara* are presented in Table 3. Weights of dough nuts were similar irrespective of the maize variety used and levels of substitution. However, weight losses increased significantly in fried doughnut with increases in the fermentation time. Maize varietal effect was expressed on the size and oil retention capacity of finished dough nut. Doughnut from CMS 8704 was significantly higher in fat content, while the amounts of oil absorbed by the dough nut increased significantly with the rate of substitution. Dough nut diameters varied with the maize variety used in the substitution and decreased significantly with increased rate of substitution.

The composite doughnut sensory scores (Table 4) were significantly affected by the level of wheat flour substitution used for its production. Scores for texture and averall acceptance decreased with increases in the amounts of wheat flour substituted. Doughnut from yellow maize (CMS 8806 and CMS 8704) significantly received higher scores for color and flavor as compared to those from white maize (CMS 8501 and CMS 9015). *Akara* containing 10 and 20% maize flour received scores which were compared with the control independently of maize variety. As the level of maize flour incorporated increased, (30 and 40% level of substitution), the doughnuts had significant respectively ($p<0.05$) lower scores for all the attributes considered. Similar results were previously reported on 20 and 30% sorghum composite bread (Foda et al., 1987; Iwuoha et al., 1997) and 40% sorghum composite bread (Keregero and Mtebe, 1994). Lower scores for flavor and texture *of Akara* made from a blend of wheat and maize (ratio: 60:40) accounted for its least acceptance. *Akara* with best sensory properties was obtained from CMS 8806 and CMS 8704 maize cultivars all of yellow colour.

DISCUSSION

Cameroon improved maize varieties were used to produce composite flour, a major ingredient for *Makala* production. Fine flour from wet milling of maize grains was more suitable to blend with wheat flour as it improved in its functional properties. The influence of maize variety and maize processing method on composite flour was highlighted. Wet milling of maize grain produced fine flour of prarticle size much smaller than those from conventional milling of dry maize. The addition of these small particles of starch in wheat accounted for the improvements on water absorption capacities. With cassava flour, the reduction of the particule size (from 400 to 250 and 180 µm) had a positive effect on bread volume, meaning that particle size index affects the functional properties of composite flours. Although, the various maize cultivars showed specificities in their chemical composition, their wet milling resulted in fine flours of similar composition, thus suggesting the possibility of using any maize cultivars for composite flour production.

The level of wheat flour substituted affected more the functional properties of composite flours as compared to maize variety. Substitution of wheat for maize flour reduced the protein contents and particularly the gluten contents well known to affect gas production and consequently swelling of the dough. Previous results have shown that the major factors responsible for the bulk of water uptake were proteins contents, quality of starch and to a lesser extent, cellulose contents (Sefa-Dedeh et al., 2001; Martin and Fitzgerald, 2002). Substituting wheat for maize affected these factors resulting in doughnut of different sensory features. In sorghum, it was observed (Fliedel et al., 1989; Njintang et al., 2001) that flour of finer particle sizes would affect the particle size index giving a larger surface area in contact with water and thus increasing WAC which plays major roles in the food preparation processes. The range of application of most flours as food ingredients is dependent to a large extent on their ability to hydrate and thus imparting their desired functional property on the food system.

The sensory analysis done was useful is assessing the effect of maize variety and substitution level in the composite doughnut. It provided a basis for determining acceptance of composite dough nut by potential consumers. Sensory evaluation showed that all the various dough nuts produced were of acceptable quality as none was graded unfit for consumption and suggesting that blends of wheat and maize (ratio between 90:10 and 60:40) can serve for *Akara* production. Wet milling selected for maize flour production is a routine practice carried out daily by village women and can be achieved through the machinery locally available. This low cost of production of one ingredient will reduce the overall cost of production of doughnut and ensure more revenue to village women who are the main producers of doughnuts. The use of composite flour is of great importance to small scale producers with very little capital for rural women. Mixing is a simple process that can easily be adopted by women producing doughnuts. Comment from panelists showed that composite dough nuts were slow digesting and therefore could sustain farmers much longer on the field for long working hours. This argument was strongly in favor of the production of doughnuts from composite flour.

Conclusion

The maize cultivars developed were found suitable for the production of fine fraction in wet milling for composite flour. The differences in the cultivars did not affect the functional properties of composite flours, thus suggesting their possible use for large scale production. The acceptability of *Akara* decreased with increasing substitution rates from 10 to 40%. These results suggest that *Akara from* composite flour containing up to 40% maize can satisfy the consumers' demand in Sub-Saharan areas, especially in North Cameroon.

ACKNOWLEDGEMENTS

The WECAMAN, CORAF and the African Development Bank are gratefully acknowledged for their technical and financial support.

REFERENCES

AOAC (1999). Official methods of analysis (13th ed). Association of Official Analytical Chemists. Arlington, VA, USA.

Carson L, Sester C, XS Sun (2000). Sensory characteristics of sorghum composite bread. International J. Food Sci. Technol. 35:465-471.

Cerdan C, Ndjouenkeu R, K Mbayhoudel (2004). Increasing the value of food crops/ impact of small-scale processing activities on the economic development of the Central African Savannahs. Cahiers Agricultures 13:85-90.

Cheong ML, XS Sun (1998). Dough improvers on dough properties and bread quality of sorghum composite flour. Cereal Foods World 43:631-639.

Delhaye C, Clement P, JP Rossi (1984). Appreciation of fermenting power. In B. Godon and W. Loisel (eds), Guide pratique d'analyses dans les industries des céréales. Tech et Doc Lavoisier, Paris, France; 685 p.
Egan H, Kirk RS, R Sawyer (1981). Pearson's Chemical Analysis of Foods. 8th ed, Churchill. Longman Group, London, UK.
Enwere NJ (1998). Foods of Plant origin: Processing and utilization. Afro-Orbis publishing Ltd, Nigeria. p. 34.
FAO (1993). Maize in human nutrition. Collection FAO: Food and nutrition. 25:174.
Fliedel G, Grenet C, Gontard N, B Pons (1989). Hardness, physico-chemical characteristics and de-hulling aptitude of sorghum. Céréales en régions chaudes. Aupelf-Uref, Eds John Libbey Eurotext, Paris.
Foda YH, Ramy A, Rasmy NM (1987). Rheological and sensory characteristics of doughs and balady bread based on wheat, sorghum, millet and defatted soy flours. Ann. Agric.Sci. Ain Shams University 32:381-395.
He H, Hoseney RC (1991). Differences in gas retention, protein solubility and rheological properties between flours of different baking quality. Cereal Chem. 68:526-530.
IRA-NCRE (1990). Annual report. Institute of Agricultural Research-National Cereal Research and Extension Projet. IRA, Nkolbisson, Yaounde Cameroon.
Iwuoha CI, Anyadike AC, OS Eke (1997). The effect of flour blending on the physico-chemical and sensory qualities of bread. J. Food Sci. Technol. India 34:311-315.
Keregero MM, K Mtebe (1994). Acceptability of wheat sorghum composite flour products: an assessment. Plant Foods Human Nutri. 46:305-312.
Lopez E, J Muchnik (1997). Petites entreprises et grands enjeux. L'Harmattan, Paris, France.
Martin M, MA Fitzgerald (2002). Proteins in rice influence cooking properties. J. Cereal Sci. 37(1):101-109.
Mbofung CMF, Njintang YN, KW Waldron (2002). Functional properties of cowpea-soy-dry red beans composite flour paste and sensorial characteristics of *Akara* (deep fat fried food): effect of whipping conditions, ph, temperature and salt concentration. J. Food Eng. 54:207-214.
Munck L (1995). New milling technologies and products: whole plant utilization by milling and separation of the botanical and chemical components. In D.A.V. Dendy (ed), Sorghum and Millets: Chemistry and Technology, American Association of Cereal Chemist, Inc, St Paul MN, USA.
Ndjouenkeu R, Mbofung CMF, FX Etoa (1989). Comparative study of some tchninic of maize processing into flour in Adamawa. Céréales en régions chaudes. Aupelf-Uref, Eds John Libbey Eurotext, Paris 179-186.
Njintang YN, Mbofung CMF, KW Waldron (2001). *In vitro* protein digestibility and physico-chemical properties of dry red bean (*Phaseolus vulgaris*) flour: effect of processing and incorporation of soybean and cowpea flour. J. Agric. Food Chem. 49:2465-2471.
Phillips RD, Chinnan MS, Branch AL, Miller J, KH Mcwatters (1988) Effect of pre-treatment on functional and nutritional properties of cowpea meal. J. Food Sci. 53(3):805-809.
Powers JJ (1984). Current Practices and Application of Descriptive Methods: In Sensory Analysis of Food. Ed J R Piggot Elsevier Applied Science Publisher, London.
Pyler EJ (1979). Baking science and technology. Vol 2, Siebel Publishing Co.
Sefa-Dedeh S, Kluvitse Y, Afoakwa EO (2001). Influence of fermentation and cowpea steaming on some quality characteristics of maize-cowpea blends. Afr. J. Sci Technol. 2(2):71-80.
SPSS (1993). Statistical Package for the Social Science for Windows. Illinois, USA..
Tiekoura R (1994). The B*ro* of maize : traditional dough from composite flours of 60% maize and 40% wheat. Acts of seminar *« Maïs prospère »* Cotonou, Bénin pp. 281-284.
Watson SA (1984). Corn and Sorghum starches: production. In R.L. Whistler, E.H. Pashall and J.N. BeMiller, (eds), Starch: Chemistry and Technology. Academic Press, Orlando Fl. USA.

Softening of dried *Vangueria infausta* (African medlar) using maltodextrin and sucrose

E. Chiau[1,2] , J. Da Cruz Francisco[2], B. Bergenståhl[1] and I. Sjöholm[1]

[1]Department of Food Technology, Engineering and Nutrition, Lund University, PO Box 124, SE-221 00 Lund, Sweden.
[2]Chemical Engineering Department, Eduardo Mondlane University, PO Box 257, Maputo, Mozambique.

Softening of dried *Vangueria infausta* using maltodextrin and sucrose was investigated experimentally. The pulp of *V. infausta* was dried in a convective dryer, the influence of drying temperature as well as relationship between water content and water activity were examined. In this study, 48 samples of *V. infausta* pulp, with and without maltodextrin or sucrose, were dried at air temperatures of 60 and 80°C, at a constant air velocity of 3 m/s, for 0-240 min, and until constant weight. The water content, water activity, hardness and toughness were evaluated throughout the drying process. Samples dried at 80°C exhibited shorter drying times than samples dried at 60°C. The water activity of fruit dried for 240 min at 80°C varied from 0.633 to 0.759, and the fruit was thus safe regarding microbiological spoilage. The hardness and toughness of the dried fruit pulp increased with the reduction in water content, and it was observed that samples of pure pulp became unacceptably hard and tough, when dried to the level needed to obtain microbiological stability. The addition of sucrose as well as maltodextrin has shown to be able to reduce the hardness as well as the toughness of the dried fruit pulp; which can be considered as a strategy to obtain dried fruit pulp with suitable consistency.

Key words: *Vangueria infausta,* drying, softener, sorption isotherm, hardness, toughness.

INTRODUCTION

In southern and eastern African regions, namely Madagascar, Tanzania, Kenya, Botswana and southern and central Mozambique, the indigenous fruit *Vangueria infausta* plays an important role in the diet of the rural population. This fruit has different names, according to different languages used in the area: *African medlar* in English (*wild medlar* is used in South Africa), *mapfilwa* in Tsonga and Shangan (South of Mozambique), *mmilo* in Sotho (South Africa), *um viyo* or *um tulwa* in Zulu (South Africa), *um vilo* in Xhosa (South Africa) and *mothwanyê* in Tswana (Botswana).

V. infausta belongs to the Rubiaceae family (Mbukwa et al., 2007; Motlhanka et al., 2008). The fruit of *V.infausta* has a diameter of about 3-6 cm, its colour varies from green to light brown when ripe, and it contains 3 to 5 seeds embedded in the soft pulp. The taste is sweet and the flavour is like that of apple (Behr, 2004) and ripens from January to May in southern

Mozambique (Prins and Maghembe, 1994; Amarteifio and Mosase, 2006). It is consumed fresh, mixed with milk (Motlhanka et al., 2008), or made into a pulp, as a substitute for apple sauce in puddings. It is also used for making juice and alcoholic beverages. A type of vinegar can also be produced from the fruit (Wild medlar, 2011). The composition of *V. infausta* has been investigated by Magaia et al. (2013). The fruit contains large amounts of total dietary fiber (on average 62 g/100g db) of which more than half was insoluble dietary fiber; Magaia et al. (2013) also argue that the *V. infausta* pulp contains organic acid.

V. infausta deteriorates rapidly after harvest, therefore, there is a great loss of the product during the period of maximum production (January to May) due to the lack of methods of preservation. Drying is one of the methods used for industrial preservation of fruit pulps, concentrating the raw material and enabling the product to be stored for long periods (Gomes et al., 2004). With drying, the fruit is preserved as microorganisms cannot grow below a certain level of water activity. The water activity is the most important factor in controlling spoilage (Jangam and Mujumdar, 2010). The water activity of different dry matters can be expressed in sorption isotherms, and this will give a possibility to predict the shelf life of dried products (Visavale, 2012; Jangam and Mujumdar, 2010; Singh and Heldman, 2009).

Drying can be achieved by different ways: convective drying, osmotic dehydration, vacuum drying, drum drying, freeze drying, microwave drying, sun drying and spray drying (Chong and Law, 2011; Fernandes et al., 2011; Hii et al., 2011; Mujumdar and Law, 2010; Santos and Silva, 2008; Marques et al., 2006). Drying causes physicochemical changes in the product, such as the Maillard reaction, caramelization, enzymatic reactions, fat oxidation, the loss of volatile components other than water, and crystallization; the effects of these changes can be an increase in aroma, as well as changes in texture of the dried product. The degree of changes in the texture of fruits and vegetables developed during drying is strongly associated with the composition and structure of the tissue fragments and cell walls (Marzec et al., 2009; Reeve, 1970). Recently, microwave assisted vacuum drying has been investigated to predict texture parameters of dried fruits and vegetables (Huang et al., 2012). In addition, the combined drying methods also lowered the energy consumption, processing time and by that reduced the drying costs (Lee et al., 2013; Huang et al., 2012; Bonazzi and Dumoulin, 2011; Mujumdar and Law, 2010).

Many fruit materials are difficult to dry. Commonly, there are difficulties with stickiness and hygroscopicity. A common source of these difficulties is the high sugar concentrations in the fruit solids (Chong and Law, 2011). A way of understanding the material properties behind these observations is by analysing the state of the dried fruit (Roos, 1995). Products with a state above the glass transition temperature tends to be soft and sticky while products below the glass transition temperature are hard and non-sticky (Lee et al., 2013; Bonazzi and Dumoulin, 2011). Jaya and Das (2004, 2009) have shown how maltodextrin and other ingredients can be used to change the state of dried fruits and thereby optimize the properties of the final product.

The purpose of this study was to evaluate the texture properties of the dried fruit pulp of *V. infausta* and to evaluate improvements on the texture, when adding maltodextrin and sucrose to the matrix. The influence of drying temperature and the relationships between water content, water activity, hardness and toughness were investigated during convective drying.

MATERIALS AND METHODS

Raw material

The *V. infausta* fruit used in this study was obtained from the Paquete district of Maputo, Mozambique (25.5794°S, 32.6646°E) during the harvest season of 2011. After harvesting, the fruit was maintained at air temperature in plastic cooling bags and transported to the Eduardo Mondlane University laboratory within 2-3 h. The fruit was graded according to size in order to obtain samples of the same degree of homogeneity, 3-4 cm in diameter and 31.0 to 61.0 g in weight. The fruit was then packed in vacuum bags containing 10-24 units of the fruit each (around 0.5 kg), and frozen in a domestic freezer at -20°C.

Commercial samples of other various dried fruits were studied for comparison purposes, namely: dried figs, Sunny Fruit, Sunfruit k.F.C. Gida Tekstil San Ith, Izmiry, (Turkey), dried pineapple, osmotically treated with sugar, supplied by Candy King, Sverige AB, Solna, (Sweden) and dried cranberries, osmotically treated with sugar, supplied by Coop Trading, Albertslund, (Denmark).

Maltodextrin and sucrose were used for softening the dried *V. infausta;* the maltodextrin was supplied by Cargill, Charlottenlund (Denmark) and the sucrose was supplied by Nordic sugar AB, Malmo (Sweden).

Sample preparation

Preparation of pulp

Each vacuum bag containing frozen fruit was temperate in a cooling room to reach 3°C, placed at room temperature for 30 min, then placed in a vessel, containing 1.5 L water of 11°C. The coming up time to heat the water up to 97°C was 15 min and the heating was maintained for 5 min and then the sealed fruit bags were taken out from the water and allowed to cool in room temperature. The fruit was then removed from the bag, peeled using a knife, and the seeds were removed. The remaining fruit was pulped using a domestic puree sieve.

Preparation of samples and drying procedure

Three sets of experiments were performed: one with pulp, second with pulp and maltodextrin, and third with pulp and sucrose. The pulp was homogenised. The commercial maltodextrin (DE 12 - 16)

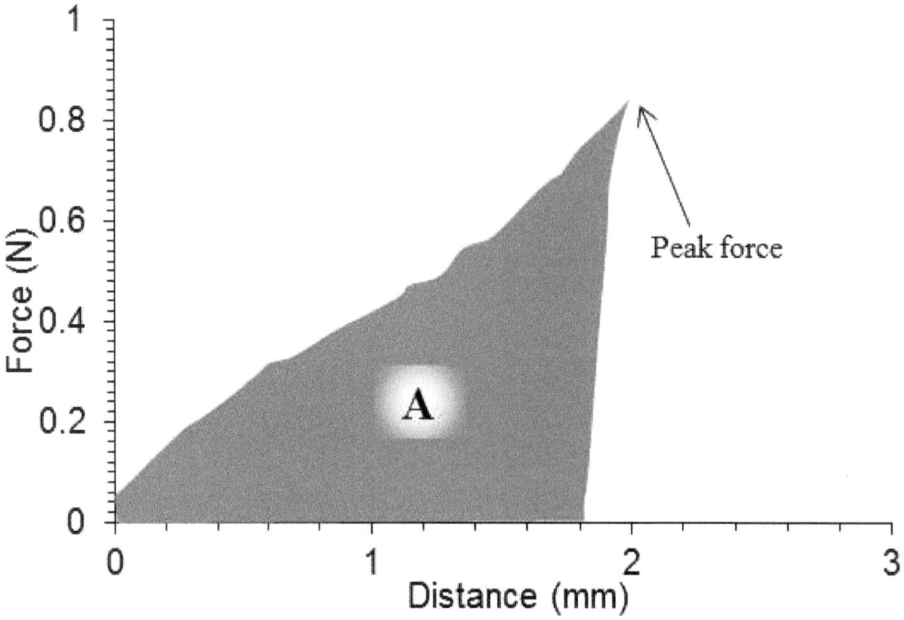

Figure 1. Hardness and toughness plot.

respectively sucrose was added to the fruit pulp at the ratio of 1:1 on weight basis, then the pulp was thoroughly mixed by spoon for 15 min. In each experiment, the pulp was distributed into eight circular samples: before drying (0 minute), six samples were placed on tray 1 (30, 60, 90, 120, 180 and 240 min) and one sample on tray 2 (dried until constant weight).

The samples were 5 mm thick, 28 mm in diameter and weight 5 g each. The samples were dried in a convective dryer as described by Skjöldebrand (1980), at air temperatures of 60 or 80°C, and air velocity of 3 m/s. During drying, the weight of the samples was recorded using a Mettler AE160 analytic weighing balance with a range of 0-162.0 g, and an accuracy of ±0.0001 g (Mettler Instrumente AG, Greifensee, Zurich, Switzerland).

The temperature of the drying air and the samples was monitored continuously using type K, 0.5 mm thermocouples. Samples were removed from the oven at time intervals and the last sample was dried up to constant weight. In addition, water content, water activity and texture were carried-out.

Analysis

pH

The pH of fresh *V. infausta* pulp was determined within 2-3 h after harvesting, using a glass electrode pH meter (Basic 20) at 25°C, standardized with buffer solutions of 4.0 and 7.0. The measurement of pH was performed in triplicate for each sample.

Water content and water activity

The water content of the initial and final samples (0.5-1.0 g) was determined by heating in a vacuum oven (Forma Scientific, Marietta, OH, USA) at 70°C for 24 h. Weighing was performed using a Mettler AE160 analytic balance. The determination was also performed in triplicate for each sample. Water activity (a_w) was measured at 20°C using an AquaLab water activity meter S 3TE, supplied by Decan Devices, Inc. (Pullman, Washington, USA), calibrated using standard salt solutions (NaCl: a_w = 0.760 M and KCl: a_w = 0.984 M), standard deviation of a_w = ±0.003.

Soluble solids

Soluble solids were determined at 20°C using a Digital Abbe WAY-S Refractometer supplied by Cometa Selecta Group, Barcelona, Spain, and expressed as °Brix (total soluble solids).

Texture analysis

The texture of *V. infausta* was determined using a texture analyser (Stable Micro Systems, version 07.13, SMS TAXT2i, Godalming, England), with stainless steel needle 2.0 mm in diameter, where the penetration force (N) was monitored. The operating conditions of the instrument were as follows: pre-test speed: 1.0 mm/s, test speed: 0.2 mm/s, post-test speed: 3.0 mm/s, 2.0 mm distance and 5.0 g trigger force. Two texture parameters (hardness and toughness) were evaluated in order to characterize the modification of the fruit resulting from drying.

In this study, hardness (N) was defined as the maximum compression force of a penetration test, while toughness (N·mm) was defined as the energy required to penetrate the sample completely (Bchir et al., 2012; Al-Said et al., 2009). Figure 1 shows an example of how the hardness and toughness were calculated. The peak force measured during the test is referred to as the hardness (Sutar and Thorat, 2011; Hii and Law, 2010; Chong et al., 2008a, b) and the area A under the curve as the toughness. Three measurements were performed on each sample obtained from each drying condition, and the final texture of the sample was recorded as the average of the three measurements.

Statistical analysis

The results were presented as mean values with the standard error. The data were analysed by analysis of variance (ANOVA) in MS Excel (Microsoft Office Excel, professional 2007).

Table 1. Properties of *Vangueria infausta* pulp.

Dry matter (%)	Experimental data			
	Water content (%)	Water activity	Soluble solid (°Brix)	pH
48.0±0.2	52.0±0.2	0.966±0.001	35.5±0.2	3.30±0.003

The values indicate mean of three measurements ± standard error. Water activity mean of two measurements ± standard error.

Table 2a. Water content (X) and water activity (a_w) of dried samples along the drying procedure at 60°C.

Drying time (min)	Pulp, 60 °C		Pulp add Md, 60 °C		Pulp add Suc, 60 °C	
	X (g/g db)	a_w	X (g/g db)	a_w	X (g/g db)	a_w
0	1.085±0.009	0.966±0.001	0.386±0.003	0.944±0.001	0.365±0.005	0.828±0.001
30	0.714±0.010	0.948±0.000	0.292±0.009	0.870±0.000	0.283±0.004	0.797±0.000
60	0.635±0.007	0.940±0.000	0.291±0.003	0.864±0.000	0.264±0.013	0.790±0.001
90	0.520±0.003	0.923±0.000	0.289±0.002	0.853±0.000	0.261±0.004	0.777±0.000
120	0.457±0.003	0.907±0.000	0.271±0.003	0.845±0.000	0.256±0.004	0.771±0.001
180	0.340±0.004	0.871±0.000	0.268±0.004	0.838±0.000	0.251±0.005	0.760±0.001
240	0.275±0.001	0.831±0.001	0.256±0.009	0.806±0.000	0.230±0.007	0.755±0.000
338 (*)	0.185±0.001	0.749±0.000				
486 (*)			0.171±0.007	0.744±0.000		
499 (*)					0.203±0.006	0.668±0.001

The values indicate mean of three measurements ± standard error, water content (X) in dry basis, drying time in minutes, a_w is water activity and (*): drying time until constant weight.

RESULTS AND DISCUSSION

Properties of *V. infausta* pulp

The characteristics of the *V. infausta* pulp used in the drying experiments is presented in Table 1. Most of the values obtained were in agreement with previous results for *V. infausta*. The dry matter of fruits in this study was high, 48%, in agreement with that reported by Magaia et al. (2013) collected in Marracuene and Manhiça, although it was higher than that previously reported by Amarteifio and Mosase (2006) (23.5%) and Saka and Msonthi (1994) (26.5%) for samples from Malawi. The dry matter varies depending on local weather conditions, and harvest time.

There is little information on fresh *V. infausta* such as on water activity, for that reason, other fruits were used for comparison purposes. The water activity of the pulp, 0.966, was comparable to values of other comparable fresh fruits such as olives, 0.96-0.97 (Fernández-Salguero et al., 1993) and figs, 0.96 (Piga et al., 2004).

Drying

Drying characteristics

The w ater content and water activity during the drying process of *V. infausta* are presented in Table 2a,b. The water content and water activity decreased with drying time for all drying processes. The final water content obtained by drying to constant weight was around 0.19 g/g db for samples dried at 60°C and 0.11 g/g db at 80°C.

The water activity of the samples dried for 240 min was between 0.633 and 0.831. The water activity of the samples dried up to constant weight was between 0.410 and 0.749. Food with water activity between 0.4 and 0.6 is considered as a dry product, whereas food with water activity between 0.65 and 0.75 is considered to be an intermediate moisture content product (Jangam et al., 2008). In addition, the water activity of 0.7 is oftcn used as a practical limit for food of intermediate moisture content. Product deteriorating bacteria cannot grow at water activity below 0.86, however, yeasts and moulds may grow to 0.62 (Hii and Law, 2010; Barbosa-Cánovas et al., 2003).

The addition of maltodextrin or sucrose at the ratio of 1:1 to fruit pulp before drying modifies the drying characteristics. The water activity of samples to which sucrose has been added was later dried for 240 min at 80°C, 0.633. This is similar to the values of 0.61-0.63 obtained by Piga et al. (2004) for dried figs. The commercial dried fruits studied here: dried figs, pineapple and cranberry, had water activities of 0.69, 0.57 and 0.55, respectively. Fruit with water activity within this range is

Table 2b. Water content (X) and water activity (a_w) of dried samples during the drying procedure at 80°C

Drying time (min)	Pulp, 80 °C		Pulp add Md,80 °C		Pulp add Suc, 80 °C	
	X (g/g db)	a_w	X (g/g db)	a_w	X (g/g db)	a_w
0	1.085±0.009	0.966±0.001	0.386±0.003	0.944±0.001	0.365±0.005	0.828±0.001
30	0.654±0.003	0.943±0.000	0.267±0.011	0.849±0.001	0.259±0.005	0.755±0.000
60	0.521±0.005	0.926±0.000	0.241±0.008	0.819±0.001	0.234±0.005	0.729±0.000
90	0.416±0.005	0.894±0.000	0.228±0.008	0.817±0.001	0.230±0.004	0.703±0.000
120	0.320±0.001	0.852±0.000	0.222±0.006	0.798±0.001	0.211±0.001	0.692±0.001
180	0.267±0.001	0.821±0.000	0.206±0.013	0.770±0.001	0.204±0.007	0.656±0.001
240	0.176±0.002	0.738±0.000	0.200±0.004	0.759±0.001	0.196±0.006	0.633±0.001
330 (*)	0.096±0.005	0.613±0.000				
385 (*)			0.130±0.011	0.668±0.001		
480 (*)					0.102±0.001	0.410±0.000

The values indicate mean of three measurements ± standard error, water content (X) in dry basis, drying time in minutes, a_w is water activity and (*): drying time until constant weight.

Table 3. Water content (X) and water activity (a_w) of dried commercial samples.

Sample	Experimental data	
	Water content(g/g db)	Water activity
Cranberry	0.173±0.005	0.550±0.001
Pineapple	0.101±0.001	0.574±0.001
Figs	0.254±0.003	0.689±0.001

The values indicate mean of three measurements ± standard error, water content in dry basis.

safe for consumption (Hii and Law, 2010; Jangam et al., 2008; Perera, 2005).

The water content and water activity of the commercial samples are presented in Table 3. The water content (0.196 g/g db) and water activity (0.633) were in agreement with the values reported for other fruits. Azeredo et al. (2006) reported the water content of 0.208 g/g db and water activity of 0.62 for dried mango, and Díaz et al. (2009) reported the water content of 0.269 g/g db and water activity of 0.692 for dried apple.

This can be compared with dried date palm fruits (*Phoenix dactylifera L.*), that was considered as safe and with better color and texture at 24 to 25% (0.32 to 0.33 db) (Chong and Law, 2011). Thus, we may confidently considered *V. infausta* as safe at the water content of 16.4% (wet basis) (0.196 db).

Sorption isotherms

The sorption isotherms of *V. infausta* pulp dried at 60 and 80°C are shown in Figure 2a,b. Each data point on the curves represents the mean of three measurements.

Different models have been used to predict the drying behaviour of fruit. One model used to predict water sorption isotherms, the GAB model, has been described by Visavale (2012) and Toledo (2007), Equation (1):

$$\frac{a_w}{X} = \alpha\, a_w^{\,2} + \beta\, a_w + \gamma \qquad \text{Equation (1)}$$

where X is the water content (db), a_w is the water activity, and α, β and γ were obtained by non-linear regression of the experimental data on water activity and water content.

The coefficients of the GAB equation are presented in Table 4 and the regression lines are shown in Figure 2a,b. The high regression coefficient of determination (R^2) shows that this model accurately describes the drying characteristics of *V. infausta* in the present study.

Texture analysis

Changes in the texture of dried *V. infausta* were determined throughout the drying process (Table 5a,b). It is obvious that the hardness and toughness increased with the increase in drying time and drying temperature. The hardness of the samples dried during 240 min increased from 0.11 to 3.10 N, and those samples dried until constant weight increased from 0.10 to 14.00 N. The toughness of the samples dried for 240 min increased from 0.2 to 3.8 N·mm, and the toughness of the fully dried samples ranged from 0.2 to 15 N·mm.

The hardness of the *V. infausta* pulp, and the pulp with added maltodextrin or sucrose was before drying: 0.17, 0.34 and 0.08 N (2.0 mm needle probe), respectively. These values are in agreement with values reported for the chempedak fruit *(Artocarpus integer)* (Chong et al., 2008a, b) (0.09 N) (2.0 mm cylindrical probe) and grape

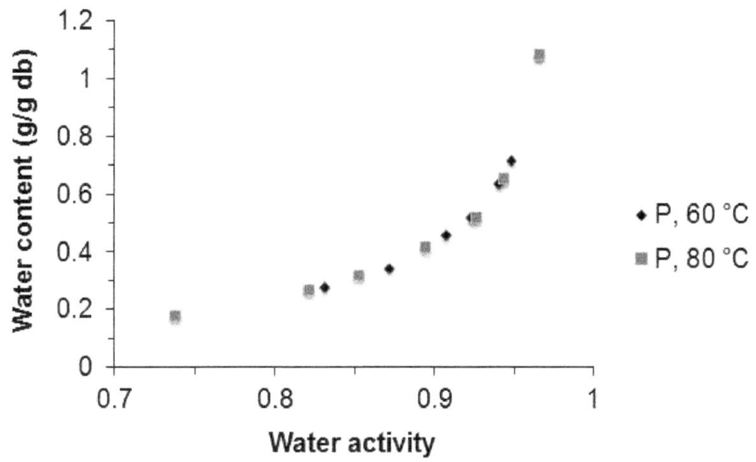

Figure 2a. Sorption isotherms of *v.A infausta* pulp dried at 60 and 80°C based on experimental data on dry basis.

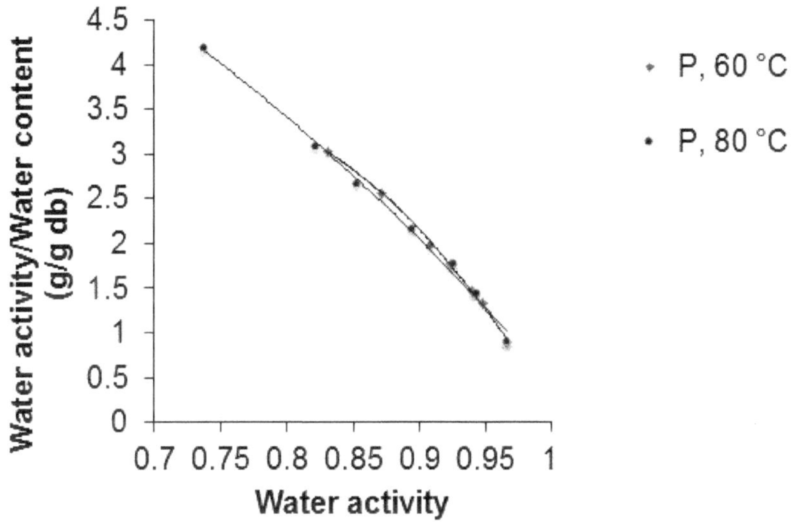

Figure 2b. Fitting GAB Equation to sorption isotherm of *V. infausta* pulp using experimental data and predicted (full line). The GAB model is described in Equation (1).

Table 4. Parameters of GAB adjustments applied to drying data

Sample (g/g db)	Drying air temperature (°C)	α	β	Ɣ	R^2
Pulp	60	-43.09	61.92	18.68	0.9983
Pulp	80	-10.96	4.96	6.48	0.9939

R^2 is coefficient of determination, α, β and Ɣ are coefficient of GAB, calculated by Equation (1), sample in dry basis, drying air temperature in degrees Celsius.

(Xiao et al., 2010) (0.62 N) (2.0 mm spherical probe). The hardness of the *V. infausta* pulp was 3 N when dried at 60°C, and 14 N when dried at 80°C (2.0 mm needle probe), which are similar to the values reported for dried

Table 5a. Toughness (P_t) and hardness (P_h) of dried samples during the drying procedure at 60°C.

Drying time (min)	Pulp, 60 °C		Pulp add Md, 60 °C		Pulp add Suc, 60 °C	
	P_t (N.mm)	P_h (N)	P_t (N.mm)	P_h (N)	P_t (N.mm)	P_h (N)
0	0.22±0.02	0.17±0.02	0.36±0.03	0.34±0.03	0.10±0.01	0.08±0.02
30	0.24±0.01	0.18±0.01	0.09±0.01	0.06±0.01	0.09±0.01	0.06±0.01
60	0.23±0.01	0.19±0.01	0.08±0.01	0.05±0.01	0.13±0.03	0.08±0.02
90	0.30±0.01	0.28±0.01	0.13±0.01	0.09±0.01	0.12±0.01	0.08±0.01
120	0.34±0.01	0.29±0.01	0.13±0.02	0.10±0.02	0.49±0.34	0.49±0.36
180	0.80±0.09	0.71±0.08	0.11±0.01	0.08±0.01	0.57±0.15	0.64±0.11
240	1.13±0.02	0.97±0.03	0.15±0.01	0.11±0.01	0.21±0.06	0.15±0.06
338 (*)	3.8±0.1	3.2±0.1				
486 (*)			1.0±0.2	1.0±0.2		
499 (*)					0.2±0.1	0.1±0.1

The values indicate mean of three measurements ± standard error, toughness (P_t) in N.mm, hardness (P_h) in N, drying time in minutes and (*): drying time until constant weight.

Table 5b. Toughness (P_t) and Hardness (P_h) of dried samples along the drying procedure at 80°C.

Drying time (min)	Pulp, 80°C		Pulp add Md, 80°C		Pulp add Suc, 80°C	
	P_t (N.mm)	P_h (N)	P_t (N.mm)	P_h (N)	P_t (N.mm)	P_h (N)
0	0.22±0.02	0.17±0.02	0.36±0.03	0.34±0.03	0.10±0.01	0.08±0.02
30	0.26±0.01	0.24±0.02	0.17±0.01	0.13±0.02	0.25±0.04	0.14±0.05
60	0.34±0.01	0.30±0.01	0.23±0.06	0.19±0.06	0.24±0.03	0.14±0.03
90	0.50±0.01	0.42±0.01	0.19±0.03	0.13±0.02	0.30±0.03	0.14±0.02
120	0.82±0.03	0.74±0.03	0.25±0.02	0.20±0.01	0.23±0.04	0.14±0.03
180	1.70±0.03	1.37±0.01	0.39±0.01	0.34±0.02	0.34±0.08	0.22±0.05
240	3.79±0.46	3.08±0.34	0.44±0.04	0.32±0.01	0.39±0.04	0.26±0.02
330 (*)	15.0±4.6	13.5±3.8				
385 (*)			8.7±1.9	8.5±2.2		
480 (*)					3.6±1.6	3.2±1.5

The values indicate mean of three measurements ± standard error, toughness (P_t) in N.mm, hardness (P_h) in N, drying time in minutes and (*): drying time until constant weight.

grapes: 10 N at 50°C, 14 N at 60°C and 17 N at 65°C (2.0 mm spherical probe) (Xiao et al., 2010) and dried chempedak fruit (Chong et al., 2008a) (3.39 N) (2.0 mm cylindrical probe). In addition to this, Chong et al. (2008a, b) reported that dried chempedak using sun drying and hot air drying has significant textural change as compared to fresh fruit. In addition, the combined drying method reported for dried mixed potato with apple chips: 7 N dried by microwave-freeze drying and 14 N dried by microwave vacuum drying (5.0 mm cylindrical probe) (Huang et al., 2011), are similar to the results reported in this study.

Adding maltodextrin or sucrose affected the hardness of the dried product. Drying at the higher temperature of 80°C resulted in a harder product: 3 N when sucrose was added and 9 N when maltodextrin was added, than drying at 60°C: 0.1 N with sucrose and 1 N with maltodextrin.

Statistical analysis using ANOVA showed that the relationship between the hardness and the toughness was linear, such that it could be expressed by the linear equation, Equation (2):

$$P_h = aP_t + b \qquad (2)$$

Where, P_h is the hardness (N), P_t is the toughness (N·mm), a is the linear coefficient and b is the ordinate-intercept.

Table 6 presents the results of the linear regression of dried *V. infausta* pulp, and the pulp to which maltodextrin or sucrose had been added, coefficient a and ordinate-intercept b, coefficient of determination (R^2), which were

Table 6. Analysis of linearity between hardness (P_h) and toughness (P_t).

Sample (g/g db)	Drying air temperature (°C)	a (mm)$^{-1}$	b (N)	R^2	ε
Pulp	60	0.851	0.005	0.9997	0.006
Pulp	80	0.901	-0.076	0.9993	0.011
Pulp-Maltodextrin	60	0.972	-0.023	0.9962	0.027
Pulp-Maltodextrin	80	0.989	-0.055	0.9999	0.004
Pulp-Sucrose	60	1.189	-0.067	0.9864	0.063
Pulp-Sucrose	80	0.927	-0.094	0.9996	0.008

a is the linear coefficient and b is the intercept of linear regression, calculated by Equation (2), R^2 is coefficient of determination and ε is the standard error.

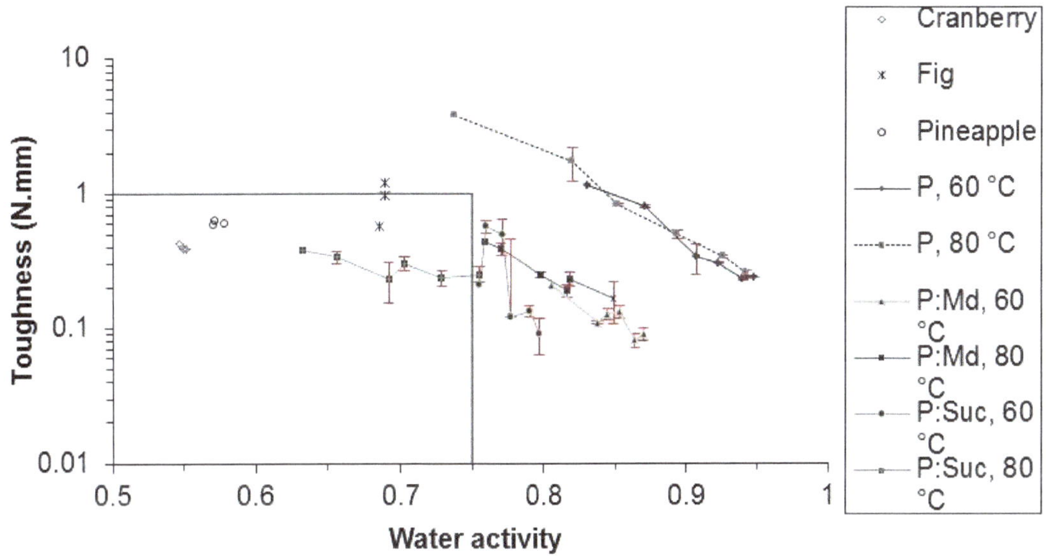

Figure 3. Toughness of dried samples as function of water activity, P - pulp, Md - Maltodextrin, Suc - Sucrose. The standard error of the commercial fruit was zero.

satisfactory and well within experimental standard errors. Kingsly et al. (2006) also reported a linear relationship between the hardness and the water content for dried pomegranate seeds (Anardana, *Punica granatum*). Thereby, it can be concluded that the critical texture can be equally well described by using either the hardness parameter or the toughness parameter.

Figure 3 shows the toughness as a function of water activity for dried *V. infausta*. We can note that each drying curve follows a trajectory with increasing toughness when the water activity is decreased. A possible commercial window can be defined if we consider the microbiological stability critical for the distribution and the toughness critical to the consumer perception of a dried fruit product. Possible limits of the commercial windows are for water activity values below 0.75 (Barbosa-Cánovas et al., 2003; Jangam et al., 2008) for toughness values below 1 N·mm.

The pure fruit samples form very tough structures before they reach a sufficiently low water activity to be distributed while the fruit samples with the added sucrose or maltodextrin become softer and display a drying trajectory passing through the commercial window. For comparative reasons, a few commercially dried products are included in the graph.

It should be noted that the sugars are added as dry material to the fruit pulp before drying. The solubilisation of the sucrose crystals as well as of the maltodextrin particles may be incomplete and the heterogeneity in the finally dried material may contribute to the softness.

Conclusions

The relationship between water content and water activity measured using the GAB model was found to be adequate to predict the dried *V.* infausta models for water sorption isotherms. The high coefficient of determination demonstrates that this model describes the drying characteristics in a reliable way.

The drying time is determined by the drying parameters such as temperature. Reduced water content lowers the water activity and create microbiologically stable products (aw < 0.75) at water contents below 0.2 g/g db. However, the drying also leads to an increase in hardness and toughness. An increase in drying temperatures resulted in high values of hardness and toughness. The hardness and toughness were linearly correlated making the use of both parameters to characterize the samples unnecessary. The toughness trajectory described by a fruit pulp sample never passes through the commercially possible window defined by an attractive texture and a suitable water activity. Thus, there is a need to use a softening component when drying the fruit of *V. infausta*. The use of maltodextrin and sucrose in order to soften the dried product resulted in drying trajectories passing through the commercial window.

Further work is required to investigate the level of sugar required for better softening of the dried *V. infausta* and for reducing water activity and water content.

ACKNOWLEDGEMENTS

The authors acknowledge the financial support of SIDA programme, Lund University (Sweden) and Eduardo Mondlane University (Mozambique).

REFERENCES

Al-Said FA, Opara LU, Al-Yahyai RA (2009). Physico-chemical and textural quality attributes of pomegranate cultivars (*Punica granatum L.*) grown in the sultanate of oman. J. Food Eng. 90:129-134.

Amarteifio JO, Mosase MO (2006). The chemical composition of selected indigenous fruits of Botswana. J. Appl. Sci. Environ. Mgt. 10:43 - 47.

Azeredo HMC, Brito ES, Moreira GEG, Farias VL, Bruno LM (2006). Effect of drying and storage time on the physico-chemical properties of mango leathers. Int. J. Food Sci. Technol. 41:635 - 638.

Barbosa-Cánovas GV, Fernández-Molina JJ, Alzamora SM, Tapia MS, López-Malo A, Chanes JW (2003). Handling and preservation of fruits and vegetables by combined methods for rural areas. Technical Manual. FAO, Rome. 39-60.

Bchir B, Besbes S, Karoui R, Paquot M, Attia H, Blecker C (2012). Osmotic dehydration kinetics of pomegranate seeds using date juice as an immersion solution base. Food Bioprocess Technol. 5:999-1009.

Behr K (2004). Vangueria infausta Burch.subsp.infausta South Africa National Biodiversity Institute. South Africa. Obtained through: http://www.plantzafrica.com. On 9/3/2013.

Bonazzi C, Dumoulin E (2011). Quality changes in food materials as influenced by drying processes, in Modern drying technology:Product quality and formulation - Volume 3, ed. Tsotsas E and Mujumdar AS. John Wiley & Sons, Weinheim, Germany 1 - 20.

Chong CH, Law CL (2011). Drying of Exotic Fruits, in Drying of Foods, Vegetables and Fruits - Volume 2, ed. Jangam, SV, Law CL and Mujumdar AS. Singapore. 1 - 42.

Chong CH, Law CL, Cloke M, Abdullah LC, Daud WRW (2008a). Drying kinetics, texture, color, and determination of effective diffusivities during sun drying of chempedak. Drying Technol. 26:1286 - 1293.

Chong CH, Law CL, Cloke M, Hii CL, Abdullah LC, Daud WRW (2008b). Drying kinetics and product quality of dried chempedak. J. Food Eng. 88:522 - 527.

Díaz EL, Giannuzzi L, Giner SA (2009). Apple pectic gel produced by dehydration. Food Bioprocess Technol. 2:194 - 207.

Fernandes FAN, Rodrigues S, Law CL, Mujumdar AS (2011). Drying of exotic tropical fruits: A comprehensive review. Food Bioprocess Technol. 4:163 - 185.

Fernández-Salguero J, Gómez R, Carmona MA (1993). Water activity in selected high-moisture foods. J. Food Compos. Anal. 6:364-369.

Gomes PM de A, Figueirêdo RMF de, Queiroz AJ de M (2004). Armazenamento da polpa de acerola em pó a temperatura ambiente. Ciênc. Tecnol. Aliment. Campinas. 24.384 - 389.

Hii CL, Law CL (2010). Product quality evolution during drying of foods, vegetables and fruits, in Drying of foods, vegetables and fruits - Volume 1, ed. Jangam SV, Law CL and Mujumdar AS. Singapore. 125 - 144. Obtained through: www.mujumdar.net78.net_on 8/26/2013.

Hii CL, Ong SP, Law CL (2011). Drying studies of tropical fruits cultivated in Malaysia: A review. J. Appl. Sci. 11:3815-3820.

Huang L-I, Zhang M, Mujumdar AS, Lim R-X (2011). Comparison of four drying methods for re-structured mixed potato with apple chips. J. Food Eng. 103:279 - 284.

Huang L-I, Zhang M, Wang L-p, Mujumdar AS, Sun D-f (2012). Influence of combination drying methods on composition, texture, aroma and microstructure of apple slices. Food Sci. Technol. 47:183 - 188.

Int. J. Food Sci. Technol. 39:793-799.

Jangam SV, Joshi VS, Mujumdar AS, Thorat BN (2008). Studies on dehydration of sapota (*Achras zapota*). Drying Technol. 26:369 - 377.

Jangam SV, Mujumdar AS (2010). Basic concepts and definitions, in Drying of foods, vegetables and fruits - Volume 1, ed. Jangam, SV, Law CL and Mujumdar AS. Singapore. 1 - 30. Obtained through: www.mujumdar.net78.net_on 8/26/2013.

Jaya S, Das H (2004). Effect of maltodextrin, glycerol monostearate and tricalcium phosphate on vacuum dried mango powder properties. J. Food Eng. 63:125-134.

Jaya S, Das H (2009). Glass transition and sticky point temperatures and stability/mobility diagram of fruit powders. Food Bioprocess Technol. 2:89-95.

Kingsly ARP, Singh DB, Manikantan MR, Jain RK (2006). Moisture dependent physical properties of dried pomegranate seeds (Anardana). J. Food Eng. 75:492 - 496.

Lee D-J, Jangam S, Mujumdar AS (2013). Some recent advances in drying technologies to produce particulate solids. KONA Powder and Particle J. 30:69 - 83.

Magaia T, Uamusse A, Sjöholm I, Skog K (2013). Dietary fiber, organic acids and minerals in selected wild edible fruits of Mozambique. SpringerPlus 2:88 - 96.

Marques LG, Silveira AM, Freire JT (2006). Freeze-drying characteristics of tropical fruits. Drying Technol. 24:457 - 463.

Marzec A, Kowalska H, Pasik S (2009). Mechanical and acoustic properties of dried apples. J. Fruit Ornamental Plant Res. 17:127 - 137.

Mbukwa E, Chacha M, Majinda RRT (2007). Phytochemical constituents of *Vangueria infausta*: their radical scavenging and antimicrobial activities. ARKIVOC. (ix) :104 - 112.

Motlhanka DMT, Motlhanka P, Selebatso T (2008). Edible indigenous wild fruit plants of eastern Botswana. Int. J. Poult. Sci. 7:457 - 460.

Mujumdar AS, Law CL (2010). Drying technology: Trends and applications in postharvest processing. Food Bioprocess Technol. 3:843 - 852. Obtained through: www.mujumdar.net78.net__on 8/26/2013.

Perera CO (2005). Selected quality attributes of dried foods. Drying Technol. 23:717-730.

Piga A, Pinna I, Özer KB, Agabbio M, Aksoy U (2004). Hot air dehydration of figs (*Ficus carica L.*): drying kinetics and quality loss.

Prins H, Maghembe JA (1994). Germination studies on seed of fruit trees indigenous to Malawi. For. Ecol. Manage. 64:111-125.

Reeve RM (1970). Relationships of histological structure to texture of fresh and processed fruits and vegetables. J. Texture Stud. 1:247 - 284.

Roos Y (1995). Phase transitions in foods. Academic Press. New York. US

Saka JDK, Msonthi JD (1994). Nutritional value of edible fruits of indigenous wild trees in Malawi. For. Ecol. Manage. 64:245 - 248.

Santos PHS, Silva MA (2008). Retention of vitamin C in drying processes of fruits and vegetables - A review. Drying Technol. 26:1421 - 1437.

Singh RP, Heldman DR (2009). Introduction to food engineering. Academic press. London. UK. 4th Edition. pp. 654 - 655.

Skjöldebrand C (1980). Convection oven frying: heat and mass transfer between air and product. J. Food Sci. 45:1354-1362.

Sutar PP, Thorat BN (2011). Drying of roots, in Drying of foods, vegetables and fruits - Volume 2, ed. Jangam SV, Law CL and Mujumdar AS. Singapore. pp. 44 - 74.

Toledo RT (2007). Fundamentals of food process engineering. Springer science. New York. 3rd Edition. pp. 431 - 473.

Wild medlar (2011). Department of agriculture. South Africa. Obtained through: www.daff.gov.za/docs/Brochures/WildMedlar. pdf on 3/9/2013.

Visavale GL (2012). Principles, classification and selection of solar dryers. In Solar drying: fundamentals, applications and innovations, ed. Hii CL, Ong SP, Jangam SV and Mujumdar AS. Singapore. 1 - 50. Obtained through: www.mujumdar.net78.net_on 8/26/2013.

Xiao HW, Pang CL, Wang LH, Bai JW, Yang WX, Gao ZJ (2010). Drying kinetics and quality of monukka seedless grapes dried in an air-impingement jet dryer. Biosyst. Eng. 105:233 - 240.

Thermal oxidative alteration of sunflower oil

Sadoudi R[1] , Ammouche A[2] and Ali Ahmed D[1]

[1]Département d'Agronomie. Faculté des Sciences Biologiques et Agronomiques. Université de Tizi-Ouzou, Algérie.
[2]Ecole nationale supérieure agronomique (ENSA, Ex. INA). Département de Technologie alimentaire et de Nutrition Humaine. Hassen badi, El-Harrach (Alger). Algérie.

Sunflower oil is extensively used in frying in Algeria as an alternative to olive oil due to its low cost. However, the high level of unsaturated fatty acids (FA) contained in sunflower oil enhances its susceptibility to oxidation. In our study, the sunflower oil was heated at 99±2°C with incorporation of 9 L of oxygen/second for 52 h continuously in the absence of foodstuff. Heating polyunsaturated fatty acids (PUFAs) in the presence of air causes a greater degree of lipid peroxidation. The oil oxidation degree was monitored through several physicochemical analyses. The products of thermal oxidation were monitored using UV- spectrophotometric method and Fourier transform infrared spectroscopy (FT-IR). Compared to fresh oil, the free-fatty acid contents, peroxide value, density and moisture of the thermally oxidized sunflower oil increased. In addition, the iodine and saponification values decreased during thermal treatment. The treatment applied had a negative effect on FA composition; the most significant effects were on C18:2, C18:1 and C16:0 contents. Analysis of chromatographic profile of thermoxidized sunflower oil showed a reduction in linoleic acid (LA) and an increase in oleic and palmitic acids; decrease of linoleic acid content is used as an indicator of lipid oxidation. Moreover, during the early stages, conjugated dienes (CDs), absorbing at 233 nm, were formed upon decomposition of hydroperoxides. In our study, the early stages of lipid oxidation were measured by UV-spectrophotometric method. Hydroperoxides broke down into secondary products and were revealed by FT-IR; these scission products are generally odoriferous by nature. The C=O stretching band at 1739-1724 cm^{-1} of the aldehydes was much more intense. Formation of conjugated double bond systems and the isomerisation of *cis* to *trans* double bonds was observed in the C=C stretching region at 980 to 965 cm^{-1}. The results obtained reveal that even fresh oil contains products of peroxidation and isomerization of C18:2, n-6; indeed, CDs can be produced during the refining process of oil. The treatment applied increased the rate of these products and conferred a marked rancid taste and a thick texture to the thermoxidized sunflower oil.

Key words: Sunflower oil, Linoleic acid, thermally oxidative treatment, alteration level.

INTRODUCTION

Sunflower oil is a high-quality edible oil. It is used in cooking, frying, and in the manufacture of margarine and shortening and considered by some as desirable as olive oil. Sunflower oil was selected in this study due to its high use in food as it is a rich source of linoleic acid. Furthermore, it is light in taste and appearance and has a high vitamin E content compared to other vegetable oils (Shahidi et al. 1992).

At high temperature and in the presence of air, many chemical reactions can be observed in oil: hydrolysis, polymerization, oxidation and isomerization (Rossell, 2001). Thus, new and unstable compounds potentially

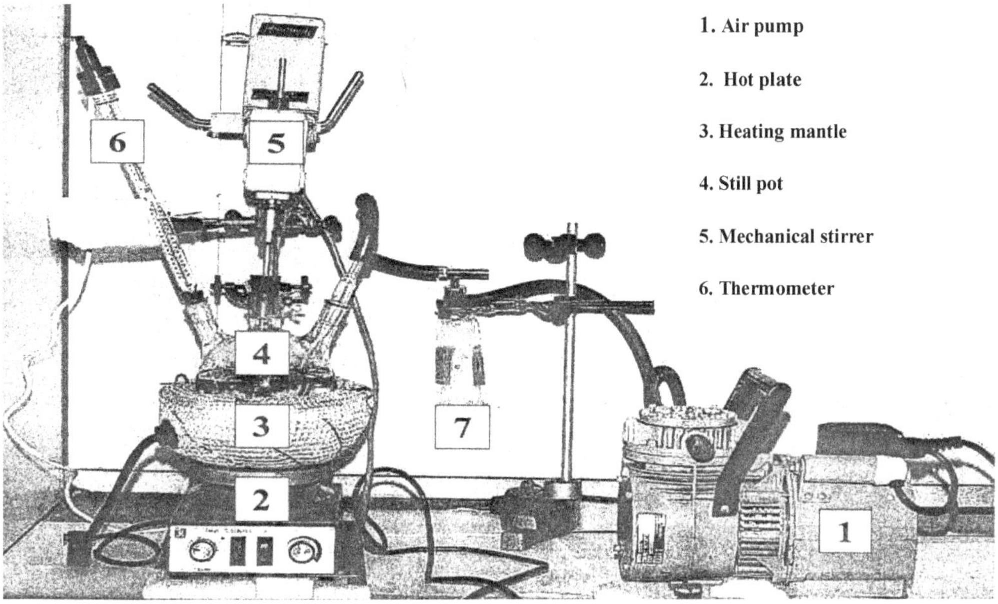

Figure 1. Mounting of thermal oxidation device.

toxic at low concentrations may be generated after destruction of the linoleic acid (Nawar, 1996; Min and Boff, 2001).

The objective of this study was to evaluate the deterioration levels of sunflower oil subjected to thermoxidative treatment (in the absence of foodstuff) according to the laboratory instrumental method developed by Drozdowski and Szukalska (1987) modified by Blanc-Gondardmary et al. (1989). There is no standard method to detect oxidetive changes during the entire process, and a combination of different analytical techniques is usually required (Gray, 1978; Frankel, 1993; Warner and Eskin, 1995). In our study, the primary oxidative products were monitored through peroxide value (PV), loss of unsaturated fatty acids, conjugated diene value, and others. Secondary changes are measured by Fourier transform infrared (FT-IR) spectroscopy, this latter gives information about the different functional groups present in the sample; thus, it is not limited to just one kind of compound like the indices mentioned above (Muik et al., 2005).

MATERIALS AND METHODS

Sunflower oil was purchased at a local market in Tizi-Ouzou, Algeria. This oil was obtained from the seeds of *Héliantalus annus linnaeus* containing 40% of oil. Sunflower oil-like most vegetable oils is composed mainly of triacylglycerols (98 to 99%), and a small fraction of phospholipids, tocopherols, sterols, and waxes (all of the latter are commonly referred to as the "unsaponifiable fraction"). Sunflower oil is characterized by a high concentration of linoleic acid, followed by oleic acid and it has a low content of palmitic acid compared with other oils (Grompone, 2005). Decrease of linoleic acid content is used as an indicator of lipid oxidation. The mechanism of lipid oxidation changes significantly at elevated temperatures and depends strongly on oxygen availability. In our study, the sunflower oil (700 ml) was heated at 99±2°C with incorporation of 9 L of oxygen / s for 52 h continuously in dark. Figure 1 show the apparatus setup used in thermally oxidative treatment.

After treatment, thermoxidized sunflower oil was kept under nitrogen before sealing airtight in a glass bottle and stored at -20°C until further analysis. The degree of oxidation of the oil was monitored through several physicochemical analyses. The following parameters were studied: free-fatty acid (FFA) contents, peroxide value (PV), refractive index (RI), humidity (H%), iodine value (IV), saponification value (SV) and density (AOCS, 1989); fatty acid (FA) composition (AOAC, 1999); specific extinction at 232 nm (*K*232) and 270 nm (*K*270) related to the content of conjugated dienes (CDs) and trienes (CTs) of linoleic acid (18:2 n-6) respectively was determined using UV spectrophotometer (IUPAC, 1987); secondary lipid oxidation products were determined using FT-IR spectroscopy.

Statistical data analysis

The average comparison was realized by the analysis of variance (ANOVA) with Stat. Box. Edition 6.4. The significance level was selected at $p < 0.05$.

RESULTS

The study shows that compared to fresh oil, the FFA content, PV, RI, density and moisture of the oxidized sunflower oil increased from 0.093 to 1.25%, 5.83 to 152.5 meq/kg, 1.461 to 1.476, 0.910 to 0.985 and 0.100 to 2.006 respectively. The IV and SV decreased from 125.84 to 80.51 g I_2/100 g and 192.60 to 183.79 mg KOH/g respectively. In addition, a reduction in linoleic acid (from 58.14 to 40.59%) and an increase in oleic acid

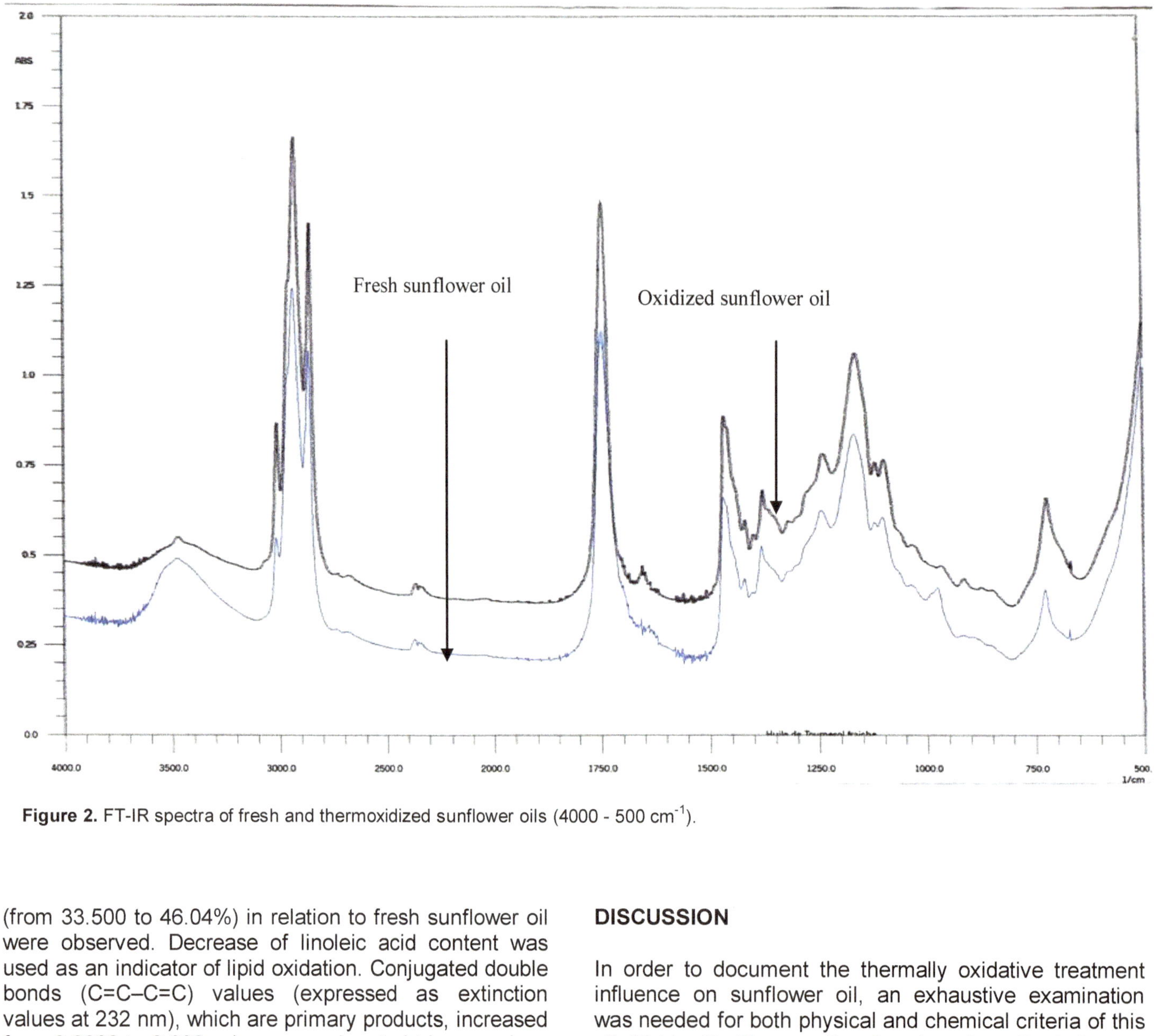

Figure 2. FT-IR spectra of fresh and thermoxidized sunflower oils (4000 - 500 cm^{-1}).

(from 33.500 to 46.04%) in relation to fresh sunflower oil were observed. Decrease of linoleic acid content was used as an indicator of lipid oxidation. Conjugated double bonds (C=C–C=C) values (expressed as extinction values at 232 nm), which are primary products, increased from 0.0002 to 0.033, whereas conjugated trienes values at 232 nm ranged from 0.423 to 0.655. Formation of aldehydes, ketones and other secondary oxidation products, revealed by FT-IR spectroscopy, in thermoxidized oil was accelerated by constantly bubbling air into the oil during heating (Figure 2). Then, flavour, aroma and taste of the oil were affected. Also, the formation of conjugated double bond systems and the isomerisation of *cis* to *trans* double bonds as observed in the C=C stretching region contributed to change of the density and viscosity of the thermoxidized sunflower oil.

Nevertheless, part of the double bonds of the linoleic acid remained at configuration *cis* (C=C–C–C=C); this essential FA was proportioned at 40.59% in oxidized oil. Our results show a strong increase of the band located at 1667 to 1639 cm^{-1} which can be assigned to *cis* double bonds in oxidized oil compared to fresh oil.

DISCUSSION

In order to document the thermally oxidative treatment influence on sunflower oil, an exhaustive examination was needed for both physical and chemical criteria of this oil (Table 1). The results of various investigated parameters led to the conclusion that the treatment applied in our study caused a high level of deterioration of this oil.

Compared to fresh oil, the thermoxidized oil contained approximately 13 times more total FFA than fresh oil (0.093 *vs.* 1.25%) and its acidity content was higher (P=0) than fresh oil (0.182 *vs.* 2.51 mg KOH/g). Formation of FFA might be an important measure of rancidity of foods. FFAs are formed due to hydrolysis of triglycerides and may get promoted by reaction of oil with moisture (Freja et al., 1999). The advanced deterioration of thermoxidized oil was due to its strong humidification. The treatment caused a very significant (P=0) rise in humidity (2% *vs.* 0.1%). Humidification of oxidized oil would be due to the formation of water and volatile compounds which constitute the final products of decomposition of hydroperoxides.

Table 1. Physico-chemical properties of fresh and thermoxidized sunflower oil.

Characteristic	Fresh oil	Thermoxidized oil
Physical state at room temperature	Fluid	Thick texture, flavor faded
Peroxide value (meq/Kg)	5.83 ± 0.76	152.5 ± 5.33
Acid value (mg KOH/g)	0.182 ± 0.04	2.51 ± 0.11
Free Fatty Acids (%)	0.093 ± 0.02	1.25 ± 0.07
Iodine value (g I_2/100g)	125.84 ± 1.59	80.51 ± 1.37
Saponification value (mg KOH/g)	192.60 ± 2.91	183.79 ± 1.32
Refraction index	1.461 ± 0.037	1.476 ± 0.023
Density at 20°C (g/ml)	0.910 ± 0.018	0.985 ± 0.010
Moisture (%)	0.100 ± 0.012	2.006 ± 0.221
C16 :0 content (%)	6.14 ± 0.451	9.48 ± 0.367
C18 :1, n-9 content (%)	33.500 ± 2.317	46.040 ± 3.939
C18 :2, n-6 content (%)	58.140 ± 2.475	40.590 ± 0.829

Values are mean±standard deviation of triplicate determinations.

Table 2. CDs and CTs extinctions values in both oils.

λ (nm)	Extinction values « K »	
	Fresh oil	Thermoxidized oil
200	-	0.112
232	-	0.036
250	0.0002	0.033
262	-	0.027
268	0.293	0.407
270	0.423	0.655

In addition, this treatment decreased the global unsaturation of thermoxidized oil. The IV is a measure of the total number of double bonds. During the oxidation process, a very significant decrease (P=0) in IV was observed (125.84 *vs.* 80.51). This reduction was more pronounced than those obtained by Blanc-Gondardmary et al. (1989). This variation was due to the more prolonged time of heating applied. Decrease in IV is an indicator of lipid oxidation (Naz et al., 2004). Our results obtained by gas chromatography (GC) showed a dramatically reduced level of unsaturated FA.

PV is a widely used measure of primary lipid oxidation indicating the amount of peroxides formed in fats and oils during oxidation (Ozkan et al., 2007). Our result show a considerable (P=0) increase (5.83 *vs.* 152.5 meq/Kg) in PV after subjecting the oil to oxidation conditions. Such increase in PV had been reported by Neff et al. (1994), Liu and White (1992).

Oxidation of polyunsaturated fatty acids (PUFA) leads to primary and secondary oxidation products. Compounds and amounts of these products vary, depending on the oxidative conditions. To evaluate the oxidation state of the oil, the following parameters have been measured in the two oils. The UV spectrophotometric absorption at 232 and 270 nm, expressed as *K*232 and *K*270, measures the formation of CD and CT, respectively. The *K*232 value increased strongly in oxidized oil reflecting that much CD was formed (Table 2). The spectrophotometric absorption at 270 nm showed for the thermoxidized oils a tendency that resembles very well that of the band located at 833 to 866 cm^{-1} for hydroperoxides, 1739 to 1724 cm^{-1} for aldehydes and 1724 to 1709 cm^{-1} for ketones.

In addition, our experimental conditions supported the peroxidation of linoleic acid and increased the polarity of oil revealed by measurement of PV and IR spectrophotometry at 3636 to 3571cm^{-1} respectively. In the frame of this study, we compared the FA composition of thermoxidized and fresh sunflower oil.

Thermal treatment led to dramatical differences in sunflower oil FA composition; the linoleic acid decreased upon heating, which explained the increase in oleic and palmitic acids. This has been previously reported in heated sunflower oil (Juaneda et al., 2003). The sensitivity of the double bounds of linoleic acid to the combined action of heat and oxygen would be responsible for this finding. Crapiste et al. (1999) demonstrated that as alteration advanced, there was a continuous decrease of unsaturated fatty acids, particularly linoleic acid, being more pronounced at the highest temperature. It resulted in an increase in the oleic acid to linoleic acid ratio (o/l), indicating a preferential use of linoleic acid in oxidation reactions. Otherwise, Marmesat et al. (2009) observed, in high linoleic sunflower oil, that the oxidation took place mainly in the linoleyl group of the triacylglycerols, while the loss of oleyl group was minimum throughout the total oxidation period. However, in high oleic sunflower oil, due to the high relative content in oleic acid, the decrease of oleic acid was similar to that found for linoleic acid.

Different FA isomers were formed during heating and higher differences were found between their contents in the thermoxidized and fresh sunflower oils. The hydroperoxides and CD, relatively stable, absorbed at 232 nm. They are quantifiable by UV spectrophotometric method (Laguerre et al., 2007). These primary products of oxidetion were formed highly in oxidized oil; an extinction of 0.073 was noted. Hydroperoxides can be decomposed into secondary products. Aldehydes and ketones have been reported as major secondary oxidation products. For oxidized oil, the value of this parameter was much higher than for the fresh oil (1.122 *vs.* 0.716). Accumulation of these secondary products in oxidized oil caused deterioration of its flavor.

FT-IR spectroscopy has been used to investigate the chemical changes taking place during lipid oxidation in several edible oils. Differences among the spectra of the fresh and oxidized oils were located at all of the bands suggested by Mazliak (1968) (Figure 2). The intensities of these bands were higher in oxidized oil than in fresh oil which reflects the high level deterioration of oil. Thermoxidized oil contained more hydroperoxides than fresh oil; part of this primary product was converted into secondary products; the C=O stretching band of the aldehydes was much more intense. Thermal oxidation led to strong increases of the bands at 833 to 866 and at 1739 to 1724 cm^{-1} respectively.

Oxidative degradation of sunflower oil was accelerated by heating at 100°C combined to oxygen insufflated. The formation of conjugated double bond systems and the isomerisation of *cis* to *trans* double bonds was observed in the C=C stretching region (980 to 965 cm^{-1}). Thermal oxidation led to strong increases of the band at 980 to 965 cm^{-1} assigned to CD. Thick texture of thermoxidized oil can be attributed to high amounts of CD and CT produced, which is also reflected in *K*270 and *K*232 of this oil; this principal parameters reflect the isomerisation of *cis* double bonds to *trans* double bonds reflecting change of the density and texture of this oil.

In addition to ketones and aldehydes, oxidized oil exhibited a very strong band at 1761 cm^{-1} which corresponds to the C=O stretching (1761 cm^{-1}) of the formic acid (HCOOH). In fact, this compound contributed to the deterioration of the organoleptic quality of oil. This acid will be converted into CO_2 and H_2O, other final products of oxidation.

Conclusion

From the results of various investigated parameters, it can be concluded that the treatment applied to the sunflower oil, which is heating at 100°C with continuous air insufflations for 52 h, caused a high level deterioration with development of oxidative rancidity. Results obtained spectrophotometrically showed very strong band at 980 to 965 cm^{-1} which corresponds to the C=C stretching vibration in conjugated systems; the C=O stretching band of the aldehydes (1739 to 1724 cm^{-1}) was much more intense in oxidized oil regardless of treatment. Overconsumption of these components can be detrimental to health. Significant losses in the essential fatty acid (linoleic acid) were also evident in sunflower oil subjected to treatment.

In further work, we plan to apply this technique to the study of related oxidation processes in biological systems that are gaining increased attention with regard to possible connection between lipid oxidation and pathological events.

REFERENCES

AOCS (1989). Official and Recommended Practices of the American oil Chemists Society, 5th eds. Champaign, I.L. pp. 48-62.

Blanc-Gondardmary P, Revol A, Pacheco H (1989). Chronical ingestion of oxidized oil in young rat. Effect on lipid composition and cytidyl transferase activity. Biomembranes et nutrition, colloque INSERM., Paris, 12-14 juin.

Crapiste GH, Brevedan MIV, Carelli AA (1999). Oxidation of Sunflower Oil During Storage. JAOCS 76 : 1437-1443.

Drozdowski B, Szukalska E (1987). A rapid instrumental method for the evaluation of the stability of fats. J. Am. Oil. Chem. Soc. 64:1008–1011.

Frankel EN (1993). In Search of Better Methods to Evaluate Natural Antioxidants and Oxidative Stability in Food Lipids. Trends Food Sci. Technol. 4:220–225.

Freja N, Mozzon M, Lercker G (1999). Effect of free fatty acids on the oxidative stability of vegetable oil. J. Am. Oil Chem. Soc. 76 : 325-329.

Gray JI (1978). Measurement of Lipid Oxidation: A Review. J. Am. Oil Chem. Soc. 55:539–546.

Grompone MA (2005). Sunflower Oil. Bailey's Industrial Oil and Fat Products, Sixth Edition, Six Volume Set. Edited by Fereidoon Shahidi. Copyright © John Wiley & Sons, Inc.

International Union of Pure and Applied Chemistry (IUPAC) (1987). Standard method for the analysis of oils, fat and derivatives; 7th revised and enlarged edn., edited by C. Paquat and A. Hautfenne, Blackwell Scientific, London.

Juaneda PSB, De la Perriere JL, Sebedio, Gregoire S (2003). Influence of heat and refining on formation of CLA isomers in sunflower oil. J. Am. Oil Chem. Soc. 80:937–940.

Laguerre M, Lopez-Giraldo LJ, Lecomte J, Pina M, Villeneuve P (2007). Outils d'évaluation in vitro de la capacité antioxydante. Oléagineux, Corps Gras, Lipides. 14 : 278-292.

Liu RH, White RJ (1992). High temperature stability of soybean oils with altered fatty acid compositions. J. Am. Oil. Chem. Soc. 69:533-537.

Marmesat S, Morales A, Velasco J, Ruiz-Méndez MV, Dobarganes MC (2009). Relationship between changes in peroxide value and conjugated dienes during oxidation of sunflower oils with different degree of unsaturation. Grasas Y Aceites 60 : 155-160.

Mazliak P (1968). Le métabolisme des lipides dans les plantes supérieures. [Ed. Masson et Cie. pp. 224.

Min DB, Boff JF (2001). Lipid oxidation of edible oil. In C. Akoh & D. B . Min edn. Food lipids. New York, Marcel Dekker. pp. 335-363.

Muik B, Lend B, Molina A, Ayora MJ (2005). Direct monitoring of lipid oxidation in edible oils by Fourier transform Raman spectroscopy. Chem. Phys. Lipids. 134:173–182.

Nawar WW (1996). Lipids. In O.R. Fennema edn. Food chemistry 3rd edn. New York: Marcel Dekker. pp. 225-319.

Naz S, Sheikh, H, Saddiqi, Sayeed SA (2004). Oxidative stability of olive, corn and soybean oil under different conditions. Food Chem. 88:253-259.

Neff WE, Mounts TL, Rinsach WM, Konishi H, Elegamy MA (1994). Oxidative stability of purified canola oil triacylglycerols with altered fatty acid compositions. J. Am. Oil Chem. Soc. 71:215-221.

Official Methods of Analysis, AOAC (1999). 16th eds. AOAC International, Gaithersburg.

Ozkan, G, Simsek B, Kuleasan H (2007). Antioxidant activity of satureja cilicica essential oil in butter and in vitro. J. Food Eng. 79:1391-1396.

Rossell JB (2001). Frying. Improving quality. In J. B. Rossell edn. Cambridge England. Woodhead Publishing Limited.

Shahidi F, Janitha PK, Wanasundara PD (1992). Phenolic antioxidants. Crit. Rev. Food Sci. Nutr. 32:67–103.

Warner K, Eskin NAM (1995). Methods to Assess Quality and Stability of Oils and Fat-Containing Foods. AOCS Press, Champaign.

Biological control of apple gray mold by mixtures of *Bacillus Subtilis* and yeast isolates

E. Zangoei, H. R. Etebarian and N. Sahebani

Plant Pathology, University of Tehran.Tehran, Iran.

In this study the effect of biocontrol agents *Candida membraniciens* (isolates A4, A5), *Pichia guilliermondii* (isolate A6) and *Bacillus subtilis* (isolates B2, B6) were evaluated individually and in combination on gray mold of apple. Our results show that the antagonists were compatible when they were tested *in vitro*. Also, results show the spore germination of *Botrytis cinerea* decreased significantly by the combination of B2+A5 more than the other treatments. In the dual culture test, the combinations of B2+A5 and B6+A6 prevented mycelial growth of pathogen and that the combined application of the two agents was more effective in control of gray mold of apple *in vivo* than the application of each one alone. In different proportions antagonist test, when the inoculum size favored A5 was 70 in the mixed, the biocontrol could be improved. This study suggested that antagonists are more effective in biological control of apple gray mold when used in proper combinations than when each one is used alone.

Key words: Gray mold, biocontrol, mixtures antagonist, *Botrytis cinnerea*.

INTRODUCTION

Fruits and vegetables suffer significant losses from fungal diseases after harvest (Filonow, 1998). Postharvest losses of fruits and vegetables are high, ranging from 10 to 40% depending on the species. Among them gray mold of apple is created by wound-invading necrotrophic *Botrytis. cinerea Pers.:Fr* (Romano et al., 1983). Currently, Thiabendazole, is the main fungicide globally used for the control of postharvest fungal fruit decays of apple (Pusey, 1989). The fungicide is applied as the drench treatment before or after the cold storage. The development of resistance in fungal pathogens to fungi-cides and the growing public concern over the health and environmental hazards associated with high levels of fungicides have resulted in a significant interest in the development of alternative nonchemical methods for control of post harvest diseases (Gullino et al., 1994). Biological control using some antagonistic microorganisms is the safe method for control of postharvest pathogens (Yu and Zheng, 2006). Puesey, (1984) and Lima et al. (1997) claimed that considerable success was achived by utilizing antagonistic microorganisms for controlling postharvest diseases. In recent years many attempts are made to develop the efficacy of existing biological agents. There are many aproaches to achieve this goal through: Enhancing efficacy by addition of fungicides (Chand-Goyal and Spotts, 1996); salt (Gholamnejad and Etebarian, 2009); nutrients (El-Ghaouth et al., 2000a); chitosan (El-Ghaouth et al., 2000b);integration with physical means (Stevens et al., 1997); and finally using mixture of antagonists (Janisiewicz, 1988). The mixture of biological control agents have many benefits: it can increase the constancy of biological control of apples in the field and later in storage (Leibinger et al., 1997), and also it can be effective to control many post hasvest diseases at the same time (Janisiewicz, 1996). Also suggested, the employment of more than one biocontrol agent caused to survive biological control in unsteady position (Guetsky et al., 2001). And, the combination of antagonists with different mechanisms of disease suppression might simultaneously suppress many pathogens (Guetsky et al., 2002). Several studies have tested different biocontrol strains in combination, for example Leibinger et al. (1997) showed that mixture of two yeast isolates and one bacterium controlled the postharvest pathogens *Pencillium expansum*, *B. cinerea* and *Pezicula malicorticis* on apple. Calvo et al. (2003) showed that mixing different isolates of yeast *Rhabdotorula* and *Cryptococus* were considerably more effective in controlling *P. expansum* and *Botrytis cinerea* on Red Delicious apple fruits. Guetsky et al. (2001) combined *Pichia guilliermondii* with *Bacillus mycoides* to improve control of gray mold of strawberries by *B. cinerea* in storage. In this study, we applied different mixture of antagonists for controlling postharvest gray mold of apples.

MATERIALS AND METHODS

Atagonists

The antagonistic yeasts *Candida membranifuciens* (A4 & A5) and *Pichia guilliermondii* (A6) were obtained from plant pathology laboratory of Abureihan Campus, University of Tehran. The isolates were identified previously in Centralbureau voor Schimmelcultures (CBS). For biocontrol experiments, yeasts were grown on PDA media for 24 to 48 h at 25°C before use, then cultures were harvested by a bacteriological loop and resuspended in sterile distilled water and cell concentrations were adjusted with sterile distilled water to 1 × 10^7 cell/ml.

The *Bacillus subtilus* isolates B2 and B6 were also obtained from plant pathology laboratory of Abureihan campus, university of Tehran. *B. subtilus* isolates were grown in 250 ml erlenmeyer flasks containing 50 ml of NYDB (*Nutrient Yeast Dextrose Broth)* on a rotary shaker (150 rpm) at 25°C. Then the cultures were harvested by centrifugation and resuspended in sterile distilled water and the population of the bacterial cells was adjusted to 10^8 cells/ml (optical densities (OD) 0/05 transmittance at 590 nm). For biocontrol experiments suspensions of the yeast and bacterium at each concentration were mixed in proportions of 50: 50 (v/v).

Pathogen

B. cinerea was isolated from decayed apple. The fresh cultures were grown on PDA plate at 25°C before use then spore suspension were prepared from the spourulating edges of 10 day old culture with a bacteriological loop, then suspending them in sterile distilled water containing 0.05% (v/v) Tween 80. Spore concentration was determined with hemacytometer and adjusted with sterile-distilled water to 1 × 10^5 spore/ml.

Plant

Golden Delicious apples were obtained from organically grown position in orchard in Damavand, Tehran.

Compatibility of biocontrol agents with each other

The compatibility of the biocontrol agents (yeast and bacteria isolates) with each other was tested following the methods of antibiosis and mixed culture tests, *in vitro*.

Antibiosis test

15 mm diameter disks from 5 days old NYDA cultures of the three yeast isolates (A4 , A5 , A6) were cut and then they were spotted onto center of one surface dry plate PDA (9 cm plate). The plates were incubated for 72 h at 25°C. Following incubation, the plates of the yeast isolates were oversprayed with a suspension (10^8 cell/ml) of *Bacillus subtilis* isolates (B2, B6). The plates were incubated for 48 h at 25°C. Then, the plates were examined for the presence of inhibition zones. All treatments consisted of four replicates (Wilson and Lindow, 1994).

Mixed culture test

To study interaction of the antagonists *in vitro*, the cultures of yeast and bacteria were resuspended in water (adjusted to 10^8, 10^7 cell/ml bacteria and yeast isolates respectively) and 100 micro liter of these suspensions were added individually to 10 ml Erlenmeyer flasks containing 5 ml of NYDB. Another set of flasks was inoculated with a combination of both organisms in the proportion of 50:50. The flasks were incubated on the shaker at 150 rpm and 24°C for 30 h. Samples from the flasks were taken at 15 and 30 h intervals and were plated on PDA or PDA supplemented with 25 mg of streptomycin per liter for recovering bacterial and yeast cells respectively. This samples were inoculated with mixture of the antagonists were plated in duplicate, once on regular PDA to recovery of the bacterial antagonist and the second on PDA supplemented with streptomycin sulfate at 25 mg/lL, which inhibited the bacterium but allowed the yeast to grow. Finally, for evaluate the yeasts population, the colonies were counted with a colony counter after incubation for 48 h. All treatments consisted of three replicates (Janisiewicz and Bors, 1995).

In vitro antagonism

The biocontrol agents were tested alone and in different combinations for their effectiveness against the mycelial growth of *B. cinerea* by the dual culture technique (Etebarian et al., 2005). Half of the agar surface were streaked with 100 µl suspension of 10^7 Cell/ml of the yeast isolates (A4, A5 and A6), 10^8 cell/ml of *B. subtilis* isolates (B2 and B6) and an equal their mixtures (50:50). After 2 days of incubation, a mycelial disc (9 mm) of *B. cinerea* of ten-day old culture was placed on the other side of each plates. For control, the medium was incubated with the pathogen alone. Four replications were carried out for each treatment; the dishes were incubated at 25°C for 7 days and the percent growth inhibition was calculated using the formula: $n = (a-b)/a \times 100$, where *n* is the percent growth inhibition, *a* is the colony area of *B. cinerea* for the control and *b* is the colony area of *B. cinerea* treated with different combinations of the yeasts or the bacteria.

Effects of alone or the combination treatments of antagonists on spore germination of pathogen were carried out in PDB culture. The 100-µl of 1 × 10^8 Cell/ml of the washed cell suspension of bacteria isolates (B2, B6) and 1 × 10^7 Cell/ml of the yeast isolates (A4, A5, A6) and the sterile distilled water as the control were added into 10 ml glass tube containing 5 ml PDB respectively. Another set of glass was inoculated with same quantity of a combination of both antagonists in the proportion 50:50. At the same time, aliquots 100 µl of spore suspensions 1 × 10^5 spores/ml of *B. cinerea* were added into each tube. After 20 h incubation at 25°C, at least 100 spores per replicate were observed microscopically to determine germination rate on a rotary shaker (150 rpm). All treatments consisted of four replicates (Droby et al., 1997).

In vivo biological control studies

Fruits were surface-disinfected with sodium hypochlorite (0.1%) for 2 min and washed two times by immersion in distilled water,Then allowed to air-dry and wounded 3 × 3 × 3 mm in three places on the equator of the fruit with a sterilized needle , then 40 µl of suspension 10^8 and 10^7 Cell/ml of bacteria and yeast isolates respectively (singly or their mixtures 50:50) was pipeted in to apple wounds. after 24 h , the treated wounds were inoculated with 20 µl of *B. cinerea* spore (1×10^5 spore / ml). The lesion diameters were determined 15 and 30 days after storage at 20 and 4°C in 95% relative humidity. Each apple constituted a single replicate and each treatment was replicated four times. Wounded apples were distributed into five sets: (1) Non-treated apples (control), (2) Apples treated with isolates yeast or bacteria alone (antagonist), (3) Apples inoculated with *B. cinerea* alone (pathogen), (4) Apples treated with isolates yeast or bacteria individual and then inoculated with *B.cinerea* (antagonist+ pathogen) (5) Apples treated with mixture of antagonists and inoculated with pathogen (mixture of antagonists + pathogen). The percentage of lesion reduction was estimated according to the formula:

$$LR = (Dc - Dt) / Dc \times 100$$

Where, Dc and Dt are, the lesion diameters of the control and treated apples respectively (Etebarian et al., 2005). Limpel's formula also was used to determine the presence of synergistic or antagonistic interactions between the antagonists in mixtures:

$$Ee = X + Y - X.Y/100$$

Where, *Ee* is the expected effect from additive responses of two inhibitory agents and *X* and *Y* are the pecentages of lesion reduction for each agent used alone. If the combination of the two agents produces any value of inhibition greater than Ee, then synergism exists, on the same basis, if the observed effect (pecentage of lesion reduction) would be less than the expected effect, then the mixture exhibits antagonism (Limpel et al., 1962).

Biocontrol test *in vivo* in the various proportions of mixture antagonists

In vivo biological control studies when two treatments as B2+A5 and B6+A6 were combined at approximately equal biomass (50:50), they were consistently compaierd to the other combinations. The apple fruits were disinfected and wounded as described above, then aliquots 40 µl suspensions at 10^7 and 10^8 cell/ml yeast and bacterium were mixed in proportions from 0:100, 30:70, 70:30, 50:50, 60:40, 40:60, 100:0. This proprtions were pipetted into wounds after 24 h, 20 ml of 10^5 spores/ml suspension of *B. cinerea* were inoculated into each wound. Treated fruits were stored at 20°C for 15 days or 4°C for 30 days. The lesion diameter was recorded afterwards. There were three replicate trials of 5 fruits per treatment with complete randomization (Janisiewicz and Bors, 1995).

Recovery of the antagonists

Two combinations B2+A5 and B2+A6 was conducted to recover the antagonists. Three wounds per fruit were made as described above, and then 40 µl of each antagonist alone and an equal mixture (50:50) was pipetted to each wound. After 24 h, the tested wounds were inculated with 20 µl of *B. cinerea* spore (1 × 10^5 spore/ml).The fruit apples were stored for 30 day at 4⁝C, then the fruit samples were taken to determine antagonist populations after 15 and 30 days of storage. The wounded area was removed with a cork borer (no. 5) and placed in a sterile mortar, and ground with pestle. For each apple, recovered cell suspensions from the three wounds were diluted 10^3. 10^4 and 10^5 Serial tenfold dilution and were plated on PDA or PDA supplemented with 25 mg of streptomycin per liter for recovering bacterial and yeast cells, respectively. Samples from flasks inoculated with the mixture of the antagonists were plated in duplicate on PDA and PDA plus streptomycin. The colonies were counted with a colony counter after incubation for 24 and 48 h for the bacteria and yeast, respectively (Teixido et al., 2000).

Data analysis

The *in vitro* and *in vivo* assays were analyzed by an analysis of variance (ANOVA) with SAS Software (SAS Institute, version 9.0, Cary, NC). All assays were carried out with three/four_replicates. Statistical significance was judged at the level $p < 0.05$. When the analysis was statistically significant, Duncan's Multiple-Range Test (SSR Test) was used to test mean separations among mean values of each treatment (Little and Hills, 1978).

RESULTS

Compatibility of biocontrol agents with each other

In the test antibiosis, the isolates which grew togother were compatible, but isolates that were separated by an inhibition zone were incompatible. From these results none of the antagonists showed antibiosis effect on each other. Both *B. subtilis* isolates (B2 and B6) and the yeast isolates grew together and it indicated that these isolates were compatible.

Interaction of the antagonists *in vitro* showed the mixtures of B2+A4, B2+A5, and B2+A6 the growth of yeast isolates wasn't effected by the presence of *B.subtilis* isolate B2 ($P \leq 0.05$). In NYDB inoculated with yeast isolate alone, the yeast population increased in all flasks after 15 h, and the cells increased until end of the experiment. However, the increases of the mixtures yeast isolate (A4, A5 and A6) with the bacteria, almost were equal with alone antagonists after 30 h (Figure 1 A). Also, in the mixtures of B6+A4, B6+A5 and B6+A6, the growth of yeast isolates was not affected by the presence of *B. subtilis* isolate B6 ($P \leq 0.05$). Results showed that the populations of the yeast isolates inoculated in mixture of *B.subtilis* increased in all flasks after 30 h. In the other hand, the application of the yeast population alone increased in all flasks after 15 h, and then it increased until end of the experiment (Figure 1 B).

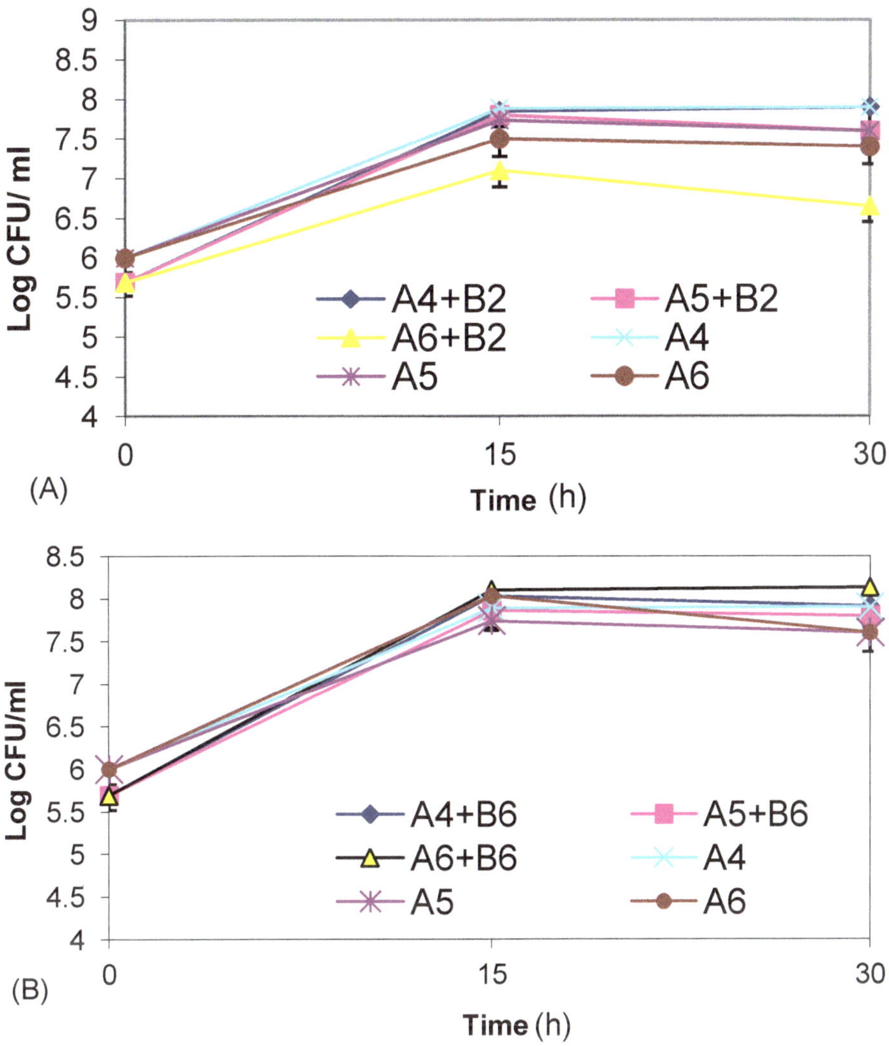

Figure 1. Population dynamics of yeast isolate (A4, A5, A6) by the presence of *B.subtilis* B2 (A), B6 (B) alone or in mixtures in NYDB. The samples were collected at 15 h intervals. Bars represent standard error.

In vitro biological control studies

Results show that all treatments reduced significantly the mycelial growth of *B. cinerea* (P≤0.05). The tretment B2+A5 had the highest percent of inhibition at 53.26%, if the single isolates inhibition A4, A5, A6, B2, B6 were 20.5, 21.28, 32.21, 40.54 and 39.73 %,respectively. The percent inhibition of *B. cinerea* of other treatments such as B2+A4, B2+A6, B6+A5 and B6+A4 was equal to, or further than the inhibition achieved by the isolates applied alone (Table 1).

Spore germination of *B. cinerea* in PDB culture was inhibited strongly in the presence of active cells of antagonists and their mixtures (P≤0.05). At the end of 20 h incubation at 25°C, the spore germination of *B. cinerea* incubated with the combination B2+A5 was 10% while the spore germination of the other combinations were equal to, or further than the isolates applied alone (Table 2).

In vivo biological control studies

The results show that wound diameter of all treated fruits were significantly lower than the untreated controls (P≤0.05). The lesion diameters of the biocontrol combinations B2+A5, B6+A6, B6+A4 on treated fruits were 5/23, 7/43, 11/66 mm, respectively, after storage at 20°C for 15 day, that provided better control than the antagonist applied alone B2, B6, A5 and A4 that had a lesion diameter of 15/06, 15/22 13/57 and 12/99 mm respectively.

On the other hand the lesion diameter, in the apple fruits treated with the combinations of yeast and bacteria, the lesion diameter of fruits with the best combination such as B2+A5, B6+A6 and B6+A5 was 3/48,3/87 and

Table 1. Effectiveness of mixture *Bacillus subtilis* and yeast isolates in inhibiting mycelial growth of *Botrytis cinerea*.

Treatment isolateds[1]	Mycelial growth of the pathogen (mm)	Inhibition over control (%)
A4	31.96$^{b\,2}$	20.5
A5	31.65^{bc}	21.28
A6	27.25^{dc}	32.21
B2	23.9^{de}	40.54
B6	24.23^{de}	39.73
B2+A4	22.94^{def}	42.64
B2+A5	18.79^{f}	53.26
B2+A6	31.48^{bc}	21.7
B6+A4	22.65^{dfe}	43.66
B6+A5	23.53^{de}	41.47
B6+A6	22.07^{fe}	45.09
Control	40.21^{a}	0.0

[1]*Bacilus subtilis* isolate B2 and B6 , *C. Membranifuciens* (A4 and A5), *P. guilliermondi* (A6). [2]Values are the means of four replications. Means followed by a common letter are not significantly different at the 5% level by Duncan's Multiple Range Test.

Table 2. Effectiveness of mixture *Bacillus subtilis* and yeast isolates in inhibiting Spore germination of *Botrytis cinerea*.

Treatment isolateds[3]	Spore germination[1] (%)
A4	59$^{ab\,2}$
A5	45.33^{cb}
A6	13^{de}
B2	14.66^{de}
B6	19.66^{de}
B2+A4	28^{dc}
B2+A5	10^{e}
B2+A6	55.33^{ab}
B6+A4	30^{cd}
B6+A5	21.66^{de}
B6+A6	19.33^{de}
Control	82.33^{a}

[1]Germination rate were measured microscopically after 20 h incubation at 25°C in PDB. [2]Each value is the mean of four experiments. Values in each column followed by different letters are statistically different according to Duncan's multiple range test at(P≤0.05). [3]*Bacilus subtilis* isolate B2 and B6, *C. Membranifuciens* (A4 and A5), *P. guilliermondi* (A6) .

5/16 mm, respectively, wheras the biocontrol isolates applied alone such as B2, B6, A6, A5 and A4, it was only 13.29, 10/83, 8/97,11/67 and 16/01 respectively, after storage for 30 day at 4°C (Figure 2 A and B).

The expected effect of Ee, accordance to the formula Limpele, showed that B2+A5 and B6+A6 compounds at 4°C and B2+A5 at 20°C had a synergistic effect in controlling gray mold. Its reason is that the calculated LR was more than Ee.In other compounds, the number of calculated Ee was more than the number of suppresions percent LR, it showed that these compounds had the antagonistic effects against gray mold (Table 3).

Biocontrol test *in vivo* in the various proportions of mixture antagonists

The results showed that the different proportion of the antagonists affected the lesion diameter, significantly (P≤0.05). Also, the results show, the mixture of B2+A5 (proportion of 30% of bacterium and the proportion of 70% of yeast), reduced the lesion diameter to 5 mm, this compounds didn't have significant difference with other mixtures (with different proportions antagonists) expect of B2(70)+A5(30) at 20°C. Also, the results showed that there is a significant difference between treatments at 4°C (P≤0.05). The treatment of B2+A5 with 50:50 ratio for yeast and bacteria, and other mode with 30:70 ratio for bacreria and yeast was more effective than other proportions and the antagonists applied individually on lesion development, respectively (Figure 3 A and B).

Recovery antagonists

The results showed that both *B. subtilis* (B2) and yeast isolates (B6) multiplied rapidly in apple wounds in the presence of *B. cinerea*. At the start of the experiment (time 0), the number of per yeast and bacterium strain was 2 × 10^5 CFU/ml for mixture of B2+A5, also the population of *C. Membranifuciens* (A5) and *B. subtilis* (B2) increased from 1.1 × 10^7 and 6.31 × 10^6 CFU/wounds, respectively at 15 d. But after 30 days incubation at 4°C, the yeast and bacteria population began to decline. The comparison between logarithm of mixed antagonists and their individual applications showed that the population of B2 and yeast A5 in mixture is more than they applied alone. Indeed the results showed that a positive interaction probably occurred between these antagonists in mixture and in this position, the quantity of both antagonist's cell is more than when they applied alone. The study of antagonist's population in mixture of B2+A6 have been showed that the population logarithm of alone antagonists as more than their mixture in all test period, this can be due to an antagonism effect among them (Figure 4 B).

DISCUSSION

In earlier studies the multiple strain mixtures of bio-

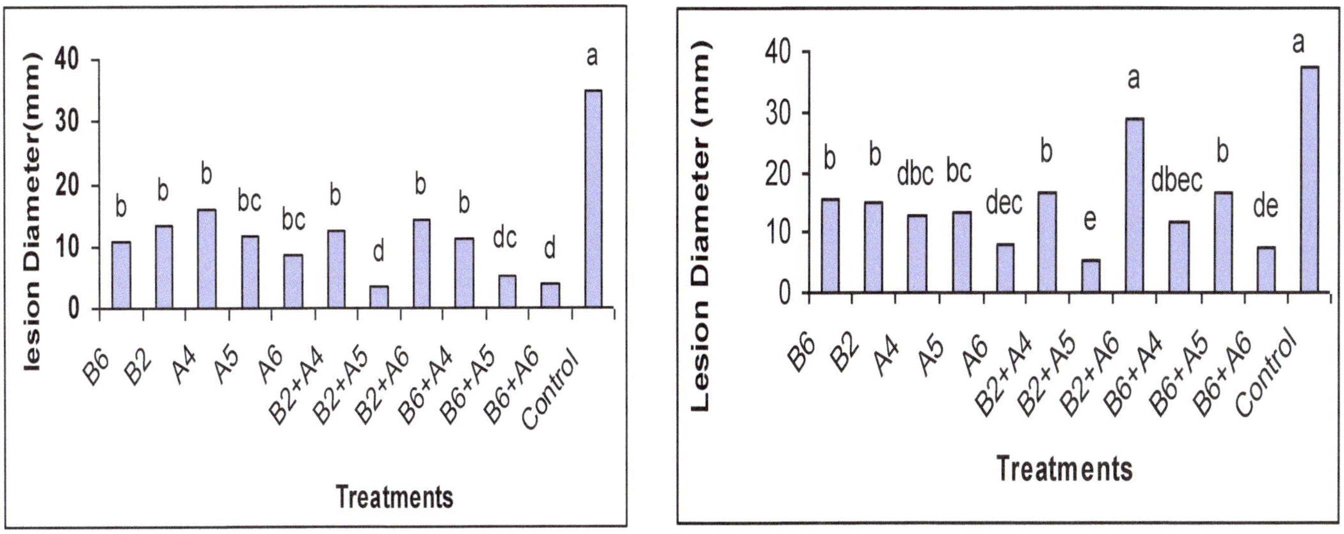

(A) (B)

Figure 2. Inhibition of gray mold in apple fruits .the lesion diameters were measured after 30 days incubation at 4°C (A), and 15 days incubation at 20°C (B). Different letters in the assay indicate significant differences between means according to Duncan's multiple range test (p ≤ 0.05).

Table 3. percentage of lesion reduction of mixture *B.subtilis* isolate (B2,B6) and yeast isolates of (A4,A5 and A6) and expected effect (Ee) of mixtures according to Limpel's formula.

	20°C		4°C	
Treatment isolateds[2]	lesion reduction (LR) %	Ee	lesion reduction (LR) %	Ee
A4	65.14^{bc1}		54.24^{c}	
A5	60.56^{bc}		66.64^{bc}	
A6	77.93^{a}		72.22^{abc}	
B2	59.63^{bc}		62^{c}	
B6	55.55^{dc}		69.09^{cb}	
B2+A4	66.59^{b}	91.09	64.38^{c}	90.58
B2+A5	86.94^{a}	85.07	90.02^{a}	87.32
B2+A6	28.41^{e}	85.92	58.96^{c}	82.61
B6+A4	59.32^{dbc}	90.18	68.32^{cb}	92.32
B6+A5	50.21^{d}	82.46	85.23^{ab}	89.66
B6+A6	80.01^{a}	84.50	88.93^{a}	85.83

[1]Each value is the mean of three experiments. Values in each column followed by different letters are statistically different according to Duncan's multiple range test at (P≤%05). [2]*Bacilus subtilis* isolate B2 and B6, *C. Membranifuciens* (A4 and A5), *P. guilliermondi* (A6) .

control agents have been employed successfully against plant pathogens (Janisiewicz, 1996). Several authors have suggested that combinations of bio-control agents have to be compatible with each other for more consistent results of biological control (Raaijmakers et al., 1995). According to *in vitro* antibiosis test the selected antagonist strains do not have any antibiotic element against each other. Interaction of the antagonists *in vitro*, also showed that the growth of yeast isolates were not affected by the presence of *B.subtilis* B2 and B6 isolates in NYDB culture. It indicated that the *B.subtilis* isolates B2 and B6 were compatible with yeast strains A4, A5 and A6. Tillugavati et al. (2007) so as to affirm the compatibility between two bacterium (*Bacillus* and *Psudomonase*) and one fungus (*Trichdema*) in mixture, used the dual culture test of these antagonists with pathogen. They concluded that these three biocontrol agents create more inhibition of growth of mycelia of *B. cinerea* than single antagonists used alone, it was suggested its reason is the compatibility between bio-

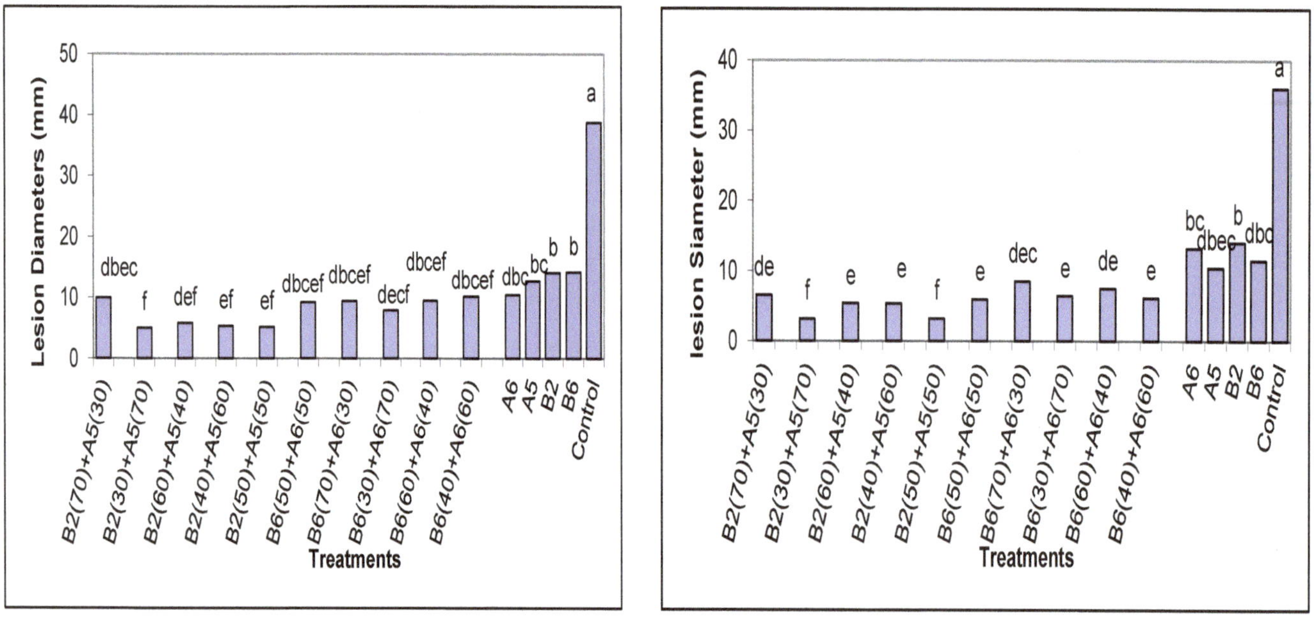

(A) (B)

Figure 3. Inhibition of grey mold in apple fruits as affected by various proportions(0:100, 30:70,70:30,50:50,60:40,40:60,100:0) treatments of the antagonist. lesion diameters were measured after 15 days incubation at 20°C (A), and 30 days incubation at 4°C (B). Values followed by different letters are significantly different according to Duncan's multiple range test at p ≤ 0.05.

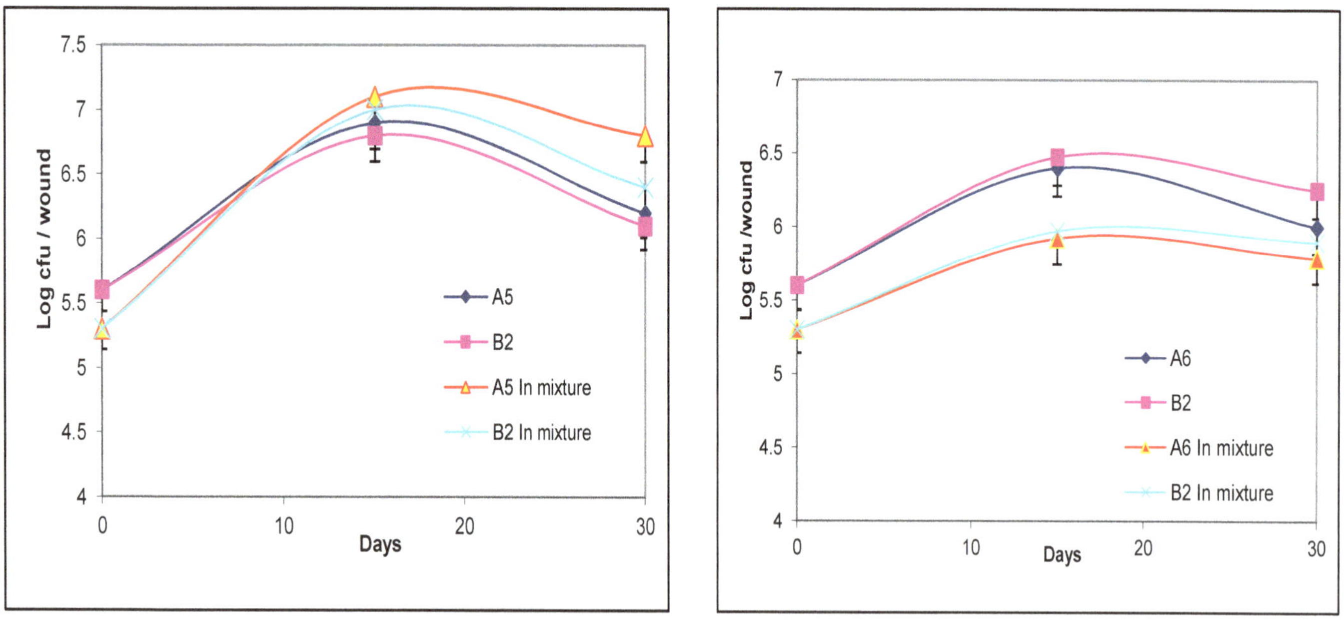

(A) (B)

Figure 4. Population dynamics in wounds of Golden Delicious apples inoculated with *B.subtilis* B2 and *C.Membranifuciens* A5 isolate (A) and *B.subtilis* B2 and *Pichia guilliermondi* A6 isolate (B) the individually or in a 50:50 mixture and incubated at 4 °C. Bars represent standard error.

control agents. In our experiments, the compounds of B2+A5 and B6+A6 created the most inhibition of the growth of mycelia and spore germination of *B. cinerea*; this can be due to the compatibility between biocontrol agents with each other. The combinations of B2+A5 and B6+A6 and a less extent B2+A4, B6+A4 and B6+A5, controlled apple gray mold more effectively when they were used alone. In addition, Limpel's formula (Limpel et al., 1962) and population size of yeasts allowed us to evaluate the interactions between yeast and bacteria isolates when they were used in mixtures. According to Limpel's formula the *B. subtilis* (B2) and *C. membranifaciens* (A5) mixture showed synergism against *B. cinerea*, because the LR% was calculated higher than Ee. Dynamic of populations, also explained that both *C. membranifaciens* A5 and *B. subtilis* B2 reached a higher cell concentration when they used in a mixture mode instead of using alone. These results agree with the findings of Janisiewicz and Bors who said carrying capacity of the wounds could be greater than the population of the single antagonists and they mentioned that another antagonist may caused to increase the population of the antagonists in the wounds to increase more improves of biocontrol (Janisiewicz and Bors, 1995). Also, they found that screening of specific amino acids may specifically enhance the growth of the antagonists but had no effect on pathogen. It was when the specific nutrient is available, the growth of the antagonist which metabolize it enhance, so both antagonists cannot compete for this nutrient and it creates desirable position for growth of both antagonists. Also, this caused even greater depletion of essential nutrients for development of *B.cinerea* and eventually better biocontrol. These results also agree with the findings of Calvo et al. (2003) and Janisiewicz (1996). In a more comprehensive study Gutsky et al. (2002) showed types of biocontrol mechanisms *B. cinerea* on strawberry by mixture of yeasts (*Pichia guilermondii*) and a bacterium (*Bacillus mycodes*) by scan of Electron Microscopy, they concluded that in the presence of yeast, some of *B. cinerea* spore cannot germinate, They indicated that *P. guilermondii* competed with *Botrytis cinerea* for glucose, sucrose, adenine, histidine, and folic acid. It was found that the half of spores of *B. cinerea* hydrolyzed fully in presence of *B. mycodes*. They suggested that these observations may be due to the presence of cell wall hydrolyzing enzymes. But when the mixture of two biocontrol agents was used, the spores were hydrolyzed with more intensity. Likewise, they expressed when both biocontrol agents were applied in a mixture, their activity reflected the sum of the biocontrol mechanisms. Most of researches on the mixtures of biocontrol agents showed that combinations of antagonists caused to improve biocontrol. But, studies show that some combinations of biocontrol agents could not improve suppression of disease compared with the separate antagonists (Thilagavathi et al., 2007; Leibinger et al., 1997). In this study, the mixture of B2+A6, had an equal or/and lower control than the separate antagonists. This may be due to an incompatibility between *B. subtilis* isolate B2 and *P.guilliermondi* isolate A6. Also, Limpel's formula showed that the interaction between them is antagonism. Morever, the dynamic of populations of B2 has a negative effect on the growth of A6 and application of isolate A6 in alone mode is more effective than in the mixture mode. Results of compatibility of biocontrol agents with each other *in vitro*, demonstrated that at least inhibitory substances (antibiosis) are not involved, probably. However many same studies showed that the basis of this incompatibility is unknown (Schisler et al., 1997).We can compare these results with Leibinger's test that saied a mixture of two yeast isolates and one bacterium controled the post-harvest pathogens *Penicillium expansum*, *B. cinerea* and *Pezicula malicorticis in vivo*. In their test the population size of yeast isolate *A. pullulans* reduced when it was applied in combination with *B. subtilis*, they reported that the basis of this incompatibility could result from the production of antibiotic compounds by *Bacillus subtilis* (Leibinger et al., 1997). From the test of different proportions of antagonist in B2+A5 mixture, we can understand that biocontrol would be improved when we use more inocoulom yeast A5 with 70%, its reason could be more colonization of yeasts in apple's wounds. It shows that yeasts, for control of the pathogens, have stronger mechanism than bacterium that include competition for nutrients and space (Leibinger et al., 1997). Calvo et al. (2003) showed that the gray mold in mixture of *Rhodotorula glutinis* SL1 and SL30 with inocoulom 1:2 for SL1, is controlled more better than proportion 2:1 for SL30. They concluded that the control of disease is depended on the size of inocoulom of antagonists in mixture. In this reserch, we showed that application of antagonists in combination mode is more effective than the application used alone, for a period of 30 days at 4°C and a period of 15 days at 20°C. The adaptation of these strains to a wide range of temperature (4 to 20 C°) provides great potential for this antagonist mixture for control of post-harvest diseases on apples in storage and transportation conditions. We hope that other biologists, physicians and agronomists work together for culturing and formulating of these agents and make commercial using of these antagonist microorganisms in the near future.

Conflict of Interests

The author(s) have not declared any conflict of interests.

ACKNOWLEDGMENTS

This paper is a part of MSc. thesis by the first author submitted to the Department of Plant Protection, College of Abureihan, University of Tehran. The authors thank the University of Tehran for funding this project. We are grateful to F. Kord and R. Ghashghaie for their assistance in the correction of this manuscript.

REFERENCES

Calvo J, Calvente VDE, Orellano M, Benazzi D, Tosetti M (2003). Improvement in the biocontrol of postharvest diseases of apple with the use of yeast mixture. Biocontrol. 48:579–593.

Chand-Goyal T, Spotts RA (1996). Control of postharvest pear diseases using natural saprophytic yeast colonists and their combination with low doses of thiabendazole. Postharvest Biology. Technol. 7:51–64

Droby ME, Wisniewski L, Cohen B, Weiss D, Touitou Y, Eilam E (1997). Influence of CaCl2 on *Penicillium digitatum*, grapefruit peel tissue, and biocontrol activity of *Pichia guilliermondii*, Phytopathol. **87**:310–315.

El-Ghaouth A, Smilanick J, Wisniewski M, Wilson CL (2000a). Improved control of apple and citrus fruit decay with a combination of *Candida saitoana* with 2-deoxy-D-glucose. Plant Disease. 84:249–253.

El-Ghaouth A, Smilanick J, Eldon GE, Ippolito A, Wilson CL (2000b). Application of *Candida saitoana* and glycolchitosan for the control of postharvest diseases of apple and citrus fruit under semi-commercial conditions. Plant Disease. 84:243–248.

Etebarian HR, Sholberg PL, Eastwell KC, Sayler RJ (2005). Biological control of apple blue mold with *Pseudomonas fluorescens*. Microbiol. 51:591-598.

Filnow AB (1998). Role of competition for sugars by yeasts in the biological control of gray mold and blue mold of apple. Biocontrol Sci. Technol. 8:243-256.

Gholamnejad J, Etebarian HR (2009). Effect of calcium chloride on the biocontrol efficacy of two antagonistic yeasts against *Penicillium expansum* on apple fruit. Phytoparastica. 37:255-267.

Guetsky R, Shtienberg D, Elad Y, Dinoor A (2001). Combining biocontrol agents to reduce variability of biological control. Phytopathol. 91:261-267.

Guetsky R, Shtienberg D, Elad Y, Fischer E, Dinoor A (2002). Improving biological control by combining biocontrol agents each with several mechanisms of disease suppression. Phytopathol. 92:976-985.

Gullino, M.L. and Kuijpers, L.A.M . 1994. Social and political implications of managing plant diseases with restricted fungicides in Europe. Annu. Rev. Phytopathol. **32**: 559–579.

Janisiewicz WJ (1988). Biocontrol of postharvest diseases of apples with antagonist mixtures. Phytopathol. 78:194-198.

Janisiewicz WJ, Bors B (1995). Development of a microbial community of bacterial and yeast antagonists to control wound-invading postharvest pathogens of fruits. Enviro. Microbiol. 61:3261-3267.

Janisiewicz W (1996). Ecological diversity, nich overlap, and coexistence of antagonists used in developing mixture for biocontrol of postharvest disease of apple. Phytopathol. 86:473-479.

Little TM, Hills FJ (1978). Agricultural Experimentation Design and Analysis, John Wiley and Sons, New York, USA.

Lima G, Ippolito A, Nigri F, Salerno M (1997). Effectiveness of *Auerobasidium pulluans* and *Candida olephia* against postharvest strawberry rots. Postharvest Biology .Technol.10:169-178 .

Leibinger W, Breuker B, Hahn M, Mendgen K (1997). Control of postharvest pathogens and colonization of the apple surface antagonistic microorganisms in the field. Phytopathol. 87:1103-1110.

Limpel L, Shuldt P, Lamont D (1962). Proc. N.E. *Weed Control Conf.* 16: 48–3.

Puesey PL, Wilson CL (1984) .Postharvest biological control of stone fruit brown rot by *Bacillus subtilis* .Plant Disease . 98:753-756.

Pusey LP (1989). Use of *Bacillus subtilis* and related organisms as biofungicides. Pestic. Sci. 27:133-140.

Romano ML, Gullino ML, Garibaldi A (1983). Evaluation of the sensitivity to several fungicides of postharvest pathogens in North-western Italy. Mededelingen van de Faculteit Landbouwwetenschappen Universiteit Gent, 48:591–602.

Raaijmakers JMI, Van dSIM, Koster PA, Bakker PJ, Schippers B (1995). Utilization of heterologous siderophores and rhizosphere competence of fluorescent Pseudomonas spp. Canadian J. Microbiol. 41:126–135.

Schisler DA, Slininger PJ, Bothast RJ (1997). Effects of antagonist cell concentration and two-strain mixtures on biological control of Fusarium dry rot of potatoes. Phytopathol. 87:177-183.

Stevens C, Khan VA, Lu JY, Wilson CL, Pusey PL, Igwegbe ECK, Kabwe K, Mafolo Y, Chalutz E, Droby S (1997). Integration of ultraviolet (UVC) light with yeast treatment for control of postharvest storage rots of fruits and vegetables. Biological Control 10:98–103.

Teixido N, Usall J, Vians I (2000).Efficacy of preharvest and postharvest Candida sake biocontrol treatments to prevent blue mould on during cold storage.Food Microbiol. 58:83-92: 50:203-210.

Thilagavathi R, Saravanakumar S, Ragupathi N, Samiyappan R (2007). A combination of biocontrol agents improves the management of dry root (*Macrophomina phaseolina*) in greengram. Phytopathol. 46:157–167.

Wilson MSL, Lindow (1994). Ecological similarity and coexistence of epiphytic ice-nucleating (Ice+) *Pseudomonas syringae* strains and non-icenucleating (Ice-) biological control agent. Appl. Enviro. Microbiol. 60:3128–137.

Yu T, Zheng XD (2006). Salicylic acid enhances biocontrol efficacy of the antagonist Cryptococcus laurentii in apple fruit. J. Plant Growth Regulator. 25:166–174.

Development and functional properties of low-cost complementary food

Getachew Addis[1] , Vasudeva Singh[2], Vishwas Pratape[2], Alok Srivastava[2], Lalitha Gowda[2], Madhavan Asha[2] and Sila Bhattacharya[2]

[1]Ethiopian Health and Nutrition Research Institute, P.O. Box 1242, Addis Ababa, Ethiopia.
[2]Central Food Technological Research Institute, Mysore-570 020, India.

Development of complementary food that contains adequate nutrients at affordable costs is quite a challenge for developing countries. The traditional complementary foods are viscous and lack essential nutrients. The objective of this study was to formulate and evaluate the physicochemical properties and acceptability of a complementary food. A preliminary study was conducted to select the appropriate processing method and estimate the nutritional composition of the ingredients. Accordingly, the complementary food was formulated from processed sorghum, pigeon pea, soybean along with skimmed milk and sucrose in the proportion of 65:15:10:5:5. The blend (95%) was mixed with malted finger millet (5%). Functional properties, sensory attributes, amino acid composition and trypsin inhibitor activity (TIA) of the product were evaluated. The product was easily dispersed in warm water and the reconstitution had a smooth and homogeneous consistency. The dry basis composition of the product was 14.4% digestible protein, 59% digestible starch, 4.2% fat, 12.8% total dietary fiber and 2.4% total minerals. Calcium, magnesium, iron and zinc level were 145.2, 94.5, 6.0 and 2.0 mg per 100 g dry basis, respectively. TIA in the product was inactivated to trace. Physicochemical properties of the formulated product were within the acceptable standard for children aged between 6 and 24 months. The newly developed complementary food may contribute to alleviating child malnutrition.

Key words: Cereals, pulses, skimmed milk, viscosity, complementary food.

INTRODUCTION

The major risk factors associated with infant and early childhood mortality and morbidity are poor infant feeding practices as well as childhood and maternal under-nutrition (WHO, 2009). Unlike developed countries where there are many children affected by obesity, most children in developing countries such as sub-Saharan Africa and Southeast Asia are conversely affected by lack of food or inadequate supply of energy and nutrients. Malnutrition is one of the major problems facing infants and young children of developing countries. Following a study conducted for 12 consecutive years (1980 to 1992), WHO has reported that more than one third of children aged less than five years in developing countries suffer from chronic malnutrition (de Onis et al., 1997). UNICEF (2009) indicated that more than 4.6 million annual deaths of children occur in Africa. In Ethiopia, 47 and 11% of under five children are stunted and wasted, respectively (Central Statistical Agency and ORC Macro, 2006). A higher rate of under five mortality due to protein-energy malnutrition is also reported in Ethiopia (Government of the Federal Democratic of Ethiopia, 2008) and Sudan (Asma et al., 2006). Furthermore, pocket studies on micronutrient deficiencies indicate higher prevalence of iron and zinc deficiency especially in young children

(Getahun et al., 2001). The situation is similar with Asian countries like India (Thathola and Srivastava, 2002). Appropriate complementary feeding of the infant and adequate nutrition are among the crucial measures required to avert infant and early childhood deaths (UNICEF, 2009).

In developing countries, children are mostly weaned on starchy, bulky gruels which are characterized by both low energy and low nutrient density. The traditional complementary food is mainly based on starch staples such as rice, sorghum, maize and wheat. Although the cereals contain sufficient sulfur containing amino acids (methionine and cysteine), which are limiting in pulses, they are deficient in lysine, which is rich in pulses (Gopalan et al., 1989). The unavailability of nutritious food and the high cost of commercial complementary foods and animal protein are the main causes of protein-energy malnutrition in children in the developing countries. Therefore, developing nutrient-dense, safe, affordable and accessible complementary food from locally produced ingredients using household or small to medium scale production technologies has been strongly recommended as a viable and sustainable approach to address the problem of malnutrition (WHO, 1998). The main objective of the present study was therefore to formulate a reduced bulk nutrient-dense complementary food for children, characterize the food and study the nutritional and sensory acceptability of the product.

MATERIALS AND METHODS

Selection and source of ingredients

Sorghum, soybean and pigeon pea were strategically selected to meet FAO/WHO/UNU complementary food quality requirements of essential amino acids (FAO/WHO/UNU, 1985). Sorghum (*Sorghum bicolor* L.), skimmed milk and sucrose were purchased from Mysore, India. Soybean (*Glycine max* L.) and pigeon pea (*Cajanus cajan* L.) were procured from the University of Agricultural Sciences, Bangalore, India. Finger millet (*Eleusine coracana* L.) malt was obtained from M/s. Christy Food Grains, Tamil Nadu. The raw materials were cleaned to free from extraneous matter and damaged grains.

Processing of ingredients for complementary food formulation

Sorghum

Grains were conditioned (equilibrated) to different moisture content ranging from ~12 to 20% and kept in tightly closed containers for 14 h at room temperature with occasional mixing. Treated grains were popped in a laboratory popcorn maker at 260°C for 30 s. Percentages of popped material as well as the bulk density of popped grains were calculated. The number of samples tested each time was five and the process was replicated twice and expressed as mean ± SD. The popped grains were then ground in a laboratory model vertical grinder to obtain flour that passed through a BSS 60 mesh with an opening of 250 micron. The temperature of the sample was always kept below 40°C during grinding by cooling the samples as well as the grinder.

Pigeon pea

Grains were soaked in water for 24 h. After decanting excess water, the grains were wrapped with gunny cloth and left for 36 h to germinate. Germinated grains were kilned at 105 – 110°C and were milled by a pulse dehusker (Ramakrishnaiah et al., 1999). It works on the principle of the dry milling technique by applying oil and heat treatment to the pulse followed by dehulling in an abrasive type unit. The dehusked kernels were then ground in the laboratory model vertical grinder to obtain flour that passed through a BSS 60 mesh.

Soybean

Soybean grains were decorticated using pulse dehusker (Ramakrishnaiah et al., 1999) and the kernels were soaked in water at room temperature (28°C) for 4 h. After draining excess water and drying surface moisture, the kernels were toasted at 200°C for 20 min and then ground in the laboratory model vertical grinder to obtain flour that passed through a BSS 60 mesh.

Finger millet

Finger millet grains were cleaned and soaked in water for 12 h. Excess water was decanted, and soaked grains were wrapped in gunny cloth and left for germination for two days. Water was sprinkled during germination to keep grains moist. Sprouted grains were kilned, rootlets were removed, decorticated and powdered as indicated by Shobana and Malleshi (2007).

Complementary food formulation

The proportion of the main ingredients (processed sorghum, pigeon pea and soybean) was determined mainly based on proximate composition as recommended by Malleshi (1995). Accordingly, the complementary food was formulated using processed sorghum, pigeon pea, soybean, skimmed milk and sucrose in the ratio of 65:15:10:5:5 and blend was mixed thoroughly until uniform distribution of particles was obtained. The formulated complementary food (95 parts) was mixed with finger millet malt (5 parts) to prepare the product. Flow chart for preparation of the formulated complementary food is presented in Figure 1.

Functional properties

Swelling and solubility studies of the complementary food and ingredients were conducted at 30, 55 and 95°C. Swelling volume was determined according to the method of Crosbie et al. (1992). Flour samples (0.4 g, dry basis, db) were mixed with 12.5 ml of water, equilibrated at the respective temperature for 5 min and then centrifuged at 5000 rpm for 15 min. The swelling volume was calculated by converting the height of the resultant gels to a volume basis, and the results were reported as g/g of dry flour. The supernatant was carefully removed and the difference of the initial and final volume due to swelling was observed for measuring solubility and expressed per gram of the initial weight of the dry flour. All experiments were repeated three times and values are presented as the mean of the three observations.

Viscosities of unprocessed and processed ingredients, as well as the complementary food formulation were measured using Brookfield viscometer. A known quantity of the respective flours was reconstituted in water at different solid concentrations ranging from 8 to 30% depending on experimental observation of the viscosity of individual test material. The slurry was heated in boiling water bath

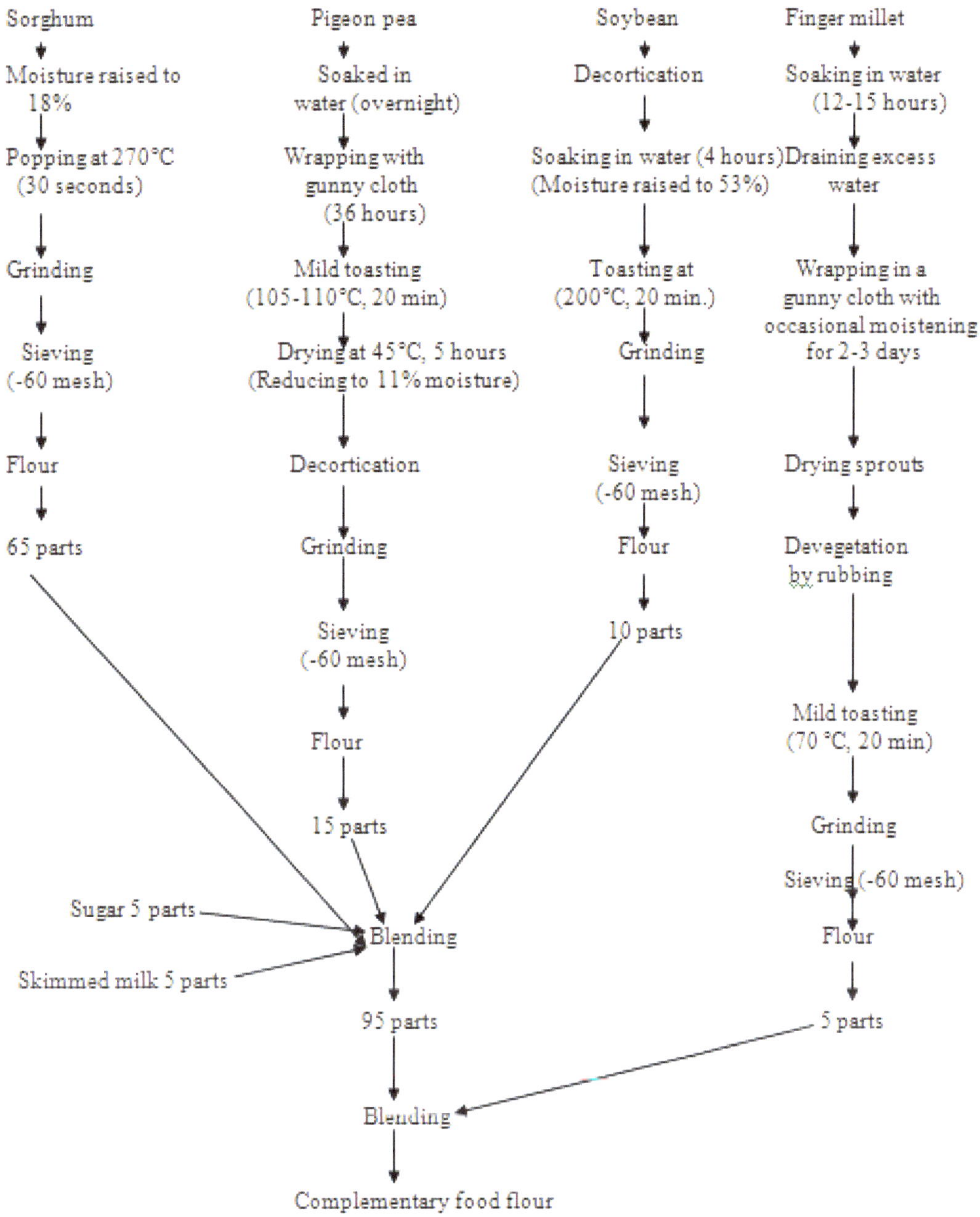

Figure 1. Flow chart for preparation of the formulated complementary food.

maintained at 95°C for 20 min with occasional stirring maintaining constant volume. The slurry was then cooled to ambient temperature and the viscosity was measured. A comparative viscosity study of processed and respective raw ingredients was also performed. The formulated complementary food with different percentage (0, 1, 2, 3, 4, 4.5 and 5%, w/v) of finger millet malt flour was also studied similarly to achieve the desired viscosity of the product.

Viscography of the complementary food and raw mix of the ingredients was also carried out following the method of Halick and Kelly (1959) and Kho and Singh (2009) by employing a Brabender visco-amylograph (Type VSK 4, Duisburg, FRG fitted with a 700-cmg sensitivity cartridge) using a dispersion containing 15% flour solids (db). The sample was heated up to 95°C at the rate of 1.5°C/min, held at 95°C for 20 min and cooled at 1.5°C/min to 50°C. Duplicate runs were made for each sample.

Nutrient composition and digestibility study

Moisture, ash, fat, crude protein and minerals, viz. calcium, iron, magnesium, zinc and copper were estimated using AOAC (2000) methods. Starch was determined following Chitra et al. (2010). Five grams (dry matter basis) of sample was boiled with 50 ml of 1.124% HCl in a boiling water bath for 15 min. The aliquot was transferred into a 100 ml volumetric flask and cooled. Three milliliters each of 15% potassium ferrocyanide and 30% zinc sulfate were added into the digested samples and incubated at 20°C and polarimetric readings were taken on the same day. Starch content was then calculated using the following equation.

$$Starch(\%, db) = \frac{Observed\,reading\ X\ 100}{185.7\ X\ 0.1\,Weight\ in\ g,\ dmb}\ X\ 100$$

The results were expressed after applying a correction factor of 2.4% (adding to each value), which was necessitated as the starch recovery using the polarographic method slightly varies from one source to the other with a mean reduction of 2.4%, db (Chitra et al., 2010). The complementary food was also analysed for amino acids (Bidlingmeyer et al., 1987), insoluble and soluble dietary fiber (Asp et al., 1983). *In vitro* protein and carbohydrate digestibility of the complementary food were also determined following the method of Walter and Mark (1964). Trypsin inhibitor activity (TIA) was estimated following Hamerstrand et al. (1981).

Sensory evaluation

Complementary food in powder form was dispersed in warm water (50°C) at 10%, and mixed well so that no lumps were present. The dispersion was boiled for 2 min and served in glass containers coded with 3-digit random numbers to minimize bias. The temperature of the sample at the time of serving to panelists was 50 to 55°C.

Sensory evaluation of the complementary food was carried out in two stages by a trained panel of 10 assessors, in sensory booths maintained at 22 ± 2°C, RH 50 ± 5%, under white fluorescent light. In the first stage of evaluation, "*Free Choice Profiling*" was used for formulating a suitable scorecard. This included listing of appropriate terminology describing individual quality attributes of the sample. Panelists were adequately trained to detect subtle differences in the perceived intensity of attributes and also to describe them. In the second stage, the "*Quantitative Descriptive Analysis*" method was employed for sensory analysis of the product. A 15 cm scale anchored at 1.25 cm on either end, representing "Recognition threshold" and "Saturation threshold," respectively was used for this purpose. Panelists were asked to mark on this scale to indicate the perceived intensity of each attribute. The scores for all the attributes were tabulated and the mean value was then taken for each attribute. These mean scores represented the panel's judgment about the sensory quality of the sample.

Statistical analysis

Descriptive statistical analysis was conducted using Microsoft ® Excel 2003 and the results were expressed as mean ± SD whenever required.

RESULTS AND DISCUSSION

Physicochemical properties of processed grains

Effect of moisture on physical properties of sorghum

Popping of sorghum reduced bulk density from 753 to 89 kg/m^3. Increasing moisture content from the 11.3 to 18% further reduced bulk density of popped sorghum from 89 to 66 kg/m^3. Popping expanded the raw sorghum to 574%, which was further improved to 767% by increasing the moisture content to 18%. When moisture content of grains was further increased to about 20%, the grain bulk density and percent expansion was decreased. Therefore, 18% moisture content was used for popping of sorghum. Higher expansion of popped grains is desirable because of enhanced food quality and also textural and economic benefits.

Viscosity

Viscosities of raw and processed flours of sorghum, pigeon pea and soybean are presented in Figure 2. The viscosity of processed flour varied considerably; at 10% slurry it was 2600 mPas for sorghum, 863 mPas for pigeon pea and was too thin for soybean to measure using the viscometer. It appears that the viscosity of the processed ingredients is highly positively correlated with the starch content of the flours. Up to 20% concentration soybean slurry was too thin and freely flowing. Sorghum at 15% concentration and pigeon pea at 20% concentration was too thick to measure the viscosity, while the viscosity of soybean at 25 and 30% concentration was only 737 and 3610 mPas, respectively.

Among the raw materials considered, sorghum showed highest viscosity at lower concentration (about 12 to 13%) both in cooked raw and cooked processed materials as it contains highest amount of starch. The viscosity of cooked processed sorghum was lower than the cooked raw sorghum, which is an indication of the change in properties of cooked starch of the processed sorghum. Similar observation was also noticed in pigeon pea but at higher concentrations; 15 to 16% in cooked raw pigeon pea and about 18% in cooked processed pigeon pea. In the case of soybean, viscosity was registered at about 30% in cooked raw material and very high concentration in cooked processed material. In the processed soybean, denaturation of protein occurs but negligible amount of starch makes the cooked slurry to assume least viscosity even at higher concentrations.

Solubility and swelling

Table 1 shows solubility and swelling of the raw and processed flours of sorghum, pigeon pea and soybean at different temperatures. At lower temperature (pregelatinization temperature), all samples have relatively less solubility and swelling. In agreement with Muralikrishna et al. (1986), both solubility and swelling of processed samples were higher than raw flour samples at temperatures of 30 and 55°C. While at cooking temperature (95°C) both solubility and swelling were decreased due to equilibration. Among the samples, raw pigeon pea has the least water holding capacity at pregelatinization temperature. At gelatinization temperature, processed soybean has the lowest solubility and swelling. Higher solubility and swellling was observed for processed flours as compared to raw flour samples, which could partly be attributed to gelatinization of starch granules during popping, kilning and toasting of the grains. Similarly, increase in solubility of all the samples at the highest temperature could be partly attributed to gelatinization of starch during cooking, and relatively lower solubility of soybean could be due to its lower content of starch.

Proximate composition and dietary fiber

The proximate, soluble and insoluble dietary fiber content

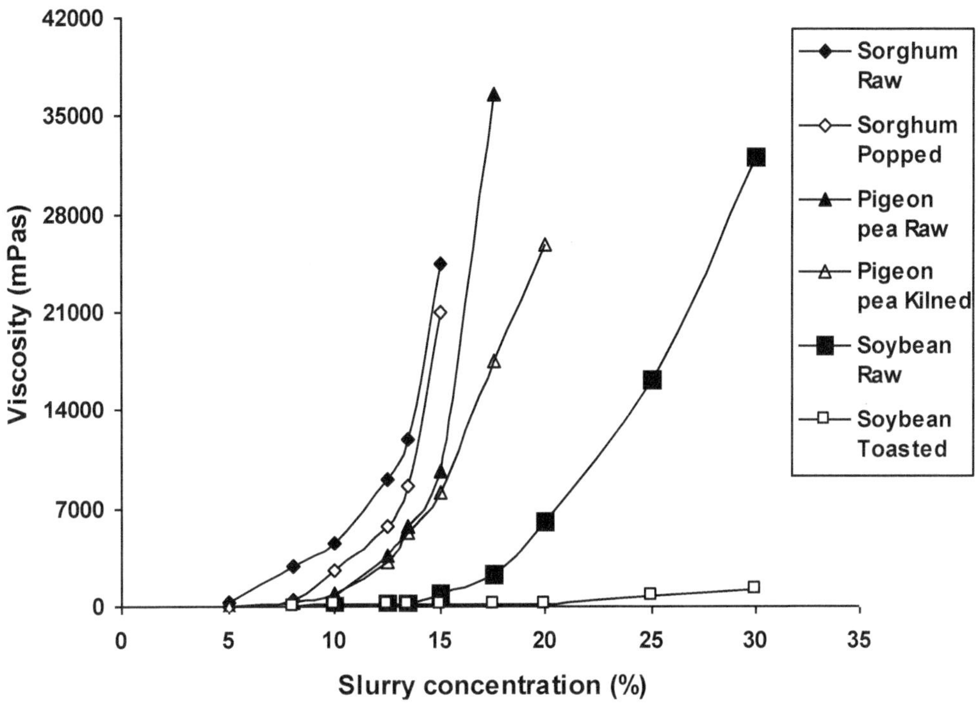

Figure 2. Viscosity of raw and processed ingredients.

Table 1. Solubility and swelling of raw and processed flours at different temperatures. Values are means of three independent measurements ± SD.

Test	Sample	30°C	55°C	95°C
Solubility (g/g)	Raw sorghum	1.1 ± 0.1	1.8 ± 0.0	5.7 ± 0.1
	Processed sorghum	1.7 ± 0.1	2.8 ± 0.0	4.8 ± 0.0
	Raw pigeon pea	1.1 ± 0.4	1.1 ± 0.1	4.9 ± 0.1
	Processed pigeon pea	1.5 ± 0.0	1.3 ± 0.0	4.6 ± 0.0
	Raw soybean	1.4 ± 0.2	1.1 ± 0.1	3.3 ± 0.4
	Processed soybean	1.5 ± 0.0	1.6 ± 0.1	2.8 ± 0.2
Swelling (g/g)	Raw sorghum	1.4 ± 0.0	1.8 ± 0.1	5.4 ± 0.2
	Processed sorghum	3.5 ± 0.0	2.6 ± 0.3	5.2 ± 0.3
	Raw pigeon pea	0.8 ± 0.0	0.9 ± 0.0	4.8 ± 0.0
	Processed pigeon pea	1.4 ± 0.1	1.7 ± 0.0	4.3 ± 0.0
	Raw soybean	1.2 ± 0.1	1.3 ± 0.2	3.0 ± 0.1
	Processed soybean	1.5 ± 0.1	2.8 ± 0.0	2.6 ± 0.1

of the ingredients of the complementary food is presented in Table 2. Raw sorghum contained the lowest quantity (9.2%) of protein and soybean had the most (41.7%, db). Protein content of pigeon pea increased from 22% in raw to 28% (db) in kilned sample which may be due to loss of moisture and leaching loss during germination. In soybean, decorticated samples contained the maximum protein level (50%, db). As compared with sorghum and pigeon pea, soybean contained the least starch content (6.3%, db). Starch content was however reduced marginally by processing (popping, toasting and kilning). Breakdown of starch to lower molecular weight compounds might have brought marginal reduction during estimation.

Total dietary fiber content varied among the different flours and processing methods ranging from 6.3% in sorghum flour (raw) to 29.7% in soybean (toasted) on dry weight basis. The insoluble to soluble dietary fiber ratio was appreciably affected by popping (sorghum), germinated and kilning (pigeon pea) and toasting (soybean). In sorghum, the insoluble to soluble dietary fiber ratio

Table 2. Nutritional composition and TIA of raw and processed flours (db). Values are means of three independent measurements ± SD.

Sample	Moisture (%)	Ash (%)	Fat (%)	Protein (protein digestibility) (%)	Starch (starch digestibility) (%)	Dietary fiber		TIA (TIU/mg)
						Insoluble (%)	Soluble (%)	
Sorghum: raw flour	14.8 ± 0.0	1.2± 0.0	3.3 ± 0.0	10.6 ± 0.1 (74.6)	80.5 ± 0.5 (83.3)	5.3 ± 0.21	1.0 ± 0.4	NC
Sorghum: popped	9.8 ± 0.1	1.1± 0.1	4.6 ± 0.4	11.9 ± 0.0 (85.4)	74.7 ± 0.2 (97.9)	7.9 ± 0.2	1.0 ± 0.3	NC
Pigeon pea: raw flour	8.6 ± 0.4	3.8 ± 0.3	1.8 ± 0.2	25.4 ± 0.3 (85.7)	56.1 ± 0.1 (58.7)	8.7 ± 0.8	3.4 ± 0.1	18.6
Pigeon pea: germinated and kilned	7.3 ± 0.2	4.0 ± 0.3	2.8 ± 0.0	28.2 ± 0.1 (87.5)	48.3 ± 0.5 (98.7)	10.0 ± 0.5	1.3 ± 0.2	4.4
Soybean: raw	11.0 ± 0.1	5.3 ± 0.3	23.0 ± 0.1	50.0 ± 0.2 (85.5)	8.8 ± 0.3 (ND)	19.7 ± 2.0	8.7 ± 0.6	88.6
Soybean: decorticated, toasted	6.1 ± 1.1	6.1 ± 0.1	23.6 ± 0.3	48.1 ± 0.4 (88.6)	6.0 ± 0.1 (ND)	24.2 ± 1.4	5.5 ± 0.1	4.7

NC= Analysis not conducted, and ND = composition not detected.

increased from 5.6:1 in raw flour to 8.1:1 of popped flour. Similarly, insoluble to soluble dietary fiber ratio increased from 2.5:1 to 7.9:1 in kilned and pulverized pigeon pea, and 2.3:1 to 4.5:1 in toasted soybean. The total dietary fiber content of each processed ingredient was higher than the raw material. Results of the present study are in close agreement with Azizah and Zainon (1997) who reported that most of the increment of dietary fiber during processing is due to insoluble fiber. According to Azizah and Zainon (1997), it is possible that thermal processing may cause production of Maillard reaction products and thus increase in the insoluble dietary fiber content. Increase in insoluble dietary fiber content by autoclaving and microwave treatment was also previously reported (Vidal-Valverde and Erias, 1991). Soluble dietary fiber mostly remained unchanged or reduced to some extent by processing. This is also in agreement with previous reports (Thed and Phillips, 1995; Azizah and Zainon, 1997).

Protein and starch digestibility

Popping of sorghum increased the starch digestibility from 83 (raw) to 98% due to gelatinization of starch during popping. Digestibility of starch in raw pigeon pea was relatively poor (59%) but increased to 65% by germination and 99% by kilning (Table 2). Muralikrishna et al. (1986) and Mouliswar et al. (1993) have also observed that toasting increases digestibility of the starch.

Protein digestibility ranged from 75% in raw sorghum to 90% in germinated and decorticated pigeon pea. An increase in the protein digestibility of sorghum by popping and pigeon pea by kilning was observed. Som et al. (1992) have reported that cooking increases digestibility of protein marginally.

Antinutritional composition

Maximum TIA was detected in raw soybean (88.6 TIU/mg). Following decortication and toasting, TIA was reduced to a safe level of 4.7 TIU/mg for soybean (Table 2). Trypsin inhibitor has been known to bind to dietary protein and inhibit its digestibility which in turn affects growth of experimental animals. Among cereals and pulses, soybean is reported to contain the maximum level of TIA. Numerous reports are available on the effect of processing on TIA. Heat treatment significantly reduces TIA. Destruction of TIA is accompanied by a concomitant increase in the nutritive value of protein and is well documented by Liener (1994).

Quality evaluation of the complementary food

Different indicators are used to evaluate the quality of complementary foods including physicochemical characteristics and sensory attributes.

Nutrient composition

Nutrient composition of the complementary food is presented in Table 3. The complementary food contained 4 fats and 16% protein (db) which are within the recommended range for complementary food development (Malleshi, 1995). The total dietary fiber content was 12.8% (db), which was slightly higher than the recommended standard for complementary food development (Malleshi, 1995). The digestibility study showed complete digestion of starch and 90% of protein. Higher digestibility of both starch and protein is an important characteristic of complementary foods. TIA was only detected in very trace quantity (2.7 TIU/mg).

The amino acid profile of the complementary food as obtained during the present study is presented in Table 3. Theoretical values from literature (Gopalan et al., 1989) of the essential amino acids (%) of the complementary food were also extrapolated from proportion of the ingredients

Table 3. Proximate and other compositions of the complementary food.

Composition/parameter	Value
Moisture (%, db)	8.4 ± 0.3
Ether extractives (%, db)	4.2 ± 0.1
Protein (%, db)	16.1 ± 0.3
Digestible protein (%, db)	14.4 ± 1.0
Amino acid (g/100 g)	
Asx*	7.38 ± 0.17
Glx**	22.12 ± 0.33
Ser	5.31 ± 0.08
Gly	4.09 ± 0.06
His	2.83 ± 0.29
Arg	6.38 ± 0.18
Thr	3.92 ± 0.06
Ala	4.97 ± 0.17
Pro	7.20 ± 0.20
Tyr	3.17 ± 0.02
Val	4.79 ± 0.05
Met	0.54 ± 0.04
Cys	0.55 ± 0.06
Ile	3.34 ± 0.14
Leu	10.80 ± 0.41
Phe	6.56 ± 0.15
Lys	6.02 ± 0.12
Starch (%, db)	59.6 ± 0.3
Digestible starch (%, db)	59.0 ± 0.8
Insoluble dietary fiber (%, db)	8.2 ± 0.2
Soluble dietary fiber (%, db)	4.6 ± 1.0
Ash (%, db)	2.4 ± 0.0
Calcium (mg/100 g, db)	145.2 ± 3.2
Magnesium (mg/100 g, db)	94.5 ± 0.02
Iron (mg/100 g, db)	6.0 ± 0.03
Zinc (mg/100 g, db)	2.0 ± 0.03
Copper (mg/100 g, db)	0.47 ± 0.08
Trypsin inhibitor activity (TIU/mg)	2.7 ± 0.1
Solubility (g/g)	
30°C	1.4 ± 0.2
55°C	1.1 ± 0.1
95°C	3.0 ± 0.4
Swelling (g/g)	
30°C	1.2 ± 0.1
55°C	1.1 ± 0.0
95°C	2.6 ± 0.3

Values are means of three independent measurements ± SD; * Asx represents both Asp and Asn, ** Glx represents both Gln and Glu.

(dry matter basis) in the complementary food to compare with the experimental results. Based on protein digestibility of the complementary food (89.71%), protein digestibility corrected amino acid score (PDCAAS) of the experimental and theoretical values were calculated and found closely similar and in very good agreement with FAO/WHO/UNU (1985) reference values (Table 4). FAO/WHO/UNU (1985) protein standard is an internationally accepted quality parameter for complementary food development. When the formulated complementary food is viewed in reference to the set standard, it appears comparable for 2 to 5 year old children. The developed complementary food also contained appreciable amount of minerals like calcium (145 mg/100 g) and magnesium (95 mg/100 g). However, it should be noted that phytic acid chelates with divalent metals including calcium, iron, magnesium and zinc thereby reducing bioavailability of the minerals (Liener, 1994). Effort should therefore be made to quantify and reduce the level of phytic acid in the complementary food as suggested by Liener (1994).

Viscosity

The cooked complementary food flour slurry exhibited high viscosity (Figure 3A) at 6% and above. This limits food intake by the child. Therefore, it was desirable to reduce the bulk in order to increase calorie and nutrient density of the complementary food. This was possible by adding 5% finger millet malt flour (Figure 3B). When the complementary food containing 5% malt was cooked, the viscosity was reduced with appreciable quantity even at 20% concentration (Figure 3B) because of partial hydrolysis of starch and reduction in its water holding capacity. The viscosity of 20% cooked food was reduced from 23733 to 450 mPas by adding 5% malt. The trend is in agreement with a previous report on complementary food (Malleshi et al., 1986). Malted complementary flour is rich in α-amylase. During cooking, alpha-amylase in the malt hydrolyzes the 1,4 α-D-glucosidic linkages of polysaccharides. Starch is therefore degraded to lower molecular weight dextrins by the enzyme, and thus, viscosity of the heated paste decreases (Pertin, 1964; Dogan, 2002). Moreover, the enzyme α-amylase is moderately stable to heat. During the heating phase (in the presence of moisture), α-amylase gets activated and liquefies starches.

Using a Brabender Viscograph, the change in pasting characteristics of the mix of raw materials (without malt) and the cooked complementary food was studied (Figure 4). The viscogram mainly indicates the behavior of starch present in the raw materials and complementary food. The curve of the raw materials was initially parallel to the X-axis. With increasing temperature, the swelling increased and the food completely gelatinized at 74.2°C. Peak viscosity (PV) and hot paste viscosity (HPV) were at 850 and 610 BU, respectively. At cooling temperatures of 50 and 30°C, the viscosity was about 1500 and 1750 BU (cold viscosity, CV), respectively, which is an indication of retrogradation or precipitation of linear molecules. The calculated set back viscosity (difference between CV and PV) was 900 BU and total set back viscosity was 1140 BU (Figure 4) which is due to retrogradation of linear molecules in the starch granule.

Figure 4 shows the paste viscosity profile of the formulated food at 15% slurry concentration. Since the materials

Table 4. Protein digestibility corrected amino acid score (PDCAAS) of formulated complementary food.

Composition	AA (mg/g)*	AA (mg/g)**	Reference pattern***	Digestibility (AA x 89.4%)*	PDCAAS*	Digestibility (AA x 89.4%)**	PDCAAS**
Histidine	31.2	28.7	19	27.89	1.47	25.66	1.35
Lysine	60.2	60.9	58	53.82	0.93	54.44	0.94
Tryptophan	12.5	ND	11	11.18	1.02	ND	-
Phenylalanine + Tyrosine	85.5	98.9	63	76.44	1.21	88.42	1.4
Methionine + Cystine	31.2	11.1	25	27.89	1.12	NCQ	-
Threonine	40.1	39.7	34	35.85	1.05	35.49	1.04
Leucine	113.3	109.5	66	101.29	1.53	97.89	1.48
Isoleucine	51.7	33.8	28	46.22	1.65	30.22	1.08
Valine	58.7	48.6	35	52.48	1.50	43.45	1.24

AA = Amino acid; * = theoretical value; ** = experimental value; *** = FAO/WHO/UNU (1985) reference pattern (mg/g); Experimental value of protein digestibility (89.4%) was used to calculate PDCAAS; ND = not detected; NCQ = not completely quantified due to degradation during hydrolysis.

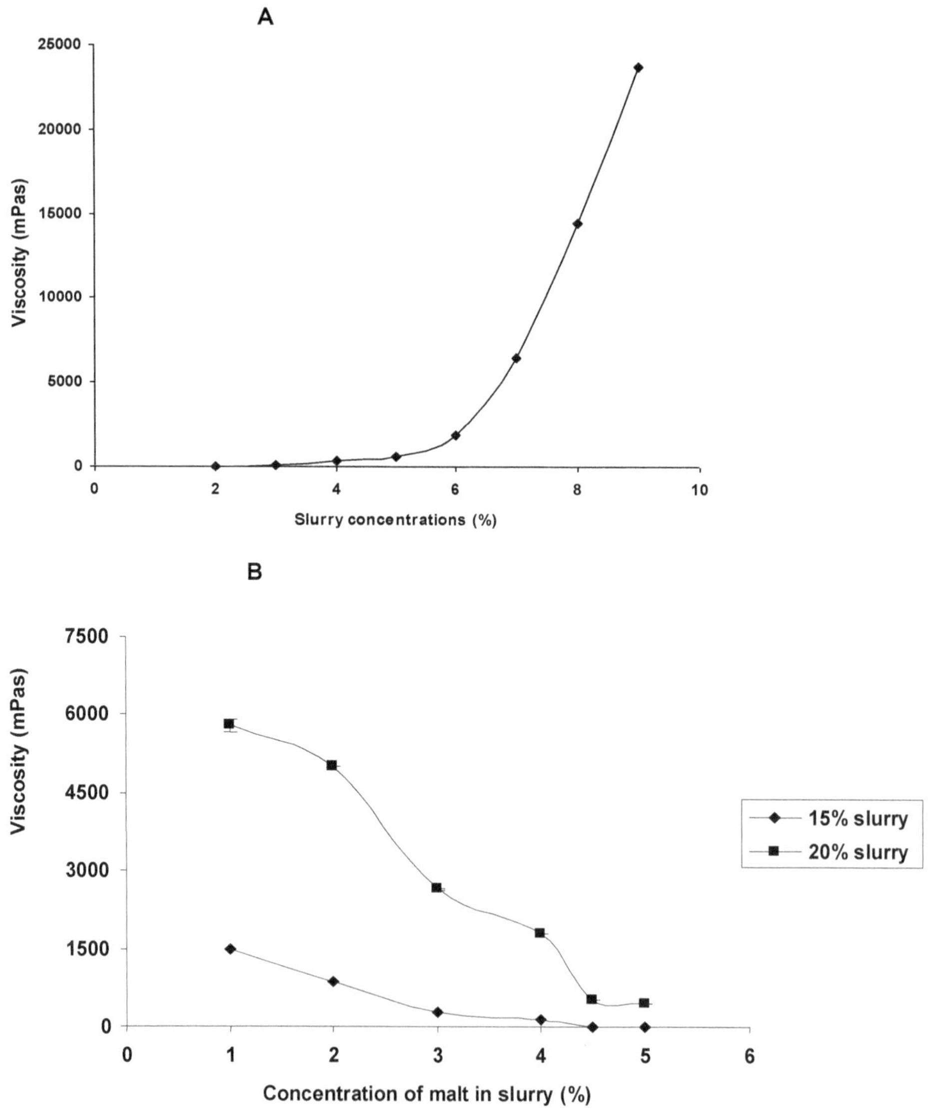

Figure 3. Effect of malt on viscosity of complementary food. A: Viscosity of complementary food without malt at different concentrations, B: Effect of malted finger millet on viscosity of complementary food at different slurry concentrations.

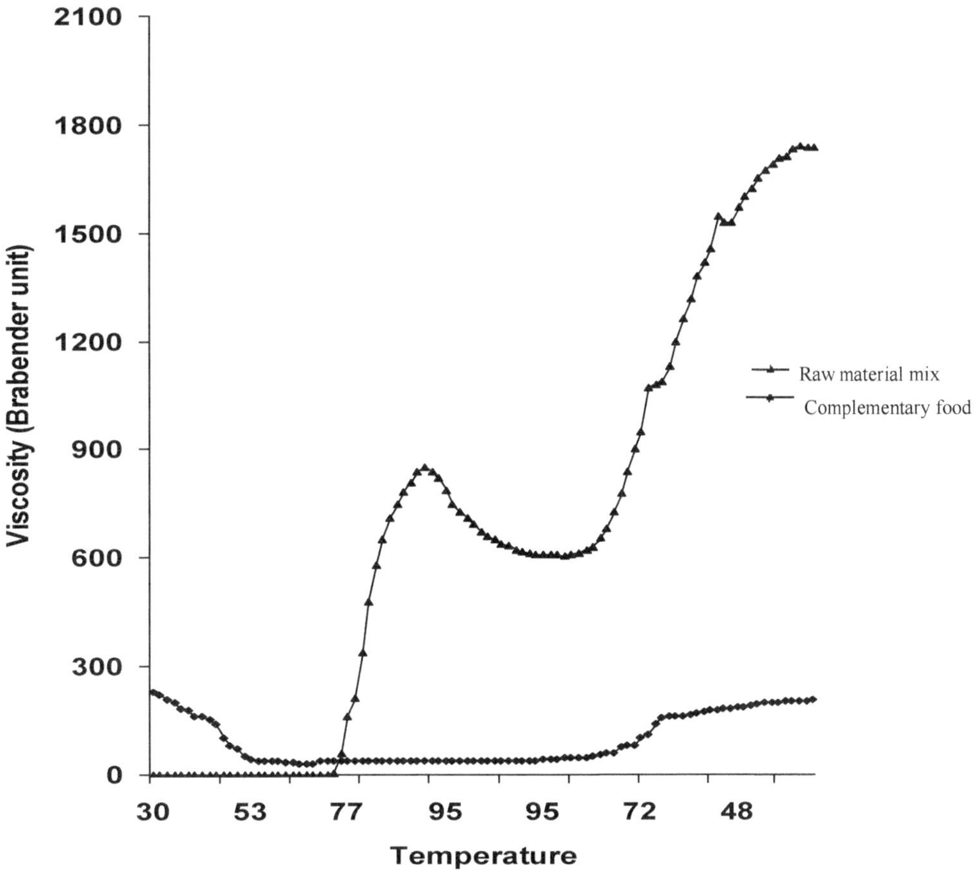

Figure 4. Paste viscosity profile of raw mix of the ingredients and formulated complementary food.

were precooked, the complementary food absorbed water soon after soaking and showed an initially higher viscosity value. During cooking temperature (95°C) in particular, the complementary food had extremely low viscosity and was a free flowing fluid. At cooling temperature (30°C), the consistency remained thin and showed a slight increase in viscosity. This is in contrast with the viscosity profile of the raw material which formed a thick gel that would probably be impossible for an infant to swallow. Previous investigators (Malleshi and Desikachar, 1988) also observed a lower viscosity of malted complementary food as compared with roller dried proprietary complementary food (PWF). The comparative study of malted complementary food (MWF) with proprietary complementary food (PWF) showed that MWF had considerably less viscosity than PWF. According to these investigators, for a comparable slurry viscosity of 55 mPas, the MWF had 50 to 100% more calories as compared to different proprietary complementary foods paste viscosity. Reduced viscosity would enable the child to take more of the product per feed or higher nutrient and calorie dense complementary food at a time. Malleshi et al. (1986) indicated that this kind of complementary food is superior over complementary foods developed by modern technologies such as roller drying.

Swelling and solubility

Severity of heat treatment during popping might cause structural changes in starch and protein resulting in increased water absorption. Although higher water absorption is desirable in some food preparations such as bakery, this 'bulking' property is disadvantageous for complementary foods where low 'dietary bulk' is desired. In the present study, the problem was solved to an appreciable extent by adding 5% of finger millet malt flour. The solubility and swelling of the formulated complementary food at three temperatures is presented in Table 3. Gelatinization temperature increased solubility and swelling of the formulated complementary food as compared to pre-gelatinization temperature. Unlike sorghum and pigeon pea, the complementary food has relatively less solubility and swelling, which could be attributed to the presence of sucrose, milk and malt. One of the important qualities of a complementary food is its water holding capacity. Since water does not have any calorific value, an increase in its level will reduce energy value of the product. Therefore, low water holding capacity is a desirable characteristic of

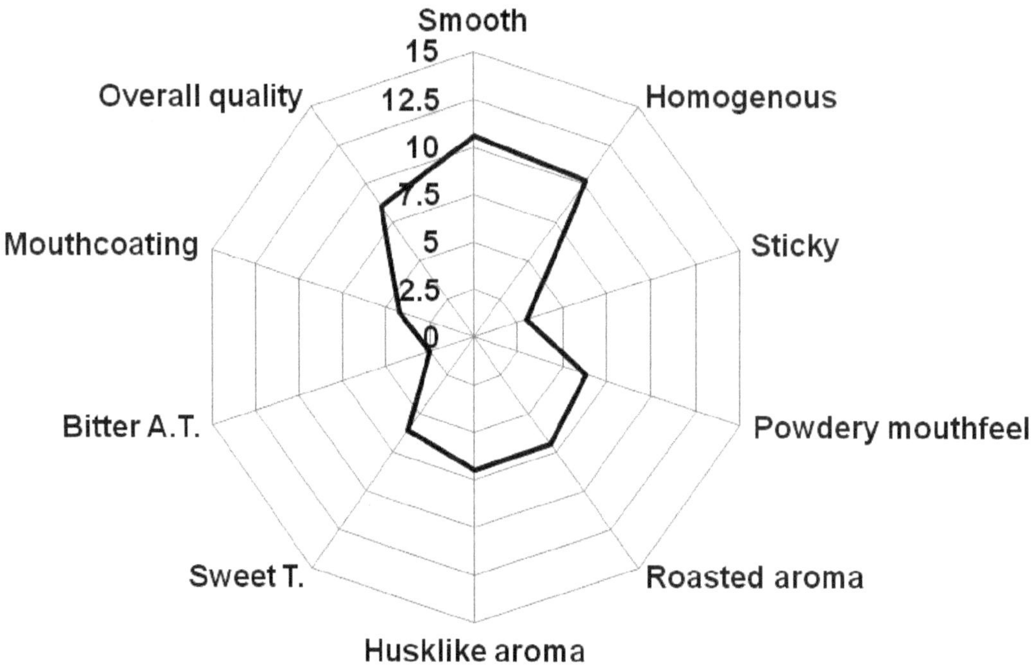

Figure 5. Sensory profile of the complementary food.

the complementary food.

Sensory evaluation

Mean scores for sensory evaluation of each attribute of the sample are graphically presented as "sensory profile" in Figure 5. It was observed that the complementary food in powder form could be easily dispersed in warm water without any lumps. The complementary food preparation was smooth and homogeneous, without any noticeable stickiness. This is a desirable property, which facilitates easy swallowing of the food by the infants, the target group for which the product has been developed. Roasted aroma and husk-like aroma were perceived by most of the panelists. Sweet taste was perceived to be at a desirable level. Overall quality was fairly high which aptly reflected the acceptability of the complementary food preparation.

Conclusions

Important aspects of complementary food formulation were studied including physical properties, quantity and quality of nutrients, antinutritional factors and sensory evaluation. *In vivo* protein bioavailability, important nutrients such as vitamins, storage and packaging characteristics await further study. The formulated complementary food can however be produced by small and large scale industries and used at the local level to alleviate child malnutrition. Depending upon the requirements, operating capacity and corresponding degree of mechanization, the scale of production canvary to suit the local needs and conditions. To produce a complementary food at industrial level, further fortification with required quality and quantity of micronutrients (premix) is required.

REFERENCES

AOAC (2000). Official methods of the AOAC International. Determination of lead, cadmium, and minerals in foods by Atomic Absorption spectrophotometry (method 999.11/985.35). Association of Official Analytical Chemists, Gaithersburg, USA.

Asma MA, Fadil EBE, Tinay AHE (2006). Development of weaning food from sorghum supplemented with legumes and oil seeds. Food Nutr. Bull. 27(1): 26-34.

Asp N-G, Johansson C-G, Hallmer H, Siljeström M (1983). Rapid enzymatic assay of insoluble and soluble dietary fiber. J. Agric. Food Chem. 31: 476-482.

Azizah AH, Zainon H (1997). Effect of processing on dietary fiber contents of selected legumes and cereals. Malaysian J. Nutr. 3: 131–136.

Bidlingmeyer BA, Cohen SA, Tarvin TL, Frost B (1987). A new, rapid, high-sensitivity analysis of amino acids in food type samples. J. Assoc. Anal. Chem. 70(2): 241-247.

Central Statistical Agency (Ethiopia) and ORC Macro (2006). Ethiopia Demographic and Health Survey 2005, Central Statistical Agency and ORC Macro, Addis Ababa, Maryland, p. 474.

Chitra C, Singh V, Ali SZ (2010). Effect of processing paddy on digestibility of rice starch by *in vitro* studies. J. Food Sci. Technol. 47(4): 414-419.

Crosbie GB, Lambe WJ, Tsutsui H, Gilmour RF (1992). Further evaluation of the flour swelling volume test for identifying wheats potentially suitable for Japanese noodles. J. Cereal Sci. 15: 271-280.

de Onis M, Monteiro C, Akr J, Clugston G (1997).The Worldwide magnitude of protein-energy malnutrition: An Overview from the WHO Global Database on Child Growth. WHO, Geneva, Switzerland.

Dogan IS (2002). Dynamic rheological properties of dough as affected by amylases from various sources. Nahrung 46: 399-403.

FAO/WHO/UNU (1985). Energy and protein requirements. WHO technical report series 724. WHO, Geneva, Switzerland. p. 209.

Getahun Z, Urga K, Ganebo T, Nigatu A (2001). Review of the status of malnutrition and trends in Ethiopia. Ethiop. J. Health Dev. 15(2): 55-74.

Gopalan C, Sastri BVR, Balasubramanian SC (1989). Nutritive value of Indian foods, National Institute of Nutrition, Indian Council of Medical Research, Hydrabad. p. 161.

Government of the Federal Democratic of Ethiopia (2008). Program Implementation Manual of National Nutrition Program (NNP) – I: June 2008 – June 2013. Federal Ministry of Health, Addis Ababa, Ethiopia. p. 128.

Halick JV, Kelly VJ (1959). Gelatinization and pasting characteristics of rice varieties as related to cooking behavior. Cereal Chem. 36(1): 91-98.

Hamerstrand GE, Black LT, Glover JD (1981). Trypsin inhibitors in soy products: Modifications of the standard analytical procedure. Cereal Chem. 58(1): 42–45.

Koh BK, Singh V (2009). Cooking behaviour of rice and black gram in the preparation of Idli, a fermented product of South Indian Origin by Viscography. J. Texture Stu. 40: 36 -50.

Liener IE (1994). Implications of nutritional components in soybean foods. Crit. Rev. Food Sci. Nutr. 34(1): 31–67.

Malleshi NG (1995). Weaning foods. Regional extension, service centre (rice milling), ministry of food processing industries, government of India. RESC Scientific Series. Central Food and Technological Research Institute, Mysore, India. pp. 8: 50.

Malleshi NGM, Desikachar HSR (1988). Reducing the paste viscosity (dietary bulk) of roller dried weaning foods using malt flour or fungal amylase. J. Food Sci. Technol. 25: 1–3.

Malleshi NGM, Rao SNR, Sreedharamurty S, Desikachar HSR (1986). Development of weaning foods based on some traditional technologies. J. Food Sci. Technol. 23: 315–318.

Mouliswar P, Kurien S, Daniel VA, Malleshi NG, Rao SV (1993). *In Vitro* digestibility of protein and starch energy food and its bulk reduction. J. Food Sci. Technol. 30(1): 36–39.

Muralikrishna G, Malleshi NG, Desikachar HSR, Tharanathan RN (1986). Effect of popping on the properties of some millet starches. Starch/stärke 38: 48–51.

Pertin H (1964). Application of the Falling number method for evaluating α-amylase activity. Cereal Chem. 41: 127-140.

Ramakrishnaiah N, Pratape VM, Narasimha HV, Shangmugam GK (1999). An Improved Mechanised Pulse Dehusker. Indian Patent Application No. 2385/Del/98.

Shobana S, Malleshi NG (2007). Preparation and functional properties of decorticated finger millet (*Eleusine coracana*). J. Food Eng. 79: 529-538.

Som JN, Mouliswar P, Daniel VA, Malleshi NG, Rao SV (1992). Digestibility of protein and starch in malted weaning foods. J. Food Sci. Technol. 29(4): 262 – 263.

Thathola A, Srivastava S (2002). Physicochemical properties and nutritional traits of millet-based weaning food suitable for infants of the Kumaon Hills, Northern India. Asia Pac. J. Clin. Nutr. Asia 11(1): 28-32.

Thed ST, Phillips RD (1995). Changes of dietary fiber and starch composition of processed potato products during domestic cooking. Food Chem. 52: 301–304.

UNICEF (2009). Press release: UNICEF highlights child survival on the Day of African Child. http://www.unicef.org/media/media_50001.html. Accessed 28 July 2009.

Vidal-Valverde C, Erias J (1991). Legume processing effects on dietary fiber components. J. Food Sci. 56(5): 1350–1352.

Walter R, Mark AS (1964). A pepsin pancreatin digest index of protein quality evaluation. J. Nutr. 83: 257-261.

WHO (1998). Complementary feeding of young children in developing countries. WHO/NUT/98.1 WHO, Geneva, Switzerland. pp. 167–170.

WHO (2009). World Health Statistics. WHO, Geneva, Switzerland. p. 149.

26

Biogenic amines and microbiological profile of egyptian cheeses

Khaled Meghawry El-Zahar

Food Science Department, Faculty of Agriculture, Zagazig University, 44511 Zagazig, Egypt.

Cheeses are among those high-protein-containing foodstuffs in which enzymatic and microbial activities cause the formation of biogenic amines (BAs) from amino acids decarboxylation. Most of the methods for amine determination in these products involve acid extraction followed by a liquid-liquid purification step to selectively separate amines and amino acids. This study aimed to describe the development of biogenic amines in Egyptian cheeses during ripening and storage regimes. Biogenic amines content in Mish, Ras and Blue cheeses were 270-1300, 340-980 and 210-700 mg/kg, respectively. The dominant biogenic amines were different. This work confirms that the main biological feature influencing amines formation is the extent of growth of microorganisms, like *Enterococci*, characterized by decarboxylase activity. It is important to note that the presence of biogenic amines due to the activities of these microorganisms is maintained within safe levels. In Egypt, reports dealing with the Egyptian cheeses (Mish, Ras and Blue) are scanty. So, the present work was carried out to fill the gap in our knowledge on its microbiological and biochemical features, focusing on hygiene and consumer health aspects.

Key words: Biogenic amines, food safety, proteolysis, ripening, *Enterococci* spp.

INTRODUCTION

Milk and milk products are very important in human nutrition and, among them; cheese is considered a good source of proteins, vitamins and minerals. However, cheese is one of the most fermented foods commonly associated with biogenic amines (BAs) contamination. These compounds are basic nitrogenous compounds formed by series of microorganisms, mainly by decarboxylation of amino acids or "*in vivo*" also by de-amination and trans-amination of aldehydes and ketones (Loizzo et al., 2012, 2013). Biogenic amines are compounds commonly present in living organisms in which they are responsible for many essential functions. They can be naturally present in many foods such as fruits and vegetables, meat, fish, chocolate and milk, but they can also be produced in high amounts by microorganisms through the activity of amino acid decarboxylases (Ten Brink et al., 1990). Excessive consumption of these amines can be of health concern because when there is are no equilibrate assumption in human organism, can generate different degrees of diseases determined by their action on nervous, gastric and intestinal systems and blood pressure (Suzzi and

Gardini, 2003). Biogenic amines are low molecular weight nitrogenous bases, they were found in fermented foods and cheese (Mohamed et al., 2013).

Also, biogenic amines are low-molecular nitrogenous compounds that are formed in foodstuffs mainly by microbial decarboxylation of the precursor amino acids (Alberto et al., 2002). The importance of observing BAs content lies in potential toxicity to human, mainly when the concentration is up to 100 mg/kg (or up to 100 mg/L). Thus, the presence of BAs significantly influences the food quality and safety (Smit et al., 2005).

The presence of relevant amounts of BAs in cheeses has been documented (Martuscelli et al., 2005; Kung et al., 2007; Pintado et al., 2008; Ladero et al., 2009; Mercogliano et al., 2010). In cheeses, BAs formation is caused by curdling and cheese decarboxylase-positive microorganisms. Histamine (HIS), tyramine (TYR), putrescine (PTR), cadaverine (CAD), spermidine (SPD), spermine (SPR), tryptamine (T), and β-phenylethylamine (PE) are frequently found in these products. Cheese is one of the fermented foods most commonly associated with BAs poisoning; mainly HIS, TYR, PTR and CAD. Indeed, the term "cheese reaction" to refer to it (Ten Brink et al., 1990). Tyramine and histamine are the most abundant and frequent BAs in cheese (Fernández et al., 2007). Consumption of food containing high levels of BAs is considered undesirable since it can be associated with several toxicological problems such as respiratory distress, headache, hyper- or hypo-tension or allergies (Ladero et al., 2010). These problems are especially severe in consumers with low levels of the enzymes involved in the detoxification system (mono and di-amine oxidases), either by genetic disorders (Caston et al., 2002) or medical treatments (Halász et al., 1994). The content of biogenic amines and polyamines significantly differed according to the technology of ripening. The cheeses unwashed during ripening had much higher contents of all observed amines and polyamines in comparison with the washed-rind cheeses. The mean content of putrescine, cadaverine and tyramine exceeded 100 mg/kg in unwashed-rind cheeses, while the other amines occurred at lower levels. The content of all detected amines was very low in washed-rind cheeses; no tryptamine, phenylethylamine and histamine were found. The effect of storage on the amine formation was not confirmed (Samková et al., 2013).

Physiologically, histamine is one of the most effective BAs; it has vasoactive and psychoactive effects (Repka-Ramirez and Baraniuk, 2002). Moreover, it is the main BAs involved in food poisoning and it is limited in some foodstuffs by law. At non-toxic doses, food borne histamine can cause intolerance symptoms such as diarrhoea, hypotension, headache, pruritus and flushes. Just 75 mg of histamine, a quantity commonly present in some meals, can induce symptoms in the majority of healthy persons with no history of histamine intolerance (Wöhrl et al., 2004).

The ability of microorganisms to decarboxylate amino acid is highly variable. Due to strain-specific, it is important to count decarboxylase-positive microorganisms to estimate; the risk of BAs food content and to prevent BAs accumulation in food products. Presence and accumulation of BAs depends on many factors such as presence of specific bacteria (*Enterococci, Micrococci, Enterobacteriaceae* and *Lactobacilli*) and enzymes, availability of free amino acids, presence of suitable cofactors, that is, pH level, water activity, temperature and salt content, type of cheese, ripening and storage period (Galgano et al., 2001). Some controversial results have been reported on the contribution of *Enterococci* sp. in BAs production in cheeses, and in particular in histamine (Sumner and Taylor, 1989). *Enterococci* have a long history of use as artisanal cultures for preparation of various types of cheeses (Izquierdo et al., 2009), they are sometimes associated with pathogenicity (Khan et al., 2010), can cause endocarditic, bacteraemia, and several infections, as well as multiple antibiotic resistances (Kayser, 2003). Although, the chemical composition and microbiological quality of cheeses in Egyptian markets have been studied extensively, little data is available on the occurrence of biogenic amines in Egyptian cheeses. Therefore, this survey was undertaken to determine the presence of BAs in commercially available cheeses during ripening and storage, also to make an assessment of the health hazard arising from the consumption of these products especially by susceptible individuals. Other studies such as Lorencová et al. (2012) and Buňková et al. (2010) deal with selection and study of microorganism (such as lactic acid bacteria), which are major producer of biogenic amine. However, these works explore the particular biogenic amine production in growth medium, where the concentration of biogenic amine could be biased optimal environment for the bacteria metabolisms. Moreover, some strains of the starter lactic acid bacteria (such as *Lactococcus lactis* subsp. *lactis*) have decarboxylase activity that was observed in model environment of growth broth. Behaviour of these strains has not been investigated in real system of the cheese and can be different in comparison with condition in growth broth.

The objective of our pilot study was to compare the BAs content and other selected parameters in Egyptian cheese and to review hypothesis that the BAs content developed during the ripening and storage period is related to the presence of decarboxylase positive strain of *Enterococci* sp.

MATERIALS AND METHODS

Cheese samples

A total 85, 49 and 44 of Mish, Ras and Blue cheeses samples were purchased from different Egyptian retail markets and small scale factories. The samples collected were 6-48 months old. They were kept in sterile plastic bags and transported to the laboratory of Food Science Department, Zagazig University (Egypt), then stored at 4 ±1°C until analyzed.

Chemical analysis

Cheeses were analyzed in triplicates for moisture by the oven drying method at 102°C (IDF, 1993), salt by titration with $AgNO_3$, and fat by Gerber method (AOAC, 2002). For pH measurement, grated cheese (10 g) was macerated with 10 mL of distilled water and the pH of the resultant slurry was measured using a digital pH meter (pH 211, Hanna Instruments, Vila do Conde, Portugal). Titratable acidity was determined as g lactic acid/100 g cheese is using the method of AOAC (2002). Total volatile fatty acids and total nitrogen (TN) were determined using the methods of AOAC (2002). All analyses were carried out in triplicates.

Assessment of proteolysis

Water-soluble nitrogen fraction (WSN) of cheese was prepared according to Kuchroo and Fox (1982) and a cheese to water ratio of 1:5 was used. 12% trichloroacetic acid soluble nitrogen-fraction (TCA-SN, that is, NPN) was obtained by mixing equal volumes of water-soluble fraction and 24% (w/w) TCA solution, followed by filtration through a white ribbon filter paper (Schleicher and Schuell, Dassel, Germany). The nitrogen content of both fractions WSN and TCA-SN, respectively, was determined by Kjeldahl method (AOAC, 2002) and expressed as percentage of TN.

Free amino acids and biogenic amines

Free amino acids (FAA) and BAs were assayed according to the method of Krause et al. (1995), modified by Pinho et al. (2001). In brief, a 4 g cheese sample was suspended in 15 mL of 0.2 M aqueous perchloric acid; the mixture was homogenized in an Ultra Turrax blender (Sotel, Warsawa, Poland) for 2 min, then kept in an ultrasonic bath (Heraeus, Osterode, Germany) for 30 min, and finally centrifuged at 4000 xg for 20min. Derivatization was carried out via dansyl chloride, at 70°C per 15 min. The reaction was quenched by placing the vials in an ice bath for 5 min. High performance liquid chromatography (HPLC, Waters 600) was used to dansylamines determination. The system was equipped with delivery system, reverse phase Cl8 Nucleosil column 250 x 4 mm, 10 µm packing (Macherey - Naggl). The detection was performed using U.V detector (Waters 486) at wavelength of 254 nm using linear program of 25 min period and 1 ml/min constant solvent flow rate. Data were integrated and recorded using a Millennium Chromatography (Waters, Milford MA 01757). Elution was carried out at a flow rate of 1 mL/ min, using a volumetric gradient of solution A, 9 mM aqueous sodium dihydrogenophosphate, 4% (w/v) dimethyl formamide and 0.1% (w/v) triethylamine (adjusted to pH 6.55 with phosphoric acid), and solution B, 80% (v/v) aqueous acetonitrile. Detection was performed by measuring absorbance at 436 nm. Quantification was carried out based on a mixture of amino acid standards: aspartic acid, glutamic acid, serine, threonine, glycine, alanine, arginine, proline, valine, methionine, isoleucine, leucine, lysine, histidine, tyrosine, cystine, tryptophan and phenylalanine; and biogenic amine standards: ornithine, tryptamine, phenyl ethylamine, putrescine, cadaverine, histamine, tyramine and spermine (Sigma Chemical). All determinations were performed in triplicates (Figure 1).

Microbiological analysis

For each cheese sample, 10 g was weighed and dispersed aseptically in 90 mL of citrate buffer (2%,w/v) and homogenized in a sterile polyethylene bag using a Stomacher (Seward Laboratory Blender Stomacher 400 Lab Blender UK) for 1.5 min. Serial dilutions were made in 0.1% sterile peptone water and all determinations were made in triplicates (Messer et al., 1985). The enumeration of total mesophilic bacteria (Plate Count Agar, Merck, Germany) at 30°C/48 h, total coliform groups (Violet Red Bile Agar, Merck, Germany) at 37°C/48 h, yeasts and moulds (Potato Dextrose Agar, Merck, Germany) at 21°C/7 days, *Lactobacilli* (MRS agar, Merck, Germany), *Lactococcus* sp. (M_{17} agar, Merck, Germany) and *Enterococci* (Azide Dextrose agar, Merck, Germany) at 28°C/48 h (Frank et al., 1993) were performed.

Statistical analysis

The effect of time of ripening on all parameters of proteolysis and on total FAA and BAs content of the cheese was assessed by analysis of variance (ANOVA) using the SPSS 10.0 for Windows software (Liu et al., 2003).

RESULTS AND DISCUSSION

Evaluation of physico-chemical parameters

The chemical compositions of Egyptian cheeses are presented in Table 1. The total solids content of cheese samples varied from 30.5 to 46.5, 47.2 to 58.3 and 41.2 to 48.8% in Mish, Ras and Blue cheeses, respectively. A significant variation of fat content was observed, 17.8-30.4; 33.8-48.3 and 25.3-38.4% in Mish, Ras and Blue cheeses, respectively. The salt content of the cheese samples fell within the range, 6.1-10.5; 5.6-6.8 and 4.5-5.7% in Mish, Ras and Blue cheeses, respectively. Whereas, the pH of cheeses ranged from 4.2 to 5.3; 4.5-5.2 and 4.9-5.8 respectively, which agrees with those reported for good quality Egyptian cheeses (Kebary et al., 1999; Ibrahim and Amer, 2010). Total nitrogen content in cheese samples was slightly higher in Ras cheese as compared to Mish and blue cheeses. Whilst, the water soluble nitrogen was lowest in Blue cheese as compared to other cheeses (Table 1). The WSN/TN ratio showed differences in the degree of ripening of the component cheeses. NPN represented more than 50% of the WSN of the tested cheeses; this may have originated from the component cheeses. WSN and NPN have been classically used as a measure of the extent of secondary proteolysis, that is, formation of small sized peptides (2–20 residues) and free amino acids (Furtado and Partridge, 1988). Total volatile fatty acids showed a significant variation among the tested cheeses, 33.5-55.4; 62.7-92.7 and 45.6-74.5 as 0.1 N NaOH/100 g, in Mish, Ras and Blue cheese, respectively. These variations indicated large differences in quality and degree of ripening in Egyptian cheeses. Production of BAs has frequently been

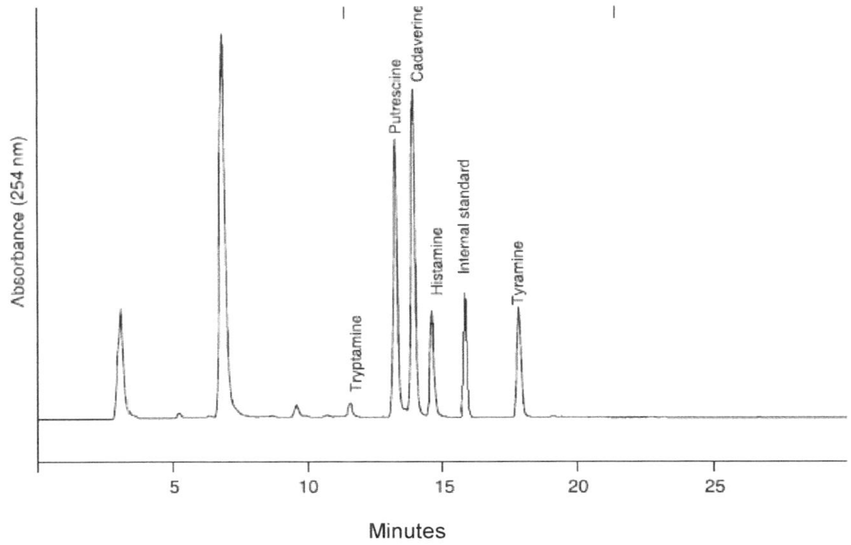

Figure 1. Chromatograms present the areas of biogenic amine standard solution derivatized with incubation at 40°C for 40 min.

Table 1. Chemical composition and proteolysis indices of Egyptian cheese samples ripened/stored at different periods.

Cheese type	Ripening/storage period (mon)	Physico-chemical parameter						Proteolysis indices			
		Total Solid (%)	Fat in solid (%)	Protein in solid (%)	Acidity *	pH	Salt in solid (%)	Total nitrogen (%)	Non-protein nitrogen (%)	Water soluble nitrogen (%)	TVFA**
Mish	6	30.5±2.0^{d}	17.8±2.5^{d}	20.1±1.07^{a}	3.31±0.71^{a}	4.2±0.1^{d}	6.1±1.5^{b}	3.15±0.98^{a}	0.22±0.4^{b}	0.29±0.5^{d}	33.5±2.2c
	12	35.6±2.1^{c}	20.8±2.2^{cd}	18.5±1.15^{b}	3.01±0.65^{b}	4.4±0.1^{cd}	7.2±1.6^{ab}	2.91±0.78^{a}b	0.31±0.6^{a}	0.34±0.6^{cd}	40.5±2.5b
	24	37.0±1.9^{b}	22.3±3.2^{bc}	17.23±1.01^{b}	2.8±0.52^{b}	4.5±0.2^{c}	8.3±2.4^{ab}	2.71±0.73^{b}c	0.33±0.6^{a}	0.41±0.4^{bc}	47.9±2.9a
	36	39.4±2.5^{b}	25.6±5.2^{b}	14.35±0.8^{c}	2.5±0.53^{c}	4.9±0.2^{b}	9.2±2.4^{ab}	2.24±0.56^{cd}	0.36±0.7^{a}	0.49±0.4^{b}	52.3±3.1a
	48	46.5±2.7^{a}	30.4±4.5^{a}	13.33±0.5^{c}	2.2±0.44^{c}	5.3±0.2^{a}	10.5±2.9^{a}	2.19±0.57d	0.39±0.7a	0.55±0.3^{a}	55.4±3.2a
Ras	6	47.2±2.9d	33.8±1.5d	24.88±1.8a	2.2 ±0.43a	4.5±0.2b	5.6±1.4b	3.91±1.29ab	0.19±0.1ns	0.31±0.6ns	62.7±3.5d
	9	50.01±3.1c	37.8±2.0c	24.44±1.3b	2.1±0.42ab	4.6±0.1b	6.7±1.5ab	3.81±1.17a	0.20±0.2ns	0.32±0.5ns	74.0±3.8c
	12	52.5±3.6b	42.5±2.4b	23.6±1.4b	2.0±0.34bc	4.8±0.2b	7.3±1.7a	3.71±1.35b	0.21±0.2ns	0.33±0.5ns	85.8±4.3b
	24	58.3±3.8a	48.3±3.1a	21.05±1.0c	1.85±0.23c	5.2±0.1a	6.8±1.9a	3.31±1.42ns	0.25±0.4ns	0.39±0.7ns	92.7±4.5a
Blue-viened	6	41.2±2.2c	25.3±2.1c	22.64±1.2a	1.9±0.22b	4.9±0.3b	4.5±1.2b	3.55±1.47ns	0.20±0.2b	0.25±0.3b	45.6±2.7c
	9	44.5±2.4b	32.1±2.9b	21.69±1.1b	1.7±0.22c	5.1±0.5b	4.8±1.5ab	3.2±1.36ns	0.21±0.3ab	0.24±0.4b	66.6±3.7b
	12	48.8±2.8^{a}	38.4±3.6^{a}	20.41±1.1^{c}	2.8±0.34^{a}	5.8±0.2^{a}	5.7±1.7^{a}	3.4±1.38^{ns}	0.24±0.4^{a}	0.34±0.5^{a}	74.5±3.9a

Averaged data of analyzed cheese samples in triplicate; Mean value ± a standard deviation * Acidity is expressed as lactic acid. ** Total volatile fatty acids is expressed as 0.1 N NaOH/100 g cheese.

Table 2. Levels of biogenic amines (mg/100 g cheese) of Egyptian cheese samples ripened/stored at different periods.

Cheese type	Ripening/storage period (month)	Biogenic amines (mg/100 g)								
		TYR	T	PTR	HIS	CAD	SPD	SPR	PE	Total
Mish	6	6 ± 0.62^{e}	4 ± 0.51^{d}	4 ± 0.45^{d}	9 ± 0.69^{e}	3 ± 0.3^{e}	1 ± 0.05^{d}	Nd^{c}	Nd^{d}	27 ± 2.2
	12	12 ± 1.5^{d}	10 ± 0.85^{c}	10 ± 0.85^{c}	14 ± 0.98^{d}	10 ± 0.5^{d}	1 ± 0.11^{d}	Nd^{c}	Nd^{d}	57 ± 3.2
	24	14 ± 0.42^{c}	17 ± 1.54^{b}	18 ± 1.5^{b}	27 ± 2.4^{c}	18 ± 1.72^{c}	2 ± 0.21^{c}	1 ± 0.08^{b}	4 ± 0.53^{c}	101 ± 4.3
	36	15 ± 0.46^{b}	21 ± 1.79^{a}	$19\pm1.7a^{b}$	29 ± 2.3^{b}	20 ± 2.1^{b}	3 ± 0.42^{b}	1 ± 0.09^{b}	7 ± 0.64^{b}	115 ± 4.6
	48	19 ± 1.55^{a}	22 ± 2.15^{a}	20 ± 2.16^{a}	31 ± 2.5^{a}	22 ± 1.97^{a}	4 ± 0.63^{a}	2 ± 0.16^{a}	12 ± 1.15^{d}	132 ± 5.3
Ras	6	3 ± 0.28^{d}	10 ± 0.93^{b}	6 ± 0.42^{d}	12 ± 0.96^{d}	Nd^{d}	1 ± 0.08^{b}	Nd^{b}	3 ± 0.19^{b}	34 ± 2.6
	9	4 ± 0.54^{c}	11 ± 1.05^{c}	8 ± 0.62^{c}	14 ± 0.91^{c}	8 ± 0.71^{c}	1 ± 0.11^{b}	Nd^{b}	3 ± 0.21^{b}	49 ± 2.8
	12	5 ± 0.61^{b}	13 ± 1.13^{b}	13 ± 0.54^{b}	23 ± 2.05^{b}	13 ± 1.26^{b}	Nd^{c}	1 ± 0.08^{a}	5 ± 0.41^{a}	73 ± 3.7
	24	14 ± 1.07^{a}	20 ± 1.86^{a}	16 ± 1.25^{a}	26 ± 2.4^{a}	20 ± 2.1^{a}	2 ± 0.21^{a}	Nd^{b}	Nd^{c}	98 ± 4.1
Blue-viened	6	Nd^{c}	11 ± 0.45^{c}	1 ± 0.15^{c}	4 ± 0.34^{c}	4 ± 0.22^{c}	Nd^{b}	Nd^{b}	1 ± 0.11^{c}	21 ± 2.1
	9	1 ± 0.1^{b}	15 ± 0.87^{b}	2 ± 0.31^{b}	9 ± 0.94^{b}	7 ± 0.65^{b}	Nd^{b}	Nd^{b}	2 ± 0.25^{b}	36 ± 2.6
	12	8 ± 0.72^{a}	17 ± 1.26^{a}	9 ± 0.93^{a}	14 ± 1.26^{a}	11 ± 1.13^{a}	3 ± 0.41^{a}	2 ± 0.17^{a}	6 ± 0.62^{a}	70 ± 3.5

Averaged data of analyzed cheese samples in triplicates, nd: not detected.

referred to as the proteolytic activity of microorganisms present in cheese during manufacture and ripening. Increases in the non-protein nitrogen fractions (WSN and NPN) often means level increase of free amino acids, which are precursors of BAs.

Amino acids and BAs

BAs content of cheese can be extremely variable and depends on the type of cheese, the ripening time, the manufacturing process and the microorganisms present (Ordonez et al., 1997). The Egyptian cheeses (Mish, Ras and Blue) examined confirmed this variability in the total content of BAs ranging from 21.0 to 130.0 mg/100 g cheese (Table 2 and Figure 2). There are significant differences among contents of the eight BAs assayed. Only the Mish cheese contained more than 100 mg/100 g cheese of the total BAs, as affected by increasing the storage period. According to Taylor (1985), the threshold of risk is 100 mg/kg total amines of cheese, if ingestion is associated with such potentiating co-factors as amine oxidase-inhibiting drugs or alcohol, or else if there are pre-existing gastrointestinal diseases (Stratton et al. 1991). Production of BAs in cheese has often been associated with non-starter lactic acid bacteria and *Enterobacteriaceae* (Joosten and Northolt, 1987), so it may be a toxicological risk associated with consumption of raw milk cheese, especially by sensitive individuals. Spanjer and van Roode (1991) suggested that the total concentration of tyramine, histamine, putrescine and cadaverine in cheese should not exceed 900 mg/kg DW^{-1}, but no upper limit for BAs in cheese has been legally enforced.

Even if no significant differences were observed in the final amounts of BAs in Blue and Ras cheeses, the dynamics of accumulation were not the same. Overall, histamine was the most prevalent amine, being found in all analyzed cheese samples. It was followed by T (98%), PTR (97%), CAD (95%), TYR (89%), SPD (73%), PE (72%) and SPR (37%). In spite of being the most frequently detected amine, SPR was present at low levels, below 2.5 mg/ 100 g cheese. Spermidine and PE were also detected at low levels (below 4 and 12mg/100 g cheese, respectively). However, HIS, CAD, T, PTR and TYR were detected at levels up to 30, 21, 20, 19 and 18 mg/ 100 g, respectively. HIS was the most prevalent amine, it was found in all analyzed cheese samples (Table 2). Higher means levels were detected for Mish, Ras and Blue cheeses (9-31, 12-26 and 4-14 mg/ 100 g, respectively). HIS levels capable of causing histamine poisoning were detected in all cheese samples. However, taking into account the concomitant presence of polyamines, it is likely that a higher percentage of cheese samples could cause HIS poisoning. TYR was present in 100% of Mish and Ras cheeses and in 60% of Blue cheese. Mish (53%) and Ras cheese (18%) contained TYR at levels capable of causing hypertensive crisis (Komprda et al., 2008). Overall, T was detected sporadically, at lower amounts as compared to HIS. Similar results were observed by Chang et al. (1985). Higher means levels of tryptamine were observed in Mish and Ras cheeses. The toxic threshold of tryptamine is not known (Joosten, 1988). PE, another amine of health significance was detected 100% in Blue, 82% in Ras and 53% in Mish cheese. The prevalence of this amine was

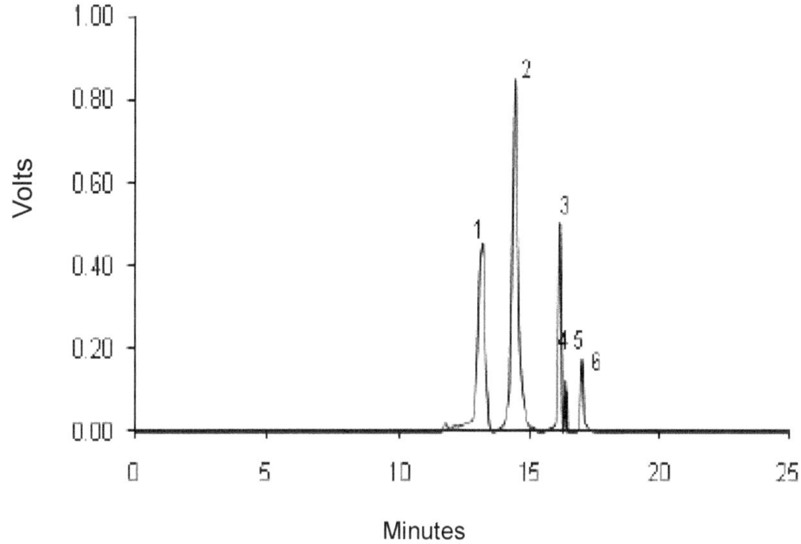

Figure 2. Chromatograms presenting the concentration of histamine, tryptamine, putrescine, tyramine, phenyl ethylamine and cadaverine in Ras cheese.

high, however the levels detected were low (≤12 mg/100 g), below its toxic threshold. The rate of CAD and T accumulation were similar for the traditional Mish and Ras cheeses and that described for the total BAs formation. The accumulation of amines increased remarkably later in the ripening and storage periods. The rate of PE accumulation was similar for Ras and Blue cheeses, whereas HIS and T values were higher in the Ras cheese and reached concentrations of 12 - 26 mg/100 g and 10-20 mg/100 g, respectively.

Similar results were obtained with heat treatment or bactofugation of the milk used for emmental production and had little effect on the TYR content (Krause et al., 1997). The rates of SPR and SPD accumulation were similar for Mish and Blue cheeses, whereas T and HIS were higher in the Mish cheeses and reached concentrations of 4-22 and 9-31 mg/100 g, respectively. According to Halász et al. (1994), Gouda cheese along with Swiss and Cheddar cheeses, which contain high levels of BAs are the most frequently incriminated cheese in histamine poisoning episodes.

Amino acid levels in cheeses types were extremely variable (Table 3). This fact was attributed to an accelerated amino acid release at the manufacturing day, when cheeses were incubated at temperatures favorable for microorganism development and activity (Bütikofer and Fuchs, 1997). Any food with free amino acids, especially tyrosine and phenylalanine, are subject to BAs formation if poor sanitation and low quality foods are used or if the food was subjected to temperature abuse or extended storage time (Schirone et al., 2011).

High variability was observed in pH, acidity, moisture and fat contents of the different analyzed cheese types. Several analyzed samples did not meet the standard of identity and quality established by Egyptian legislation. With regard to quality parameters, pH, moisture and fat content and acidity correlated significantly (P≤ 0.05) with formation and accumulation of some BAs. These results suggest that, among quality parameters evaluated, acidity influenced amine formation in several cheese types. These results are supported by the theory that the formation of BAs is a protective mechanism of bacteria against acidic environments (Maijala, 1994). The presence of micro-organisms with high decarboxylase activity has been reported as the main factor for BAs production in cheese. Moreover, some strains have proteolytic activity, which can affect the accumulation of BAs in cheese (Galgano et al., 2001). For the production of amines, the enzymatic activity of proteases derived from micro-organisms, or from another origin, is important from a qualitative point of view, that is, in relation to the type of amino acids provided to the amino acid decarboxylating microflora. The bacteriological composition of milk could be critical to define the amine profile in cheese; therefore, large amounts of amines in cheese could indicate unsuitability from a hygienic point of view, and the milk used for cheese making. Moreover, the results emphasize the necessity of controlling the indigenous bacterial population responsible for high production of BAs and the use of competitive adjunct cultures is suggested.

Microbiological evaluation

Microbiological analyses of the Egyptian cheeses were examined throughout ripening/storage period (Table 4). Lactic acid bacteria did not show any substantial change

Table 3. Total amount of free amino acids content (mg/g dry weight of cheese) of Egyptian cheese ripened/stored at different periods.

Amino acid	Mish cheese					Ras cheese				Blue cheese		
	Ripening/storage period (mon)											
	6	12	24	36	48	6	9	12	24	6	9	12
Threonine	1.38±0.1	1.36±0.13	1.38±0.13	1.20±0.11	1.18±0.1	1.18±0.1	1.32±0.1	1.38±0.1	1.28±0.1	1.28±0.1	1.18±0.1	1.12±0.1
Serine	2.26±0.2	2.09±0.23	2.06±0.23	1.96±0.23	1.70±0.2	1.62±0.2	1.88±0.2	1.88±0.3	1.81±0.3	1.79±0.2	1.72±0.2	1.44±0.1
Glutamic	8.52±1.3	8.21±1.2	8.19±1.2	6.50±1.1	5.76±1.2	9.96±1.6	6.63±1.1	5.54±0.9	5.46±1	8.66±1.3	8.60±1.3	5.34±1
Proline	3.34±0.3	3.62±0.34	3.6±0.34	2.86±0.12	2.80±0.2	2.68±0.24	2.60±0.3	3.04±0.3	3.00±0.3	2.75±0.3	2.70±0.3	2.44±0.3
Glycine	0.92±0.1	0.79±0.11	0.77±0.11	0.80±0.1	0.86±0.09	0.92±0.11	0.80±0.1	0.72±0.08	0.71±0.08	0.85±0.1	0.82±0.1	0.74±0.08
Alanine	1.56±0.1	1.55±0.19	1.54±0.19	1.34±0.13	1.40±0.1	1.14±0.13	1.44±0.1	1.38±0.1	1.31±0.1	1.42±0.1	1.40±0.1	1.38±0.1
Cysteine	0.18±0.03	0.18±0.03	0.18±0.03	0.16±0.01	0.14±0.01	0.16±0.01	0.17±0.01	0.16±0.01	0.16±0.01	0.26±0.03	0.21±0.03	0.15±0.02
Valine	1.72±0.2	1.61±0.22	1.61±0.22	1.75±0.17	1.45±0.12	1.31±0.12	1.64±0.15	1.09±0.1	1.04±0.1	1.59±0.1	1.55±0.1	1.32±0.11
Methionine	0.98±0.1	0.95±0.14	0.95±0.14	0.97±0.11	0.81±0.07	0.95±0.1	0.77±0.1	0.85±0.07	0.84±0.07	1.11±0.15	1.01±0.15	0.85±0.1
Isoleucine	2.48±0.22	2.35±0.31	2.35±0.31	2.20±0.15	2.08±0.2	2.07±0.21	2.19±0.2	2.02±0.17	1.95±0.17	2.41±0.3	2.33±0.3	2.20±0.19
Leucine	3.02±0.3	2.45±0.3	2.45±0.3	2.63±0.23	2.40±0.16	2.22±0.23	2.48±0.22	2.18±0.15	2.11±0.15	2.73±0.3	2.66±0.3	2.35±0.18
Tyrosine	2.06±0.2	1.87±0.2	1.85±0.2	1.42±0.14	1.67±0.1	1.52±0.13	1.74±0.12	1.55±0.1	1.53±0.1	1.98±0.2	1.94±0.2	1.74±0.1
Phenylalanine	1.72±0.2	1.88±0.22	1.86±0.22	1.17±0.11	1.48±0.1	1.31±0.12	1.53±0.1	1.39±0.11	1.36±0.11	1.73±0.2	1.66±0.2	1.54±0.1
Histidine	1.80±0.2	1.94±0.23	1.91±0.23	1.55±0.25	1.83±0.2	1.47±0.11	1.68±0.12	1.57±0.11	1.56±0.11	1.82±0.2	1.79±0.2	1.60±0.1
Lysine	1.12±0.1	0.99±0.17	0.95±0.17	0.77±0.09	0.85±0.1	0.66±0.02	0.78±0.06	0.80±0.08	0.78±0.08	0.92±0.1	0.87±0.1	0.81±0.1
Tryptophan	0.94±0.1	0.98±0.16	0.95±0.16	0.83±0.1	0.79±0.08	0.73±0.11	0.79±0.07	0.85±0.1	0.82±0.1	0.99±0.11	0.95±0.11	0.91±0.1
Arginine	1.10±0.1	1.11±0.16	1.01±0.16	1.15±0.13	0.78±0.09	0.82±0.12	0.75±0.1	0.78±0.1	0.75±0.1	0.83±0.09	0.79±0.09	0.69±0.05
Total	35.10±3.3	33.93±2.9	33.61±2.9	29.26±2.7	27.98±2.6	30.08±1.9	29.19±2.5	27.18±2.9	25.74±3.0	33.12±3.1	32.3±3.1	26.62±2.8

Averaged data of analyzed Cheese samples in triplicates.

during storage period, while the number of *Enterobacteriaceae* remained high during the ripening/storage period, despite a slight decrease at the end of ripening period. All bacterial groups except for coliforms were maximum in young cheeses. Numbers of *Lactococcus* sp. were slightly higher than those of *Lactobacilli* and total mesophilic bacteria. The difference of *Lactococci* counts from the other groups was maximum three log units. The predominance of *Lactococcus* sp. during the early stages of raw milk cheeses ripening was reported (Manolopoulou et al., 2003). Lactic acid bacteria (*Lactococci, Lactobacilli* and *Enterococci*) were quantitatively the dominant groups, and change of their viable numbers was significant ($P \leq 0{:}01$) throughout the ripening period. Numbers of *Enterococcus* sp. in all samples of Ras cheese were almost the same in Blue cheese. The presence of *Enterococci* sp. in high numbers could be due to their tolerance to a wide range of environmental conditions such as low temperature, high salt content and acidity (Lorencová et al., 2012; Buňková et al., 2010). Because of these properties, although all microorganisms were effected from salt significantly ($P \leq 0{:}05$), *Enterococcus* sp. were not. *Enterococci* are a group of microorganisms that may influence the ripening process due to their proteolytic and lipolytic activities and their ability to stimulate acid production by some *Lactococci* (Sarantinopoulos et al., 2001). Total mesophilic aerobic bacteria increased reaching their highest numbers during a 45 day ripening period at cold storage, and then rapidly declined. Numbers of microorganisms indicative of the hygienic quality, such as coliforms, *Enterococcus* sp. and *Lactococcus* sp. were present in cheese at relatively high levels. These counts suggest that contamination was very high in raw milk. Numbers of coliforms and

Table 4. Means counts of microorganisms in Egyptian market cheeses.

Cheese type	Ripening/ storage period (month)	Aerobic mesophilic bacteria (10^6 cfu/g)	Coliform group (10^2 cfu/g)	Moulds and Yeasts (10^3 cfu/g)	Lactococci (10^4 cfu/g)	Enterococci (10^4 cfu/g)	Lactic acid bacteria (10^6 cfu/g)
Mish	6	7.59± 0.13^{b}	4.98± .15^{ef}	3.89 ± 0.21^{f}	7.13± 0.19^{a}	6.45± 0.22^{a}	10.24±0.09^{a}
	12	7.37±0.42b^{c}	5.07± .44^{de}	4.05± 0.20^{ef}	6.94± 0.31^{b}	6.38± 0.87^{b}	9.66±0.12^{b}
	24	6.84± 0.06^{fg}	5.27± .59^{cde}	4.33± 0.63^{de}	6.75± 0.18^{c}	6.33± .44^{bc}	9.50±0.17^{bc}
	36	6.28± 0.04^{i}	5.34± .57^{cd}	4.74± 0.90^{c}	6.66± .06^{cde}	6.13± 0.54^{c}	9.23±0.08^{def}
	48	6.09± 0.18^{i}	5.53± .35^{bc}	4.82± 0.42^{c}	6.34± 0.37^{f}	5.88± 0.30^{d}	9.15±0.13^{ef}
Ras	6	7.13± 0.20^{de}	4.5± 0.23^{g}	4.05± 0.20^{ef}	6.67±0.20^{cd}	5.82±0.18^{de}	9.65±0.20^{b}
	9	6.95± 0.24^{ef}	4.63± 0.21^{g}	4.37± 0.43^{d}	6.58± .17^{de}	5.78± .39^{de}	9.55±0.13^{bc}
	12	6.65± 0.21^{gh}	4.72± .42^{gf}	4.79± 0.61^{c}	6.50± .13^{ef}	5.73± .47^{de}	9.30±0.11^{de}
	24	6.51± 0.18^{h}	3.93± 0.23^{h}	5.01± 0.39^{bc}	6.40± 0.19^{f}	5.63± 0.38^{e}	9.13±0.07^{ef}
Blue-viened	6	7.92± 0.24^{a}	5.84± 0.25^{a}	4.82± 0.42^{c}	4.51± 0.14^{g}	5.88± 0.41^{d}	9.36±0.011^{cd}
	9	7.37± 0.21^{bc}	5.65± .26^{ab}	5.23± 0.16^{ab}	4.38± 0.21^{g}	5.78± .44^{de}	9.30±0.06^{de}
	12	7.22± 0.58^{ed}	5.27± .19^{cde}	5.37± 0.39^{a}	4.00± 0.09^{h}	5.73± .30^{de}	9.08±0.08^{f}

Microbiological composition of Egyptian market cheeses (means ± SD). Means log counts in triplicates.

Enterococcus sp. were not reduced significantly (P≤0:05), while numbers of *Lactobacilli* sp. were also reduced significantly depending on the ripening time (P≤0:01), but they remained alive. This can be explained by the pH levels and the quantity of lactic acid. Counts of yeasts and moulds were in Mish cheese similar to other findings in Ras cheeses. During the ripening/storage period, the numbers were not significantly decreased (P≤0:01), and they had relatively high counts in Blue cheeses. Yeasts were present at various levels among the distinct cheeses, grouped from dairy markets, the differences in numbers may be due to the distinct pH and salt concentrations found between the corresponding cheeses, although no significant correlations resulted. Occurrence of yeasts in cheeses was variable, because they have been associated with the production of flavour compounds as a result of their relatively strong proteolytic and lipolytic activities. However, scant information is available regarding the contribution of yeasts to synthesis of BAs in foods: a histidine-decarboxylase activity was found in yeasts of the genera Debaromyces and Candida isolated from fermented meat (Montel et al., 1999) and such an activity was actually above that observed in lactic acid bacteria. Macedo et al. (1995) found that the presence of yeasts was closely related to lactic acid utilization, while their contribution to the ripening process was due to their proteolytic and lipolytic activities. In this study, the number of microorganisms such as yeasts, moulds and coliforms causing spoilage of cheeses by their putrefactive effects were decreased slightly. Formation of basic compounds from proteolysis could be as a resulting of changing of pH and a decrease of acidity. As recommendation, the permissible level of biogenic amines stipulated by Egyptian Organization for Standardization and Quality Control (EOS, 1996) should be modified to meet the more safe standard adopted by Food Drug Administration (FDA, 2001) and their levels can be lowered by using good quality raw milk and maintaining hygiene standards during manufacturing and storage processes.

Conclusions and recommendations

The main feature influencing the BAs formation is the extent of growth of microorganisms, like *Enterococci* sp. characterized by decarboxylase activity. The presence of high contents of BAs in Mish and Ras cheeses could be related to the enzymatic activity of proteases derived from microorganisms, or from another factor, that is important from a qualitative point of view, that is, in relation to the type of amino acids provided to the amino acid decarboxylating microbiota, in particular tyrosine. Therefore, a large amount of BAs in cheese reflects the bad hygienic conditions under which they are produced and stored. Accordingly, the levels of biogenic amines in different cheeses should be in accordance with the safe permissible limit recommended by FDA to ensure human safety.

Conflict of Interests

The author(s) have not declared any conflict of interests.

REFERENCES

Alberto MR, Arena ME, Manca de Nadra MC (2002). A comparative survey of two analytical methods for identification and quantification of biogenic amines. Food Control 13: 125–129.

AOAC (2002). Association of Official Analytical Chemists. Official methods of analysis (17th Ed). Arlington, Virginia, USA: Association of Official Analytical Chemists, International, Inc.

Buňková L, Bunka F, Mantlová G, Čablová A. Sedláček I, Švec P, Pachlová V, Krác S (2010). The effect of ripening and storage conditions on the distribution of tyramine, putrescine and cadaverine in Edam cheese. Food Microbiology. 27: 880-888.

Bütikofer U, Fuchs D (1997). Development of free amino acids in cheese. Le Lait 77: 91–100.

Caston JC, Eaton CL, Gheorghui BP, Ware LL (2002). Tyramine induced hypertensive episodes, panic attacks in hereditary deficient monoamine oxidase patients: case reports. J.S.C. Med. Assoc. 98: 187-192.

Chang SF, Ayres JW, Sandine WE (1985). Analysis of cheese for histamine, tyramine, tryptamine, histidine, tyrosine and tryptophan. J. Dairy Sci. 68: 284-2846.

EOS (1996). Egyptian Organization for Standardization and Quality Control. Detection of poisons and control, Report. p. 1796.

FDA (2001). Food Drug Administration. Food and drug administration hazards and controls. Guidance, 3rd ed. Center of food safety and Nutrition, Washington, USA.

Fernández M, Linares D, Del Río B, Ladero V, Alvarez MA (2007). HPLC quantification of biogenic amines in cheeses: correlation with PCR-detection of tyramine-producing microorganisms. J. Dairy Res. 74:276–282.

Frank JF, Christen GL, Bullerman LB (1993). Tests for groups of microorganisms. In: Marshall, R (Ed.), Standard Methods for the Examination of Dairy Products, 16th ed. American Public Health Association, Washington DC, pp. 271–286.

Furtado MM, Partridge JA (1988). Characterization of nitrogen fractions during ripening of a soft cheese made from ultra filtration retentates. J. Dairy Sci. 71: 1379–1400.

Galgano F, Suzzi G, Favati F, Caruso M, Martuscelli M, Gardini F, Salzano G (2001). Biogenic amines during ripening in "Semicotto Caprino" cheese: role of enterococci. Int. J. Food Sci. Technol. 36: 153–160.

Halász A, Barath A, Simon-Sarkadi L, Holzhapeel W (1994). Biogenic amines and their production by microorganisms in food. Trends Food Sci. Technol. 5: 42–46.

Ibrahim A, Amer A(2010). Comparison of biogenic amines levels in different processed cheese varieties with regulatory specifications. World J. Dairy Food Sci. 5: 127-133.

IDF (1993). International Dairy Federation. Determination of nitrogen content. Standard 20 B. Brussels, Belgium.

Izquierdo E, Marchioni E, Aoude-Werner D, Hasselmann C, Ennahar S (2009). Smearing of soft cheese with *Enterococcus faecium* WHE81, a multi bacteriocins producer, against *Listeria monocytogenes*. Food Microbiol. 26: 16-20

Joosten HMLJ (1988). Conditions allowing the formation of biogenic amines in cheese: 3 Factors influencing the amounts formed. Neth. Milk Dairy J. 41: 329–357.

Joosten HMLJ, Northolt MD (1987). Conditions allowing the formation of biogenic amines in cheese. 1. Decarboxylative properties of some non-starter bacteria. Neth. Milk Dairy J. 41: 259-280.

Kayser FH (2003). Safety aspects of enterococci from the medical point of view. Int. J. Food Microbiol. 88: 255-262.

Kebary KK, El-Sonbaty AH, Badawi RM (1999). Effects of heating milk and accelerating ripening of low fat Ras cheese on biogenic amines and free amino acids development. Food Chem. 64 (1): 67-75.

Khan H, Flint S, Yu PL (2010). Enterocins in food preservation. Int. J. Food Microbiol. 141: 1-10.

Komprda T, Burdychova R Dohnal V, Cwikova´ O, Sla´dkova´ P, Dvora´ckova H (2008). Tyramine production in Dutch-type semi-hard cheese from two different producers. Food Microbiol. 25: 219–227.

Krause I, Bockhardt A, Klostermeyer H (1997). Characterization of cheese ripening by free amino acids and biogenic amines and influence of bactofugation and heat-treatment of milk. Le Lait 77: 101-108.

Krause I, Bockhardt A, Neckermann H, Henle T, Klostermeyer H (1995). Simultaneous determination of amino acids and biogenic amines by reversed-phase high performance liquid chromatography of the dabsyl derivatives. J. Chromatogr. A 715: 67-79.

Kuchroo CN, Fox PF (1982) Soluble nitrogen in Cheddar cheese: comparison of extraction procedures. Milchwissenschaft 37: 331-335.

Kung HF, Lee YH, Chang SC, Wei CI, Tsai YH (2007). Histamine contents and histamine-forming bacteria in sufu products in Taiwan. Food Control 18: 381–386.

Ladero V, Calles-Enríquez M, Fernández M, Alvarez MA (2010). Toxicological effects of dietary biogenic amines. Curr. Nutr. Food Sci. 6: 145-156.

Ladero V, Fernández M, Alvarez MA (2009). Effect of post-ripening processing on the histamine and histamine-producing bacteria contents of different cheeses. Int. Dairy J. 19: 759-762.

Liu RX, Kuang J, Gong Q, Hou XL (2003). Principal component regression analysis with *spps*. Comput. Methods Programs Biomed. 71: 41-47.

Loizzo MR, Menichini F, Picci N, Puoci F, Spizzirri G, Restuccia D (2012). Technological aspects and analytical determination of biogenic amines in cheese. Trends Food Sci. Technol. xx: 1-18.

Loizzo MR, Menichini F, Picci N, Puoci F, Spizzirri G, Restuccia D (2013). Technological aspects and analytical determination of biogenic amines in cheese. A review. Trends Food Sci. Technol. 30: 38-55.

Lorencová E, Buňková L, Matoulková D, Dráb V, Pleva P, Kubáñ V. Buňka F (2012). Production of biogenic amines by lactic acid bacteria and bifidobacteria isolated from dairy products and beer. Int. J. Food Sci. Technol. 47: 2086-2091.

Macedo AC, Malcata FX, Hogg TA (1995). Microbiological profile in Sera ewe's cheese during ripening. J. Appl. Bacteriol. 79: 1–11.

Maijala R (1994). Histamine and tyramine production by a Lactobacillus strain subjected to external pH decrease. J. Food Protect. 57: 259-262.

Manolopoulou E, Sarantinopoulos P, Zoidou E, Aktypis A, Moschopoulou E, Kandarakis IG, Anifantakis EM (2003). Evolution of microbial populations during traditional Feta cheese manufacture and ripening. Int. J. Food Microbiol. 82: 153–161.

Martuscelli M, Gardini F, Torriani S, Mastrocola D, Serio A, Chaves-Lopez C, Schirone M, Suzzi G (2005). Production of biogenic amines during the ripening of Pecorino Abruzzese cheese. Int. Dairy J. 15: 571-578.

Mercogliano R, De Felice A, Chirollo C, Cortesi ML (2010). Production of vasoactive amines during the ripening of Pecorino Carmasciano cheese. Vet. Res. Commun. 34: 175–178.

Messer JW, Behney HM, Leudecke LO (1985). Microbiological count methods. In: Richardson, GH (Ed.), Standard Methods for the Examination of Dairy Products, 15th Ed. APHA, Washington, DC, USA. pp. 133–149.

Mohamed AG Deabes MM, Fatma AM Hassan A, Enab K, Abou- Arab AAK (2013). Biogenic amines and chemical composition of different formulations used for manufacture of processed cheese. J. Appl. Sci. Res. 9 (3): 1477-1483.

Montel MC, Masson F, Talon R (1999). Comparison of biogenic amine content in traditional and industrial French dry sausages. Sci. Aliments 19: 247–254.

Ordonez JA, Ibanez FC, Torre P, Barcina Y (1997). Formation of biogenic amines in Idiazabal ewe's-milk cheese: effect of ripening, pasteurization, and starter. J. Food Protect. 60: 1371-1375.

Pinho O, Ferreira IO, Mendes E, Oliviera BM, Ferreira M (2001). Effect of temperature on evolution of free amino acid and biogenic amine contents during storage of Azeitao cheese. Food Chem. 75: 287-291.

Pintado AIE, Pinho O, Ferreira IO, Pintado M, Gomes A, Malcata F (2008). Microbiological, biochemical and biogenic amine profiles of Terrincho cheese manufactured in several dairy farms. Int. Dairy J. 18: 631-640.

Repka-Ramírez MS, Baraniuk JN (2002). Histamine in health and disease. J. Allergy Clin. Immunol. 17: 1–25.
Samková, E, Dadáková E, Pelikánová T (2013). Changes in biogenic amine and polyamine contents in smear-ripened cheeses during storage. Eur. Food Res. Technol. 237: 309-314.
Sarantinopoulos P, Andrighetto C, Georgalaki MD, Rea MC, Lombardi A, Cogan TM, Kalantzopoulos G, Tsakalidou E (2001). Biochemical properties of *enterococci* relevant to their technological performance. Int. Dairy J. 11: 621–647.
Schirone M, Tofalo R, Mazzone G, Corsetti A, Suzzi G (2011). Biogenic amine content and microbiological profile of Pecorino di-Farindola cheese. Food Microbiol. 28: 128-136.
Smit G, Smit BA, Engels WJM (2005). Flavour formation by lactic acid bacteria and biochemical flavour profiling of cheese products. FEMS Microbiol. Rev. 29: 591-610.
Spanjer MC, Van Roode BASW (1991). Towards a regulatory limit for biogenic amines in fish, cheese and sauerkraut. De Ware (n)-Chemicus 21: 139-167.
Stratton SS, Hutkins RW, Taylor SL (1991). Biogenic amines in cheese and other fermented foods: a review. J. Food Prot. 54: 460-470.
Sumner SS, Taylor SL (1989). Detection method for histamine producing dairy-related bacteria using di-amine oxidase and leucocrystal violet. J. Food Prot. 52: 105–108.
Suzzi G, Gardini F (2003). Biogenic amines in dry fermented sausages: a review. Int. J. Food Microbiol. 88: 41-54.
Taylor SL (1985). Histamine Poisoning Associated with Fish, Cheese and Other Foods. World Health Organization, Geneva, Switzerland. pp. 1-47.
Ten Brink B, Damink C, Joosten HMLJ, Huisint-Veld JHJ (1990). Occurrence and formation of biologically active amines in foods. Int. J. Food Microbiol. 11: 73–84.
Wöhrl S, Hemmer W, Focke M, Rappersberger K, Jarisch R (2004). Histamine intolerance-like symptoms in healthy volunteers after oral provocation with liquid histamine. Allergy Asthma Proc. 25: 305–311.

Cyanide and selected nutrients content of different preparations of leaves from three cassava species

M. G. Umuhozariho[1,2], N. B. Shayo[1], J. M. Msuya[1] and P. Y. K. Sallah[2]

[1]Department of Food Science and Technology, Faculty of Agriculture, Sokoine University of Agriculture, P.O. Box 3006, Morogoro, Tanzania.
[2]College of Agriculture, Animal Sciences and Veterinary Medicine, University of Rwanda, P.O. Box 117, Huye, Rwanda.

Cassava leaves are largely consumed as vegetable in African, but contain a toxic compound, cyanide. To ascertain their safety and contribution to human nutrition, after a number of pre-treatments preceding their boiling in water, cyanide, vitamin C, β-carotene, crude protein, iron, calcium, phosphorus, potassium and zinc contents were assessed in leaves from bitter, sweet and wild cassava species, boiled for 15 and 30 min after differently processed by: (1) pounding un-dried (UND), (2) drying before pounding (DBP) and (3) drying after pounding (DAP). Blanching headed drying was done in a tunnel solar dryer. Results showed that cassava species, processing procedures, and boiling time significantly ($p < 0.05$) reduced cyanide and the nutrients. However, except vitamin C, eliminated to almost nil, other nutrients were retained at considerable levels. Sensibly decreased by drying and/or boiling, cyanide levels ranged from 32 - 50 mg HCN/kg (dry matter basis) after boiling for 30 min. These levels, above the recommended level (10 mg HCN/kg) for foods, were safe with regard to cyanide toxicity based on the fact that the vegetable is served in small quantities as side food. consumed quantities of relishes as side foods. Nevertheless, it was advisable not to make them the everyday foods, especially to lower body weight such as children, and to extend time of cooking.

Key words: Cassava leaves, cassava species, processing procedures, cyanide and nutrients, Rwanda

INTRODUCTION

Cassava is a very important crop in the tropics and a staple food for over 800 million people (Nassar et al., 2007), growing over a range of climates and altitudes and on a wide variety of soils (FAO/IFAD, 2005). In Africa, Rwanda included, cassava is primarily used as food, consuming roots as starchy food (Nweke et al., 2002), but leaves are also largely consumed as green vegetables (Achidi et al., 2005).

Producing a valuable and safe food from cassava and cassava leaves involve certain challenges. In fact, cassava and cassava leaves contain cyanide, in the form of cyanogenic glucosides, primarily linamarin and small lotaustralin (Uyoh et al., 2007). The cyanogenic glucosides are distributed throughout the cassava plant, with highest levels in leaves (Etonihu et al., 2011). Under high temperature, pressure, and use of enzyme (linamarase), or mineral acids, cyanogenic glucosides are decomposed into acetone cyanohydrins which, at pH above 5 or temperatures above 35°C, is broken down, spontaneously, into poisonous compound, hydrogen cyanide (HCN) (Siritunga and Sayre, 2004). Cyanohydrins are the most dangerous form of the cyanide because at the elevated pHs and temperatures present in the human body, it rapidly decomposes to release the poisonous hydrogen cyanide. The continued consumption of high dietary cyanogens has been linked with a number of chronic health disorders, and occasionally death, depending on the level of cyanogens, frequency of cyanogens exposure and, quality and quantity of protein intake status (Cliff et al., 2011; Nhassico et al., 2008; CCDN, 2007). For the human body detoxification, unbound cyanide is converted to less toxic thiocyanate (SCN) and is excreted in the urine. The synthesis of thiocyanate requires sulphur-containing amino acids, as a consequence of protein intake (CCDN, 2007).

Despite the presence of the poisonous component, numerous publications have provided evidence on potential contribution of cassava leaves in human nutrition by providing protein, minerals and vitamins, depending on preparation techniques (Akinwale et al., 2010; Mulokozi et al., 2007; Ayodeji, 2005). Faber and Van Jaarsveld (2007) revealed that improved handling, such as optimizing time of thermal treatment, drying process and preliminary preparations can preserve quality of treated food. In Rwanda, leaves from three species of cassava, bitter (*Manihot utilissima),* sweet (*Manihot dulcis)* and wild (*Manihot glaziovii)* are valued and highly utilized as green vegetables. They are usually cooked freshly harvested, but preservation by sun-drying to extend their shelf life is also reported by Umuhozariho et al., (2011). The direct exposure to sunlight is known to reduce the quality (colour and vitamin contents) of the final product (MMA, 2008). However, solar drying is reported among strategies to combating nutrients losses in processed food stuffs and extending the availability of the nutrient-rich foods, beyond the season in which they are in abundance (Oguche Gladys, 2011; Thompson and Amoroso, 2011).

The present study was undertaken to improve drying of cassava leaves by using tunnel solar dryer and evaluate levels of cyanide and nutrients in the dried and un-dried leaves, from the three cassava species and after preparation as human food.

MATERIALS AND METHODS

Collection of cassava leaves

In June, 2012, tender cassava leaves, the first matured up to leaf position five were harvested from three species of cassava, bitter (*Manihot utilissima*), sweet (*Manihot dulcis*) and wild (*Manihot glaziovii*). Varieties named "Seruruseke" (5280), ISAR 1961 and "Igicucu" were chosen for sweet, bitter and wild, respectively. In order to minimize the effects of age, environment and soil type on chemical composition, leaves samples of the same age were selected from the same field, Rwanda Agricultural Board (RAB)'s field at the Karama Research Station, in Bugesera District of Eastern Province of Rwanda.

Sample preparation

Samples were collected in the field and transported in closed polyethylene bags, which were stored in a cool box containing ice. Each sample was divided into two portions, first portion was analyzed in fresh condition and for the second portion analysis was done after blanching and drying. Blanching was done by submersion in boiling water for 4-5 min, and then immediately cooled in tap water at ambient temperature as described by Kendall et al. (2010).

Three different preparation procedures were conducted, namely: (1) Un-dried (UND) (2) dried before pounding (DBP) and (3) dried after pounding (DAP) leaves. Pounding was done using wooden mortar and pestle, while drying was done using a tunnel solar dryer at Sokoine University of Agriculture. The products obtained by the three different preparation procedures (Figure 1) were chemically analyzed, un-boiled and boiled for 15 and 30 min. Moisture, cyanide, protein and minerals (Ca, Fe, K, P and Zn) were determined. The first four analyses were conducted at Sokoine University of Agriculture laboratories, while vitamins (Ascorbic acid and β-carotene) analyses were done at the Tanzania Food and Drug Authority (TFDA), in Dar-Es-Salaam. All chemical analyses were carried out in quadruple.

Drying procedures

After blanching, pounded and un-pounded leaves from bitter, sweet and wild cassava were dried using a tunnel solar dryer. Temperatures inside the dryer were recorded at 8 a.m., noon and 8 p.m. each day, averaging 38°C. The complete drying was when the samples were dried until they became brittle. The dried samples were immediately packed in plastic materials, sealed and transported to laboratories, in opaque cartons to avoid light effect before analysis.

Cooking procedures

The cooking consisted of boiling for 15 and 30 min, in distilled water (1:2) and (1:9) respectively for un-dried and dried samples as volume of sample by volume of water, in stainless steel and without cover. The dried samples were first soaked in water for about 5 min before starting the fire. Un-dried and cooked samples were kept frozen before analysis.

Cyanide (HCN) and nutrients determination

Cyanide (HCN) levels in the samples were determined by alkaline titrating method as described by AOAC (1995), official method 915.03B. Moisture content of samples was determined as outlined by AOAC (1995), official method 934.01. For minerals, sample ashes and solutions were obtained respectively by official methods 965.09 and 982.23 described by AOAC (1995). Total phosphorus (P) was obtained using ascorbic acid blue color procedure and by reading the absorbance at a wavelength of 884 nm on a UNICAM 5625 UV/visible spectrometer (Okalebo et al., 1993). Calcium (Ca) and potassium (K) were measured by flame photometry, reading their absorbance at 422.7 and 766.5 nm respectively on a Cole-Parmer instrument, model 2655-00 Digital flame Analyzer. Iron (Fe) and zinc (Zn) were determined by reading their absorbance at 248.3 and 213.9 nm, respectively on a UNICAM 919 Atomic Absorption Spectrometer (AAS) using Hollow Cathode lamps (Okalebo et al., 1993). Crude protein content was determined by using the micro-Kjeldahl method (AOAC, 1995), official method 920.87. Vitamin C (ascorbic acid) content was determined as outlined by ISO (1984), method 6557/2. B-carotene was measured using a high performance liquid chromatography (HPLC), equipped with a Photodiode Array (PDA) detector fitted with a 436 nm wavelength. For sample preparation, aliquots were extracted by solvent n-Hexane (Priadi et al., 2009; Tee Siong and Lam, 1992). Further extraction and clean-up was done using a dispersive Solid Phase Extraction (dSPE) technique as described in AOAC (2007), official method 2007 0.1.

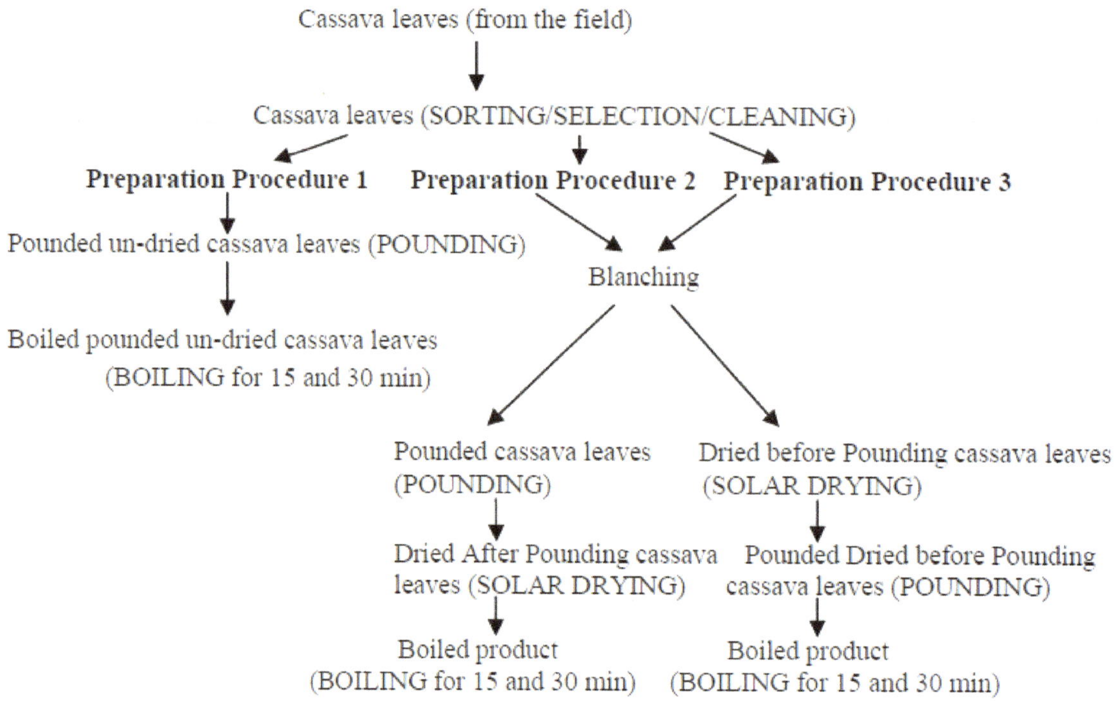

Figure 1. Flow diagram illustrating preparation procedures of cassava leaves.

Statistical data analysis

Data from the results of chemical analyses of samples were subjected to statistical analysis, using SAS 9.2 (SAS Institute, 2008). Kolmogorov-Smirnov test was first carried out to assess the normality of the data (Kutner et al., 2005). Multiple ways (species with three levels specifically bitter, sweet and wild, processing methods with three levels namely un-dried, dried before pounding and dried after pounding, boiling time with 2 levels, that is, 15 and 30 min) analysis of variance (ANOVA) was applied after assuming the normal distribution of the data. Where the treatments had statistical significant effect on the response variables of interest, Fisher's least significant differences (LSD) test was used to separate the means. The treatments were judged statistically significantly different at $p < 0.05$.

RESULTS AND DISCUSSION

Cyanide, ascorbic acid, β-carotene, protein, iron, calcium, phosphorus, potassium and zinc were chemically determined in cassava leaves from three species (bitter, sweet and wild). Overall effect of species and processing procedures on contents are shown in Table 1. From the results in the table, it was noticeable that for un-dried and dried samples, before boiling, wild species had the highest concentrations in all determinations (cyanide, ascorbic acid, β-carotene, protein, iron, calcium, phosphorus, potassium and zinc). Bitter species had also high protein content as the wild. Sweet species was less concentrated in cyanide, β-carotene, crude protein, calcium, phosphorus and potassium than wild and bitter. Bitter and sweet species had similar concentrations of ascorbic acid and iron. Bitter species was less concentrated in zinc than wild and sweet.

The presence of cyanide in all the studied leaves confirmed the earlier reports that all cassava cultivars contain cyanogenic glucoside, in a wide variation according to varieties (CIAT, 2007). The levels of cyanide in fresh leaves of the studied species were 1905, 1480 and 2179 mg HCN/kg respectively, for bitter, sweet and wild. The values were in the ranges reported by earlier researchers, from 189 to 2466 mg HCN/kg fresh weight basis by Fukuba et al. (1982) and 800 to 3200 mg HCN/kg dry matter by Ravindran (1995). The high values in the studied leaves were not surprising as cyanide is known to be distributed throughout the cassava plant, but with highest levels in leaves (Etonihu et al., 2011). However, depending on the varieties, moderate and low cyanide content cassava leaves have been discovered by Burns et al. (2012). As continued consumption of high dietary cyanide has been linked with a number of chronic health disorders (Nhassico et al., 2008; CCDN, 2007), it is evident that, cassava leaves in the present study need to be properly processed to reduce their cyanide before consumption by humans.

Nutrients varied significantly ($p< 0.05$) according to cassava species. Sarkiyayi and Agar (2010) revealed significant differences in protein and mineral contents when investigating sweet and bitter roots and it may be the same phenomenon in the leaves. In un-dried samples, the average values of crude protein (35-36%), iron (230-278 mg/kg), calcium (7373-8822 mg/kg), phosphorus (4413-4907 mg/kg), potassium (15110-17119 mg/kg) and

Table 1. Mean levels of cyanide and selected nutrients of cassava leaves according to species and processing procedure.

Parameter		Cyanide (mg/kg)	Vitamin (mg/100g)		Crude protein (%)	Mineral (mg/kg)				
PP	CS	HCN	AA	B-C	CP	Fe	Ca	P	K	Zn
	Bitter	1905.0±25.0^{b}	8.41±0.03^{b}	51.2±0.5^{b}	36.6±0.5^{a}	230.4±1.6^{b}	7441.0±60.3^{b}	4607.7±48.8^{b}	16309.9±102.9^{b}	64.2±0.7^{c}
UND	Sweet	1480.5±18.9^{c}	8.39±0.02^{b}	40.6±0.3^{c}	35.2±0.8^{b}	230.4±1.3^{b}	7373.5±63.3^{c}	4413.3±45.6^{c}	15110.4±116.8^{c}	67.0±0.6^{b}
	Wild	2179.7±29.1^{a}	13.27±0.09^{a}	80.4±0.3^{a}	36.6±0.2^{a}	278.3±2.1^{a}	8222.5±61.8^{a}	4907.4±50.2^{a}	17119.9±109.5^{a}	76.2±0.7^{a}
	Bitter	562.8±24.2^{b}	0.00075±0.00^{b}	43.7±0.8^{b}	36.8±0.7^{a}	222.5±0.6^{b}	6956.1±65.0^{b}	4590.4±43.5^{b}	15520.6±139.4^{b}	61.9±0.7^{c}
DBP	Sweet	467.8±27.6^{c}	0.00075±0.00^{b}	39.8±0.2^{c}	35.1±0.9^{b}	221.8±1.9^{b}	6587.0±60.3^{c}	4212.0±49.7^{c}	14358.1±150.2^{c}	65.5±1.2^{b}
	Wild	873.3±27.3^{a}	0.00090±0.00^{a}	65.9±0.2^{a}	36.1±0.1^{a}	243.0±0.3^{a}	8070.2±62.1^{a}	4877.4±48.6^{a}	16157.4±133.0^{a}	73.2±0.8^{a}
	Bitter	413.6±21.7^{b}	0.00070±0.00^{b}	39.1±0.1^{b}	36.6±0.1^{a}	215.2±0.1^{b}	6360.2±68.7^{b}	4454.6±48.6^{b}	14693.3±156.2^{b}	59.7±1.3^{c}
DAP	Sweet	352.4±23.6^{c}	0.00070±0.00^{b}	29.1±0.0^{c}	35.2±0.1^{b}	215.3±0.2^{b}	6124.7±69.9^{c}	3957.2±52.3^{c}	12922.2±151.0^{c}	62.1±0.2^{b}
	Wild	684.9±21.6^{a}	0.00080±0.00^{a}	63.5±0.2^{a}	36.3±0.1^{a}	228.9±0.5^{a}	7898.8±63.6^{a}	4607.2±48.7^{a}	15765.7±118.7^{a}	69.5±1.6^{a}

Values are means and SE of nine independent determinations, dry matter basis in quadruple. Means within columns superscript by similar letter are not significantly different from each other ($P < 0.05$) by Fisher's least significant difference (LSD). PP = Processing procedures CS = cassava species, HCN = hydrogen cyanide, AA = ascorbic acid, β-C = β-carotene, CP = crude protein, Fe = iron, Ca = calcium, P = phosphorus, K = potassium, Zn = zinc, UND = un-dried, DBP = dried before pounding and DAP = dried after pounding

zinc (64-76 mg/kg) agreed with values in earlier reports (Dada and Owuru, 2010; Ravindran, 1995). Values of β-carotene (406-804 mg/kg) were in agreement with values mentioned by Priadi et al. (2009). Considering these nutrient levels, it can be said that cassava leaves are good source of β-carotene, protein and minerals and similar observation has been written by Achidi et al. (2005) and Ayodeji (2005). Mulokozi et al. (2007) and Akinwale et al. (2010) noticed also the potential contribution of cassava leaves, especially in vitamin A and suggested to properly prepare them for more profit from their present nutrients.

The cassava leaves were differently processed before cooking: (1) "Un-dried (UND)", (2) "Dried before pounding (DBP)" and (3) "Dried after pounding (DAP)". DBP and DAP leaves were blanched before solar drying, principally for inactivating potentially deleterious enzymes. The leaves were dried to brittle and on average, water content of the leaves varied from 83.5% for fresh to 4.7% for dried. As it is mentioned by James and Kuipers (2003), green vegetables contains less sugar, and thus, they can be dried to brittle and water content, 4-8%, depending on the type of vegetable. Un-dried samples were more concentrated in cyanide, ascorbic acid, β-carotene, iron, calcium and potassium than dried, but crude protein seemed not to be sensibly affected by drying. Dried before pounding and dried after pounding samples were the first in protein and second in iron and potassium. Dried un-pounded (DBP) samples retained more cyanide and nutrients than the dried pounded leaves (DAP). Drying reduced deeply the ascorbic acid content of the samples. The differences between un-dried and dried leaves were due to the combination effects of blanching and drying, because before drying leaves were blanched. It has been revealed that blanching and drying reduce the poisonous compound, cyanide, but unluckily accompanied by nutrients losses (Oguche Gladys, 2011; Anhwange et al., 2011; Eze, 2010; Abah Idah et al., 2010; Udofia et al., 2010). From the same table (Table 1), it was observable that pounding promoted cyanide and nutrients removal. The decrease, probably due to leaching or solubility in evaporated or drained water, was facilitated by small sized particles of dried after pounding products.

After the solar drying, the residual cyanide was still high (684-873 mg HCN/kg dry matter). An additional treatment was necessary for safety of the foodstuffs. Cooking by boiling in water is well known to reduce sensibly cyanide (Gernah et al., 2012; Ubi et al., 2008). Therefore, the UND, DBP and DAP were boiled, in distilled water, using stainless materials, for 15 and 30 min. In addition to cyanide reduction, findings in this study showed a significant ($p < 0.05$) decrease in protein, vitamins and minerals with cooking time and it has been the same observation in earlier study of Gernah et al. (2012). The concern was to assess a state of the cooked cassava leaves in regards to cyanogens and nutrients. Means of cyanide, ascorbic acid, β-carotene, protein, iron, calcium, phosphorus, potassium and zinc of the vegetables,

Table 2. Mean levels of cyanide and selected nutrients of boiled cassava leaves for 15 minutes.

Parameter		Cyanide (mg/kg)	Vitamin (mg/100g)		Crude protein (%)	Mineral (mg/kg)				
PP	CS	HCN	AA	B-C	CP	Fe	Ca	P	K	Zn
UND	Bitter	571.2±5.6^{b}	5.78±0.33^{b}	34.1±1.9^{b}	36.8±0.7^{b}	144.7±0.4^{b}	4230.4±11.9^{c}	3644.0±30.6^{c}	8917.0±87.9^{b}	45.8±0.8^{c}
	Sweet	501.9±1.4^{c}	5.39±0.05^{c}	33.1±2.6^{c}	35.5±0.4^{c}	142.5±1.4^{c}	4622.7±34.1^{b}	3872.9±33.9^{b}	8905.2±149.4^{c}	58.8±0.9^{b}
	Wild	696.5±3.3^{a}	11.49±0.04^{a}	49.1±3.1^{a}	36.9±0.8^{a}	210.4±1.8^{a}	6729.3±19.8^{a}	4533.0±31.6^{a}	8984.3±119.6^{a}	68.7±0.8^{a}
DBP	Bitter	532.8±4.2^{b}	-	35.9±0.2^{b}	36.8±0.7^{b}	92.5±0.6^{c}	4186.2±25.0^{c}	3199.4±23.5^{b}	8520.6±139.4^{b}	45.7±2.3^{c}
	Sweet	467.8±4.6^{c}	-	33.0±4.8^{c}	36.7±0.2^{b}	97.8±1.9^{b}	4487.0 ±20.3^{b}	3212.0±20.7^{b}	8858.1±150.2^{a}	58.1±0.2^{b}
	Wild	543.3±4.3^{a}	-	39.8±0.2^{a}	37.2±0.9^{a}	113.0±0.3^{a}	6370.2±32.1^{a}	4346.3±33.4^{a}	8857.4±133.0^{a}	63.5±1.6^{a}
DAP	Bitter	303.6±3.7^{b}	-	34.2±0.1^{b}	35.3±0.2^{b}	92.2±0.1^{c}	4160.2±28.7^{b}	3177.4±28.6^{b}	8493.3±156.2^{c}	45.1±0.1^{c}
	Sweet	292.4±3.6^{c}	-	29.1±0.1^{c}	35.2±0.1^{b}	95.3±0.3^{b}	4424.7±16.9^{b}	3157.2±42.3^{b}	8822.2±151.0^{b}	58.3±0.2^{b}
	Wild	314.9±3.6^{a}	-	35.5±0.2^{a}	36.7±0.1^{a}	112.9±0.5^{a}	6298.8±36.6^{b}	4307.2±38.7^{a}	8855.7±118.7^{a}	86.3±0.1^{a}

Values are means and SE of nine independent determinations, dry matter basis in quadruple. Means within sub-columns superscript by similar letter are not significantly different from each other ($P < 0.05$) by Fisher's least significant difference (LSD). PP = processing procedures, CS = cassava species, HCN = hydrogen cyanide AA = ascorbic acid, βC = β-carotene, CP=Crude protein, Fe = iron, Ca = calcium, P = phosphorus, K = potassium, Zn = zinc, UND = un-dried, DBP = dried before pounding and DAP = dried after pounding.

boiled for 15 and 30 min are given respectively in Tables 2 and 3.

After boiling for 15 min (Table 2), depending on species and processing procedure, the residual cyanide levels ranged from 209 to 696 mg HCN/kg and remained high so that an extension of cooking time was highly indispensable. After boiling for 30 min (Table 3), the remaining levels of cyanide, across species and processing procedures, varied from 32 to 50 mg HCN/kg. From Tables 2 and 3, processing procedure that excluded blanching (UND) was more effective in removing cyanide by heating than those procedures that included blanching (DBP and DAP). This was attributed to the action of endogenous linamarase on cyanogenic glucosides, following the intimate contact in the finely-divided tissues, during pounding, between linamarin and the hydrolyzing enzyme, linamarase, which promotes rapid breakdown of cyanogens glucosides into a free form, hydrogen cyanide (HCN) (White et al., 2003), while blanching inactivated enzymes in the dried samples and limited easier hydrolysis of cyanogenic glucosides into hydrogen cyanide. The hydrogen cyanide is known to be easily removed by heat during boiling. Similar to drying, after boiling, it was visible that pounding and then drying (DAP) improved cyanide reduction. The reason may be the same for drying, small sized particles of boiled products.

Cassava leaves as a safe human food

After cooking for 30 min, moisture content of the called "relishes" was on average 87%. Therefore, one kilogram (dry weight basis) is equivalent to about 4 kg of the cooked vegetable as it is eaten (relishes). Under normal circumstances, this volume is shared by many persons in one meal considering an adult person can eat up to 100-200 g of the vegetable relish. Furthermore, different studies reported that an acute oral lethal dose of hydrogen cyanide (HCN) is proportional to body weight (WHO, 2004). But a large variation of the doses showed a lack of precision. For example, levels ranging from 30-210 mg of HCN for a 60 kg adult have been recorded by Montgomery (1980). Committee of experts in codex standards concluded that a cyanide level of up to 10 mg HCN/kg of cassava flour is not associated with acute toxicity (FAO/WHO, 1993) and the level became recommended by FAO/WHO (1991) as safe for human foods. Therefore, the cassava leaves in this study, un-dried and solar dried, after being boiled for 30 min, can be said to be safe for human consumption in regards to cyanide toxicity, based on the acute oral lethal doses, but also by considering that the quantities of green vegetables are usually small by serving, as a side relish for the starchy based food. Besides, relish from the leaves was found as source of protein, and the nutrient is known to be helpful in cyanide human body detoxification (Nhassico et al., 2008; CCDN, 2007).

Potential contribution of cassava leaves to human nutrition

For the nutrients, the nine vegetable relishes

Table 3. Mean levels of Cyanide and selected nutrients of boiled cassava leaves for 30 minutes.

Parameter		Cyanide (mg/kg)	Vitamin (mg/100g)		Crude Protein (%)	Mineral (mg/kg)				
PP	CS	HCN	AA	B-C	CP	Fe	Ca	P	K	Zn
UND	Bitter	35.4±2.0^{b}	0.24±0.01^{b}	33.0±1.0^{b}	33.7±0.6^{b}	199.4±1.4^{b}	3939.8±27.8^{c}	3564.9±16.5^{c}	8849.3±102.6^{b}	45.4±0.4^{c}
	Sweet	32.8±1.5^{c}	0.20±0.03^{c}	32.6±0.4^{c}	33.8±0.4^{b}	182.7±0.8^{c}	4391.4±24.7^{b}	3795.5±28.2^{b}	8834.1±159.1^{b}	57.9±0.1^{b}
	Wild	40.7±1.5^{a}	0.30±0.01^{a}	43.2±1.9^{a}	34.3±0.2^{a}	219.6±2.5^{a}	5939.3±27.5^{a}	4321.7±23.2^{a}	8960.3±173.5^{a}	68.4±1.2^{a}
DBP	Bitter	47.7±1.1^{b}	-	33.5±0.6^{b}	35.8±0.1^{a}	144.5±2.8^{b}	3509.7±26.3^{c}	3120.9±35.5^{c}	8216.6±124.3^{c}	45.1±0.2^{c}
	Sweet	45.2±0.8^{c}	-	32.6±0.8^{c}	35.0±1.2^{b}	143.9±1.4^{c}	3654.3±28.3^{b}	3200.4±23.6^{b}	8521.8±169.7^{b}	54.4±0.7^{b}
	Wild	50.4±0.4^{a}	-	37.7±0.3^{a}	35.9±0.6^{a}	205.9±3.5^{a}	5436.0±25.6^{a}	4166.9±4.2^{a}	8861.1±180.9^{a}	63.0±1.2^{a}
DAP	Bitter	38.0±2.9^{b}	-	33.2±0.2^{b}	35.3±0.9^{a}	87.1±3.6^{c}	3221.6±33.3^{c}	2982.9±25.0^{b}	7840.7±181.6^{b}	44.5±0.7^{c}
	Sweet	36.4±1.8^{c}	-	29.0±1.8^{c}	35.4±0.5^{a}	95.0±0.6^{b}	3400.2±22.6^{b}	2737.8±25.1^{c}	7008.9±171.6^{c}	51.9±0.4^{b}
	Wild	41.8±2.1^{a}	-	34.3±0.1^{a}	35.0±0.4^{b}	103.9±1.1^{a}	4888.4±18.9^{a}	3898.3±16.9^{a}	8746.5±168.0^{a}	61.7±0.6^{a}

Values are means and SE of nine independent determinations, dry matter basis in quadruple. Means within sub-columns superscript by similar letter are not significantly different from each other ($P < 0.05$) by Fisher's Least Significant Difference (LSD). PP = Processing procedures, CS = cassava species, HCN = hydrogen cyanide AA = ascorbic acid, βC = β-carotene, CP = crude protein, Fe = iron, Ca = calcium, P = phosphorus, K = potassium, Zn = zinc, PM = processing method, SP = species, UND = un-dried, DBP = dried before pounding and DAP = dried after pounding.

retained appreciable levels of the studied nutrients, except vitamin C, for which levels were too small to be considered as traces in relishes from dried and cooked cassava leaves. The severe reduction of ascorbic acid may be related to the fact that it is thermo-labile at mild heating and very sensitive to blanching, drying and cooking (Faber and Van Jaarsveld, 2007). To understand the contribution of the relishes to human nutrition, their content levels of β-carotene, protein, iron, calcium, phosphorus, potassium and zinc were compared with Recommended Dietary Allowances (RDAs). For β-carotene, because the body converts all dietary sources of vitamin A into retinol, it is explained as retinol activity equivalent (RAE) and believing that 1 μg of retinol is equal to 6 μg of β-carotene (Food and Nutrition Board, 2001), the β-carotene mean values were calculated into RAE before being compared to RDA. Results of the comparison are shown by Table 4.

From the Table 4, the amounts in grams (dry matter basis) of relish from un-dried (UND) leaves (128, 50, 190, 205, 202, 263, and 14 g) to meet respectively protein, iron, calcium, phosphorus, potassium, zinc and β-carotene RDAs, were less than the amounts of relishes from DBP (128, 60, 208, 228, 210, 272 and 15 g), and DAP (132, 105, 228, 249, 228, 287 and 16 g), needed to meet the respective nutrients (protein, iron, calcium, phosphorus, potassium, zinc and β-carotene) RDAs.

The results showed that relish from un-dried leaves (UND) provides more nutrients than relish from dried leaves. This was attributed to blanching and drying, indispensable treatments for quality and storability (Oguche Gladys, 2011; Anhwange et al., 2011; Eze, 2010). Moreover, comparing the dried samples, drying before pounding (DBP) procedure provides more nutrients than pounding before drying treatment (DAP), but the latter contains less cyanide and then is safer for human consumption. B-carotene and iron are adequately contributed by the cassava leaves, considering the slighter required quantities of the greens to meet their RDAs (Table 4).

Conclusion and recommendations

Cassava leaves of bitter, sweet and wild species, when un-dried or solar dried, have potential to contribute to vitamin A, protein and mineral requirements. Vitamin C is very low in cooked (un-dried and dried) cassava leaves that a complement vitamin C rich-food is necessary to accompany cassava leafy meal. Leaves from wild species are the richest in nutrients, followed by bitter while sweet is the least. Fresh (un-dried) leaves give the more safe and rich foods, but for preservation purposes, solar drying is efficient, and pounding cassava leaves before they are dried is a more recommended drying method for cyanide reduction. Cyanide levels are significantly different in leaves from different cassava species, but after sufficient cooking, the difference is small that vegetable relishes from all the cassava species can be judged to be safe for human consumption. Blanching, drying and prolonged cooking treatments (30 min) reduce sensibly cyanide levels

Table 4. Average levels of nutrients in processed and cooked (30 min) cassava leaves and required amount of the relishes to meet the RDAs.

PP	CP (%)	Fe (mg/kg)	Ca (mg/kg)	P(mg/kg)	K (mg/kg)	Zn (mg/kg)	B-C (µgRAE/g)
UND	35.3±0.4	200.7±3.9	4756.5±59.9	3893.0±37.2	8881.2±180.4	56.6±1.1	61.5
DBP	35.5±1.3	164.1±2.5	4199.4±24.8	3495.5±28.8	8532.4±162.5	54.8±0.8	58.1
DAP	33.3±1.0	95.3±1.4	3836.7±32.6	3205.3±33.5	7864.7±165.4	52.5±1.3	54.4
RDA(mg)	45,000	10	800	800	1800	15	900 (µg RAE)
Required amount of the vegetable relishes (g) to meet the RDAs (dry matter basis)							
UND	128.5	50.2	190.5	205.4	202.6	263.6	14.6
DBP	128.3	60.7	208.0	228.7	210.5	272.3	15.4
DAP	132.3	105.2	228.6	249.6	228.6	287.7	16.5

Values are means and SE of six independent determinations, dry matter basis in quadruple. CP = crude protein, Fe = iron, Ca = calcium, P = phosphorus, K = potassium, Zn = zinc, PP = processing procedures, UND = un-dried, DBP = dried before pounding, DAP = dried after pounding, RDAs = Recommended Dietary Allowances by day and RAE = retinol activity equivalent.

(32-50 mg HCN/kg dry matter basis), but not at recommended level for human foods (10 mg HCN/kg dry matter basis). However, considering the small quantities by serving of green vegetables as side food, the protein level in cassava leaves, important nutrient in cyanide human body detoxification, and the acute oral lethal doses of hydrogen cyanide by bodyweight, the cassava leaves food can be said to be safe for human consumption. However, it is not advisable to consume cassava leaves as an everyday vegetable relish or in large quantities. In addition, cassava leaves meals may be limited for lower body weights such as children. The frequency may be reduced by promoting other greens such as amaranths, spinach and cabbage, even in arid areas, where cassava leaves are highly utilized as human food because other leafy vegetables cannot grow well in the present conditions. Cassava varieties with low levels of cyanide in leaves should be released for leaves consumption purposes to alleviate nutrient deficiencies, especially β-carotene and iron.

In general, amounts of nutrients retained after cooking un-dried and dried cassava leaves (after 30 min) are significant that cassava leaves as food are judged to contribute nutritionally to human health, especially vitamin A and iron.

Protein content is also of interest, and as cassava leaves are affordable by even poor people whose access to protein rich-foods such as milk, meat and fish is hard, the leaves can be helpful. But time of cooking may be extended to at least 30 min to improve reduction of cyanide level in the cassava leaves relishes.

ACKNOWLEDGEMENTS

This study was supported by grant from the Swedish International Development Agency (SIDA), through the Research Directorate of the University of Rwanda. They are acknowledged. The authorizations for using Sokoine University of Agriculture's solar drying station and Rwanda Agricultural Board's cassava field are gratefully acknowledged.

REFERENCES

Abah Idah AP, Musa JJ, Olaleye ST (2010). Effect of temperature and drying time on some nutritional quality parameters of dried tomatoes. Aust. J. Technol. 14 (1):25-32.

Achidi AU, Olufunmke A, Bokanga M, Dixin BM (2005). The use of cassava leaves as food in Africa. Ecol. Food Nutr. 44 (6): 423-435.

Akinwale MG, Aladesanwa RD, Akinyele BO, Dixon AGO, odeyi AC (2010). Inheritance of β-carotene in cassava (*Manihot esculenta* crantz). Inter. J. Genet. Mol. Biol. 2 (10): 198-201.

Anhwange BA, Asemave K, Ikyenge BA, Oklo DA (2011). Hydrogen content of *Manihot Utilissima, Colocasia, Esculenta, Dioscora Bulbifera and Dioscorea Domentorum* tubers found in Benue State. Int. J. Chem. 3 (4): 69-71.

AOAC (1995). Official Methods of Analysis. 16th Ed. Association of Official Analytical Chemists. Washington, D.C., USA.

AOAC (2007). Official Method 2007 0.1. Association of Official Analytical Chemists. AOAC International.

Ayodeji OF (2005). Nutrient composition and processing effects on cassava leaf (*Manihot esculenta*, Crantz). Pak. J. Nutr. 4 (1): 37-42.

Burns AE, Gleadow RM, Zacarias AM, Cuambe CE, Miller RE, Cavagna TR (2012). Variations in the chemical composition of cassava (*Manihot esculenta* Crantz) leaves and roots as affected by genotypic and environmental variation. J. Agric. Food Chem. 60: 4946-4956.

Cassava Cyanide Diseases and Neurolathyrism Network (CCDN) (2007). Working together to eliminate cyanide poisoning, konzo and Tropical Ataxic Neuropathy (TAN). CCDN News. Issue No. 9. p. 4
Cliff J, Muquingue H, Nhassico D, Nzwalo H, Bradbury JH (2011). Konzo and neurolathyrism: similarities and dissimilarities between these crippling neurodegenerative diseases. Toxic 49 (3): 631-635.
Dada OA, Owuru OO (2010). Mineral and nutrient leaf composition of two cassava (*Manihot esculenta* crantz) cultivars defoliated at varying phonological phases. Notulae Scientia Biologicae 2 (4): 44-48.
Etonihu AC, Olajubu O, Ekanem EO, Bako SS (2011). Titrimetric evaluation of cyanogens in parts of some Nigerian cassava species. Pak. J. Nutr. 10 (3): 260-263.
Eze JI (2010). Evaluation of the efficacy of a family sized solar cabinet dryer in food preservation. Am. J. Sci. Ind. Res. 1(3): 610-617.
Faber M, Van Jaarsveld PJ (2007). The production of provitamin A-rich vegetables in home-gardens as a means of addressing vitamin A deficiency in rural African communities. J. Sci. Food Agric. 87: 366–377.
FAO/IFAD (2005). A review of cassava in Africa. Case studies on Nigeria, Ghana, the United Republic of Tanzania, Uganda and Benin. FAO, Rome, Italy.
FAO/WHO (1991). Joined FAO/WHO Food Standard Programme, Codex Alimentarius Commission, XII, Supplent 4. Rome, Italy.
FAO/WHO (1993). Toxicological evaluation of certain food additives and naturally occurring toxicants. Joint FAO/WHO expert committee on food additives. 39th meeting. Geneva, Suitzerland.
Food and Nutrition Board (2001). Dietary Reference Intakes for vitamin A, vitamin K, arsenic, barn, chrnium, copper, iodine, iron, manganese, molybdenum, nickel, silicon, vanadium and zinc. Washington, DC, National Academy Press.
Fukuba H, Igarashi O, Briones CM, Mendoza EMT (1982). Determination and detoxification of cyanide in cassava and cassava products. J. Crop Sci. 7 (3): 170-175.
Gernah DI, Ega BM, Umoh UE (2012). Effect of boiling time on the quality of 'Zogale': A snack food produced from peanut (Arachis hypogeal) cake and boiled Moringa leaves. Afr. J. Food Sci. 6 (10): 287-293.
International Centre for Tropical Agriculture (Centro Internacional de Agricultura Tropical, CIAT) (2007). Improved cassava for the developing World. Annual report. p. 39.
ISO/655-7-2 (1984). Fruits, vegetable and derived products determination of ascorbic acid content. Part 2: Rutine methods. International Organization for Standardization, case postale 56, CH-1211, Geneve 20, Switzerland. p. 4.
James IF, Kuipers B (2003). *Preservation of fruit and vegetables*. 4th edition. Agromisa Foundation. Wageningen. p. 86.
Kendall P, DiPersio P, Sofos J (2010). Drying Vegetables. Colorado State University Extension Ltd. Fact sheet no. 9.308.
Kutner MH, Nachtsheim CN, Neter J, Li W (2005). Applied linear statistical models. 4th edition. Mc Graw-Hill.
MMA (Match Makers Associates limited) (2008). Dried fruits and vegetables for urban and export markets. Sub sector and value chain analysis. Tanzania. Final report.
Montgomery RD (1980). Cyanogens. In *Toxic constituents of plant foodstuff*. Academic Press, New York. pp. 143-160.
Mulokozi G, Mugyabuso J, Modaha F (2007). Potential of cassava and sweet potato leaves to contribute to the vitamin A requirements. In: Proceedings of the 13th ISTRC symposium, 755-762. Arusha. Tanzania.
Nassar N, Vizzotto CS, Schwartz CA, Pires-Junior OR (2007). Cassava diversity in brazil: the case of carotenoid-rich landcases. Genet. Mol. Res. 6(1): 116-121.
Nhassico D, Muquingue H, Cliff J, Cumbana A, Bradbury JH (2008). Rising African cassava cyanide initake and controle measures. J. Sci. Food Agric. 88: 2043-2049.
Nweke FI, Spencer DSC, Lynam JK (2002). The cassava transformation: Africa's best kept secret. Michigan University Press. USA. p.188.
Oguche Gladys HE (2011). Effect of drying methods on chemical composition of Spinach "Aieifo" (Amaranthus aquatic), and Pumpkin leaf (Telfairia occidentalis) and their soup meals. Pak J. Nutr. 10 (11): 1061-1065.
Okalebo JR, Gathua KW, Woomer PL (1993). Fractionation of organic matter by particles size. In: Laboratory Method of Soil and Plant Analysis: Working Manual. Nairobi, Kenya. pp. 66-68.
Priadi D, Permana D R, Dona SE, Hartati S, Sudarmonowati E (2009). Selection of Indonesia cassava (*Manihot esculenta* Crantz) genotype as source of β-carotene. Biodiversitas 10 (1): 6-11.
Ravindran V (1995). Preparation of cassava leaf products and their use as animal feeds. Food Chem. 27: 299-309.
Sarkiyayi S, Agar TM (2010). Comparative analysis on the nutritional and anti-nutritional contents of sweet and bitter cassava varieties. Adv. J. Food Sci. Technol. 2 (6): 328-334.
SAS Instute (2008). Software 9.2. University of Notre Dame-T&R, Cary, NC, USA.
Siritunga D, Sayre R (2004). Engineering cyanogen synthesis and turnover in cassava (*Manihot esculenta*). Plant Mol. Biol. 56: 661–669.
Tee Siong E, lam L C (1992). Analysis of carotenoids in vegetables by HPLC. Asian Food J. 7(2): 91-99.
Thompson B, Amoroso L (Eds) (2011). Combating micronutrient deficiencies: food-based approaches. FAO, Rome, Italy/CABI, Wallingford, Uk. p. 432.
Ubi BE, Ibiaml A, Efisue AA, Odu EM, Udehl KI, Egesi CN (2008). Varietal differences in leaf cyanide content. J. Agric. Biotechnol. Ecol. 1: 63-69.
Udofia US, Obizoba IC (2010). Effect of traditional processing techniques on leafy vegetable and starchy staple and their uses in traditional soups and dishes as consumed in Akwa Ibom State. J. Biochem. Invest. 3: 1-5.
Umuhozariho MG, Shayo NB, Msuya JM, Sallah PYK (2011). Utilization of cassava leaves as a vegetable in Rwanda. Rw. J. Agric. Sci. 24 (Series E): 15-27.
Uyoh EA, Udensio, Natwi V, Urea I (2007). Effect of different processing methods on cyanide content of garri from cultivars of cassava. J. Food Agric. Environ. 3&4: 105-107.
White WLB, McMahon JM, Sayre RT (2003). Regulation of cyanogenesis in cassava. Food Chem. Toxicol. 41 (8): 1193-1197.
WHO (2004). Hydrogen cyanide and cyanides: Human health. Concise International Chemical Assessment Document 61. Geneva, Switzerland.

Comparison between gamma irradiation and plasma technology to improve the safety of cold sliced chicken

Ahmed A. Aly[1] and G.M.El-Aragi[2]

[1]Home Economic Department, Faculty of Specific Education, Benha University, Egypt.
[2]Plasma Physics and Nuclear Fusion Department, Nuclear Research Center, AEA, PO 13759 Cairo, Egypt.

The aim of this study was to determine the impact of gamma irradiation and plasma treatment on quality (appearance, odor and texture) and safety (total bacteria count, psychrophilic bacteria, spore forming bacteria, total molds and yeasts) of cold sliced chicken. Pulsed atmospheric-pressure plasma jet (PAPPJ) device is used for the treatment of fresh sliced chicken with different exposure time. Also, another sliced chicken samples were subjected to different gamma irradiation doses. The results illustrate that gamma irradiation and plasma treatment had no real effect on the chemical composition of sliced chicken and the effects of these treatments on the microbial load are reduced the counts of total bacteria, psychrophilic bacteria, total molds and yeasts count. Gamma irradiation and plasma treatment had no effects on the sensory characteristics for appearance, odor and texture of all sliced chicken samples. Finally, gamma irradiation and plasma treatment had no adverse effects on the sensory properties of sliced chicken samples, therefore, their use as safety technological treatments in food preservation and extending shelf-life.

Key words: Gamma irradiation, plasma, chicken, microbial load.

INTRODUCTION

Poultry carcasses are commonly contaminated with enteric pathogens such as Salmonella, Campylobacter and *Listeria monocytogenes* (Jacobsreitsma et al., 1994; Murphy et al., 2004); the possibility of cross-contamination of poultry carcasses post slaughter is high. Decontamination of poultry carcasses is therefore desirable. Various decontamination technologies have been proposed including the use of various chemical agents such as alkali (Rodriguez et al., 1996), physical methods such as steam treatment (James et al., 2007) and biological control with bacteriophages (Carvalho et al., 2010), but only treatment with water supplemented with chlorine or a chlorineting agent is used commercially. The effects of such decontaminating treatments are limited (Oyarzabal, 2005; Russell and Axtell, 2005).

Plasma is matter that contains partially or wholly ionized gas with a net neutral charge and is often referred to as the fourth state of matter as it shares properties similar to both those of gases and liquids. Plasma is created under energy deposition into gaseous mixture. Gas turns into plasma due to ionization, dissociation and excitation of the bound states of atoms and molecules of the background gas. Therefore, plasma consists of gaseous mixture of charged particles (free electrons and ions) and neutral activated species including gas molecules, free radicals, metastables and ultraviolet photons. Energetic electrons generate intensively numerous chemical active species due to collisions with atoms and molecules. In the gas mixture containing oxygen and water vapor, most of the primary radicals are O and OH.

Numerous components of plasma such as for example reactive oxygen or nitrogen species, charged particles, electric fields and even UV light are involved in these effects. Hence, they are widely used in a broad spectrum of applications ranging from low temperature plasma chemistry, decomposition of gaseous pollutants, light sources, surface modification to medical sterilization and microbial decontamination (Kogelschatz, 2004; Becker et al., 2005; Foest et al., 2005; Ehlbeck et al., 2008; Massines et al., 2012; Ehlbeck et al., 2011).Cold plasma produces (gaseous) activated ions, pho-tons, electrons and free radicals, collectively termed 'plas-ma', that exert their effects at 30 to 60°C; hence the term 'cold' or non-thermal.

Plasma may inactivate both vegetative cells and bac-terial endospores. Synergistic effects between these possible mechanisms of inactivation can be expected, depending on the operational conditions and the design of the plasma generator.

Recent investigations which include the evaluation of the inactivation of food-borne pathogens seeded on to thin films of agar, treated with a glow discharge plasma (Kayes et al., 2007) and also sprayed onto the surface of heat sensitive polyethylene terephthalate (PET) foils, exposed to a dielectric barrier discharge (Muranyi et al., 2007; Yu et al., 2006), reported that high loadings of *Escherichia coli* on polycarbonate membranes adversely affected the penetration of plasma species and the effec-tiveness of microbial inactivation. Cold atmospheric plasmas (CAP) or non-thermal plasmas have been repor-ted to be very effective against a wide range of micro-organisms, including biofilm-formers and bacterial spores (Montie et al., 2000; Deng et al., 2005; Vleugels et al., 2005).

Implementation of an emerging technology such as CAP, in food preservation, depends on reliable estima-tions of its efficacy against pathogenic and food spoilage microorganism and to reach this objective it is necessary to understand the factors which affect microbial resis-tance (Fernández et al., 2012). Among these factors, the effect of initial cell concentration is known to be one of the most relevant in food surface treatment strategies (Deng et al., 2005; Moreau, 2000).

In general, during inactivation of microorganisms on surfaces, the rate of inactivation is inversely proportional to the initial cell concentration (Shintani, 2000). Food irra-diation is being considered an important tool, in ensuring safety and extending shelf-life of fresh meat and poultry (Yoon, 2003). Thus irradiation can eliminate food-borne pathogenic microorganisms in meat. Furthermore, the use of gamma irradiation as a safety techno-logical treatment in food preservation has now become legally accepted in many countries of the world (Abdel-Daium, 2007).

Therefore, in this work, we examined the efficacy of cold atmospheric plasma to inactivation of microbial load in sliced chicken compared to gamma radiation.

MATERIALS AND METHODS

Sliced chicken were purchased from local market (Benha, Qaliobia governorate, Egypt). All samples were transported to our laboratory food irradiation unit, Nuclear Research Center in ice-box (0°C) and surveyed for microbiological counts for counts of total bacteria, psychrophilic bacteria, spore forming bacteria, total molds and yeasts. Then, sliced chicken samples were packed in tightly sealed polyethylene pouches and divided into seven groups and stored in freezing till irradiation treatments.

Gamma irradiation treatments

Four bags from each of sliced chicken were gamma irradiated at 0, 2, 4, and 6 kGy doses using cobalt-60 gamma chamber (1.367 kGy/h) in Cyclotron Project, Nuclear Research Center Atomic Energy Authority, Inshas, Cairo, Egypt. After irradiation, all samples were stored at 4±1°C.

Plasma treatments

Character of exposure machine

The plasma generator consisted of a negative dc source, a Blumlein-type pulse-forming network (E-PFN), and a dynamic spark gap switch. A triggered spark gap switch was used as a closing switch of E-PFN. E-PFN had four stages of LC ladder, which were composed of 5 nF of capacitor and 3 μH of inductor. The characteristic impedance ($2\sqrt{L/C}$) and the pulse width ($2N\sqrt{LC}$) of E-PFN, calculated from capacitance (C) and inductance (L) of the LC ladder, and number (N) of LC ladder stages were approximately 49 Ω and 1.0 μs, respectively.

A charging resistance value of 50 kΩ was chosen in the present case which corresponds to a charging RC time constant of 1 ms, which is 40 times faster compared to the repetition rate of the pulse.

A schematic of the pulsed atmospheric-pressure plasma jet (PAPPJ) device for generating high voltage pulsed, cold atmospheric plasma jets is shown in Figure 1. The high voltage (HV) wire electrode, which is made of a copper wire, is inserted into a hollow barrel of a syringe. The distance between the tip of the HV electrode and the nozzle is 0.5 cm.

When HV pulsed, DC voltage (amplitudes up to 25 kV, repetition rate up to 25 Hz), was applied to the HV electrode and helium gas was injected into the hollow barrel. This device was made using medical syringe (made out of an insulating material cylinder). The gas was fed into the system via flow meter.

The applied voltage to and the discharge current through the discharge chamber were measured using a voltage divider (Homemade), which was connected between the two electrodes, and a current monitor, which can be located upon returning to the ground. The signals from the voltage divider and the current monitor were recorded in a digitizing oscilloscope (Lecroy, USA) with a 200-MHz bandwidth.

The high voltage pulses are applied between the needle electrode positioned inside a dielectric cylinder (a simple medical syringe) and a metal ring placed on the exterior of this cylinder. In order to obtain electric discharges at atmospheric pressure, a high voltage pulses (tens of kV) which have limited duration (hundreds of nanoseconds) and are repeated (tens of pulses per second), in addition to an inert gas (argon) is introduced in the cylinder.

The gas flows were in the range 0.5-10 l/min. The discharge takes place between the metallic needle top and a metallic ring fit on the outer surface of the syringe. Under optimal conditions, plasma is emitted as centimeter-long jets, just millimeters in

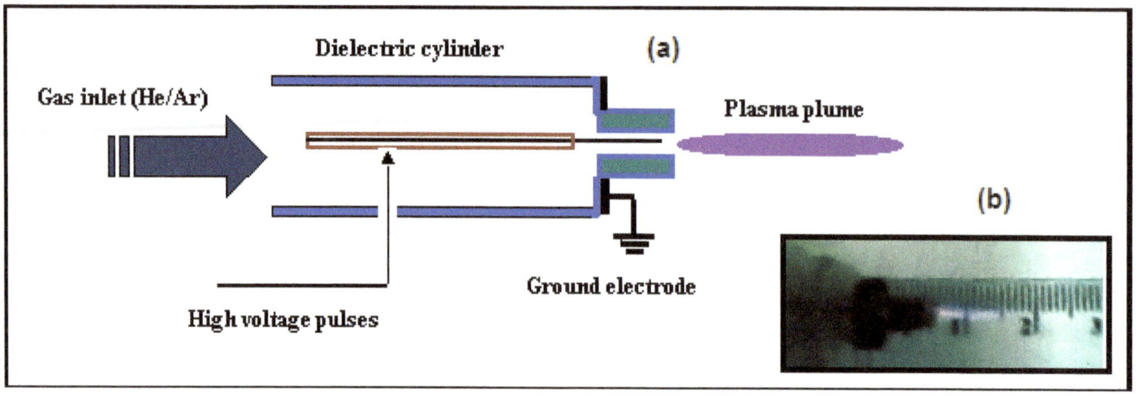

Figure 1. Schematic diagram of the plasma jet generator. (b) Image of the plasma jet with argon gas.

Table 1. Effect of gamma irradiation and plasma treatment on chemical composition of sliced chicken.

Chemical composition %	Treatments						
	Gamma irradiation doses				Exposure time of plasma (min)		
	0 kGy	2 kGy	4 kGy	6 kGy	0.5	1.0	1.5
Moisture	76.133	76.133	76.367	76.200	75.833	75.567	76.267
Total Protein	87.633	86.273	86.530	86.387	86.723	87.006	86.340
lipid	5.190	5.6133	5.353	5.550	5.410	5.453	5.320
Ash	5.5167	5.800	5.800	5.317	5.967	5.627	5.390

diameter or even smaller.

The working gases are supplied by high-pressure cylinders. Gas pressure regulators are used to reduce the pressure of gases to a workable level. Then, gas flow controllers deliver the gases with the desired flow. For voltage amplitudes of 15-18 kV, the plasma jet is very weak. The plasma jet disappears for voltage amplitudes lower than 15 kV.

When argon is injected from the gas inlet and high voltage pulses, 26 kV voltages is applied to the electrode, the plasma jet is generated and a plasma plume reaching length of 21 mm is launched through the end of the tube and in the surrounding air. The length of the plasma plume can be adjusted by the gas flow rate and the applied voltage.

Three bags from each of sliced chicken were exposured to plasma at 0.5, 1.0 and 1.5 min in Plasma Physics and Nuclear Fusion Department, Nuclear Research Center, Atomic Energy Authority, Inshas, Cairo, Egypt. After the exposure time of plasma, all samples were stored at 4±1°C.

Microbial analysis

Colony forming units for total bacterial count were counted by plating on plate count agar medium and incubated at 30°C for three to five days (APHA, 1992). Total molds and yeasts were counted on oxytetracycline glucose yeast extract agar medium according to Oxoid, (1998). psychrophilic and spore forming bacteria count according to (FDA, 2002).

Statistical analysis

The statistical evaluation of the mean data was compared using one-way analysis of variance (ANOVA) according to Zar (1984). The chosen level of significance was $P \leq 0.05$.

RESULTS AND DISCUSSION

Chemical composition

The effect of various gamma irradiation doses on the chemical composition of sliced chicken was studied and the data and its analysis of variance are illustrated in Tables 1 and 2. From Table 1, it could be noticed that the moisture total protein, lipid and ash contents were 76.133, 87.633, 5.190 and 5.5167% for control (0 kGy) sliced chicken respectively.

From data in Tables 1 and 2, it could be noticed that there was no significant changes ($P \leq 0.05$) in the chemical analysis of sliced chicken irradiated with gamma doses (2, 4, and 6 kGy) and exposed to plasma at 0.5, 1.0 and 1.5 min compared to non irradiated sample.

Microbial evaluation

Bacterial aspects of untreated and treated sliced chicken samples with gamma irradiation and plasma during cold storage at (4±1°C) were enumerated and the data are recorded in Table 3. The initial total bacteria count, psychrophilic bacteria, spore forming bacteria, total molds

Table 2. Analysis of variance for the effect of gamma irradiation and plasma treatment on chemical composition of sliced chicken.

Chemical composition	Source of variation	SS	Degree of freedom	Main square values	F	P-value	F crit
Moisture	Between groups	1.382	6	0.231 NS	0.039	1	2.848
	Within groups	82.26	14	5.876			
	Total	83.642	20				
Total protein	Between groups	4.212	6	0.702 NS	0.399	0.867	2.848
	Within groups	24.611	14	1.758			
	Total	28.823	20				
Lipid	Between groups	0.3674	6	0.061 NS	0.242	0.955	2.848
	Within groups	3.542	14	0.253			
	Total	3.9092	20				
Ash	Between groups	1.0193	6	0.170 NS	0.694	0.659	2.848
	Within groups	3.4271	14	0.245			
	Total	4.446	20				

NS, Non significant.

and yeasts of non-treated sliced chicken were high counts (5.6×10^2, 9.0×10^2, 71×10^1 and 6.3×10^2 cfu/g, respectively) while treatment sliced chicken by gamma irradiation at doses (2, 4 and 6 kGy) and plasma treatment (0.5, 1, 1.5 min) reduced the counts of microbial loads.

The best antimicrobial treatments which com-pletely eliminated psychrophilic bacteria were noticed in sliced chicken samples treated by gamma irradiation at dose 4 kGy followed by plasma at 1.5 min. Furthermore, the use of these doses extended the shelf-life of sliced chicken to 27 and 23 days, respectively compared to seven days for untreated samples under cold storage (4±1°C).

Sensory evaluation

Sensory evaluation is a common and very useful tool in quality assessment of processed meat products. It makes use of the senses to evaluate the general acceptability and quality attributes of the products.

Tables 4 and 5 show the changes in sensory evaluation of sliced chicken as affected by gamma irradiation and plasma treatment during storage period at 4±1°C temperature. From these results, it is noticed that gamma irradiation and plasma treatment had no remarkable effect on sensory properties (appearance, odor and texture) of sliced chicken under investigation at zero time of storage period.

Tables 4 and 5 also show the appearance, odor and texture during storage temperature (4±1°C). At the beginning of cold storage (4±1°C) (zero time), no significant difference (P≤0.05) was found between the samples which were irradiated with gamma rays at 0, 2, 4, 6 kGy and exposed to plasma treatment at 0.5, 1.0, 1.5 min. A significant difference (P≤0.05) was shown in all treated samples after four days.

It was observed that the reference sample were rejected being samples that had been contaminated on day 8 of storage while irradiated with gamma rays (2 kGy) and exposure to plasma treatment (0.5 min.) were scored as poor and rejected on day 12. It has been found that irradiated samples at 6 kGy and those exposed to plasma treatment at 1.5 min were better than other samples.

This may be due to the effect of gamma irradiation dose (6 kGy) and exposure to plasma treatment (1.5 min) improves the appearance, odor and texture by inhibition of the microbial load and discoloration of the surface. It was noticed that a small spots of mold growth appeared on the surface of irradiated samples at doses of 4 and 6 kGy and were rejected on day 16 and 28 of storage respectively.

On the other hand, sliced chicken samples were exposed to plasma treatment at 0.5 and 1 min and were rejected being samples that had been contaminated on day 16 and 24 of storage respectively.

Gamma irradiation and plasma treatments of sliced chicken samples have the same extended shelf-life up to 27 and 23 day respectively compared to non- treated samples during cold storage (4±1°C).

Conclusion

Cold or non-thermal plasma has several advantages over

Table 3. Effects of gamma irradiation and plasma treatment on microbial load in sliced chicken during cold storage (4±1°C).

Microbial decontamination (cfu/g)	Storage period (days)	Treatment						
		Gamma irradiation dose				Time of cold atmospheric plasma (min)		
		0 kGy	2 kGy	4 kGy	6 kGy	0.5	1	1.5
Total bacteria counts	0	5.6×10^2	2.5×10^2	6.2×10^1	5.1×10^1	2.9×10^2	2.8×10^2	1.9×10^2
	4	1.1×10^3	4.2×10^2	7.1×10^1	6.8×10^1	8.5×10^2	4.8×10^2	4.0×10^2
	8	8.0×10^3®	5.0×10^2	7.9×10^1	7.5×10^1	6.8×10^3	5.9×10^3	5.2×10^2
	12		1.7×10^3®	1.9×10^2	1.3×10^2	8.8×10^3®	6.2×10^3	5.9×10^3
	16			6.9×10^3®	4.0×10^3		8.9×10^3®	8.0×10^3
	20				8.5×10^3			9.3×10^3
	24				9.5×10^3			9.0×10^4®
	28				9.8×10^4®			
Psyhcrophilic bacteria	0	9.0×10^2	7.4×10^2	4.3×10^2	Nil	8.0×10^2	3.9×10^2	Nil
	4	7.1×10^3	5.9×10^3	8.1×10^2	Nil	6.1×10^3	4.5×10^3	Nil
	8	5.7×10^4®	8.7×10^3	2.1×10^3	Nil	9.8×10^3	6.3×10^3	Nil
	12		2.0×10^4®	4.2×10^3	Nil	3.0×10^4®	9.2×10^3	Nil
	16			2.2×10^4®	Nil		1.0×10^4®	Nil
	20				Nil			Nil
	24				Nil			Nil ®
	28				Nil ®			
Spore forming bacteria	0	71×10^1	21×10^1	12×10^1	0.9×10^1	32×10^1	19×10^1	1.3×10^1
	4	98×10^1	33×10^1	20×10^1	1.0×10^1	46×10^1	31×10^1	3.5×10^1
	8	99×10^1®	50×10^1	32×10^1	2.0×10^1	70×10^1	45×10^1	7.0×10^1
	12		95×10^1®	64×10^1	4.5×10^1	98×10^1®	59×10^1	18×10^1
	16			97×10^1®	21×10^1		100×10^1®	30×10^1
	20				50×10^1			79×10^1
	24				89×10^1			89×10^2®
	28				99×10^2®			
Total molds and yeasts	0	6.3×10^2	4.1×10^2	3.2×10^2	1.3×10^2	5.3×10^2	4.8×10^2	1.5×10^2
	4	8.9×10^2	6.9×10^2	5.4×10^2	3.6×10^2	6.8×10^2	5.1×10^2	3.2×10^2
	8	5.9×10^3®	9.3×10^2	7.8×10^2	5.6×10^2	9.3×10^2	5.4×10^2	4.3×10^2
	12		8.1×10^3®	9.9×10^2	8.3×10^2	3.2×10^3®	6.3×10^2	7.1×10^2
	16			2.1×10^3®	23×10^2		5.1×10^3®	8.0×10^2
	20				9.0×10^2			9.3×10^3
	24				7.1×10^3			8.4×10^4®
	28				7.1×10^4®			

®: At this point, samples were rejected.

competing preservation technologies such as irradiation (gamma rays), chemical sterilization (ethylene oxide) or disinfection treatments (chlorine). These can have detrimental sensory effects on treated food products or they are being phased out because of safety concerns.

Although cold plasma technology is not yet used commercially on a large scale in the food industry, it has been successfully used for decades in many other industries. Cold plasma systems for treating foods at atmospheric pressure are currently being developed and tested.

Table 4. Changes in the sensory attributes of sliced chicken as affected by gamma irradiation and plasma treatment during cold storage temperature (4±1°C).

Sensory attribute	Storage period (days)	Treatments						
		Gamma irradiation dose				Exposure time of plasma(minute)		
		0 kGy	2 kGy	4 kGy	6 kGy	0.5	1	1.5
	0	9.73	9.70	9.69	9.72	9.74	9.63	9.72
	4	8.63	7.60	8.82	9.17	7.60	8.53	8.94
Appearance	8	4.37®	6.67	7.10	8.37	6.63	7.63	8.45
	12		4.27®	6.35	7.72	4.23®	6.37	7.72
	16			4.22®	7.02		3.50®	7.57
	20				6.15			5.75
	24				5.84			4.14®
	28				4.19®			
Odor	0	9.65	9.58	9.63	9.67	9.25	9.60	9.68
	4	8.58	7.55	8.75	9.12	7.55	8.47	8.88
	8	4.10®	6.58	7.05	7.82	6.58	7.58	8.40
	12		3.77®	6.30	7.68	4.18®	6.07	7.65
	16			3.83®	6.95		3.45®	7.50
	20				6.12			5.70
	24				5.80			3.97®
	28				3.92®			
Texture	0	9.08	8.95	8.97	8.98	9.00	8.92	8.98
	4	8.03	6.80	8.02	8.37	6.80	7.73	8.13
	8	3.47®	5.77	6.20	7.47	5.73	6.73	7.55
	12		3.37®	5.45	6.82	3.33®	5.47	6.82
	16			3.22®	6.02		3.00®	6.57
	20				5.75			5.35
	24				5.27			3.52®
	28				3.48®			

®: At this point samples were rejected.

Table 5. Analysis of variance between treatments for sensory evaluation of sliced chicken during storage temperature (4±1°C).

Storage period (days)	D.F	F crit	F value and level of significance					
			Appearance		Odor		Texture	
			MS	F	MS	F	MS	F
0	6	2.37	0.01 NS	0.80	0.14 NS	1.16	0.02 NS	1.04
4	6	2.37	2.36***	27.21	2.35***	31.86	2.46***	28.36
8	6	2.37	11.55***	129.55	11.62***	42.14	11.55***	129.55
12	5	2.53	14.65***	172.20	16.70***	192.84	14.65***	172.21
16	3	3.10	24.39***	177.39	26.13***	215.65	20.62***	149.94
20	1	4.96	0.48***	10.21	0.52***	18.06	0.48***	10.21
24	1	4.96	9.19***	254.03	10.08***	270.10	9.19***	254.03

D.F, Degree of freedom; ***Significant at 0.05. NS, Non significant.

REFERENCES

Abdel-Daium MH (2007). Manufacturing of low-fat Chicken sausage and keeping its quality by gamma irradiation. Arab J. Nucl. Sci. Appl. 40: 296-304.

APHA (1992). Compendium of Methods for the Microbiological Examination of Foods,(2nd ed.), American Puplic Heath Association, Washinton DC.

Becker K, Koutsospyros A, Yin SM, Christodoulatos C, Abramzon N, Joaquin JC, No GBM (2005). Environmental and biological applications of microplasmas Plasma Phys. Control. Fusion 47, B513-B523.

Carvalho CM, Gannon BW, Halfhide DE, Santos SB, Hayes CM, Roe JM, Azeredo J (2010). The in vivo efficacy of two administration routes of a phage cocktail to reduce numbers of Campylobacter coli and Campylobacter jejuni in chickens. BMC Microbiol. 10:232.

Deng XT, Shi JJ, Shama G, Kong MG (2005). Effects of microbial loading andsporulation temperature on atmospheric plasma inactivation of Bacillus subtilisspores. Appl. Phys. Lett. 87:153901.

Ehlbeck J, Brandenburg R, von Woedtke T, Krohmann U, Stieber M, Weltmann KD (2008). PLASMOSE - antimicrobial effects of modular atmospheric plasma sources. *GMS Krankenhaushygiene Interdisziplin•ar* 3(1):1-12

Ehlbeck J, Schnabel U, Polak M, Winter J, von Woedtke T, Brandenburg R, von dem Hagen T, Weltmann K-D (2011). Low temperature atmospheric pressure plasma sources for microbial decontamination. J. Phys. D: Appl. Phys. 44:18.

FDA, Food and Drug Administration (2002). Bacteriological Analytical Manual. 9th Ed.,AOAC Int., Arlington,VA,USA.

Fernandez A, Shearer N, Wilson DR, Thompson A (2012). Effect of microbial loading on the efficiency of cold atmospheric gas plasma inactivation of Salmonella enterica serovar Typhimurium International. J. Food Microbiol. 152:175-180.

Foest R, Kindel E, Ohl A, Stieber M, Weltmann KD (2005). Non-thermal atmospheric pressure discharges for surface modification. Plasma Phys. Control. Fusion 47:B525-B536.

Jacobsreitsma WF, Bolder NM, Mulder RWAW (1994). Cecal Carriage of Campylobacter and Salmonella in Dutch broiler Flocks at slaughter - A one-Yearstudy. Poult. Sci. 73:1260-1266.

James C, James SJ, Hannay N, Purnell G, Barbedo-Pinto C, Yaman H, Araujo M, Gonzalez ML, Calvo J, Howell M, Corry JEL (2007). Decontamination of poultry carcasses using steam or hot water in combination with rapid cooling,chilling or freezing of carcass surfaces. Int. J. Food Microbiol. 114:195-203.

Kayes MM, Critzer FJ, Kelly-Wintenberg K, Roth JR, Montie TC, Golden DA (2007). Inactivation of foodborne pathogens using a one atmosphere uniform glowdischarge plasma. Foodborne Pathog. Dis. 4(1):50-59.

Massines F, Sarra-Bournet C, Fanelli F, Naude N, Gherardi N (2012). Atmospheric Pressure Low Temperature Direct Plasma Technology: Status and Challenges for Thin Film Deposition. Plasma Process. Polym. 9:1041-1073.

Montie TC, Kelly-Wintenberg K, Roth JR (2000). An overview of research using the one atmosphere uniform glow discharge plasma (OAUGDP) for sterilizationof surfaces and materials. IEEE Trans. Plasma Sci. 28:41-50.

Moreau S (2000). Using the flowing afterglow of a plasma to inactivate Bacillussubtilis spores: Influence of the operating conditions. J. Appl. Phys. 88(2):1166-1174.

Muranyi P, Wunderlich J, Heise M (2007). Sterilization efficiency of a cascade dielectric barrier discharge. J. Appl. Microbiol. 103:1535-1544.

Murphy RY, Osaili T, Duncan LK, Marcy JA (2004). Thermal inactivation of Salmonella and Listeria monocytogenes in ground chicken thigh/leg meat and skin. Poult. Sci. 83:1218-1225

Oxoid,"The oxoid Manual" (1998). (8th ed.)Oxid ltd., Wade Road, Basingstoke, Hampshire, R G24 8PW, Engand.

Oyarzabal OA (2005). Reduction of Campylobacter spp. by commercial antimicrobialsapplied during the processing of broiler chickens: a review from the United States perspective. J. Food Prot. 68:1752-1760.

Rodriguez De Ledesma AM, Riemann HP, Farver TB (1996). Short-time treatmentwith alkali and/or hot water to remove common pathogenic and spoilage bacteria from chicken wing skin. J. Food Prot. 59:746-750.

Russell SM, Axtell SP (2005). Monochloramine versus sodium hypochlorite as antimicrobial agents for reducing populations of bacteria on broiler chickencarcasses. J. Food Prot. 68:758-763.

Shintani H (2000). The reason for the dependency of D value on the initial concentrationof microorganisms. J. Antibacterial Antifungal Agents 28:680.

Vleugels M, Shama G, Deng XT, Greenacre E, Brocklehurst T, Kong MG (2005). Atmospheric plasma inactivation of biofilm-forming bacteria for food safetycontrol. IEEE Trans. Plasma Sci. 33:824-828.

Yoon KS (2003). Effect of gamma irradiation on the texture and microstructure of chicken breast meat. Meat Sci. 63:273.

Yu H, Perni S, Shi JJ, Wang DZ, Kong MG, Shama G (2006). Effects of cellssurface loading and phase of growth in cold atmospheric gas plasma inactivation ofEscherichia coli K12. J. Appl. Microbiol. 101:1323-1330.

Zar JH (1984). Biostatistical analysis. Prentice Hall,Englewood, N.J. pp. 718.

Physicochemical characterization of some multifloral honeys from honeybees *Apis mellifera* collected in the Algerian northeast

Assia Amri and Ali Ladjama

Department of Biochemistry, Faculty of Sciences, University of Annaba, BP 12 - 23000 - Annaba, Algeria.

The characterization of some Algerian northeast honey was carried out on the basis of their physico-chemical properties: moisture, hydroxymethylfurfural, diastase activity, pH, free, total and lactonic acidity, electrical conductivity, mineral and proline content. Studied samples are found to be low in moisture and therefore safe from fermentation, low in HMF level and high in diastase activity. Additionally, the diastase activity and the HMF content are widely recognized parameters indicating the freshness of honey. The quality of honey samples differs on account of various factors such as season, packaging and processing conditions, floral source, geographical origin and storage period. It is important that precautions should be taken to ensure standardization and rationalization of beekeeping techniques, manufacturing procedures and storing processes to improve honey quality.

Key words: Honey, physico-chemical characterisation, HMF, diastase activity.

INTRODUCTION

Honey is a very important energy food and is used as an ingredient in hundreds of manufactured foods, particularly in cereal-based products, for its sweetness, colour, flavour, caramelisation and viscosity (Rodrigez and Ferrer, 2004). The composition of honey depends on the plant species visited by the honeybees and the environmental, processing and storage conditions (Bertoncelj et al., 2007; Guler et al., 2007). Honey produced by the honeybee is a natural super saturated sugar solution, which is mainly composed of a complex mixture of carbohydrates. In addition, to carbohydrates, it also contains certain minor constituents, including proteins, enzymes (invertase, glucose oxidase, catalase, phosphatases), amino and organic acids (gluconic and acetic acids), lipids, vitamins (ascorbic acid, niacin, pyridoxine), volatile chemicals, phenolic acids, flavonoids and carotenoid-like substances and minerals (Blasa et al., 2006). Fresh honey is usually heated in order to facilitate processing and to maintain good quality. However, excessive heat treatment leads to the formation of hydroxymethylfurfural (HMF) and reduced honey quality. HMF value is virtually absent or very low in fresh honey and is high in honey that has been heated, stored in inadequate conditions, or adulterated with invert syrup (Nozal et al., 2001). Chemical properties of honey such as pH, mineral content and total acidity also affect HMF content. The presence of organic acids and low water activity also favours the production of HMF (Kalabova et al., 2003). The Codex Alimentarius (2001) and International Honey Commission (2002) set the maximum concentration of HMF to 40 mg/kg for honey from non-tropical regions and 80 mg/kg for honey from tropical regions.

Extremely high (>500 mg/kg) HMF values demonstrate adulteration with invert syrup (Coco et al., 1996). There is a vast number of floral species characteristic of Algeria

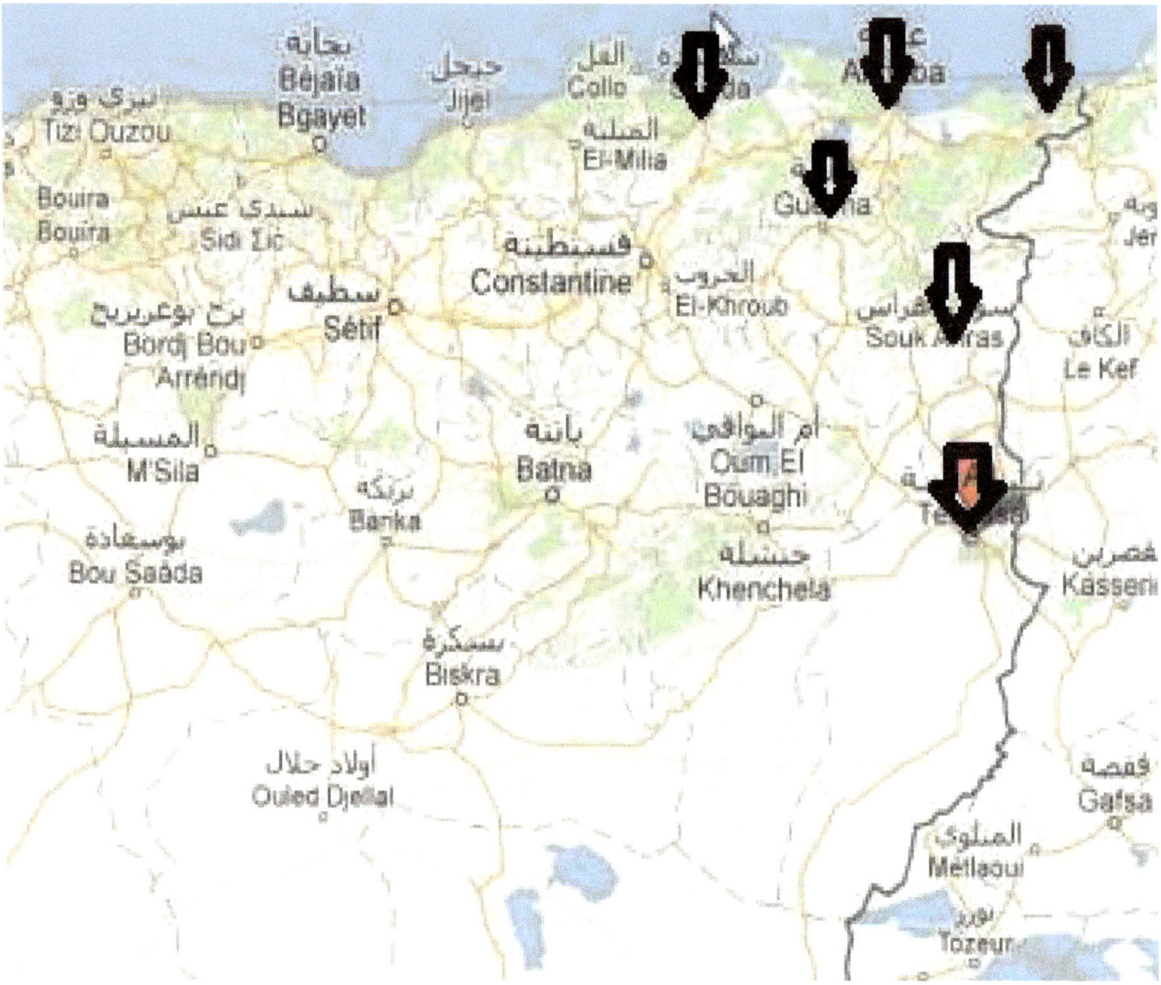

Figure1: Maps of the geographical origin of the honey samples.

visited by the honeybees. However, reports of the chemical composition of Algerian honeys are scarce (Makhloufi et al., 2007; Ouchemoukh et al., 2007). The potential honey production is very high in the Algerian northeast; this is related to the botanical wealth and the biodiversity of this region. Therefore, this study was carried out to evaluate the physicochemical quality of twenty Algerian Northeast honey samples from *Apis mellifera*.

MATERIALS AND METHODS

Honey samples

Twenty blossom honeys from various regions of Algerian Northeast were used. The samples of honey collected from bee-keepers were classified according to their botanical origin using the method established by the International Commission of Bee Botany described by Louveaux et al. (1978). The result shows that all samples are multifloral honey (no dominant pollen) with the presence of Secondary pollen (16 to 45%): *Trifolium* sp, *Eucalyptus*, *Hedysarium coronarium*, *Echium* and *Rosaceae*. Minor pollen (3 to 15%): Prunus / *Pyrus*, *Fabaceae*, *Carduus* and *Erica arborea*. Pollen very minor or isolated (<3%): *Allium* spp, *Apiaceae*, *Lavandula asphodelus*, *Lavandula stoechas*, *Lamiaceae*, *Brassicaceae*, *Taraxacum*, *Euphorbiaceae*, *Rhamnaceae*, *Asteraceae*, *Prunus / Pyrus*, *Mimosa pudica*, *Helianthus*, *mimosa bimucromata*, *Arctium* and X (indeterminate species). All the samples were collected in six different region of Northeast Algeria during the years 2004 to 2007 (Figure 1 and Table 1).

Physico-chemical analysis

Water content was determined by refractometry, pH was assessed in a 10% (w/v), total acidity, free acid, lactone and proline contents were measured in fresh honey samples using AOAC methods (Helrich, 1990). Proline content was determined by the measurement of the absorbance at 510 nm of the resulting product between proline and ninhydrin. Mineral content and electrical conductivity were determined by the methods of Bogdanov et al. (1997). Ash content was determined by heating 5 g of honey at 625°C in a muffle furnace. Electrical conductivity was measured in a 20% (w/v) solution of honey. The protein content was determined by the method of Bradford (1976), the quantity of proteins was estimated at 595 nm in relation to bovine serum albumin standard. HMF and diastase activity were measured as recommended by the

Table 1. Botanical and geographical origins of honey samples

Samples	Botanical origin	Geographic origin	Year collected
S1	multifloral	El taref	2006
S2	multifloral	Tebessa	2007
S3	multifloral	Tebessa	2007
S4	multifloral	El taref	2007
S5	multifloral	Annaba	2007
S6	multifloral	El taref	2004
S7	multifloral	Annaba	2004
S8	multifloral	El taref	2004
S9	multifloral	Skikda	2004
S10	multifloral	Skikda	2004
S11	multifloral	El taref	2004
S12	multifloral	SoukAhras	2004
S13	multifloral	Tebessa	2004
S14	multifloral	Guelma	2004
S15	multifloral	SoukAhras	2004
S16	multifloral	Annaba	2005
S17	multifloral	Guelma	2005
S18	multifloral	SoukAhras	2005
S19	multifloral	Tebessa	2006
S20	multifloral	Annaba	2006

Codex Alimentarius Commission (CAC, 1989).

RESULTS

Table 2 shows the mean values, standard deviations and ranges of the data of the different physico-chemical parameter analyses. The parameters indicating product maturity had values that fell into the limits set by current European Community requirements.

DISCUSSION

Moisture content

The moisture content in the investigated samples ranged from 13.7 to 18.09% (Table 2). All tested Algerian honeys had moisture content below 20%, which is the maximum prescribed limit for the moisture content as per the Codex standard for honey (Codex Alimentarius, 2001). The different moisture content of honey depends on harvest season, the degree of maturity reached in the hive and moisture content of the plant nectar source (Finola et al., 2007).

pH

The pH values of honey are of great importance during extraction and storage, since acidity can influence the texture, stability and shelf life of honey (Terrab et al., 2003). All honeys are acidic, having a pH in the range of 3.33 to 4.6 (Table 2). The acidity of honey is due to the presence of organic acids, particularly gluconic acid; and inorganic ions such as phosphate and chloride (Nanda et al., 2003). These results agreed with data reported by Azeredo et al. (2003). In general, honey is acidic in nature irrespective of its variable geographical origins. The pH values of Algerian, Brazilian, Spanish, Turkish and Indian honeys have been found to vary between 3.49 to 4.53, 3.10 to 4.05, 3.63 to 5.01, 3.67 to 4.57 and 3.49 to 4.43, respectively (Azeredo et al., 2003; Ouchemoukh et al., 2007; Kayacier and Karaman, 2008; Sudhanshu et al., 2010).

Free acidity of honeys

Free acidity of all twenty samples fell within the permitted range proposed by Codex Alimentarius (2001) of not more than 50 milliequiv acid/kg. The free acidity of honey samples in this study ranged from 25.763 to 44.884 milliequiv acid/kg (Table 2).

Lactonic acidity of honeys

Lactonic acidity of the honey ranged from 7.063 to 10.9 milliequiv acid/kg (Table 2).

Total acidity of honeys

Total acidity of the honey ranged from 34.363 to 54.317

Table 2. Physico-chemical characteristics of some Algerian honey samples.

Samples	Moisture %	pH	Electric Conductivity (µs/cm)	Mineral content %	Free acidity (meq/Kg)	Lactonic acidity (meq/Kg)	Total acidity (meq/Kg)	HMF (mg/Kg)	DN Schade unite	Proline content mg/Kg
S1	15.4 ±0.23	4.4 ±0.02	309 ±1	0.128 ±0.001	25.763±0.37	8.6±0.01	34.363±0.38	4.1 ±0.45	198.0 ±0.45	454.3±0.01
S2	13.8 ±0.453	4.6 ±0.03	686 ±2	0.441 ±0.084	28.098±0.05	9.073±0.4	37.171±0.45	9.7 ±1.15	117.23 ±0.12	110.7 ±0.52
S3	14.1 ±0.125	4.1 ±0.03	314 ±2	0.35 ±0.067	41.231±0.50	10.9±0.75	52.131±1.25	10.3 ±0.16	100.33 ±2.67	100 ±0.388
S4	17.7 ±0.651	4.1 ±0.035	1050.5±1.5	0.518 ±0.013	32.7±0.35	7.983±0.00	40.683±0.35	7.4±0.91	154.1 ±2.11	364 ±0.632
S5	13.7 ±0.587	4.4 ±0.023	326.5 ± 0.5	0.111 ±0.054	35.7±0.18	10.873±0.50	46.573±0.68	4.0 ±0.33	202.123 ±3.56	502.5 ±0.752
S6	15.21 ± 0.685	3.33 ± 0.351	349 ± 5.13	0,37 ± 0.036	29.9±1.00	7.063±1.00	36.963± 2	10.217 ± 0.529	44.45 ± 4.0	403.36 ± 0.557
S7	16.63 ±0.34	4.027 ±0.006	499.5 ± 1.5	0,44 ± 0.095	44.884±2.517	9.10±0.90	54.317 ±1.303	11.26 ± 0.276	51.82 ± 4.06	305.67 ±0.58
S8	16.18 ±0.16	3.47 ± 0.334	269.33 ± 1.15	0,233 ± 0.030	28.023±0.70	8.0±1.05	36.02± 1.75	1.12 ± 0.11	94.23 ± 3.6	407.593 ± 0.713
S9	17.177 ±0.26	3.8 ±0.025	312.67 ± 1.15	0,283 ± 0.015	27.02±2.00	10.02±0.02	37.04 ± 2.02	2.115 ± 0.094	105.63 ± 4.41	203.99 ± 0.165
S10	17.14 ± 0.140	3.5 ±0.055	286 ±2.00	0,263 ± 0.032	33.3±1.00	10.067±1.017	43.367±2.017	7.276 ± 0.216	76.85 ±1.61	201.133 ±1.0
S11	16.243±0.084	4.34 ±0.270	147.97 ±0.55	0,11 ± 0.01	40.0±0.481	10.38±1.00	50.38± 1.481	4.483 ± 0.379	83.81 ± 4.81	206.98 ± 0.076
S12	15.043 ± 1.092	4.2 ±0.396	266.67 ±9.02	0,213 ± 0.020	31.68±1.87	11.033±1	42.713± 2.87	6.217 ± 0.758	86.08± 4.01	506.86 ± 1.091
S13	15.633 ±1.041	4.0 ±0.215	307 ±1.07	0,363 ± 0.028	42.033±0.475	9±1.50	51.033 ± 1.975	3.747 ± 0.357	76.96 ± 4.95	309.5 ±0.608

Table 2. Contd.

S14	16.07 ± 0.113	3.84 ± 0.064	328.83±3.69	0,234 ± 0.173	27.5±0.5	9.5±0.5	37 ± 1.00	14.153 ± 0.055	29.28 ± 1.95	156.167 ± 1.258
S15	18.09 ± 0.131	4.05 ± 0.055	67.83 ±2.57	0,019 ± 0.001	35.3±0.20	8.033±1.32	43.333 ±1.52	8.599 ± 1.678	43.67 ± 1.10	104.56 ± 0.833
S16	15.81 ± 0.685	3.45±0.151	449 ± 1.13	0,27 ± 0.006	28.9±1.0.11	8.063±0.00	36.963±0.11	5.217 ± 0.629	54.45 ± 3.0	303.3 ± 0.56
S17	16.633 ± 0.349	4.02 ± 0.001	599.5 ± 1.0	0,45 ±0.015	44.884±0.193	9.10±0.40	54.317 ±0.59	11.0 ± 0.176	52.82 ± 4.06	405.67 ±0.5
S18	17.18 ± 0.101	4.47 ± 0.134	369.33 ± 0.15	0,238 ± 0.003	27.023±0.485	9.0±1.5	36.02± 1.985	1.12 ± 0.117	95.23 ± 2.6	307.003 ± 0.813
S19	16.17 ± 0.164	3.91 ± 0.125	412.67 ± 1.25	0,383 ± 0.015	28.02±.00	9.02±2.02	37.04 ± 2.02	1.11 ± 0.004	100.3 ± 0.41	103.99 ± 0.275
S20	16.14 ± 0.150	4.50 ± 0.055	487 ± 2.00	0.283 ± 0.012	38.3±2.420	9.067±0.03	43.367±2.25	6.36 ± 0.116	76.00±1.61	301.18±1.003

milliequiv acid/kg (Table 2). The acidity of honey is due to the presence of organic acids, particularly gluconic acid, in equilibrium with their lactones or esters, and inorganic ions such as phosphate and chloride (El-Sherbiny and Rizk, 1979; Al-Khalifa and Al-Arify, 1999).

Electrical conductivity of honeys

The electrical conductivity as measured in most honeys is less than 700 µS/cm (Table 2). The sample S4 has the highest conductivity (1050.5 µS/cm). This parameter depends on the mineral, organic acid, protein, complex sugars and polyol content and varies with botanical origin (Terrab et al., 2003).

Mineral content

The mineral content of the honey ranged from 0.019 to 0.518% (Table 2). These differences in mineral content are dependent on the type of soil in which the original nectar bearing plant was located (Anklam, 1998).

The percentage mineral content is considered as a quality criterion indicating the possible botanical origin of honey. The variability in the mineral content of honeys could be due to harvesting processes, beekeeping techniques and the material collected by the bees while foraging on the flora (Finola et al., 2007).

Proline content

Proline comes mainly from the salivary secretions of *A. mellifera* during the conversion of nectar into honey (Bergner and Hahn, 1972). The concentrations of proline ranged between 103.66 and 506.86 mg/kg (Table 2).

Amylase activity

The amylase activity is usually expressed as diastase number, symbol DN, also known as Gothe units. The amylase activity ranged from 29.28 to 202.123 Gothe units (Table 2) and the remaining DN was still above the minimum limit (8 DN). Consequently, it has been recommended that other quality indicators, such as invertase activity, which is more heat-sensitive than amylase, should be used (Oddo et al., 1999).

HMF content

Since HMF can be formed either by Maillard reaction (heating of reducing sugars in the presence of proteins), or by dehydration under acidic conditions. The HMF values were low (<15 mg/ kg) (Table 2). The diastase activity and the

HMF content are widely recognized as parameters indicating the freshness of honey (Mendes et al., 1998; Terrab et al., 2002).

Conclusion

The result of this study indicated that honey samples purchased from Northeast Algeria, were mostly of good quality. The research describes the variability of some physicochemical characteristics of twenty Algerian honeys. The results obtained agreed with requirements of European Community Directive. Honey samples differ in quality on account of various factors such as season, the origin of honey, the activity of the bee, the food of the bee, the period and technique of extraction of honey, conditions of storage and the freshness of honey. The comparison of these honey varietals with published international standards allowed us to have an idea on the quality of our honeys produced in the Algerian Northeast.

ACKNOWLEDGEMENT

We are grateful to beekeepers that collaborated with us in providing honey samples.

REFERENCES

Al-Khalifa AS, Al-Arify IA (1999). Physicochemical characteristics and pollen spectrum of some Saudi honeys. Food Chem. 67:21-25.

Anklam E (1998). A review of the analytical methods to determine the geographical and botanical origin of honey. Food Chem. 63:549-563.

Azeredo LDC, Azeredo MAA, De Souza SR, Dutra VML (2003). Protein content and physicochemical properties in honey samples of *Apis mellifera* of different floral origins. Food Chem. 80:249-254.

Bergner KG, Hahn H (1972). Zum Vorkommen und zur Herkunft der freien Aminosa¨uren in honig. Apidologie 3(1):5-34.

Bertoncelj J, Dobersek U, Jamnik M, Golob T (2007). Evaluation of the phenolic content, antioxidant activity and colour of Slovenian honey. Food Chem. 105:822-828.

Blasa M, Candiracci M, Accorsi A, Piacentini MP, Albertini MC, Piatti E (2006). Raw Millefiori honey is packed full of antioxidants. Food Chem. 97:217-222.

Bogdanov S (1997). Nature and origin of the antibacterial substances in honey. Lebensmittel Wissenchard und Technology 30:748-753.

Bradford MM (1976). A rapid and sensitive for the quantitation of microgram quantitites of protein utilizing the principle of protein-dye binding. Anal. Biochem. 72:248-254.

Coco FL, Valentini C, Novelli V, Ceccon L (1996). High-performance liquid chromatographic determination of 2-furaldehyde and 5-hydroxymethyl-2-furaldehyde in honey. J. Chromatogr. A 95-102.

Codex Alimentarius Commission (1989). Codex standard for honey (worldwide standard). FAO-WHO, CAC Vol 3, Ed 1 Supp 2, Rome.

Codex Alimentarius Commission (2001). Revised codex standard for honey. Codex Standard 12-1981. Rome: FAO and WHO FOOD STANDARDS PROGRAMME.

El-Sherbiny GA, Rizk SS (1979). Chemical composition of both clovers and cotton honey produced in A.R.E Egypt. J. Food Sci. 7: 69-75.

Finola MS, Lasagno MC, Marioli JM (2007). Microbiological and chemical characterization of honeys from central Argentina. Food Chem. 100:1649-1653.

Guler A, Bakan A, Nisbet C, Yavuz O (2007). Determination of important biochemical properties of honey to discriminate pure and adulterated honey with sucrose (Saccharum officinarum L.) syrup. Food Chem. 105(111):9-1125.

Helrich K (1990). Oficial methods of analysis (15th ed.). Arlington, VA: Association of O�cial Analytical Chemists, Inc.

Kalabova K, Borkovcova I, Smutna M, Vecerek V (2003). Hydroxymethylfurfural in Czech honeys. Czechoslovak J. Anim. Sci. 48: 551-557.

Kayacier A, Karaman S (2008). Rheological and some physicochemical characteristics of selected Turkish honeys. J. Texture Studies 39:17-27.

Louveaux J, Maurizio A, Vorwohl G (1978). Methods of melissopalynology. Bee World 59:139-157.

Makhloufi C, Schweitzer P, Azouzi B, Persano Oddo L, Choukri A, Laaredj H et al (2007) .Some properties of Algerian honeys. Apiacta 42:73-80.

Mendes E, Brojo PE, Ferreira IMPLVO, Ferreira MA (1998). Quality evaluation of Portuguese honey. Carbohydr. Polym. 37(3):219-223.

Nanda V, Sarkara BC, Sharma HK, Bawa ASV (2003). Physico-chemical properties and estimation of mineral content in honey produced from different plants in Northern India. J. Food Composit. Anal.16:613-619.

Nozal MJ, Bernal JL, Toribio L, Jimenez JJ, Martin MT (2001) .Highperformance liquid chromatography determination of methylanthranilate, hydroxymethylfurfural and related compounds in honey. J. Chromatogr. A 917:95-103.

Oddo LP, Piazza MG, Pulcini P (l999). Invertase activity in honey. Apidologie 30:57-65.

Ouchemoukh S, Louaileche H, Schweitzer P (2007). Physicochemical characteristics and pollen spectrum of some Algerian honeys. Food Control 18:52-58.

Rodriguez GO, Ferrer BS, Ferrer A, Rodriguez B (2004). Characterization of honey produced in Venezuela. Food Chem. 84:499-502.

Sudhanshu S, Satyendra G, Arun S (2010) .Physical, biochemical and antioxidant properties of some Indian honeys. Food Chem.118: 391-397.

Terrab A, Diez MJ, Heredia FJ (2002). Characterization of Moroccan unifloral honeys by their physicochemical characteristics. Food Chem. 79:373-379.

Terrab A, Diez MJ, Heredia FJ (2003). Palynoloical, physico-chemical and colour characterization of Moroccan honeys: I. River red gum (Eucalyptus camaldulensis Dehnh) honey. Int. J. Food Sci. Technol. 38:379-386

Preparation and physicochemical analysis of some Ethiopian traditional alcoholic beverages

Tadele Yohannes, Fekadu Melak and Khalid Siraj

Department of Chemistry, College of Natural Sciences, Jimma University, P.O. Box 378, Jimma, Ethiopia.

Ethiopian traditional alcoholic beverages namely *tella, tej* and *areki* are very common drinks in the country. Ten (10) alcohol vending houses were considered in the study by considering two from each five sub-cities, purposely for 'filtered' *tella* sampling and five for the unfiltered *tella, tej* and *areki* samples. The alcoholic contents of *filter-tella, tej* and *areki* were measured and found in the range of 3.84 to 6.48, 8.94 to 13.16 and 33.95 to 39.9% v/v ethanol, respectively. Difference in pH values and ethanol levels among all samples was significant ($p<0.05$). Variations within samples of each vending houses, coefficient of variation (CV>10%) among all samples were significant, though the variation in pH of the alcohols studied were analyzed and found at drinking range or according to European Brewery Convention. Finally, sensory responses which were taken from the community (people drinking these alcohols), indicated that they are high in aroma (bitterness for *tella* than other drinks reported).

Key words: Filter-tella, tej and areki, traditional alcoholic beverages.

INTRODUCTION

In nearly all areas of the world, some types of alcoholic beverage native to their region are prepared and consumed (Steinkraus, 1986). Indigenous fermented alcoholic beverages from different parts of the world are described (Abegaz et al., 2003; Kabak and Dobson, 2011). In Ethiopia, some of the fermented beverages include *tella, tej, borde* and *shamita* which are very popular traditional drinks. The latter two, *borde* and *shamita* are mainly prepared in central and southern Ethiopia (Alemu et al., 1991; Ghebrekidan, 1992) whereas considering popularity, traditional alcoholic beverages namely, *tella* and 'katikala'(*areki)* is very common in northern part of Ethiopia. Currently, even bottled 'Gojam *areki*' from north Ethiopia is in the market by gathering from different rular vending houses.

Among fermented foods, alcoholic beverages have been widely consumed since prehistoric times by people around the world. Fermented products can play an important role, contributing to the livelihoods of rural and perturbing dwellers (Cadaga et al., 1999; Jeyaram et al., 2009). In developing countries, traditional fermentation serves many purposes. It can improve the taste of food, enhance the digestibility of a food, preserve food from degradation by noxious organisms, and increase nutritional values. Further, it is used for medical reasons, recreational purposes, in marriages, in religious and non-religious ceremonies (Kohajdova and Karovicova, 2007; Anteneh et al., 2011), at festivals and social gatherings, at burial ceremonies and as food substitutes (Hall and Sharples, 2008).

Fermentation methods for traditional alcohols are inexpensive and adaptable at household level in traditional communities (Paredes-López and Harry, 1988). Its process involves chemical change in bringing organic substrates through the action of biochemical catalysts, called enzymes. In brewing industry: Starch substrate ↦ glucose ↦ ethanol + carbon dioxide + energy.

Consumption of alcohols from different brewing products posses several damaging effects such as, mental problems, job trouble, loss of control, impaired judgment in humans and irresponsible behaviors (Coleman and Cater, 2005; Bhargav et al., 2008; Sui et al., 2009). Depending on the dose and the regularity of its consumption, alcohol is a more problematic causing heart disease, cancer diabetes, acute respiratory failure or death (Guidot and Hart, 2005; Gun et al., 2006). There are considerable evidences in some African countries that produced alcoholic drinks are known to have toxic components (Reilly, 1976; Sanni, 1993).

Various works have been reported in fermented food analysis (Aloys and Angeline, 2009; Satish Kumar et al., 2012), particularly alcoholic beverages in different parts of Ethiopia including the data with regards to methanol, fuel oil and ethanol contents of *tella, areki and tej* (Urga et al., 1997; Bekele et al., 2001). The ethanol and methanol content were reported seemingly from 'unfiltered' *tella* of Ethiopia though not clearly indicated (Desta, 1977). In light of this, 'filtered' *tella* physic-chemical parameters like ethanol level, pH, turbidity, sensory evaluation and community perception has been seldom addressed in Ethiopia. Hence, this study aimed to analyze ethanol level, pH and turbidity of some traditional alcoholic drinks accustomed in Ethiopia which include *tella*, *tej* and *areki*. In the case of *tella*, both 'filtered' and 'unfilterd' products were considered for analysis. Moreover, sensory evaluation and community perception were taken into consideration with freshly homemade *tella*.

MATERIALS AND METHODS

Sampling and sampling site

Ten (10) samples of 'filtered' *tella*, five samples of *tej* and *areki* vending houses were considered from different sub-cities of Jimma, namely: Ajjip, Qochi, Markato, Menahera and Matrik Sefer. Moreover, five 'unfiltered' *tella* samples were monitored starting from preparation and taken for analysis from each five sampling sites. All those vending houses were selected purposively, and *tella* samples were collected using polyethylene bottles for analysis.

Preparation of traditional alcoholic beverages: A case in Jimma, Ethiopia

Tella

Tella is one of the Ethiopian traditional beverages, which is prepared from different ingredients. It is brewed from various grains and different cereals which include; barely, corn, wheat and sorghum and also from teff and maize, although in some regions, millet and *Rhamnus prinioid* (Fite et al., 1991).

Pre-preparation steps were similar to that described by Sahle and Gashe (1991). The clay container (*insera*) was washed with *Grawa* (Vernonia amygdalina) and water several times and after that smoked with wood from *weyra* (*Olea eurapaea* subsp. Cuspidate) in order to find good fragrance or flavor and to be neat. During the preparation of *tella*, an average duration of six to eight days was required by the *tella* producers at winter season. The preparation of *tella* has four major phases.

Tej

Tej is a home processed, fermented alcoholic beverage; it is also commercially available honey wine. It is prepared from honey, sugar, water and leaves of *Gesho* (*Rhamnus prepoides*). Preparation procedure was similar to those reported by Bahiru et al. (2001).

Areki

Areki is a distilled alcoholic beverage. It is a colorless, traditional alcoholic beverage which is distilled from fermentation products prepared in almost the same way as *tella* except that fermentation mass in this case is more concentrated.

Chemicals and apparatus

Abbee Refractometer (ATAGO, USA) pH meter, density meter (Compaq Evo D500, USA) were used for the experimental work. Distilled water was utilized. All experimental process was performed according to standard procedure.

Experimental

The collected samples of *tella* ('filtered'-case) and 'unfiltered' *tella* monitored during the preparation process plus *tej* were primarily filtered with 150 μm sieve and kept in a refrigerator until experimental work. The density meter was adjusted to determine the desired parameters.

pH values of the samples

The pH of the samples was measured by dipping the electrode of a digital pH meter into the samples after a proper calibration of the meter with standard solution.

Determination of the refractive index

All the refractive indices was measured by using Abbee refractrometer on a water bath and thermostat. The function of the water bath and the thermostat is to keep the temperature of the system constant (20°C). Before and after measurement of the refractive index of each sample, the refracto-meter was calibrated with distilled water.

Determination of the specific gravity

The specific gravity of the samples was determined using a pycnometer (long neck pycnometer). The weights of an empty pycnometer, pycnometer with distilled water, and pycnometer with the samples were measured using a balance.

$$\text{Specific Gravity} = \frac{\text{Weight of pycnometr with a sample} - \text{Weight of empty pycnometer}}{\text{Weight of distilled water}}$$

Table 1. Experimental results of physico-chemical parameters of 'filtered'-*Tella*.

S/N	Collection area	pH values	Refractive index (RI)	Specific gravity (SG)	Alcohol content (%v/v)
1	Qochi1	4.00	1.3380	1.0080	4.82
2	Qochi2	4.53	1.3378	1.0038	4.00
3	Ajip1	4.99	1.3374	1.0047	3.98
4	Ajip2	4.74	1.3372	1.0045	3.84
5	Matrik 1	4.80	1.3379	1.0047	6.26
6	Matrik 2	4.75	1.3353	1.0045	6.48
7	Markato 1	4.68	1.3377	1.0053	6.15
8	Markaro 2	4.81	1.3378	1.0052	6.19
9	Menahara1	4.60	1.3405	1.0097	5.46
10	Menahara2	4.76	1.3511	1.0087	4.51

Table 2. Ethanol level (%v/v) and original gravity measurements in each sample of 'unfiltered' *tella*.

Sample type	Alcohol (%v/v)	Original gravity (%)
Menahara	3.89	12.69
Matrik	4.52	13.74
Markato	4	12.69
Qochi	3.84	12.66
Ajip	3.5	13.00

Determination of ethanol level of the samples

The ethanol level of *tella, tej and areki* was determined in the quality control laboratory of Bedele Brewery, Bedele. It is first filtered using a fluted filter paper. Then, the ethanol level determination involved measurement of refractive index and specific gravity. The temperature of the filtrate was then set at 20°C. The refractive index and the specific gravity of the samples were measured at 20°C temperature. The alcoholic content of the samples were calculated using Dave Miller (Miller, 1988) who estimates the alcohol by volume using a formula: Alcohol by volume = (initial gravity - final gravity) / 0.75.

RESULTS AND DISCUSSION

Physicochemical properties of alcoholic beverages

Significant variations in pH values of the *tella* within the sampling sites and pH value ranged from 4.00 to 4.99 (Table 1). There were also significant variations in the alcohol content of *tella* samples observed where the values ranged from 3.98 to 6.48 (% v/v) and averaged 5.17 (%v/v). Wide variations in results were obtained and this might be attributed to different methods of preparation of the beverages at the domestic level and the constituent variations are also suggested in literatures (Vogel and Gobezie, 1983). As these fermentations are natural and, thus, uncontrolled, alcohol produced during the fermentation methods are spontaneous, and variability of the product is inevitable (Zvauya et al., 1997). Statistical one-way ANOVA analysis indicated that at the 0.05 (p = 0.028) level, the means of the ethanol (% v/v) of 'filterd' and 'unfiltered' *tella* are significantly different (Tables 1 and 2). The Original gravity reported for 'unfilterd' tella was found in the range of EBC (European Beer Convention).

In Table 3, the pH value of *tej* and *areki* ranged from 3.56 to 4.45, and 4.30 to 4.51, respectively. Moreover, alcohol contents of *tej* and *areki* samples ranged from 8.94 to 13.16 and 33.95 to 39.90 with average values of 11.47 and 37.22 %v/v, respectively. There is a significant variation in the alcoholic content of *tej* and *areki* samples collected from different area as it is supported by Gizaw (2006) through the study conducted at different location. The alcohol contents of the sub-cities reported (Figure 1) for each alcoholic drink is as indicated: Tella case; matrik> merkato> menaheria> Qochi> Ajip. For Areki the order of alcoholic content is matrik> merkato> qochi> menaheria> ajip. For tej: the order in alcoholic content is like merkato> qochi> menaheria> ajip> matrik.

In Jimma town, the variation in ethanol contents of traditional alcoholic beverages like *areki* and *tej* might be due to differences in preparation and fermentation as it is reported by Bahiru et al. (2001) and Susan and Lewis (1993). Conditions such as temperature, aeration and actions of the microorganisms also obviously affect the level of the alcohol in the samples (Sahle and Gashe,

Table 3. Experimental results of pH and ethanol level of Ethiopian traditional alcoholic drinks, *tej* and *areki*.

S/N	Collection area	pH value		Alcohol content (%v/v)	
		Tej	Areki	Tej	areki
1	Qochi	3.66	4.30	12.98	36.99
2	Ajjip	3.92	4.40	10.72	33.95
3	Matrik	3.56	4.51	8.94	39.90
4	Markato	4.45	4.49	13.16	38.96
5	Menihara	4.15	4.48	11.56	36.30
6	Average	3.948	4.436	11.472	37.22

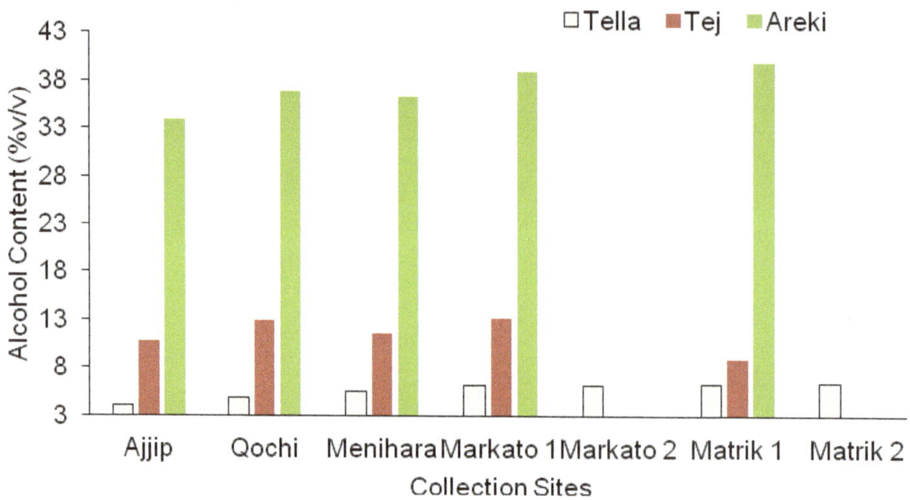

Figure 1. Alcoholic content versus study site variation of alcoholic drinks.

1991). The average pH values of *tej* and *areki* in this study show considerable variation for each sample as indicated in the results obtained. In comparing the mean alcohol contents of *tej* from other studies in Ethiopia, deviation was observed such that the result is more than that reported by Desta (1977), but similar to that reported by Gizaw (2006).

Sensory evaluation of *tella*

A panel of ten people afterwards called judges who have had experiences in drinking alcohols were selected from the community on the basis of interest and availability. They were initially trained in descriptive analysis of the mouth feel attributes of *tella*. Most of the judges provide a sensory evaluation reflection of bitterness in the 'unfiltered' *tella* samples prepared under the supervision of the researcher. But for the cases of 'filtered' *tella* collected from five study sites, the judges dictated that most of the samples >70% have medium sourness by stating that they feel sweet taste and sour homogenized. Some of the mouth feels such as bitterness, sweet, sour, soapy and aromatic used by these judges were reported by Langstaff and Lewis (1993) though they used a term softness instead of soapy or CO_2 (Kebede et al., 2002). The judges emphasized that the reason for bitterness might be the higher use of *gesho* in the preparation. In view of this, it seems evident (Motarjemi and Nout, 1996) that the raw materials utilized for the production of the beverages as well as the type of yeast during fermentation conditions played the major role in defining the mouth feel parameters. In sensory evaluation, tella were found to be strongly affected by the method of preparation, which in turn played an important role in the perception of the people in the ingredient at the initial. The judges were also interviewed on the perception of the community on alcoholic beverages in general about *tella*, *areki* and *tej* and responded as follows: there are assumptions that *tej* is to be consumed by peasants living near the town or by the aged people in the town. All the judges agreed that on the perceptions or importance, they

they accepted because they contribute to medicine, it makes them to get fatter in short periods, and there is eye-watering as length of consumption period increased. Other serious cases those judges suggest when asked to provide their last words, was about the secrets things done to increase the alcoholic content of tella in the community which is adding cement and the use of certain plants. Market value or the business of venders is not as expected due this related secret perceptions and trigs-judges prediction.

Conclusion

The study indicates that the alcoholic content (% v/v) of traditional alcoholic drinks (*tella, tej* and *areki*) in Jimma town varied considerably by the type of alcoholic drink considered. Similarly, there was a significant variation in pH values of the three alcoholic drinks (*tella, Tej and Areki)* though it was on drink range. Not only the variation in alcoholic drinks but also ethanol level were reported with significant variation from each experimental site studied. The alcoholic content variability in local alcoholic beverages could be attributed to the spontaneous fermentation differences that the producers build.

ACKNOWLEDGEMENT

We thankfully acknowledge the Department of Chemistry, College of Natural Sciences, Jimma, University, Ethiopia, for providing necessary facilities to carry out this work.

REFERENCES

Abegaz K, Beyene F, Langsrud T, Narvhus JA (2003). Indigenous processing methods and raw materials of borde, an Ethiopian traditional fermented beverage. J. Food Technol. Afr 7:59-62.

Alemu F, Amha-Selassie T, Kelbessa U, Elias S (1991). Methanol, fuel oil and ethanol contents of some Ethiopian traditional alcoholic beverages. SINET:Ethiop. J. Sci. 14:19-27.

Aloys N, Angeline N (2009). Traditional fermented foods and beverages in Burundi. Food Res. Int. 42:588-594.

Anteneh T, Tetemke M, Mogessie A (2011). Antagonism of lactic acid bacteria against foodborne pathogens during fermentation and storage of borde and shamita, traditional Ethiopian fermented beverages. Int. Food Res. J. 18:1189-1194.

Bekele B, Tetenike M, Mogessie A (2001). Chemical and nutritional properties of "tej", an indigenous Ethiopian honey wine: variations within and between production units. Food Technol Afr. 6:104-108.

Bhargav S, Panda BP, Ali M, Javed S (2008). Solid-state Fermentation : An Overview. Chem. Biochem. Eng. Quart. 22:49-70.

Coleman LM, Cater SM (2005). A qualitative study of the relationship between alcohol consumption and risky sex in adolescents. Arch. Sex. Behav. 34:649-661.

Desta BA (1977). Survey of the alcoholic contents of traditional beverages. Ethiop. Med. J. 15:65-68.

Fite A, Tadesse A, Urga K, Seyoum E (1991). Methanol, fusel and ethanol contents of some Ethiopian traditional alcoholic beverages. SINET. Ethiopia JSCi. 14:19-27.

Gadaga TH, Mutukumira AN, Narvhus JA, Feresu SB (1999). A review of traditional fermented foods and beverages of Zimbabwe. Int. J. Food Microbiol. 53:1-11.

Ghebrekidan H (1992). The effect of different chemical and physical agents on the viability of Cysticercus bovis:a preliminary report. Ethiop. Med. J. 30:23-31.

Guidot DM, Hart CM (2005). Alcohol abuse and acute lung injury:epidemiology and pathophysiology of a recently recognized association. J. Investig. Med. 53:235-245.

Gun, RT, Pratt N, Ryan P, Gordon I, Roder D (2006). Tobacco and alcohol-related mortality in men:estimates from the Australian cohort of petroleum industry workers. Aust. N. Z. J. Public Health 30:318-324.

Hall CM, Sharples L (2008). Food and Wine Festivals and Events Around the World:Development, Management and Markets.

Jeyaram K, Singh TA, Romi W, Devi AR, Singh WM(2009). Traditional fermented foods of Manipur. Knowledge Creation Diffusion Utilization. 8:115-121.

Kabak B, Dobson ADW (2011). An introduction to the traditional fermented foods and beverages of Turkey. Crit. Rev. Food Sci. Nutr. 51:248-260.

Kebede A, Fekadu B, Thor Langsrud JA (2002). Indigenous processing methods and raw materials of borde, an Ethiopian traditional fermented beverage. J. Food Technol. Afr. 7:59-64.

Kohajdova Z, Karovicova J (2007). Fermentation of cereals for specific purpose. J. Food Nutr. Res. 46:51-57.

Miller DG (1988). The Complete Handbook of Home Brewing. Storey Publishing, LLC, Bowker.

Motarjemi Y, Nout MJ (1996). Food fermentation: a safety and nutritional assessment. Joint FAO/WHO Workshop on Assessment of Fermentation as a Household Technology for Improving Food Safety. Bull. World Health Organ. 74(6):553-559.

Paredes-López O, Harry GI (1988). Food biotechnology review:traditional solid-state fermentations of plant raw materials--application, nutritional significance, and future prospects. Crit. Rev. Food Sci. Nutr. 27:159-187.

Reilly C (1976). Cancer and the consumption of home-produced alcoholic drinks in Zambia: a possible correlation. Afr. J. Med. Med. Sci. 5:191-194.

Sahle A, Gashe BA (1991). The microbiology of Tella fermentation. SINET: Ethiop. J. Sci. 14: 81-92.

Sanni AI (1993). The need for process optimization of African fermented foods and beverages. Int. J. Food Microbiol. 18:85-95.

Satish Kumar R, Kanmani P, Yuvaraj N,. Paari KA, Pattukumar V, Arul V (2012). Traditional Indian fermented foods:a rich source of lactic acid bacteria. Int. J. Food Sci. Nutr. 1-14.

Steinkraus KH (1986). Fermented foods, feeds, and beverages. Biotechnol. Adv. 1:219-243.

Sui D.-Y, Liu T.-Q., Jiang X.-M., Ge D., Li X.-Q., Mu C.-F., Sun X.-Y. (2009). Ethanol fermentation process in series with gel immobilized cells and integrated with membrane. Gao Xiao Hua Xue Gong Cheng Xue Bao. J. Chem. Eng. Chin. Univ. 23:80-86.

Susan AL, Lewis MJ (1993). The mouth fcel of beer - a review. J. Inst. Brew. 99:31-37.

Urga K, Fite A, Biratu E (1997). Effect of natural fermentation on nutritional and antinutritional factors of tef (Eragrostis tef). Ethiop. J. Health Dev. 11:61-66.

Vogel S, Gobezie A (1983). Ethiopian "Tej" In: Hand book of indigenous fermented Foods.

Zvauya R, Mygochi T, Parawira W (1997). Microbiological and biochemical changes occurring during production of masvusvu and mangisi, traditional zimbabulean beverages. Plant Foods Hum. Nutr. 51:43-51.

The effect of *Withania coagulans* as a coagulant on the quality and sensorial properties of Tofu

Reyhaneh Sarani[1] , Javad MohtadiNia[2] and Mohammad Asghari Jafarabadi[3]

[1]Nutrition Faculty, Tabriz University of Medical Sciences, Tabriz, Iran.
[2]Department of Food Science and Technology, Tabriz University of Medical Sciences, Tabriz, Iran.
[3]Department of Statics and Epidemiology, Tabriz University of Medical Sciences, Tabriz, Iran.

Tofu is a nutritional, gel-like soy food. The present study was carried out to investigate the effects of *Withania coagulans* extract on the soymilk coagulation for producing tofu. For this purpose, soymilk was coagulated by *W. coagulans* extract, and the properties of the prepared tofu were analyzed. The results indicate that the extract of *W. coagulans* as a coagulant significantly ($p<0.05$) reduced yield and moisture content when compared with calcium sulfate tofu, but fat and protein contents of *W. coagulans* tofu and calcium sulfate tofu were the same. Results of sensory evaluation show that the sensorial properties of *W. coagulans* tofu are not different from that of calcium sulfate tofu. In this way, it was concluded that the extract of *W. coagulans* fruits could be used as a natural coagulant in coagulation processes of soymilk.

Key words: *Withania coagulans* soymilk, tofu, coagulant, syneresis.

INTRODUCTION

Soybean is a major source of protein, having a protein content of 35-40% on a dry basis, making it a relatively inexpensive source of protein for human consumption (Derbyshire et al., 1976). Soybeans are processed into different kinds of products such as tofu one of the most important and valued soy foods throughout the world (Mujoo et al., 2003), especially in East and South Eastern Asian countries due to their inexpensive and high quality protein (Birthal et al., 2010). Use of tofu and soy-products in the Western countries are in an increasing rate (Atkinson et al., 2002).

Tofu preparation generally includes soaking and grinding of soybeans in water, filtering, boiling and coagulation of soymilk. The yield, quality and texture of tofu are influenced by many factors such as variety of soybeans (Shen et al., 2006), processing methods (Kyoko, 1979; Shih et al., 2006) and type and concentration of coagulants (Lim et al., 2006; Shen et al., 2006; Sun and Breene, 1991).

Usually, calcium sulfate and glucono -δ-lactone are used more than other coagulants on an industrial scale for the tofu making (Lim et al., 2006; Sun and Breene, 1991; Tsai et al., 2006; Wang and Hesseltine, 1982). Tofu is generally known as a salt or acid coagulated soy protein gel, with soya lipids and proteins and other constituents trapped in its networks (Kohyama et al., 1995).

Most studies about tofu are available in the literature and used chemical coagulants for preparation of tofu (Beddows and Wong, 1987; DeMan et al., 1987; Schaefer and Love, 2007; Shen et al., 2006; Sun and Breene, 1991; Wang and Hesseltine, 1982). The extract of *Withania coagulans* is considered as a natural coagulant (Dastur et al., 1949).

W. coagulans Dunal (Family: Solanaceae) is in the form of a shrub and it is usually recognized as Indian cheese maker. *W. coagulans* grows in Pakistan, Afghanistan and India as well as in southern regions of Iran. This vegetative coagulant has been used for preparation of traditional cheeses from raw cow's milk (Dastur, 1949; Roseiro et al., 2003). The main components of the berries other than the milk-coagulating enzyme are esterases, free amino acids, fatty oils (an essential oil), alkaloids and withanolides (Atta-ur-Rahman et al., 2003).

The aim of this study was to examine the effects of Vegetative coagulant from *W. coagulans* on the quality and yield of tofu making, in an attempt to use this enzyme as a new local source of coagulant in preparation of tofu.

MATERIALS AND METHODS

Soybean used for tofu making was purchased from a local market (city of Mamaghan, Iran). Calcium sulfate and other chemicals were purchased from Merck Company (Darmstadt, Germany). All chemical reagents were in analytical grade.

Dried fruits of *W. coagulans* were collected from wild plants growing in the Southeast of Iran (Sistan and Baluchestan Province).

Enzyme extraction

According to Naz et al. (2009), dried fruits of *W. coagulans* powdered by grinder (Bel-Art Products, Pequannock, NJ, USA) and enzyme extract was obtained by soaking the powder in 0.85% saline solution at 4°C for 24 h. The extract was filtered to obtain crude extract. This coagulant was kept in 4°C for future uses.

Preparation of soymilk

Soymilk was prepared according to the method of Noh et al. (2005) with a few modifications. Soybeans were soaked in water at room temperature for a period of 10 h, rinsed and drained, then ground with water at a bean: water ratio of 1:4 by using a waring blender (Bel-Art Products, Pequannock, NJ, USA) for 5 min at high speed. After grinding, the resultant slurry was cooked under stirring condition and held at 95°C for 15 min. The slurry was squeezed handy with a muslin cloth and pressed to remove the soy residue (okara), to obtain a soymilk of 12–14° Brix. A single batch of soymilk was made from the soybean and was used for the preparation of tofu by using each coagulant.

Preparation of tofu

According to Liu et al. (2004), 100 ml portion of prepared soymilk (75°C) was poured in a 250 ml beaker, and food grade 0.5% calcium sulfate solution was added to the solution with constant stirring. Stirring was stopped after complete coagulation (10 min) and content was kept without stirring for 15 min at room temperature. Tofu gel was formed and cooled to 20°C without moving from the beaker, and then kept at 4°C until the next day for analysis and yield rate measurement.

For preparation of *W. coagulans* tofu, different concentrations of prepared *W. coagulans* extract (0.25, 0.5, 1 and 1.5%) were added in a 100 ml of soymilk (37°C) with a little stirring. The content of beaker was kept for 10, 20, 30 and 40 min at 35, 37, 40 and 45°C and the same procedure was used for calcium sulfate tofu.

Analysis

Yield, moisture, protein, pH and fat analyses

Before analysis, when the tofu reached room temperature, tofu carefully was removed out from the beaker and at the same time, whey was removed and fresh tofu-gel was weighted.

Yield of tofu was calculated as fresh weight of tofu-gel obtained from a specified amount of the soybean used for its making.

Moisture content was determined by drying 5 g of fresh tofu at 105°C in an oven to constant weight (Tsai et al., 2006). Total protein was determined by the micro Kjeldahl method (AOAC, 1995). The factor N×6.25 was used to convert nitrogen to protein. The fat content of tofu was determined by the Soxhlet method (AOAC, 1995). pH of the tofu samples was measured with a digital pH meter with a glass electrode (Metrohm AG, Switzerland).

Syneresis rate

After equilibrating in the room temperature (20°C) for 1 h, the tofu-gel was carefully removed wholly from the beaker and cut cross-wise with a sharp knife to pieces with diameter of 1.5 cm.

Syneresis was evaluated by employing the modified method of Armstrong et al. (1994). After cutting, six pieces of tofu-gel samples were put on 20 mesh stainless steel grid in a plastic box. The grid was supported by four sticks. The exuded liquid was allowed to be separated from tofu-gel. The box was covered with plastic to prevent evaporating. Since the liquid was released slowly, six pieces of tofu-gel samples were stored in the box for 10 h in refrigerator. The liquid quantity exuded during the 10 h was used to calculate syneresis rate:

$$\text{Syneresis}(\%) = \frac{\text{Weight of water exuded from the sample in 10h}}{\text{Weight of sample}} \times 100$$

Sensory evaluation

Thirty-two (32) untrained panelists, composed of adult males and females, scored the sensory characteristics of tofu-gel. Tests on overall acceptability, color, flavor and texture were conducted using a 5-point hedonic scale (5 = excellent, 1 = not good) for each attribute. All samples were coded and presented in a randomized arrangement.

Statistical design and analysis

Experiments were based on a randomized complete block design. All extraction experiments and analyzes were performed seven times and values are reported as means ± SD as obtained. Analysis of variance was conducted for each data collected, using SPSS version 17.0 (SPSS Inc., Chicago, IL, USA). Duncan's multiple range tests was used to determine differences between the two prepared tofus. The Statistical tests were conducted at the 5% probability level.

RESULTS AND DISCUSSION

W. coagulans tofu coagulated only when 1% prepared *W. coagulans* extract was added in a 100 ml of soymilk

Table 1. Syneresis, yield and contents of moisture, protein, fat, pH of prepared tofu with two different coagulants.

Sample	Yield					
	Soybean (kg/kg)	Moisture (%)	Protein (%, d.b.)	Fat (%, d.b.)	pH	Syneresis (%)
Calcium sulfate tofu	3.13[a]	86.75±0.16[a]	52.70±1.41[a]	28.73±0.63[a]	5.44±0.25	15.4±0.3[a]
Withania coagulans tofu	2.19[b]	85.98±0.33[b]	52.25±1.53[a]	28.44±0.81[a]	5.08±0.12	17.0±0.5[b]

Values are means ± standard deviation. Means with the same letter within columns are not significantly different from each other (P < 0.05).

Table 2. Effect of *Withania coagulans* on sensory characteristics of tofu.

Sample	Flavor	Color	Texture	Overall acceptability
Calcium sulfate tofu	2.15[a]	2.9[a]	2.84[a]	2.28[a]
*Withaniacoagulans*tofu	2.53[a]	2.87[a]	2.81[a]	2.44[a]

Means with the same letter within columns are not significantly different from each other (P < 0.05).

(37°C) with a little stirring and then it was kept for 20 min at 37°C. These results are in the line with the results of Naz et al. (2009). The main objective of their study was to purify the protease from *W. coagulans* and partially characterize it, and they concluded that 37°C and 15 $\mu L mL^{-1}$ of enzyme would be the best concentration for coagulation.

Yield and composition of tofu

The composition and yield of tofu samples are shown in Table 1.

No significant differences were found between protein and lipid content of tofu coagulated by these two coagulants. The formation of tofu gels entangles proteins, fats, hydrocarbonates, and air within the gel networks. The aggregated soy protein networks which are full of air bubbles, scaffold the tofu gel. The networks form numerous boundaries among the media (Kohyama et al., 1995; Saowapark et al., 2008).The yield of *W. coagulans* tofu was lower than that of calcium sulfate tofu. The lower yield of *W. coagulans* tofu was reflected by the lower moisture content because tofu yield and moisture contents are highly correlated (Cai et al., 1997). These results are in the line with the results of Noh et al. (2005). They studied quality of tofu as affected by freezing treatment of soybeans and reported that the lower yield and moisture content of tofu from frozen soybeans may be ascribed to the denser and more compact structure, which made water easily release from the curd during pressing. The variation in the moisture content of tofu prepared with different coagulants is probably due to the differences in the gel network affected by the ionic strengths of the coagulants and/or the effect of the different anions on the water-holding capacity of the soy-protein gels (Wang and Hesseltine, 1982).The pH of *W. coagulans* tofu was lower than that of calcium sulfate tofu; these results are in line with the results of Pezeshki et al. (2011) which reported that pH was significantly (p< 0.05) lower in cheeses made with *W. coagulans*as coagulant than pH of cheeses with animal or fungi rennets. A decrease in pH was described as essential for the coagulation of soy proteins by many previous researchers (Beddows and Wong, 1987; Lu et al., 2006). The lower pH of *W. coagulans* tofu may reflect the isoelectric precipitation of soy-proteins by the release of protons from δ-lactone (Smith, 1978) because Withanolides are a group of ergostanolides, generally having a δ-lactone in the side chain (Abraham et al., 1968).

Syneresis rate

It is recognized that *W. coagulans* significantly (p<0.05) increased the tofu's syneresis rate. Probably due to an unspecific proteolitic activity of *W. coagulans* enzymes (Pezeshki et al., 2011), however, increase of syneresis from the curd could be due to increase in bonding occurring during storage, making the protein matrix more dense and compacted (Sun and Breene, 1991) and also differences between microstructures may be responsible for different physical properties between the two kinds of tofu gels (Liu et al., 2004).

Sensory evaluation

Table 2 shows the results of sensory evaluation of tofu for color, flavor, texture and overall acceptance. The results were expressed on a 5-point hedonic scale.

There were no significant differences between sensory characteristics of tofu coagulated by these two coagulants. This means that extract of *W. coagulans* had few effect on the sensory attributes of tofu.

Conclusion

This study has confirmed the possibility of replacing calcium sulfate by extract of *W. coagulans* for the coagulation of soybean milk in the preparation of tofu. Furthermore, the study has demonstrated that *W. coagulans* tofu had fat and protein content similar to calcium sulfate tofu but both yield and moisture content were lower. In addition, using *W. coagulans* could reduce heating process needed for coagulation of soymilk, without impairing the quality of tofu, and the cost of tofu preparation is reduced because of the cheaper price of *W. coagulans* than that of calcium sulfate

ACKNOWLEDGEMENTS

This research was supported by Tabriz University of Medical Sciences, Research Center (Iran). We thank the participants of this study for their enthusiastic support.

REFERENCES

Abraham, A, I Kirson, E Glotter, D Lavie (1968). A chemotaxonomic study of Withania somnifera (L.) dun. Phytochemistry. 7:957-962.

AOAC (1995). Official methods of analysis.16th ed. Arlington, VA.

Armstrong, Helen J, Sandra E Hill, Peter Schrooyen, John R Mitchell (1994). A comparison of the viscoelastic properties of conventional and Maillard protein gels. J. Texture Study. 25:285-298.

Atkinson, Charlotte, Heather E Skor, E Dawn Fitzgibbons, Delia Scholes, Chu Chen, Kristiina Wähälä, Stephen M Schwartz, Johanna W Lampe (2002). Overnight urinary isoflavone excretion in a population of women living in the United States, and its relationship to isoflavone intake. Cancer Epidemiol. Biomarkers Prev. 11:253-260.

Beddows, CG, Wong J (1987). Optimization of yield and properties of silken tofu from soybeans. IJFST. 22:15-21.

Birthal, Pratap Singh, Parthasarathy Rao P, Nigam SN, Bantilan CS, Bhagavatula S (2010). Groundnut and soybean economies in Asia: Facts, trends and outlook: ICRISAT.

Cai, TD, Chang KC, Shih MC, Hou HJ, Ji M (1997). Comparison of bench and production scale methods for making soymilk and tofu from 13 soybean varieties. Food Res Int. 30:659-668.

Dastur, Nosher N (1949). Milk clotting enzymes from plants. Indian Farming. 9:451-455.

DeMan L, DeMan JM, Buzzell (1987). Composition and properties of soymilk and tofu made from RI Ontario light hilum soybeans. Can. Inst. Food Sci. Technol. J. 20:363-367.

Derbyshire E, Wright DJ, Boulter D (1976). Legumin and vicilin, storage proteins of legume seeds. Phytochemistry.15:3-24.

Kohyama, Kaoru, Yoh Sano, Etsushiro Doi (1995). Rheological characteristics and gelation mechanism of tofu (soybean curd). J. Agric. Food Chem. 43:1808-1812.

Kyoko, Saio (1979). Tofu-relationships between texture and fine structure. CFW. 24:342-345.

Lim, BT, DeMan JM, DeMan L, Buzzell RI (2006). Yield and quality of tofu as affected by soybean and soymilk characteristics. Calcium sulfate coagulant. J. Food. Sci. 55:1088-1092.

Liu, Zhi-Sheng, Sam Chang KC, Li-Te Li , Eizo Tatsumi (2004). Effect of selective thermal denaturation of soybean proteins on soymilk viscosity and tofu's physical properties. Food Res. Int. 37:815-822.

Lu, JY, Eloise Carter, Chung RA (2006). Use of calcium salts for soybean curd preparation. J. Food Sci. 45:32-34.

Mujoo, Rajni, Dianne T. Trinh , Perry K. W. Ng (2003). Characterization of storage proteins in different soybean varieties and their relationship to tofu yield and texture. Food Chem. 82:265-273.

Naz, Shehla, Tariq Masud , Malik Adil Nawaz (2009). Characterization of milk coagulating properties from the extract of Withania coagulans. Int. J. Dairy Technol. 62:315-320.

Noh, EJ, Park SY, Pak JI, Hong ST, Yun SE (2005). Coagulation of soymilk and quality of tofu as affected by freeze treatment of soybeans. Food Chem. 91:715-721.

Pezeshki, A, J Hesari, A Ahmadi Zonoz , B Ghambarzadeh (2011). Influence of Withania coagulansProtease as a Vegetable Rennet on Proteolysis of Iranian UF White Cheese. JAST. 13:567-576.

Roseiro, Luisa Bivar, Manuela Barbosa, Jennifer M Ames , R Andrew Wilbey (2003). Cheesemaking with vegetable coagulants-the use of Cynara L. for the production of ovine milk cheeses. Int. J. Dairy Technol. 56:76-85.

Saowapark, Suteera, Arunee Apichartsrangkoon , Alan E Bell (2008). Viscoelastic properties of high pressure and heat induced tofu gels. Food Chem. 107:984-989.

Schaefer MJ, J Love (2007). Relationships between soybean component and tofu texture. J. Food Qual. 15:53-66.

Shen CF, L Man, RI Buzzell , JM DE MAN (2006). Yield and Quality of Tofu as Affected by Soybean and Soymilk Characteristics: Glucono-delta-lactone Coagulant. J. Food Sci. 56:109-112.

Shih MC, Hou HJ, Chang KC (2006). Process optimization for soft tofu. J. Food Sci. 62:833-837.

Smith, Allan Kay (1978). Soybeans: chemistry and technology. Volume 1. Proteins: Avi Publishing Company Inc.

Sun, Nong, William M Breene (1991). Calcium sulfate concentration influence on yield and quality of tofu from five soybean varieties. J. Food Sci. 56:1604-1607.

Tsai S-J, Lan CY, Kao CS, Chen SC (2006). Studies on the yield and quality characteristics of tofu. J. Food Sci. 46:1734-1737.

Wang HL, Hesseltine CW (1982). Coagulation conditions in tofu processing. Process Biochem. 17:7.

Studies on physical, chemical and rheological characteristics of pasta dough influenced by inulin

Reza Afshinpajouh, Soodabeh Heydarian, Mehdi Amini, Ehsan Saadatmand and Matin Yahyavi

Zar Macaron Company, Iran.

The objective of this research was to evaluate the effects of different amounts (0, 1, 2.5 and 5%) of inulin on the physical, chemical and rheological properties of pasta dough. The results show that increasing the percentage of inulin, caused decrease in dough development time, dough stability, farinograph quality number, energy, resistance to extension as well as dough elasticity, while water absorption increased. The swelling index and the length of alveograph increased upon increasing the percentage of inulin. Increasing inulin addition to 5% caused the reduction in amount of proteins and ashes of samples, but this changes were not statistically significant (>0.05). 1 and 2.5% were generally selected as a superior treatment because they have fewer effects on dough rheological characteristics. While pasta dough containing 1% inulin was not optimum due to not having functional effects, 2.5% was accepted as a best level of inulin added to pasta dough.

Key words: Pasta, dough, inulin, rheological characteristics.

INTRODUCTION

Inulin is an indigestible or low digestible sugar composition (from the type of oligosaccharides), which is found in more than 30,000 various plants (Kim et al., 2001). Chicory root is one of these plants that is cultivated, produced and exported by Belgium, Netherlands and French (Silva, 1996). The main reason of using chicory root as a source of inulin is the existence of glucose and long chains of fructose (Ozer et al., 2005). Inulin is a water soluble composition and its solubility depends on temperature. At 10°C, the solubility of inulin is 6%, while at 90°C, it is ~35% (Kim et al., 2001). Inulin, after getting to intestinal environment, as a source of carbon or energy causes the development and/or activity of prebiotics (beneficial intestinal bacteria containing Lactobacillus and Bifidobacterium) (Oliveira et al., 2009).

As a food additive with high nutritional value, prebiotics are considered as factors of increasing beneficial intestinal bacteria (Bifidobacterium and Lactobacillus) activity (Probert et al., 2004). Some health effects of prebiotics include: reducing the blood cholesterol level (Schrezenmeir and de Vrese, 2001), preventing the constipation and dysentery (Ozer et al., 2005), resisting infections such as salmonella (Oliveira et al., 2009), increasing the water absorption of mineral elements (manganese, magnesium and calcium) (Coudray et al., 1997), reducing the risk of osteoporosis, preventing the colonic cancer, reducing the blood glucose level (Gibson et al., 2004), treating the intestinal infections and reducing the intestinal bacteria and thus intestinal pH (Schrezenmeir and de Vrese, 2001).

In recent years, formulations of food stuffs with functional compositions especially prebiotics have been of great consideration (Fuad and Prabhasankar, 2010). Pasta has been the subject of interest of researchers because of its ease of use, low glycemic index and long shelf life (Jenkins et al., 1988). In order to enhance the

Table 1. Chemical composition and physical specifications of inulin composition.

Composition ingredients of inulin	Amount of compositions
Moisture	Less than 3.5%
Dry matter	96±1%
Ashes	Max 0.3%
Carbohydrates	Min 99.7%
Dietary fiber	Min 90%
Sugars (glucose, fructose and sucrose)	Max 10%
Color	White
Taste	A little sweet
pH	6
Degree of polymerization	10
Aggregation	Less than 500 µ
Stability	Heat stable

Table 2. Pasta production formulation with different percentage of inulin composition.

Formula	Flour (%)	Dry gluten (%)	Inulin (%)	Total (%)
Control pasta	95.9	4.1	-	100
1	94.9	4.1	1	100
2	93.4	4.1	2.5	100
3	90.9	4.1	5	100

nutritional value of pasta, it is possible to add dietary fibers into pasta (Brennan et al., 2004). Findings of Peressini and Sensidoni (2009) showed that 5% inulin addition into flour caused a principal impact on customer attractiveness. Brennan and Tudorica (2008) found that incorporation of 10% inulin in pasta had no significant effect on pasta cooking loss; on the other hand it reduced water absorption and firmness. Manno et al. (2009) concluded inulin caused the change of starch-protein matrix in pasta structure.

According to Silva (1996) findings, inulin is a heat resistant prebiotic which can tolerate high temperatures up to 120°C. Therefore, the temperatures of pasta processing (80°C) have no significant effect on reduction of inulin content. Similar activity within flour pH is another inulin specification. The pH between 6-7 will have the best effect on inulin activity through the pasta production process. The objective of this study was to evaluate the effect of inulin on physical, chemical and rheological characteristics of pasta.

MATERIALS AND METHODS

Common wheat flour from red wheat (variety N-80-19) with ~0.650% ash, 14.3% moisture, 10.82% protein and 49.3% gluten index (provided by Karaj Zar Flour, Iran) and inulin extract of chicory root by Belgium Cosucra Company with fibruline instant brand, were prepared and stored in the warehouse for sensitive ingredients of factory with 15°C temperature until test. The chemical and physical properties of inulin are shown in Table 1. Pasta inulin with 1, 2.5 and 5% formulations were used to make prebiotic pasta dough (Table 2). Farinograph and Extensograph Brabender (with AACC approved method number (54-10) and (54-27) respectively), Chopin Alveograph (standard method (54-30A)) and ashes and proteins test (standard number (08-01) and (46-30) respectively) were used to measure the rheological properties of pasta dough (AACC, 2010). These experiments in random complete block designed by 4 replications were done, the variance of data obtained by SAS9.1 software was analyzed and the obtained mean results were compared by Duncan multiple range test (Der and Everitt, 2002).

RESULTS

The effect of inulin addition on the rheological properties of pasta dough in Farinograph are listed in Table 3. According to the results, by increasing the percentage of inulin in pasta dough, dough resistance or stability against mixing, dough development time, dough softening degree and finally, the correlation between the quality number and the rheological properties of pasta dough were decreased. The results showed that water absorption was increased by adding inulin. The effect of different percentages of inulin on the rheological properties of pasta dough in extensograph are listed in

Table 3. Result of analyzing the variance of inulin effects on pasta dough properties in farinograph.

Inulin	Water absorption	Dough stability	Degree of dough softening (after 10 min)	Degree of dough softening (after 12 min)	Farinograph number
Pasta inulin 5%	64.03±2.03[a]	5.625±0.02[c]	42.42±1.14[c]	27.88±0.12[c]	66.72±1.23[c]
Pasta inulin 2.5%	62.47±1.78[ab]	6.9±0.14[b]	55.78±1.25[c]	31.83±0.26[b]	94.7±1.31[b]
Pasta inulin 1%	61.13±2.32[ab]	7.925±0.31[a]	62.58±0.63[b]	37.95±0.41[ab]	107.3±1.72[b]
Control pasta	59.85±1.34[b]	7.6±0.26[ab]	70±1.2[a]	45.58±0.09[a]	138.1±1.64[a]

The same letter is not significantly different at $P < 0.05$.

Table 4. The effect of different inulin percentages on dough properties in extensograph.

Inulin	Resistance to extension (after 5 min)	Energy	Extensibility	Maximum resistant to extension	Ratio number	Maximum ratio number
Pasta inulin 5%	331± 2.41[c]	79±0.14[c]	127.3±1.52[c]	454.8±3.28[c]	2.6±0.71[b]	3.55±0.45[b]
Pasta inulin 2.5%	395.5± 3.25[b]	86.25±0.14[b]	139±1.62[b]	482±2.76[bc]	2.85±0.92[b]	3.475±0.62[b]
Pasta inulin 1%	428.5± 3.76[ab]	92.25±0.14[bc]	147.8±1.98[a]	515.8±4.16[ab]	2.925±0.56[ab]	3.475±0.31[b]
Control pasta	473.5± 3.92[a]	102.3±0.14[a]	144.5±2.21[ab]	552.5±3.81[a]	3.275±0.18[a]	3.8±0.08[a]

The same letter is not significantly different at $P < 0.05$.

Table 5. The effect of different inulin percentages on dough properties in alveograph.

Inulin	P value	L value	G value	W value	P/L value	I_e	P_{200}
Pasta inulin 5%	85.75± 1.17 [c]	75.25±1.51[a]	19.75±0.67[a]	204.8±0.14[b]	1.15±0.02[c]	47.42±1.08[b]	40.65±0.47[c]
Pasta inulin 2.5	105.5± 1.26 [b]	71.25±1.62[a]	17.75±0.43[ab]	219±0.14[ab]	1.5±0.02[c]	53.28±1.16[a]	56.13±0.93[b]
Pasta inulin 1%	128.3± 1.92 [a]	47.75±2.84[b]	15.5±0.77[bc]	224±0.14[a]	2.7±0.06[b]	41.72±0.87[c]	53.45±1.04[b]
Control pasta	142± 2.31 [a]	40.75±2.63[b]	14.25±0.58[c]	222.8±0.14[a]	3.525±0.08[a]	47.15±1.33[b]	66.93±1.61[a]

P: Pressure, L: length, G: swelling index, W: energy, P/L: pressure/length, I_e: extensibility index, P_{200}: the pressure in the first 4 cm. The same letter is not significantly different at $P < 0.05$.

Table 4. By increasing of inulin addition, the resistance to extension, dough energy, extensibility, maximum resistance to extension, ratio number were reduced. Evaluating the effect of different percentages of inulin on maximum ratio number shows that this parameter was decreased in all treatments as compared to the control pasta, by increasing of the amount of inulin.

The effect of Inulin addition on the rheological properties of pasta dough in Alveograph are listed in Table 5. Pressure value (P) and pressure/length index (P/L) were reduced by increasing the amount of inulin. On the other hand, length value (L) and swelling index (G) were enhanced. Evaluating energy (W) showed that although the pasta inulin 1% had more energy than the control pasta, this difference was statistically insignificant. The extensibility index (I_e) showed a significant decrease as compared to the control pasta by 1% inulin addition. The amount of P_{200} was reduced by increasing the percentage of inulin.

The effect of inulin addition on the pasta dough dry matter

According to Figure 1, although the content of protein is reduced by increasing the amount of inulin, this reduction is not statistically significant. Figure 2 indicates that addition of different percentages of inulin did not have a significant effect on the ash content of pasta dough. Although the amount of final product had some changes in the test samples, none of these changes were significant.

DISCUSSION

Evaluations indicate that existence of hydroxyl groups in inulin fiber structure will cause more hydrogen bonds, which will in turn, increase the water absorption (Chimrov et al., 1981). Inulin addition has a significant effect on

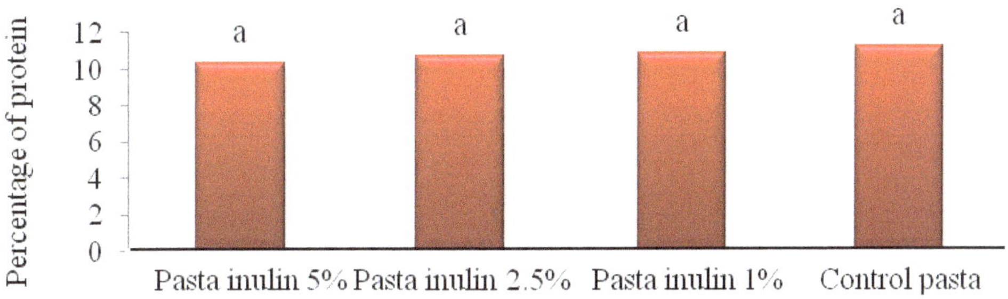

Figure 1. The effect of different percentages of inulin on pasta protein.

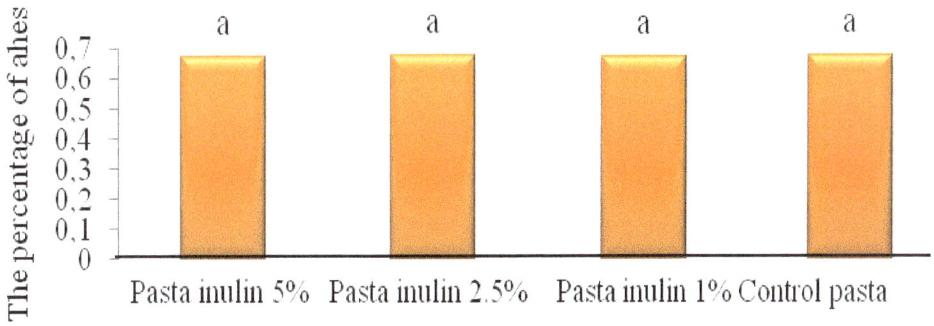

Figure 2. The effect of different percentages of inulin on pasta ashes.

pasta dough structure and would also make changes in pasta dough protein-starch bonds (Probert et al., 2004). Using the high percentages of inulin in pasta dough causes some specific changes in its organoleptic properties and consumer acceptability. Inulin addition by 1 and 2.5% percentages would cause little changes in pasta dough starch texture, while adding high percentages would cause many harsh changes in its structure and also weaken the protein-starch bonds in it. Addition of optimum percentages of inulin would cause the balance of protein-starch dehydration in the flour. Moreover, it will increase consumer attractiveness, reduce the calorie, increase the amount of indigestible fiber in it and finally increase the product's organoleptic properties (Ozer et al., 2005).

Although the addition of high percentages of more than 10% inulin cause significant decrease in pasta protein (Ozer et al., 2005), researches have shown that inulin addition by 1 to 5% will have no significant effect on reducing the pasta protein. For pasta, the minimum amount of protein is 10.5, which is also decreased by replacing the high percentages of inulin instead of flour as this reduction is not permitted below the standard range (10.5). Among the three pasta inulins of 1, 2.5 and 5%, pasta Inulin 5% is not appropriate in terms of the quality related to pasta dough properties in farinograph, because of some problems such as dough softening and reduction in dough resistance against mixing. According to the scientific articles, suggesting that pasta inulin 1% would not have functional specifications, inulin 2.5% was selected as a superior treatment among the other three experimented treatments.

The results obtained from extensograph indicated that pasta inulin (2.5%) has the highest capability among the three applied treatments and was more optimum in these properties. The results of alveograph, showed a significant correlation between the increase of the amount of inulin and rupture time, in other words, the higher the dough swelling index, the sooner it will rupture (Silva, 1996). So addition of inulin percentages more than 2.5% would cause the dough to rupture sooner, thus the high percentages are not optimum in terms of physical properties. Therefore, pasta inulin was selected as the best optimum treatment, although pasta inulin (1%) had better qualification than other percentages in alveograph.

Conclusion

Since the purpose of producing the pasta with inulin is primarily to increase its nutritional properties and then to have a high quality product, so in this research, the formulation that retains both aspects of nutritional properties and product quality according to the optimization of product process was used. The results showed that 2.5% treatment is more superior to the others. Therefore, we recommend the use of 2.5% inulin in the production formulation.

ACKNOWLEDGEMENT

The authors gratefully acknowledge Zar Industrial Group (Karaj/Iran) for providing the equipments and laboratory facilities and the financial support for this research.

REFERENCES

AACC (2010). AACC Approved Methods. St. Paul, Minnesota, USA: AACC, American Association of Cereal chemists, Inc.

Brennan CS, Kuri V, Tudorica CM (2004). Inulin-enriched pasta: effects on textural properties and starch degradation. Food Chem. 86(2):189-193.

Brennan CS, Tudorica CM (2008). Evaluation of potential mechanisms by which dietary fibre additions reduce the predicted glycaemic index of fresh pastas. Int. J. Food Sci. Technol. 43:2151-2162.

Coudray C, Bellanger J, Castiglia-Delavaud C, Remesy C, Vermorel M, Rayssignuier Y (1997). Effect of soluble or partly soluble dietary fibres supplementation on absorption and balance of calcium, magnesium, iron and zinc in healthy young men. Eur. J. Clin. Nutr. 51(6):375-380.

Der G, Everitt BS (2002). A Handbook of Statistical Analyses using SAS Second Edition Chapman &Hall/CRC.

Fuad T, Prabhasankar P (2010). Role of ingredients in pasta product quality: A review on recent developments. Crit. Rev. Food Sci. Nutr. 50(8):787-798.

Gibson GR, Probert HM, Van Loo J, Rastall RA, Roberfroid MB (2004). Dietary modulation of the human colonic microbiota: updating the concept of prebiotics. Nutr. Res. Rev. 17(2):259-275.

Jenkins,DJ, Wolever TM, & Jenkins AL (1988). Starchy foods and glycemic index. Diabetes care, 11(2), 149-159.

Kim Y, Faqih MN, Wang SS (2001). Factors affecting gel formation of inulin. Carbohydrate Polymers 46(2):135-145.

Manno D, Filippo E, Serra A, Negro C, De Bellis L, Miceli A (2009). The influence of inulin addition on the morphological and structural properties of durum wheat pasta. Int. J. Food Sci. Technol. 44(11):2218-2224.

Oliveira RP, Florence AC, Silva RC, Perego P, Converti A, Gioielli LA, Oliveira MN (2009). Effect of different prebiotics on the fermentation kinetics, probiotic survival and fatty acids profiles in nonfat symbiotic fermented milk. Int. J. Food Microbiol. 128(3):467-472.

Özer D, Akin S, Özer B (2005). Effect of Inulin and Lactulose on Survival of Lactobacillus AcidophilusLA-5 and Bifidobacterium Bifidum BB-02 in Acidophilus-Bifidus Yoghurt. Food Sci. Technol. Int. 11(1):19-24.

Peressini D, Sensidoni A (2009). Effect of soluble dietary fibre addition on rheological and breadmaking properties of wheat doughs. J. Cereal Sci. 49(2):190-201.

Probert HM, Apajalahti JH, Rautonen N, Stowell J, Gibson GR (2004). Polydextrose, lactitol, and fructo-oligosaccharide fermentation by colonic bacteria in a three-stage continuous culture system. Appl. Environ. Microbial. 70(8):4505-4511.

Schrezenmeir J, de Vrese M (2001). Probiotics, prebiotics, and synbiotics- approaching a definition. Am. J. Clin. Nutr. 73(2):361s-364s.

Silva RF (1996). Use of inulin as a natural texture modifier. Cereal Foods World 41(10):792-794.

Physicochemical properties and consumer acceptability of soft unripened cheese made from camel milk using crude extract of ginger (*Zingiber officinale*) as coagulant

Yonas Hailu[1], Eyassu Seifu[2] and Zelalem Yilma[3]

[1]School of Animal and Range Sciences, Haramaya University, P.O. Box: 138, Haramaya, Ethiopia.
[2]Department of Food Science and Technology, Botswana College of Agriculture, Private Bag 0027, Gaborone, Botswana.
[3]Netherlands Development Organization (SNV), EDGET/EADD-ET Project, Country Project Mobilization Coordinator, Churchill Road, Tracon Building, 6th Floor, P.O. Box P.O. Box 40675, Addis Ababa, Ethiopia.

This study was conducted to analyze the quality attributes of soft unripened cheese (SUC) made using ginger (*Zingiber officinale*) crude extract (GCE) by comparing it with cheese made using camel chymosin (CC). SUC made using GCE had higher acidity and ash but lower fat, total solids and protein as compared to cheese made using CC. The camel chymosin resulted in higher cheese yield as compared to cheese made using GCE. Although, lower than the values for cheese made using CC, sensory scores of cheese made using GCE is in line with literature values for camel milk cheese. The results indicated that GCE can be used to coagulate camel milk and thus help make cheese from camel milk.

Key words: Camel milk, cheese, ginger crude extract, physicochemical properties, sensory properties.

INTRODUCTION

Dromedary camels (*Camelus dromedarius*) produce more milk of high nutritional value that can be kept for a longer period of time as compared to other species in a hostile environment (Khan and Iqbal, 2001). Although camel milk has been consumed fresh in traditional pastoral systems for several centuries, processing of camel milk into more shelf stable value added milk products is not well developed and camel milk products are not common (Mehaia, 2006; Farah and Bachmann, 1987). Although coagulation of camel milk is reported to be difficult without addition of commercial milk clotting enzymes (FAO, 2001), possibilities of cheese-making from camel milk were reported by several authors (Khan et al., 2004; Mehaia, 2006; Ahmed and El Zubeir, 2011).

Recently, camel chymosin was developed using recombinant DNA technology by Danish scientists (Kappeler et al., 2006). However, it is not easily available for pastoralists or small camel milk processers; when available, it is not affordable. Generally, cheese making is reportedly curtailed by factors such as high cost and limited availability of calf rennet (Jacob et al., 2010; Hashim et al., 2011a). This calls for the need to search for alternative coagulants from easily available and cheap

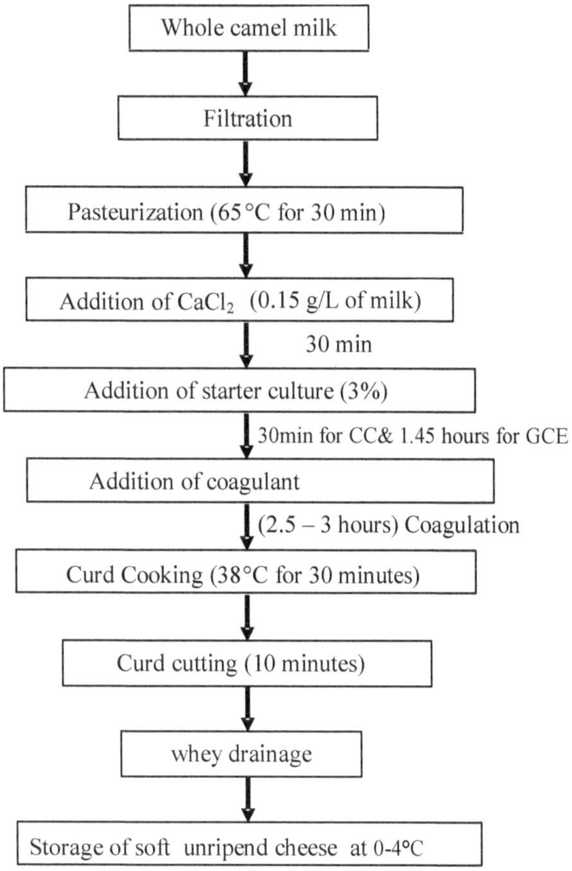

Figure 1. Flow diagram for the manufacture of soft unripened cheese from camel milk.

resources such as plant extracts to coagulate camel milk and make cheese out of it. Ginger extract was used to coagulate cow milk in which the milk clotting activity was caused due to the proteolytic activity of protease enzymes (Huang et al., 2011; Hashim et al., 2011a, b). Further, Hashim et al. (2011a) revealed that ginger extracted enzyme had high specificity for *α*-casein followed by *β*-casein and κ-casein and exhibited a similar affinity for κ-casein, α-casein and *β*-casein. Its higher specificity for κ-casein with increasing temperature was also reported (Huang et al., 2011). The milk clotting activity of ginger crude extract was reported to be higher than that of calf rennet and papain but lower than mucor rennet (Hashim et al., 2011a). This study was, therefore, conducted to assess the physicochemical properties and consumer acceptability of soft unripened cheese made by coagulating camel milk using ginger crude extract.

MATERIALS AND METHODS

Sample collection and physicochemical properties of camel milk

Camel milk samples were obtained from pastoralists living in the Erer valley in Eastern Ethiopia and delivered to Haramaya University Dairy Technology Laboratory by placing it in an ice box. The milk was collected by directly milking into sterile bottles. The extraction of ginger crude extract was performed according to the procedure described by Hashim et al. (2011a) with some modification as follows. Briefly, fresh ginger rhizomes were peeled, chopped, washed with de-ionized water and frozen at -23°C, then homogenized using a blender (Model 38BL40, Blender 8010E, Christiano Scientific Equipment, USA) with five parts of cold acetone (w/v) (-23°C) and kept at 4°C for about 15-20 min. The homogenate was filtered through cotton cloth and the precipitate was further washed with cold acetone followed by air drying. The air dried material was made into powder using a food grinding mill (Model M20, KIKA®WERKW, Germany). The powder was then homogenized in 20 mM phosphate buffer (pH 7.0) for 2.4 minutes (Hashim et al., 2011b) and the extract was filtered using a muslin cloth. The filtrate was centrifuged at 12,000 g for 20 min (Model 1020, D Centrifuge, Centrurion Scientific LTD) and the supernatant was considered as crude extract. Liquid camel Chymosin (Chy-Max® M, Christian Hansen, Denmark) with an activity of 1000 IMCU/ml at a concentration of 0.1 ml/L was also used for manufacturing of soft unripened cheese for comparison purpose. A yoghurt culture (*Streptococcus thermophilus*) (Micro-Milk, Italy) was used for acidification of milk. The physicochemical properties (pH, titratable acidity, total solids, ash, protein and fat) of raw camel milk were analyzed using standard procedures (AOAC, 1995).

Manufacturing of soft unripened cheese

Soft unripened cheese was manufactured from camel milk according to the method described by Mehaia (2006). Three different batches of cheese were made. At the end of each manufacturing process, the cheese samples were collected in sterile containers and kept in refrigerator at 4°C for further analysis. Ten liters of camel milk brought to the laboratory was divided into two lots of five liters each (ginger crude extract was used to coagulate one of the lots, added at a pH value of 5.2 and camel chymosin was used to coagulate the second lot and added at a pH value of 6.2, which served as a control). Both lots were inoculated simultaneously with the yoghurt starter culture, *S. thermophilus* (3%). Calcium chloride (0.15 g/L of milk) was added into the milk 30 min prior to the addition of the coagulants as recommended by FAO (2001). The other subsequent steps were maintained the same during cheese-making for both treatments (Figure 1).

The physicochemical properties (pH, titratable acidity, total solids, ash, protein and fat) of soft unripened cheese (SUC) were analyzed following standard procedures (AOAC, 1995). pH and titratable acidity were measured in the same day the cheese was manufactured; however, the other parameters were analyzed within 48 h of manufacture of the cheese. The cheese yield was calculated as suggested by Mehaia (2006) and was expressed as kg/100 kg of milk. Cheese yield = weight of cheese/weight of the milk sample.

Sensory evaluation

Thirty voluntary panelists were selected based on the criteria suggested by Hashim (2002): age between 18 and 34 years, and usual consumers of camel milk or fermented camel milk and cheese from milk of other species to rate the acceptability of the cheese based on its color, appearance, aroma, taste, texture and overall acceptability using a 7-point hedonic scale (1 = dislike extremely, 2 = dislike moderately, 3 = dislike slightly, 4 = neither like nor dislike, 5 = like slightly, 6 = like moderately and 7 = like extremely). Each batch of cheese was coded with three digit numbers and provided to panelists after 24 h of manufacturing. Fifteen grams of soft

Table 1. Physicochemical properties of camel milk samples used for making soft unripened cheese.

Variables	Values
pH	6.64 ± 0.02
Titratable acidity (% lactic acid)	0.15 ± 0.01
Total solids (%)	11.6 ± 0.27
Ash (%)	0.94 ± 0.03
Fat (g/100ml)	3.0 ± 0.10
Protein (g/100 ml)	2.9 ± 0.13

Table 2. Physicochemical properties of soft unripened cheese made from camel milk using crude extracts of ginger and chymosin as coagulant.

Variables	Cheese made using ginger crude extract	Cheese made using chymosin
pH	4.87 ± 0.04^{b}	5.27± 0.02^{a}
Titratable acidity (% lactic acid)	0.81 ± 0.06^{a}	0.57 ± 0.06^{b}
Total solids (%)	35.40 ± 0.03^{b}	39.90 ± 1.76^{a}
Ash (%)	2.10 ± 0.04^{a}	1.98 ± 0.00^{b}
Fat (g/100 g)	12.90 ± 0.14	13.40 ± 0.59
Protein(g/100 g)	16.40 ± 0.54^{b}	19.70 ± 0.89^{a}
Actual yield (kg/100 kg milk)	8.70 ± 1.22^{b}	11.4 ± 0.36^{a}

Means with the same superscript letters within a row are not significantly different (P > 0.05); Values in the table are mean ± SD of three replications.

unripened cheese was allotted to each judge. Sensory evaluation was conducted in a well-ventilated and illuminated room.

Data analysis

The physicochemical and sensory score data were analyzed using the ANOVA technique of completely randomized design. Means were separated using least significant difference method.

RESULTS AND DISCUSSION

Physicochemical properties of soft unripened cheese made from camel milk

The whole camel milk used for cheese making had pH 6.64 ± 0.02, titratable acidity 0.15± 0.01, percent total solids 11.6 ± 0.27, percent ash 0.94 ±0.03, fat 3.0 ± 0.10 g/100ml and protein 2.9 ± 0.13 g/L (Table 1). Significant difference (P<0.05) was observed in physicochemical properties of cheese samples made from camel milk using ginger crude extract (GCE) and camel chymosin (CC) (Table 2).

In the present study, lower pH values were observed for SUC made from camel milk as compared to values reported in literature. As indicated by El-Aziz et al. (2012), pickling of cheese from buffalo milk using ginger extract resulted in a reduced pH (from pH of 5.86 to 4.94). Inayat et al. (2003) reported a comparable pH value (5.23) for SUC made from camel milk as that made using CC in the present study. The higher titratable acidity recorded in the present study for SUC made using GCE, might be due to the type of coagulant added, temperature used and the time that elapsed in the cheese-making process that has an effect on the acidification process.

Higher total solids were observed for SUC made using CC as compared to SUC made using GCE (Table 2). Khan et al. (2004) and Mehaia (1993) indicated higher values of total solids (44.4 and 45.5% respectively) for cheese made from whole camel milk as compared to the results of the present study. Lower ash content was observed for SUC made using CC when compared with SUC made using GCE (Table 2). Lower values of ash content were reported for soft white cheese manufactured from camel milk by Khan et al. (2004) (1.53%), Ahmed and El Zubeir (2011) (1.53%) and Mehaia (1993) (1.85%) when compared with that observed in the current study.

Several factors can be responsible for these differences that, among others, include the nature of ingredients used in the cheese-making process and the ash content of the raw material used. Mehaia (2006), for instance, indicated that addition of salt during cheese manufacturing resulted in higher ash content of Domiati cheese.

Higher protein and cheese yield values were observed for SUC made using CC as compared to SUC made using GCE (Table 2). Inayat et al. (2003) reported a

Table 3. Mean sensory scores of camel milk cheese made using crude extract of ginger and camel chymosin as coagulants (n= 30).

Sensory attributes	Cheese made using ginger crude extract	Cheese made using camel chymosin
Color	5.20 ± 0.96^{b}	6.50 ± 0.68^{a}
Appearance	4.70 ± 1.26^{b}	6.17 ± 0.83^{a}
Aroma	5.93 ± 0.79^{a}	5.20 ± 1.27^{b}
Taste	5.00 ± 1.34^{b}	5.70 ± 1.09^{a}
Texture	4.83 ± 1.12^{b}	5.67 ± 1.47^{a}
Overall acceptability	5.23 ± 0.94^{b}	6.47 ± 0.68^{a}

Means with the same superscript letters within a row are not significantly different (P > 0.05); n = total number of sensory panelists.

comparable value of protein (19.64%) for soft unripened cheese made from camel milk as to the cheese made using chymosin in the present study. Hashim et al. (2011a) and Huang et al. (2011) indicated high proteolytic activity of ginger extracts on milk casein. Therefore, the variations in protein contents between the cheese types in the present study could have occurred due to the hydrolytic activity of the coagulant used for the cheese-making. Bansal et al. (2009), for example, reported that Cheddar type cheese made from camel chymosin resulted in low proteolysis as compared to cheese made with calf chymosin. El-Aziz et al. (2012) also reported that cheese proteolysis could be enhanced by the use of ginger crude extract as a coagulant for cheese making from buffalo milk.

Weak firmness of curd was observed for milk treated with crude extracts of ginger and resulted in smaller curd particles, which were not well coagulated; thus escaped out with the whey through the pores of the muslin cloth. This might be the reason for low cheese yield observed for milk treated with GCE. According to the report of Kamoun and Bergaouin (1989), cheese yield of 12 kg was obtained from 100 kg of camel milk. Cheese yield of 13.22 and 14.9% were also reported by Khan et al. (2004) and Mehaia (2006), respectively for fresh white cheese made from camel milk.

Sensory evaluation of soft unripened cheese made from camel milk

Significant differences (P<0.05) were observed for all sensory attributes between cheeses made using GCE and CC (Table 3). Cheese made using GCE had lower score for color than that made using CC (Table 3). According to Delahunty and Drake (2004), appearance of cheese is a function of the interaction between cheese color and texture, and coagulant type used.

On the other hand, in the present study, cheese made using GCE had higher score for aroma as compared to cheese made using CC (Table 3). This variation in cheese aroma might be attributed to the inherent property of the crude extract of ginger rhizome (aromatic plant used as a spice) that provides a pleasant aroma and enhances the flavor of the cheese (El-Aziz et al., 2012).

In the present study, cheese made using CC had higher scores for texture and overall acceptability than cheese made using GCE (Table 3). Bansal et al. (2009) suggested the less primary proteolysis by camel chymosin to be a possible reason for acceptable texture of Cheddar cheese made using camel chymosin. Proteolytic activity of ginger contributed to textural softening during ripening of Iranian ultra-filtrate white cheese (Fathollahi et al., 2010). El-Aziz et al. (2012) indicated that fortification of buffalo milk with ginger extract caused an increase in cohesiveness and a decrease in firmness of the curd, which resulted in more softness and smoothness during storage of cheese but high score for overall acceptability of cheese.

Although, the sensory score values for most of the sensory attributes were lower for cheese made using GCE than that made using CC, the values for sensory attributes observed in cheese made using GCE are higher than sensory scores reported by earlier researchers for camel milk cheese. Benkerroum et al. (2011) reported mean sensory scores of 4.30, 4.15, 4.25 and 4.50 for color, texture, flavor and acceptability for camel milk cheese obtained using chymosin (Chy.Max™) as a clotting agent.

Conclusions

The results of the current study revealed the possibility of making SUC from camel milk by coagulating it using locally available ginger crude extract. Although, the values for most of the quality parameters of SUC made using GCE were lower than the corresponding values for cheese made using the commercial camel chymosin, the observed result is promising and shows possibilities of improving the quality of such cheese in the future. Production of cheese from camel milk using easily available and cheap coagulants such as ginger crude extract is of paramount importance in adding value to camel milk. Further studies are needed to confirm the result and improve the outcomes using pure enzyme from ginger.

ACKNOWLEDGEMENTS

The authors are grateful to the Ministry of Education and Haramaya University for the study grant.

REFERENCES

Ahmed NAA, El Zubeir IEM (2011). Effect of salt level on some physical and chemical properties and acceptability of camel milk cheese. J. Camelid Sci. 4:40-48.

AOAC (1995). Official Methods of Analysis, 17th ed. Washington, DC: Association of Official Analytical Chemists.

Bansal N, Drake MA ,Piraino P, Broe ML, Harboe M, Foxa PF, McSweeney PLH (2009).Suitability of recombinant camel (*Camelus dromedarius*) chymosin as a coagulant for Cheddar cheese. Int. Dairy J. 19:510-517.

Benkerroum N, Dehhaoui M, El Fayq A, Tlaiha R (2011). The effect of concentration of chymosin on the yield and sensory properties of camel cheese and on its microbiological quality. Int. J. Dairy Technol. 64:232-239.

Delahunty CM, Drake MA (2004). Sensory character of cheese and its evaluation. In: Cheese: Chemistry, Physics and Microbiology. Volume 1.General *Aspects*, 3rd ed, Fox PF, McSweeney PLH, Cogan TM, Guinee TP, eds. Amsterdam: Elsevier Academic Press. pp. 455-487.

El-Aziz MA, Mohamed SHS, Seleet FL (2012). Production and evaluation of soft cheese fortified with ginger extract as a functional dairy food. Polish J. Food Nutr.Sci.62:77-83.

FAO (2001). The Technology of Making Cheese from Camel Milk *(Camelus dromedarius).* Animal Production and Health Paper, Vol. 113.Rome: Food and Agriculture Organization of the United Nations.

Farah Z, Bachmann MR (1987). Rennet coagulation properties of camel milk. *Milchwissenschaft* 42:689-692.

Fathollahi I, Hesari J, Azadmard S, Oustan S (2010). Influence of proteolysis and soluble calcium levels on textural changes in the interior and exterior of Iranian UF white cheese during ripening. World Acad. Sci. Eng. Technol. 66:844-849.

Hashim BI (2002). Acceptance of camel milk among elementary school students in Al Ain city, United Arab Emirates. Emir. J. Food Agric. 14(1): 54-59.

Hashim MM, Iqbal MF, Xiaohong C, Mingsheng D (2011b). Impact of processing conditions on the milk clotting activity of crude protease extracted from Chinese ginger. In: International Conference on Food Engineering and Biotechnology. 9:327-331.

Hashim MM, Mingsheng D, Iqbal MF, Xiaohong C (2011a). Ginger rhizome as a potential source of milk coagulating cysteine protease. J. Phytochem. 72:458-464.

Huang XW, Chen LJ, Luo YB, Guo HY, Ren FZ (2011). Purification, characterization and milk coagulating properties of ginger proteases. J. Dairy Sci. 94:2259-2269.

Inayat S, Arain MA, Khaskheli M, Malik, AH (2003). Study of the effect of processing on the chemical quality of soft unripened cheese made from camel milk. Pak. J. Nutr. 2:102-105.

Jacob M, Jaros D, Rohm H (2010). Recent advances in milk clotting enzymes. Int. J. Dairy Technol. 64:14-33.

Kamoun M, Bergaouin R (1989). Un essai de production et de transformation de lait de dromadaire en tunisie. Rev. Elev. Med. Vet. Pays Trop. 42:113-115.

Kappeler SR, Van den Brink HJM, Rahbek Nielsen H, Farah Z, Puhan Z, Hansen EB, Johansen E (2006).Characterization of recombinant camel chymosin reveals superior properties for the coagulation of bovine and camel milk. Biochem. Biophys. Res. Commun. 342:647-654.

Khan BB, Iqbal A (2001). Production and composition of camel milk: Review. Pak. J. Agric. Sci. 38:64-68.

Khan H, Athar IH, Aslam M (2004). Evaluation of cheese prepared by processing camel milk. Pak. J. Zool. 36:323-326.

Mehaia MA (1993). Fresh soft white cheese (Domiati type) from camel milk: Composition, yield and sensory evaluation. J. Dairy Sci. 76: 2845-2855.

Mehaia MA (2006). Manufacturing of fresh soft white cheese (Domaiti type) from dromedary milk using ultra-filtration process. J. Food Technol. 42:06-212.

34

Traditionally fermented alcoholic beverages of high altitude Bhotiya tribal community in Kumaun Himalaya, India

Kamal Kishor[1] , Lalit M. Tewari[1], Ashish Tewari[2] and Brij M. Upreti[1]

[1]Department of Botany, D.S.B. Campus, Kumaun University, Nainital-263 001, Uttarakhand, India.
[2]Department of Forestry and Environmental Science, D.S.B. Campus, Kumaun University, Nainital-263 001, Uttarakhand, India.

***Bhotiya* is a tribal community in the high altitude hills of Uttarakhand state. *Bhotiya* community is known for its ethnic knowledge globally. The *Bhotiya* have old age tradition of preparing fermented beverages like *Jaan* and *Kacchi* with traditionally prepared starter culture called *Balam*. These beverages play an important role in the socio-economic and cultural life of *Bhotiya* tribe of Munsyari region in Johar valley. This paper deals with the detailed account of two important alcoholic beverages prepared and consumed by the *Bhotiya* tribal community, inhabiting the Munsyari region of Kumaun Himalaya.**

Key words: *Bhotiya* tribe, traditional knowledge, *Balam, Jaan*, *Kacchi*, Kumaun Himalaya.

INTRODUCTION

Bhotiya is a hilly tribal community, which is mainly found in high altitudes of Kumaun and Garhwal regions of Uttarakhand state of India. In Kumaun region, Bhotiyas are mainly found in Pithoragarh district and some parts of Bgeswar district. *Bhotiya* tribes are known for their traditional knowledge of sustaining life in high altitudes all over the world. The tribes are mainly depends on the forests and meadows for food, medicine and shelter, etc. Preparation and consumption of traditionally fermented beverages are common with all tribal peoples of India (Kumar and Rao, 2007). A fermented food is defined as an edible product prepared from raw or cooked material of plant or animal origin by micro organism(s) either spontaneously or by adding mixture or pure culture (Hansen, 2002; Hulse, 2004). Fermented foods provide the bio-nutrients, minerals and are fortified with bio-actives compounds, enhancing the flavour and aroma and exert health-promoting beneficial substances (Darby, 1997; Campbell-Platt, 1994; Steinkraus, 1996; Tamang, 2009). *Bhotya* community of Johar Valley also have tremendous knowledge of making such alcoholic beverages for sustaining their life in extreme conditions. *Jaan* and *Kacchi*, two alcoholic beverages are integral part of socio-economic life of *Bhotya* tribes. *Jaan* is the rice beer containing low concentration of alcohol while *Kacchi* is distilled liquor that contains higher concentration of alcohol. *Jaan* is a soft drink (local beer) prepared by fermentation of boiled rice with traditionally prepared starter culture called *Balam*. While *Kacchi* is distilled beverage prepared by fermentation of boiled barley grains and distillation of ferment. Beverages are not only popular drink of *Bhotya* tribe but also considered to be pure drink and used also by Bhotiya women during various festivals and ceremonies. It is a tradition to offer these beverages to guest and relatives. Hence, this study aimed to explore the fermentation technology and its socio-economic impact on Bhotiya tribal community and commercial exploration of traditional fermentation technology and knowledge.

MATERIALS AND METHODS

The study was carried out during 2011-2012 in Munsyari developmental block of Pithoragarh district in Kumaun Himalaya. The house head and village elders were requested to share the traditional knowledge on various aspect of topic such as fermentation process, preparation of starter culture, *Balam* and plants used in it, distillation process of fermentation and traditional uses of beverages and implications on the socio-economic culture of Bhotiya tribes. Collected information was confirmed from elderly persons and women who are often engaged in distillation process and preparation of *Balam* (yeast culture).

Preparation of *Balam* (starter culture)

The starter culture, *Balam* is prepared by using wheat (*Triticum* spp.) as substrate and mixed with number of herbs and spices. For preparation of *Balam,* wheat grains are washed first by water, dried in sunlight, ground to flour and then roasted in pan over fire till it turns brown colour. The roasted flour is mixed with the plant *Cinnamomum zeylanicum* Breyn., *Ammomum subulatum* Roxb., *Piper longum* Linn. and *Ficus religiosa* Linn. seeds (Das and Pandey, 2007). The mixture is mixed thoroughly and required amount of water is added and rolled into a thick paste open to semidry. The semi-dried mixture is then pressed between palms to make *Balam* balls of required size. These balls are dried in shade, ready to use and stored.

Flow sheet of *Balam* preparation

Selection of wheat grain
↓
Washed in water
↓
Sun dried
↓
Ground to flour
↓
Roasted to brown
↓
Mix thoroughly with essential plant parts and water
↓
Mixture rolled into thick paste and left to dry
↓
Made *Balam* balls (starter culture prepared)

Preparation of *Jaan* (local beer)

Jaan is prepared by fermentation of boiled rice (*Oryza sativa* Linn.). At first, appropriate amount of cleaned rice is boiled, dewatered and left to moderate cool. After that, starter culture (*Balam*) mixed thoroughly with it and this mixture was transferred to an earthen pot or container. The partially air tight earthen pot or containers are left for four to seven days in dark places for further fermentation. The ferment is filtered and filtrate is called *Jaan*. The filtered rice is used in making a traditional dish Sej.

Flow sheet of *Jaan* preparation

Cleaned and washed rice grains
↓
Boiled and dewatered
↓
Left to moderate cool
↓
Mixed with starter culture (*Balam*)
↓
Transfer the mixture in earthen pot or container made partially air tight
↓
Fermented for 3-7 days in dark place
↓
Ferment filtered, filtrate is called *Jaan* (ready to drink)
↓
Rice left as residue is used in making traditional dish (*Sej*)

Preparation of *Kacchi* (distilled liquor)

Kacchi is the barley (*Hordeum vulgare* Linn.) based distilled beverage of *Bhotiya* tribe. First of all barley seeds are soaked in water, half boiled and allowed to moderate cool. After that the cooled seeds are mixed thoroughly with starter culture (*Balam*) in appropriate proportion and transferred to earthen pot or container, made partially airtight and left in dark for 4-7 days for fermentation. The ferment is then transferred to a distillation pot (Taula) for distillation (Photo 1(A)).

Then the pot is kept in fire for distillation. On this pot, a wooden frame is fixed which are perforated at middle and above this another copper plate (Parat) is placed in which cold water is filled. The vapour passing through the middle perforated wooden frame (Jokhal) strikes the bottom of copper pot in which cold water is filled. The vapour cools and droplets are collected in middle wooden frame. The water of upper copper plate is regularly changed to keep it cool. The strength of *Kacchi* depends on temperature of water of upper copper plate (Parat).

Flow sheet of *Kacchi* preparation

Collection of barley seeds
↓
Soaked in water
↓
Water soaked seeds are half boiled and left to moderate cool
↓
Mixed well with starter culture (*Balam*)
↓
Transfer the mixture in earthen pot or container made partially air tight
↓
Fermented for 3-7 days in dark place
↓
Ferment is transferred to traditional distillation apparatus
↓
Distil is collected in wooden pots
↓
Collected distilled liquor is called *Kacchi* (ready to drink)

RESULTS AND DISCUSSION

Investigation among *Bhotiya* people of Munsyari reveals that preparation of these two beverages is very common among the *Bhotiya* tribe in Johar Valley and prepared in almost every house. Both beverages play important role in the social life of *Bhotiya* community. These beverages

Photo 1. (A) A Traditional distillation apparatus and pots for storage of *Jaan* (B). Starter culture of *Balam* balls hanged in a wooden pot (C). Serving pots made with wood of Yew tree (*Taxus baccata* L.). *Courtesy:* Tribal Heritage Museum, Munsyari.

act as binding agent among this tribal community. In every function, they join together and drink the beverages. *Jaan* is generally prepared during festivals and ceremonies. Women and children are also fond of these beverages but consume them in small quantity during festivals and ceremonies. In marriage ceremonies *Kacchi* is offered by the girl to the guests. *Bhotiyas* not only consume these beverages for amusement but these beverages also play supplementary role in the nutrition of these people. In small quantity, these beverages are also used as medicine for treating different ailments and diseases. *Jaan* is also used to treat fever, dysentery, cough and cold and stomach ailments. *Kacchi* is used in stomach-ache, dysentery, cold and fatigue. *Balam* are also used in cholera and treating weakness of cattle. *Sej* is prepared with residual rice in *Jaan* preparation, eaten as mind booster and treatment of physical and mental fatigue.

These beverages also give financial support to the poor tribal people. They sell *Kacchi* in local areas at cost of 60-80 rupees per litre and *Balam* balls cost is 40-50 rupees per kilograms, *Jaan* cost is rupees 40-60 per litre. It gives financial support to significant number of families residing in remote areas depending on it. The present study also reveals that *Balam* is the potential yeast medium known to the *Bhotiyas* since time immemorial. The commercial exploitation of *Balam* may be explored.

ACKNOWLEDGEMENTS

The authors are thankful to UGC-New Delhi for providing financial support. Thanks also due to *Bhoitya* people of Munsyari for sharing their traditional knowledge with us.

REFERENCES

Campbell-Platt G (1994). Fermented foods- a world perspective. Food Res. Int. 27:253-257.

Darby WJ (1979). The nutrient contribution of fermented beverages in Fermented food Beverages. In Nutrition, edited by C P Gastineaw, W J Darby and T B Turner, Academic press, New York. pp. 61-79.

Das CP, Pandey A (2007). Fermentation of traditional beverages prepared by *Bhotiya* community of Uttaranchal Himalaya. Ind. J Trad. Knowledge. 6(1):136-140.

Hansen EB (2002). Commercial bacterial starter cultures for fermented foods of the future. Int. J. Food Microbiol. 78:119-I31.

Hulse JH (2004). Biotechnologies: past history, present state and future prospects. Trends Food Sci. Tech. 15:3-18.

Kumar V, Rao RR (2007). Some interesting indigenous beverages among the tribal of Central India. Ind. J. Trad. Knowledge. 6(1):141-143.

Steinkraus KH (1996). Handbook of Indigenous Fermented Food, 2nd edition. Marcel Dekker, Inc., New York.

Tamang JP (2009). *Himalayan* fermented foods: Microbiology, nutrition and ethnic value. CRC Press Taylor and Francis Group, USA. pp. 1-216.

Physicochemical analysis of desi yoghurts produced by the local community in Gilgit District

Muhammad Shahnawaz[1] , Saghir Ahmed Sheikh[2] and Zahid Ali Akbar[1]

[1]Department of Agriculture and Food Technology, Karakoram International University, Gilgit.
[2]Institute of Food Sciences and Technology, Sindh Agriculture University, Tandojam.

This study was carried out in the Department of Food Technology, Karakoram International University Gilgit, Pakistan. The desi yoghurt samples were collected from various narrow villages of Gilgit District. The desi yoghurt is prepared with pure cow milk by a hygienic and organic process. Results indicate that yoghurt of Oshikandas showed high moisture content and total soluble solids; whereas, Gilgit town yoghurt showed low moisture content and high titratable acidity. Likewise, Danyoure yoghurt showed low total soluble solids as compare to other yoghurts. This study concludes that Gilgit town yoghurt is recommended to be followed for manufacturing since its shelf life is high due to having low moisture content than other two, furthermore; these yoghurts are as per standard of high quality. Yoghurt being nutritious, having healing characteristics and curing gastro intestinal problems, is very beneficial for human's health. Therefore, uses of yoghurt may be considered mandatory in the daily diet.

Key words*:* Desi yoghurt, different villages, Gilgit-Baltistan.

INTRODUCTION

Yoghurt is a fermented milk product. The word yoghurt originates from Turkish word 'Yogen' meaning thick. Yoghurt is a partially solid fermented milk product which originated centuries ago in Bulgaria. Its attractiveness has grown and is at this time used in most parts of the world (Lee and Lucey, 2010). It is one of the most unique dairy products (Ebenezer and Vedamuth, 1991). Yoghurt is considered highly nutritious for having significant concentration of Ca++ and many other bioactive compounds to carry probiotics to the lower part of the intestine, which can significantly influence the intestinal microflora (Coisson et al., 2005). Preservation of milk by lactic acid bacterial through the acidification process is being practiced since centuries and use of different starter cultures, manufacturing techniques and milk types considerably change the characteristics of fermented milk products (Thapa, 2000). Under normal milk processing industry, special lactic starter cultures are used to ferment milk during preparation range of fermented dairy products (Tamime and Robinson, 1999). Quality and shelf life of fermented dairy products greatly depends upon the quality of raw milk, low total bacterial counts, absence of antibiotics and bacteriophages (Thapa, 2000).

The quality of industrial yoghurt can vary up to some extent due to difference in chemical composition of yoghurt milk, manufacturing process and distribution. Technology of yoghurt, microbiology of starter culture and quality appraisal is the prime importance of any type of yoghurt (Yadav et al., 1993). Yoghurt is highly nutritious and an easily palatable diet due to the predigested nutrients by bacterial starters. It is perishable in view of its unused lactose content (Durga et al., 1986). There is an apparent need for a valuable preservation method to control acid-tolerant spoilage yeasts and molds

Table 1. Determination of percent chemicl characteristics.

Sample	Moisture	pH	Acidity	TSS
Gilgit yoghurt (S1)	79.5C	5.3A	0.81A	12.6B
Danyour yoghurt (S2)	80.2B	5.29A	0.75B	11.3C
Oshikandas yoghurt (S30	81.5A	4.91B	0.68C	13.7A

*Means followed by same letter in a column are not significantly different at 0.05.

in yoghurt. Micotoxigenic fungi and pathogenic bacteria are able to grow at refrigeration temperature to numbers, which can result in an infection (Potter and Hotchkiss, 1995). Physical properties of yoghurt are changed by milk ingredients and manufacturing conditions, variables affect physical properties such as heat treatment which apply to milk, acidity, additives, protein content, culture, homogenization mechanical handling of coagulum and presence of stabilizers (Rasic and Kurman, 1978). Changes in the chemical, physical and microbiological composition of yoghurt determine the storage and shelf life of the product (Sofu and Ekinci, 2007). Keeping in view the importance of yoghurt and variable manufacturing processes, this study was conducted to access the physicochemical characteristics of desi yoghurt made from cow milk either meeting the international standard of yoghurt or not.

MATERIALS AND METHODS

Determination of moisture content

The moisture content was determined according to AOAC (2000). In this regard, the sample materials were taken in a flat-bottom dish (pre-weighed) and kept overnight in an oven at 100 to 110°C and weighed. The loss in weight was regarded as a measure of moisture content which was calculated by the following formula.

$$\text{Moisture (\%)} = \frac{\text{Weight of fresh sample} - \text{Weight of dry sample}}{\text{Weight of fresh sample}} \times 100$$

Determination of titratable acidity

Titratable acidity as tartaric acid was determined according to the method of AOAC (2000). Each sample of the products was treated with 0.1 N NaOH solution using titration kit; of which three to five drops of phenolphthalein indicator were used. The volume of alkali used was noted and calculated using following formula.

$$\text{Titratable acidity (\%)} = \frac{1 \text{ x Eq. Wt. of acid} \times \text{Normality of NaOH} \times \text{titer} \times 100}{10 \text{ x Weight of sample (g)}}$$

Determination of TSS

The total soluble solids (TSS) were determined as per method described by Mazumdar and Majumder (2003) using Digital-Bench-Refrectometer. Before use, the instrument was cleaned and adjusted to zero at 20°C using distilled water. An appropriate quantity of sample of each product prepared was placed on the prism-plate of the refractometer with the help of a glass rod and folding back the cover. For each sample, the instrument was calibrated using distilled water. The reading appeared on the screen was directly recorded as total soluble solids as brix.

Determination of pH

For determination of pH (hydrogen ion concentration) in the products, a method of AOAC (2000) was adopted and digital pH meter was used. Sample solution was taken in the beaker and directly inserted the electrode into the solution. When the first reading was completed, the electrode was wiped with distilled water and dried-up with tissue paper. Similarly, as a continue series, all other samples were determined accordingly.

RESULTS AND DISCUSSION

In Table 1, results at column one show that Oshikandas yoghurt has comparatively higher moisture content (81.5%) whereas Gilgit yoghurt has lower moisture content (79.5%). The results are in agreement with Ammara and Imran (2010) who analyzed the moisture content during storage by using different starter cultures and recorded 84.66, 85.32, 85.16, 86.29, 86.22 and 86.40%. pH of Gilgit, Danyour and Oshikandas yoghurts were 5.30, 5.29 and 4.91, respectively. Gilgit yoghurt has high pH value as compared to Danyour and Oshikandas yoghurts. Our results relatively correspond to Rodrigues et al. (2010) who concluded that pH of yoghurt is 4.30 to 5.08. The low pH of yoghurt is due to environment for the growth of acid tolerant yeasts and molds (Banaquio et al., 1981; Spillmann and Geiges, 1983). Furthermore, results indicate that mean acidity of all three yoghurt is 0.75%. However, comparatively, Oshikandas yoghurt has the lower acidity, whereas Gilgit yoghurt has higher acidity. This result is also in agreement with Younus et al. (2002) who analyzed the quality evaluation of market yoghurt/dahi and recorded 0.89, 0.87 and 1.13 titratable acidity. Our findings are also in accordance with the findings of Saleh et al. (2007) who concluded that the titratable acidity of stirred yoghurt is 0.93. Likewise, TSS was determined as 12.6, 11.3 and 13.7 brix for Gilgit, Danyour and Oshikandas yoghurts, respectively. To show whether TSS of these three yoghurts meets the standard or not, another study was reviewed (Zekai, 2003) which is agreement to d value of TSS found (14.58 Brix)

Conclusion and recommendation

From the research it is concluded that the Oshikandas yoghurt has high moisture content and total soluble solids. The Gilgit yoghurt has low moisture content but high titrable acidity; while Danyoure yoghurt have low total soluble solids as compared to other yoghurts. Thematically, all three yoghurts are strongly recommended to be used because these yoghurts have all quality characteristics of international standard. Yoghurt being nutritious, having healing characteristics and curing gastro intestinal problems is very beneficial for human's health.

REFERENCES

Ammara K, Imran P (2010). Determination of moisture content in stored yoghurt with different starter culture. Pak. J. Nutr. 20(2):221-227.

Banaquio VL, Publico CB, Calisay OG (1981). Keeping quality of yogurt. Philippine 1. Biol. 10(1):109-112.

Coisson JD, Travaglia F, Piana G, Capasso M, Arlorio M (2005). Euterpeoleracea juice as a functional pigment for yogurt. Food Res. Int. 38:893-897.

Durga LC, Sharda D, Sastry MP (1986). Effect of storage conditions on keeping riboflavin and niacin of plain and fruit yoghurt. Indian J. Dairy Sci. 39(4):404-409.

Lee WJ, Lucey JA (2010). Formation and Physical Properties of Yogurt. Asian-Aust. J. Anim. Sci. 23(9):112 -1136.

Mazumdar BC, Majumder K (2003). Methods on physico-chemical analysis of fruits. Practical Manual Book. Metropolitan New Delhi, India. neural network modeling. J. Dairy Sci. 90(7):3118-3125.

Mohammed Saleh, Khursheed Ahmed, Taws Khan (2007). Effect of wheat germ and wheat bran on the fermentation activity, phytic acid content and colour of tarhana, a wheat flour–yoghurt mixture. J. Food Eng. 78(2):681-686.

Potter NF, Hotchkiss JH (1995). Food Science 5th Edition, Chapman and Hall (Routledge), Florence, KY.quality, riboflavin and niacin of plain and fruit yoghurt. Indian J. Dairy Sci. 39(4): 404-409.

Rasic J, Kurmann JA (1978). Yoghurt: Scientific Grounds ,Technology Report of the FAO E-mail conference on small-scale milk collection and processing in developing countries.

Sofu A, Ekinci FY (2007). Estimation of storage time of yogurt with artificial neural network modeling. J. Dairy Sci. 90(7):3118-25.

Rodrigues LA, Ortolani MBT, Nero LA (2010). Microbiological quality of yoghurt commercialized in Viçosa, Minas Gerais, Brazil. Afr. J. Microbiol. Res. 4(3):210-213.

Spillmann DK, Geiges M (1983). Recent Trends in Development of Fermented Milks. Curr. Nutr. Food Sci. 3:91-108.

Thapa TB (2000). Small - scale milk processing technologies. Discussion paper. 23(9):1127-1136.

Yadav JS, Grover S, Batish VK (1993). A Comprehensive dairy microbiology. Metropolitan New Delhi, India.

Younus S, Tariq M, Tariq A (2002). Quality Evaluation of Market Yoghurt /Dahi. Pak. J. Nutr. 1(5):226-230.

Zekai T (2003). Studies on effect of different varieties of date palm paste incorporation on quality characteristics of yoghurt. Carpathian J. Food Sci. Technol. 3(1):28-38.

Some traditional Algerian products from durum wheat

R. Kezih, F. Bekhouche and A. Merazka

Institut de Nutrition, de l'Alimentation et des Technologies Agro-alimentaires, Université1, 7ème km, route de Sétif, Constantine, Algérie.

Foods derived from durum wheat play an important role in the diet of the Algerian people, this is attributed to the semiarid climate favourable for the cultivation of this staple cereal crop. Flat bread, pasta, couscous and frik are the most common products made from durum wheat in Algeria. It is also used to produce various kinds of traditional cakes. All these products are the major products made from durum wheat in Algeria and most of them are scientifically unknown. This work is a general review on currents works on some traditional products such as flat bread, pasta, traditional cake and others. It describes for each product, raw materials used, manufacturing process and end products quality criteria.

Key words: Durum wheat, semolina, traditional product, pasta, pastry, bread, frik.

INTRODUCTION

Algeria is located in northwest Africa, in an area known as the Maghreb, "Sunset" in Arabic. The center of the country is a semi-arid highland region where the climate is conducive for the cultivation of durum wheat, making this staple cereal crop and its derived foods an important element in the Algerian people's diet (Winget and Chalbi, 2004). Flat bread, pasta, couscous and frik are the most common products made from durum wheat in Algeria. It is also used to produce various types of traditional cakes (Abecassis et al., 2013).

This paper is a summary review on some traditional Algerian products from durum wheat such as flat breads, pasta products, pastries and frik.

DURUM WHEAT FLAT BREADS

In Algeria, housewives produce two types of durum flat bread:

Kesra

This means fraction, and designate four kinds of home-made flat bread. Kesra includes four kinds of flat bread: Mathlouaa, maadjouna, rakhsis and harcha. These appellations vary throughout Algeria but recipes and flow charts are alike.

Khobz eddar

This literally mean home bread, differs from kesra by its recipe (fat, eggs and some additives) and by its flow chart for production (Dagher, 1991).

Kesra types

Mathlouaa

Mathlouaa (Figure 1a) (which means leavened or risen) is one of the most consumed flat breads in Algeria especially during fasting month of Ramadan. It is leavened bread made from fine durum semolina. Usually the leavening agent is dry yeast. It is baked on a clay pan with little spikes all over the bottom called tajine (Figure 1b). The bread is light and spongy, with typical tenacity from the semolina. The bread has a short shelf life. It is good

Figure 1. Mathlouaa, a leavened flat bread. (a): whole and parts of bread. (b): Mathlouaa baking on tajine (clay pan with little spikes all over the bottom).

Figure 2. Maadjouna, unleavened flat bread (a): whole bread. (b): Smooth clay pan for baking maadjouna and other products.

only the day it is made (Kezih, 1998; Cheriet, 2000).

Maadjouna

Maadjouna (Figure 2a) (which means kneaded), the recipe for this type of bread is simple (fine semolina + water + salt) as its preparation (kneading, shaping, baking) is quick and easy. Households make maadjouna when they are too busy. It is baked on a smooth clay pan (smooth tajine (Figure 2b). The bread is chewy and dense, with a pleasant taste from the burned blisters on its faces. The bread has a very short shelf life. It goes to stale quickly. It is good only the hours it is made. It serves as an edible spoon (Kezih, 1998; Cheriet, 2000).

Rakhsis

Rakhsis (Figure 3) (which means soft), this bread has also a simple recipe (fine semolina + fat + water + salt) and its preparation is easy and quick (no fermentation). Rakhsis is thinner than mathloua. It is also baked on a clay pan with little spikes. The end product has a homogenous surface without blisters because fat prevent their development. With a firm crust and a compact crumb, the product is soft. Rakhsis has a long shelf life because the addition of fat retards staling. Rakhsis is full of flavor (Kezih, 1998; Cheriet, 2000).

Harcha

Harcha (Figure 4) (which means rough) is a kind of bread "kesra" which is prepared a little differently from the above kinds. Coarse semolina, fat with high rate, salt, and water are mixed and undergo a moderate kneading to form unrefined dough which lacks visco-elastic properties. The dough is fragile and must be handled with care. Then it is shaped into squares or lozenges which are baked on a smooth clay pan (Figure 2b). The end product has a rough surface, a long shelf life and it is tasty (Cheriet, 2000).

Khobz eddar

This traditional bread is prepared during major celebrations such as weddings and religious feasts. It is more

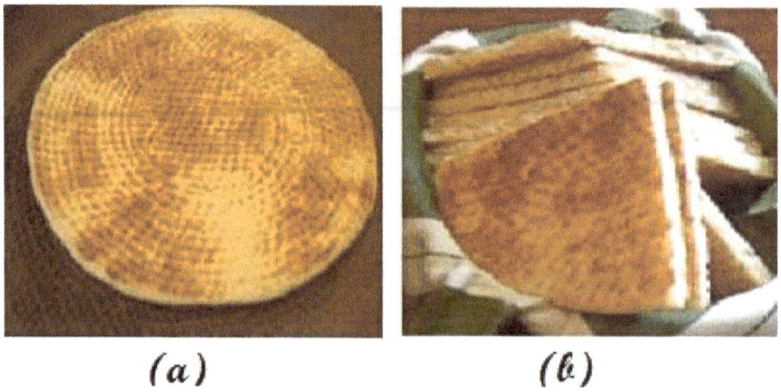

Figure 3. Rakhsis: unleavened flat bread with fat (a): hole bread. (b): slices of bread.

Figure 4. Harcha, unleavened flat bread with high fat content.

Figure 5. Khobz eddar, traditional leavened flat bread.

Figure 6. Two types of Algerian traditional pasta. (a): trida, (b): rechta.

challenging to make because its sophisticated recipe (fine semolina + fat + milk + salt + yeast + egg + sesame bread is baked in a bakery or kitchen oven. The end product has a crispy crust and a soft crumb and delicious (Figure 5) (Cheriet, 2000).

PASTA PRODUCTS AND COUSCOUS

Pasta products

Semolina and water are the greatest components of all seeds + water), also it required more steps than kesra: knead, flatten and set on the baking pan to leaven. The traditional pasta, which is processed by sheeting and cutting the dough into strips, squares or some other shapes (Figure 6). Pasta is consumed either fresh, or undergoes drying and stocked for a later use. Usually traditional Algerian pasta is steam cooked. Good pasta should have the following criteria: yellow amber color, smooth aspect, non-sticky, and when cooked should swell well, do not disintegrate and have a good taste (Namoune et al., 2000; Namoune et al., 2007).

Couscous

Couscous (Figure 7), the Algerian national dish is made from mixing semolina with water. Couscous preparation requires five steps: Blinding, agglomeration, shaping, steaming and drying. However in some regions of Algeria couscous does not undergo steaming and it is dried immediately. Characteristics of good couscous are: amber yellow color, uniform particle size, high sauce absorption capacity, ability to keep its integrity during steaming or sauce application, and particles are non-sticky, good taste (Derouiche et al., 2002).

PASTRIES DURUM WHEAT PRODUCTS

In Algeria, durum semolina is widely used to produce a

Figure 7. Couscous undergoing preparation in a wooden pan.

Figure 8. Some traditional Algerian pastries from durum wheat. (a): zlabia, (b): tamina, (c): braje, (d): makroud.

Figure 9. Frik whole grains (a); ground and sieved grains (b); frik soup (c).

range of pastries products (Figure 8), like zlabia, tamina, bradje, makroud and many others. Production and consumption of these pastries increase in Islamic holidays and in summer where families' celebrations are frequent. The recipes of these pastries are various and their manufacturing is different. They share a characteristic: sweet or must be eaten sweet (Ouelhi et al., 2004).

NON-PASTE DURUM WHEAT PRODUCT

Frik

Throughout Algerian eastern regions, immature durum is processed into frik (Figure 9). This is not specific to Algeria but there are similar traditions in other parts of the world. The grains are harvested during the interval between the "milk stage," and the "soft dough" stage. Wheat is cute with a sickle and the spikelets are stacked on a large sheet metal. The spikelets are then roasted. After which the grains are threshed and winnowed. The fire imparts a smoky flavor to the grain, and the heating stops the maturation of the endosperm and stops the degradation of the chlorophyll. The ideal frik grain is green with just the tip charred. The smoky, sweet and grassy quality of the grain adds a unique taste to frik dish. These green grains are more nutritious than mature grains, high in dietary fiber and low in phytic acid. Frik grains are ground and sieved and used to prepare a soup often consumed in the Eastern regions of Algeria especially in the month of Ramadan (Ouelhi et al., 2002; Bouza and Nacer, 2006).

CONCLUSION

Traditional products cited in this work are the major products of durum wheat in Algeria. The introduced products are not the only products produced; there are many others products not cited in the work. The review of these products shows that they constitute an interesting research field. Study of these products shows: rational use of durum wheat; research and definition of specific quality criteria for these products; improving the manufacturing process in order to have products with required qualities.

REFERENCES

Abecassis J, Boggini G, Cuq B, Namoune H (2013). Other Traditional Durum Derived Products, in:M.J. Sissons, M. Carcea, M. Marchylo, J. Abecassis, Durum Wheat: Chemistry and Technology, seconde éd. AACC International, in press.

Bouza S, Nacer F (2006). Essai de formulation et fabrication d'un mixe: soupe instantanée à base de frik. Mémoire d'Ingénieur, INATAA, Université, Mentouri, Constantine, Algérie, p. 76.

Cheriet G (2000). Etude de la galette: différents types, recettes et mode de fabrication. Thèse de Magister INATAA, Université Menrouri, Constantine, Algérie, p. 99.

Dagher MS (1991). Traditional foods in the Near East, FAO, Food Nutr. 50:161.

Derouiche M, Namoune H, Kezih R (2002). Appréciation de la qualité culinaire de deux types de couscous. Troisièmes Journées Scientifique sur le Blé. Laboratoire de Génétique, Biochimie et Biotechnologies Végétales, Université Mentouri, Constantine, Algérie, 9 pages

Kezih R (1998). Galettes: Mise au point de méthodes de préparation et d'évaluation de la qualité. Thèse Magister, Option Sciences alimentaires, INATAA, Université Mentouri, Constantine, Algérie, p. 116.

Namoune H, Abbaz C, Abdelaziz N, Kezih R (2000). Technique de fabrication, mode de caisson et qualité d'une pâte traditionnelle algérienne:chekhchoukha: Proc. Sec. Sci. Conf. Agric. Sci., Assiut University, Faculty of Agriculture, Assiut, Egypt. Vol. 2, pp. 1029-1039.

Namoune, H, Kezih, R., Othmani Marabout, Z, Cheriet G (2007). Pâtes alimentaires traditionnelles algériennes: diagrammes de fabrication, de préparation et qualités culinaires: Séminaire, INATAA, Université Mentouri, Constantine, Algérie, 12- 13 Novembre 2007.

Ouelhi G, Namoune H, Kezih R (2002). Soup consumption made with durum wheat (frik) at Constantine in Algeria. The Second Symposium on Scientific Research Outlouk in the Arab World. Scientific Research and Technology Development in the Arab World 24- 27 March, Shrajah, UAE, p. 5.

Ouelhi G, Namoune, H, Messad, N (2004). The traditional cake (makroud) in the Constantine city.The 4th Scientific Conference of Agricultural Sciences, Assiut, Egypt, p. 7.

Winget M, Chalbi H (2004). Cooking the North African Way. Ed. Lerner Publications Company, A division of Lerner Publishing Group 241 First Avenue North Minneapolis, MN 55401 U.S.A. p. 73.

Antioxidative and flavouring effects of *Aframomum danielli* on biscuits

Afolabi M. O.[1] and Adegoke G. O.[2]

[1]Department of Food Science and Technology, Bowen University, Iwo, Osun State, Nigeria.
[2]Department of Food Technology, University of Ibadan, Ibadan, Oyo State, Nigeria.

Antioxidative and flavouring effects of *Aframomum danielli* seed crude ethanolic extract on biscuits were studied. The degree of oxidation during storage was monitored by determining the acidity of the extracted fat from the biscuit samples using the standard method. Biscuits were baked with *A. danielli* seed extract at the following levels of addition: 100, 200, 300, 400, 500 ppm and a sample using butylated hydroxytoluene (BHT) at a level of 200 ppm, with a synthetic flavour (vanilla) added at a level of 50 ppm. A control sample was baked without the inclusion of *A. danielli*, BHT or synthetic flavour. The crude ethanolic extract of *A. danielli* seed at a concentration of 400 ppm was very effective as an antioxidant in the biscuit samples within the period of10 weeks of monitoring at a temperature of 25°C. Based on taste, aroma and acceptability scores, the flavour of *A. danielli* extract was not noticeable in biscuit until it was 500 ppm.

Key words: Antioxidative, flavouring effects, *Aframomon danielli,* biscuits.

INTRODUCTION

The term biscuit refers to a thin flat baked product made from flour, salt, sweetening agents, fat and food additives (Coup and Sanderson, 1987). Serious consideration is given to the shelf life of biscuits, so far as palatability is concerned (Hoseney et al., 1998). Many manufacturers are prepared to remove from the shelves, products which are ninety days old, thereby accepting this period as the maximum shelf life.

It is thought that the ninety days period is much too low and could be extended, especially if oxidative rancidity is prevented by including antioxidants in the recipe (Buck, 1991; Smith, 1972). Better results is obtained if the moisture content of the finished product is low (2.5%) and the biscuit wrapper is hermetically sealed (Smith, 1972).

The use of food additives such as antioxidant will eliminate the problems of flavor deterioration and loss of nutritional value during storage, distribution and hawking by the retailers especially where added fat/oil levels is high in the recipe (up to 20%) as it is in the case of cookies (Gisslen, 1985). Presently, in the biscuit industry, synthetic antioxidants like BHT/BHA are the antioxidants being used to prevent oxidation and flavour deterioration during shelf life of biscuits. There has been an increasing concern about possible adverse affects of synthetic food additives, especially butylated hydroxyl anisole (BHA) and butylated hydroxyl toluene (BHT) on baked products (Alice, 1998; SON, 1987). This has stimulated the research for a natural food additive that could serve as an antioxidant in such products. Example of the additives that may be used as substitutes is *Aframomon danielli*. *A. danielli* has been reported to have antioxidant and antimicrobial properties in addition to its flavour (Adegoke and Gopala Krishna, 1998; Adegoke et al., 2000a, b). The objective of this study was to investigate the possibility of replacing synthetic antioxidants (BHA and BHT) in baked biscuits with *A. danielli* seeds extract.

MATERIALS AND METHODS

A. danielli seeds, soft wheat flour and other ingredients were procured from a local market in Ibadan, Oyo state, Nigeria. Specially refined palm oil sample (free from any antioxidant and anti microbial agent) were obtained from a local manufacturer of vegetable oil and the equipment and facilities of a local bakery in Ibadan were used to bake the biscuit samples.

Spice preparation

A. danielli seed were sorted, cleaned, dried and pulverized in attrition mills, after which the powdered spice were packed in polythene bags. Then after, 25.1 g of the powdered spices was weighed into flask and refluxed in the soxhlet apparatus for 3 h, using 95% ethanol as the solvent (Adegoke et al., 2000). Solvent was evaporated off at 80°C to give the antioxidant extract.

Biscuit ingredients and baking

Biscuit recipe

The biscuits were produced using the method described by Akpapunam and Derbe (1994). The basic recipe was 100 g flour, 31 g vegetable oil, 10 g sugar, 2 eggs, 1 g baking powder. The dried ingredients were weighed and mixed manually till well blended.

Vegetable oil was added and rubbed in until uniform. The egg was added and the dough formed was kneaded manually on a flat table until a smooth consistency was obtained. The dough was cut out into round shapes using biscuit cutter. The dough pieces were transferred into greased bakery trays and baked in a hot oven at 160°C for 15 min. The biscuits were cooled on racks, packaged in sealed cellophane film and stored at 25°C. Various biscuit samples were baked as follows:

1. A control sample was baked with no anti oxidant (BHT), synthetic flavour and *A.danielli*.
2. The second set of samples were baked with *A. danielli* at the following level of addition: 100, 200, 300, 400 and 500 ppm.
3. The last sample was baked with BHT at a level of 200 ppm with a synthetic flavour (vanilla) added at a level of 50 ppm.

Analysis of samples

Chemical analysis

The acidity of extracted fat was determined using the method of Onwuka (2005) as follows: About 25 g of the powdered biscuits was weighed into flask and refluxed in the soxhlet apparatus for 3 h, using 95% ethanol as the solvent. Solvent was evaporated off at 80°C to give the extracted oil from the biscuits and the weight of the extracted oil was determined. To the oil in the flask were added mixture of 25 ml diethyl ether, 25 ml 95% ethanol and 1 ml phenolphthalein solution (1%), carefully neutralized with 0.1 M NaOH. The resulting solution was titrated with aqueous 0.1 M NaOH with constant shaking until a pink colour which persisted for 15 s was obtained.

The acidity (as palmitic acid) = Titre (ml) x 5.61 / weight of sample of biscuit used

The moisture content of baked biscuit samples were determined using AOAC (1984) method. The energy value of biscuit samples was determined using the oxygen bomb calorimeter as described by Anon (1960).

Sensory evaluation

The sensory evaluation of the freshly baked biscuit samples was conducted using multiple comparison test (Larmond, 1977). A panel of 20 judges assessed the biscuit samples for aroma, taste and overall acceptability. The samples were scored using a 6 point hedonic scale where 6 represented excellent and 1 represented very poor. The results were analyzed using analysis of variance (Snedecor, 1956).

RESULTS AND DISCUSSION

Effects of moisture content on biscuits at ambient condition

The moisture content values obtained for the baked biscuits ranged between 2.5±0.01 and 2.7± 0.72%. The Standard Organization of Nigeria (1987) set a standard of 5.0% max and with lower value, the condition for safe keeping being enhanced. The moisture content is fairly low for each sample from week 0 to week 10, an indication that the samples were well preserved against moisture (Table 1).

Effects of treatment on energy value under ambient condition

Table 2 shows the data obtained for energy value for various samples. There were no appreciable differences in energy value over the period of examination among the treatment except for control and sample with 100 ppm level of addition of *A. danielli*.

This was probably due to the fact that there had been no appreciable reduction in protein, carbohydrate and fat level due to activities of micro organisms and enzymes reaction that might have degraded the nutrients.

Effects of treatment on the sensory score of the biscuit samples

The result of the physical assessments (Table 3) conducted by the panel of 10 on the taste, aroma, and overall acceptability indicated that the assessors rated the aroma of the biscuit sample without any significant

Table 1. Moisture content (%) of sample during storage at ambient condition.

Week	Control	100 ppm *A. danielli*	200 ppm *A. danielli*	300 ppm *A. danielli*	400 ppm *A. danielli*	500 ppm *A. danielli*	200 ppm BHA/BHT
0	2.5±0.01	2.5±0.05	2.5±1.20	2.5±0.03	2.5±0.16	2.5±0.60	2.6±0.34
2	2.5±0.10	2.5±0.03	2.5±0.84	2.5±0.05	2.5±0.63	2.5±0.71	2.5±0.76
4	2.5±0.03	2.6±0.40	2.6±0.02	2.6±0.71	2.6±0.14	2.6±0.89	2.5±0.59
6	2.6±9.10	2.6±0.39	2.5±0.11	2.6±0.83	2.5±0.09	2.5±0.67	2.5±0.52
8	2.6±0.20	2.6±0.07	2.6±0.29	2.7±0.12	2.7±0.14	2.7±0.50	2.6±0.43
10	2.6±0.30	2.7±0.11	2.7±0.31	2.7±0.99	2.6±0.97	2.6±0.67	2.7±0.72

*Mean of three readings ± standard deviation.

Table 2. Effects of treatment with *A.danielli* on energy value of biscuits under ambient condition.

Week	Control	100 ppm	200 ppm	300 ppm	400 ppm	500 ppm	BHA/BHT
0	670±23	680±30.	700±20	675±10	650±25	660±50	690±20
2	650±10	680±10	690±30	650±10	675±25	695±35	680±15
4	600±25	660±20	670±20	660±20	685±40	645±10	690±25
6	580±33	650±27	685±17	670±15	690±23	670±22	700±18
8	590±22	640±21	685±19	650±19	720±13	675±33	670±43
10	570±13	620±32	710±10	670±27	690±17	665±17	650±37

*Mean of three readings ± standard deviation.

Table 3. Effects of treatment on the sensory scores of biscuits sample.

Sensory attribute	Control	100 ppm	200 ppm	300 ppm	400 ppm	500 ppm	200 ppm BHA/BHT 500 ppm vanilla
Taste	2.7^{c}	3.9^{b}	3.4^{b}	2.5^{c}	2.6^{c}	4.2^{a}	4.3^{a}
Aroma	3.9^{a}	3.8^{a}	3.8^{a}	3.4^{a}	3.4^{a}	3.3^{a}	3.3^{a}
Overall acceptability	2.5^{c}	2.6^{c}	3.6^{c}	2.8^{c}	2.6^{c}	4.3^{a}	4.3^{a}

Means within a row followed by the same letters are not significantly different at 5% level of probability of Turkey's test.

difference for all the samples, and the sensory evaluation for taste result indicated that *A. danielli* (500 ppm) and the vanilla flavoured biscuit were most preferred. The result of overall acceptability show that the sample with 500 ppm *A. danielli* was *the* most acceptable sample while the control and sample with 100 ppm *A. danielli* was not acceptable at 5% level of significant.

Effects of treatment on acidity of extracted fat of the biscuit

Figure 1 gave the acidity of the extracted fat of the biscuits baked with varying levels of *A. danielli* antioxidant extracts (100-500 ppm), BHT at 200 ppm and the control (with the exclusion of *A. danielli*, BHT and any synthetic flavour).

The result shows a steady increase in acidity of extracted fat with time in the control sample. *A. danielli* was more effective at higher concentration. The critical addition point of *A. danielli* was 400 ppm. At this concentration, the effectiveness to control lipid oxidation was comparable to that of BHT.

The antioxidative properties of the seeds extract are due to the presence of antioxidant compounds in the extract which have the ability to scavenge the lipid peroxyl radicals generated in the biscuits during storage. Such compounds are phenolic compounds that are naturally present in the seeds (Adegoke et al., 2000; Afolabi et al., 2011).

Conclusion

From this study, it was found that the inclusion of the extract of *A. danielli* in biscuit recipe at a concentration of 400 ppm was found to be effective as an antioxidant, it was however discovered that the flavour of the extract is

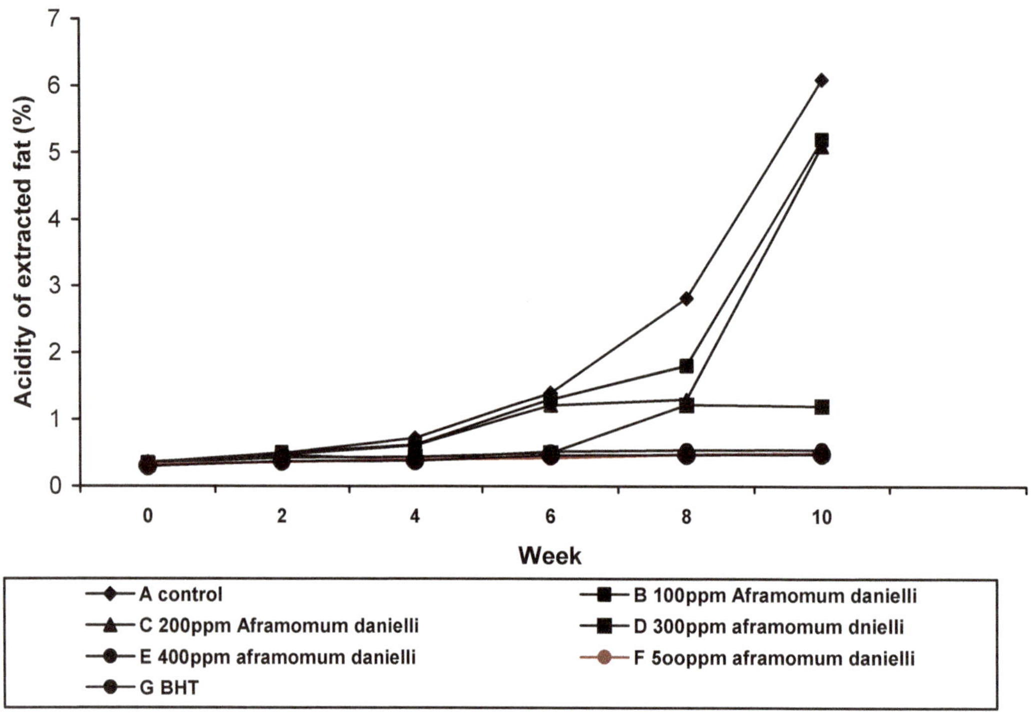

Figure 1. Effects of treatment with *A. danielli* on acidity of extracted fat of biscuit samples.

not noticeable in biscuit until it was 500 ppm. To enhance the effectiveness of the extract at lower concentration, the flavour can be purified and concentrated.

Conflict of Interests

The author(s) have not declared any conflict of interests.

REFERENCES

Adegoke GO, Fasoyiro SB, Skura B (2000). Control of Microbial Growth Browning and Lipid Oxidation by the spice *Aframomum danielli.* Food Res. Technol. 211:342-345.

Adegoke GO, Gopala Krishna AG (1998). Extractionand identification of antioxidants from the spice *Aframomum danielli.* (Family: Zingiberaceae). J. Amer. Oil Chem. Soc. 75:104-105.

Adegoke GO, Iwahashi H, Komatsu Y, Obuchi K, Iwahashi Y (2000). Inhibition of food spoilage yeast and aflatoxigenic mould by monoterpenes of the spice *Aframomum danielli.* Flavour Fragr. J. 15 (3):147-150

Afolabi MO, Adegoke GO, Mathooko FM (2011). Phytochemical characterization of the extracts of *Aframomum danielli* flower, leaf, stem and root. Afr. J. Agric. Res. 6(1):97 - 101.

Akpapunam MA, Derb JW (1994). Chemical compositionand functional properties of blends of maize anbambara groundnut flours for cookies production. Plant Foods Hum. Nutr. 46:147-155S.

Alice A (1998). Antioxidants. Go ask Alice. A Guide to Good Physical Sexual and Emotional Health.Publication of the Trustee of Colombia University, Colombia University Health Education Programme.

Anon (1960). Oxygen Bomb Calorimetry and Combination Method Technical Manual 1300. Parr Instrument Co. Molume, III, USA.

AOAC (1984).Officialmethod of analysis. Association of official Analytical Chemist, 13th edition. Washington D. C. p. 278.

Buck DF (1991). Food Antioxidants-Application and uses in Snack Foods, Cereals Foods World 29. 301-303.

Coup RO, Sanderson WB (1987). Functionality of Dairy ingredients in bakery Products. Food Technol. 1(10):86-90.

Gisslen W (1985). 'Professional Baking' John Wiley and Sons, New York.

Hoseney RC, Wade P, Finialy TW (1998). Soft Wheat Products. Wheat Chemistry and Technology, 3rd ed. Vol. 11 Pomeranz, y. (ed). American Association of Cereal Chemist. St. Paul Minnesota.

Larmond E (1977). Methods for Sensory Evaluation of Food.Food Research Central Experiment Farm, Lanana, Department of Agric, Iowa.

Onwuka GI (2005). Food Analysis and Instrumentation: Theory and Practice. Naphthali Prints. Lagos, Nigeria. pp. 93-94.

Smith WH (1972). Biscuits Crackers and Cookies; Technology, Production and Management, Applied Science Publisher, London.

Snedecor GW (1956). Statistical Methods 5th ed, Iowa State College Press, Ames Iowa p. 534.

Standard Organization of Nigeria (SON) (1987). Technical Committee Report on Revised Standard for biscuit in Nigeria.

Permissions

All chapters in this book were first published in AJFS, by Academic Journals; hereby published with permission under the Creative Commons Attribution License or equivalent. Every chapter published in this book has been scrutinized by our experts. Their significance has been extensively debated. The topics covered herein carry significant findings which will fuel the growth of the discipline. They may even be implemented as practical applications or may be referred to as a beginning point for another development.

The contributors of this book come from diverse backgrounds, making this book a truly international effort. This book will bring forth new frontiers with its revolutionizing research information and detailed analysis of the nascent developments around the world.

We would like to thank all the contributing authors for lending their expertise to make the book truly unique. They have played a crucial role in the development of this book. Without their invaluable contributions this book wouldn't have been possible. They have made vital efforts to compile up to date information on the varied aspects of this subject to make this book a valuable addition to the collection of many professionals and students.

This book was conceptualized with the vision of imparting up-to-date information and advanced data in this field. To ensure the same, a matchless editorial board was set up. Every individual on the board went through rigorous rounds of assessment to prove their worth. After which they invested a large part of their time researching and compiling the most relevant data for our readers.

The editorial board has been involved in producing this book since its inception. They have spent rigorous hours researching and exploring the diverse topics which have resulted in the successful publishing of this book. They have passed on their knowledge of decades through this book. To expedite this challenging task, the publisher supported the team at every step. A small team of assistant editors was also appointed to further simplify the editing procedure and attain best results for the readers.

Apart from the editorial board, the designing team has also invested a significant amount of their time in understanding the subject and creating the most relevant covers. They scrutinized every image to scout for the most suitable representation of the subject and create an appropriate cover for the book.

The publishing team has been an ardent support to the editorial, designing and production team. Their endless efforts to recruit the best for this project, has resulted in the accomplishment of this book. They are a veteran in the field of academics and their pool of knowledge is as vast as their experience in printing. Their expertise and guidance has proved useful at every step. Their uncompromising quality standards have made this book an exceptional effort. Their encouragement from time to time has been an inspiration for everyone.

The publisher and the editorial board hope that this book will prove to be a valuable piece of knowledge for researchers, students, practitioners and scholars across the globe.

List of Contributors

S. A. O Adeyeye
Department of Food Science and Technology, Federal University of Agriculture, Abeokuta, Nigeria

O. B. Oyewole
Department of Food Science and Technology, Federal University of Agriculture, Abeokuta, Nigeria

A. O. Obadina
Department of Food Science and Technology, Federal University of Agriculture, Abeokuta, Nigeria

A. M. Omemu
Department of Hospitality and Tourism Management, Federal University of Agriculture, Abeokuta, Nigeria

O. P. Sobukola
Department of Food Science and Technology, University of Agriculture, P.M.B. 2240, Abeokuta, Ogun State, Nigeria

O. E. Kajihausa
Department of Food Science and Technology, University of Agriculture, P.M.B. 2240, Abeokuta, Ogun State, Nigeria

V. I. Onwuka
Department of Food Science and Technology, University of Agriculture, P.M.B. 2240, Abeokuta, Ogun State, Nigeria

T. A. Esan
Department of Food Science and Technology, University of Agriculture, P.M.B. 2240, Abeokuta, Ogun State, Nigeria

I. A. Adesokan
Department of Science Laboratory Technology, Faculty of Science, The Polytechnic, Ibadan, Oyo State, Nigeria

O. P. Abiola
Department of Science Laboratory Technology, Faculty of Science, The Polytechnic, Ibadan, Oyo State, Nigeria

M. O. Adigun
Department of Science Laboratory Technology, Faculty of Science, The Polytechnic, Ibadan, Oyo State, Nigeria

O. A. Anifowose
Department of Science Laboratory Technology, Faculty of Science, The Polytechnic, Ibadan, Oyo State, Nigeria

H. O. Agu
Department of Food Science and Technology, Federal Polytechnic, Bauchi, P.M.B. 0231 Bauchi, Bauchi State, Nigeria

G. C. Ezeh
Department of Food Science and Technology, Federal Polytechnic, Bauchi, P.M.B. 0231 Bauchi, Bauchi State, Nigeria

A. I. O. Jideani
Department of Food Science and Technology, University of Venda, Thohoyandou, 0950 Limpopo province, South Africa

Oladeji, Babatunde Stephen
Department of Food Science and Technology, Obafemi Awolowo University, Ile-Ife, Nigeria

Akanbi, Charles Taiwo
Department of Food Science and Technology, Obafemi Awolowo University, Ile-Ife, Nigeria

Gbadamosi, Saka Olasunkanmi
Department of Food Science and Technology, Obafemi Awolowo University, Ile-Ife, Nigeria

Ashu Walia
Department of Food Engineering and Technology, Sant Longowal Institute of Engineering and Technology, Longowal - 148106 India

H. N. Mishra
Department of Agricultural and Food Engineering, Post Harvest Technology Centre, Indian Institute of Technology, Kharagpur 721302 India

Pradyuman Kumar
Department of Food Engineering and Technology, Sant Longowal Institute of Engineering and Technology, Longowal - 148106 India

Clarisse S. Compaoré
Département Technologie Alimentaire (DTA/IRSAT/CNRST), Ouagadougou 03 BP 7047, Burkina Faso
Institut du Développement Rural/Université Polytechnique de Bobo, BAMSB, 01 BP 10 91 Bobo-Dioulasso, Burkina Faso

Lars B. Jensen
Technical University of Denmark (DTU), National Food Institute, Division of Food Microbiology, Mørkhøj Bygade 19, 2860 Søborg, Denmark

Bréhima Diawara
Département Technologie Alimentaire (DTA/IRSAT/CNRST), Ouagadougou 03 BP 7047, Burkina Faso

Georges A. Ouédraogo
Institut du Développement Rural/Université Polytechnique de Bobo, BAMSB, 01 BP 10 91 Bobo-Dioulasso, Burkina Faso

Mogens Jakobsen
Department of Food Science, Faculty of Science, University of Copenhagen, Rolighedsvej 30, 1958 Frederiksberg C, Denmark

Labia I. I. Ouoba
London Metropolitan University, FLSC/SHS, Microbiology Research Unit, 166-220 Holloway Road, London N7 8DB, United Kingdom
Independent Senior Research Scientist-Consultant, London, UK

N. O. Kabuo
Department of Food Science and Technology, Federal University of Technology, Owerri, Nigeria

J. O. Uzuegbu
Department of Food Science and Technology, Federal University of Technology, Owerri, Nigeria

C. N. Ubbaonu
Department of Food Science and Technology, Federal University of Technology, Owerri, Nigeria

E. U. Onyeka
Department of Food Science and Technology, Federal University of Technology, Owerri, Nigeria

O. E. Adedeji
Department of Food Science and Technology, Federal University, Wukari, Taraba State, Nigeria

O. D.Oyinloye
Department of Human Nutrition, University of Ibadan, Oyo State, Nigeria

O. B. Ocheme
Department of Food Science and Technology, Federal University of Technology, Minna, Niger State, Nigeria

Oludare Olumuyiwa Adekoyeni
Food Science and Technology, Federal University Dutsinma, Katsina State, Nigeria

Rahman Akinoso
Food Technology, University of Ibadan, Oyo State, Nigeria

Akintunde Stephen Fagbemi
National Agency for Food and Drug Administration and Control (NAFDAC), Abuja, Nigeria

Djadouni Fatima
Faculty of Sciences, Mascara University, Algeria

Kihal Mebrouk
Laboratory of Applied Microbiology, Department of Biology, Faculty of Sciences, Es-Senia University, Oran, Algeria

Md. Asaduzzaman
Department of Food Processing and Preservation, Hajee Mohammad Danesh Science and Technology University, Dinajpur, Bangladesh

Md. Emdadul Haque
Department of Food Processing and Preservation, Hajee Mohammad Danesh Science and Technology University, Dinajpur, Bangladesh

Jiaur Rahman
Department of Food Processing and Preservation, Hajee Mohammad Danesh Science and Technology University, Dinajpur, Bangladesh

S. M. Kamrul Hasan
Department of Food Processing and Preservation, Hajee Mohammad Danesh Science and Technology University, Dinajpur, Bangladesh

M. A. Ali
Department of Food Science and Nutrition, Hajee Mohammad Danesh Science and Technology University, Dinajpur, Bangladesh

Mst. Sorifa Akter
Department of Food Science and Technology, State University of Bangladesh, Dhaka, Bangladesh

Maruf Ahmed
Department of Food Processing and Preservation, Hajee Mohammad Danesh Science and Technology University, Dinajpur, Bangladesh

A. O. Oyebanji
Nigerian Stored Products Research Institute, Km 3 Asa-Dam Road, PMB 1489, Ilorin, Kwara State, Nigeria

A. O. Ajani
Nigerian Stored Products Research Institute, Onireke-Dugbe, P.M.B. 5044, Dugbe, Ibadan, Oyo State, Nigeria

A. Adeize
Nigerian Stored Products Research Institute, Km 3 Asa-Dam Road, PMB 1489, Ilorin, Kwara State, Nigeria

M. O. Oyelakin
Nigerian Stored Products Research Institute, Onireke-Dugbe, P.M.B. 5044, Dugbe, Ibadan, Oyo State, Nigeria

I. Ikotun
Nigerian Stored Products Research Institute, Onireke-Dugbe, P.M.B. 5044, Dugbe, Ibadan, Oyo State, Nigeria

A. A. Agboola
Nigerian Stored Products Research Institute, No, 1 Birikisu-Iyede St., Yabba, Lagos, Nigeria

J. P. Alimi
Nigerian Stored Products Research Institute, Onireke-Dugbe, P.M.B. 5044, Dugbe, Ibadan, Oyo State, Nigeria

T. Awoite
Nigerian Stored Products Research Institute, Onireke-Dugbe, P.M.B. 5044, Dugbe, Ibadan, Oyo State, Nigeria

J. E. Obiegbuna
Department of Food Science and Technology, Nnamdi Azikiwe University, Awka, Anambra State, Nigeria

P. I. Akubor
Department of Food Science and Technology, University of Nigeria, Nsukka, Enugu State, Nigeria

C. N. Ishiwu
Department of Food Science and Technology, Nnamdi Azikiwe University, Awka, Anambra State, Nigeria

J Ndife
Department of Food Science and Technology, Kaduna Polytechnic, Kaduna State, Nigeria

O. O. C. Chukwu
Departments of Molecular Biology National Veterinary Research Institute, Vom, Nigeria
Medical Microbiology, Federal College of Veterinary and Medical Laboratory Technology,
National Veterinary Research Institute, Vom, Nigeria

A. A.Chukwuedo
Medical Microbiology, Federal College of Veterinary and Medical Laboratory Technology,
National Veterinary Research Institute, Vom, Nigeria

P. Otalike
Medical Microbiology, Federal College of Veterinary and Medical Laboratory Technology,
National Veterinary Research Institute, Vom, Nigeria

I. D. Chukwu
Central Diagnostic Laboratory, National Veterinary Research Institute, Vom, Nigeria

I. D. Echeonwux
Medical Microbiology, Federal College of Veterinary and Medical Laboratory Technology,

National Veterinary Research Institute, Vom, Nigeria

J. G. Bitrus
Medical Microbiology, Federal College of Veterinary and Medical Laboratory Technology,
National Veterinary Research Institute, Vom, Nigeria

S.I. Akubo
Medical Microbiology, Federal College of Veterinary and Medical Laboratory Technology,
National Veterinary Research Institute, Vom, Nigeria

Gopinath Lakshmipathy
Department of Industrial Biotechnology, Dr. MGR Educational and Research University, Maduravoyal, Chennai - 600095, India

Arunkumar Jayakumar
Department of Industrial Biotechnology, Dr. MGR Educational and Research University, Maduravoyal, Chennai - 600095, India

Meera abhilash
Department of Industrial Biotechnology, Dr. MGR Educational and Research University, Maduravoyal, Chennai - 600095, India

Shantha Prema Raj
Department of Industrial Biotechnology, Dr. MGR Educational and Research University, Maduravoyal, Chennai - 600095, India

Edith U. Madukwe
Department of Home Science, Nutrition and Dietetics, University of Nigeria, Nsukka, Nigeria

Okoye
Department of Home Science, Nutrition and Dietetics, University of Nigeria, Nsukka, Nigeria

Vivian
Department of Home Science, Nutrition and Dietetics, University of Nigeria, Nsukka, Nigeria

Ayogu
Department of Home Science, Nutrition and Dietetics, University of Nigeria, Nsukka, Nigeria

N. B. Rufina
Department of Home Science, Nutrition and Dietetics, University of Nigeria, Nsukka, Nigeria

Ikemefuna C. Obizoba
Department of Home Science, Nutrition and Dietetics, University of Nigeria, Nsukka, Nigeria

Calvin Onyango
Food Technology Division, Kenya Industrial Research and Development Institute, Kenya

Eunice A. Mewa
Department of Food Technology and Nutrition, University of Nairobi, Kenya

Anne W. Mutahi
Department of Food Technology and Nutrition, University of Nairobi, Kenya

Michael W. Okoth
Department of Food Technology and Nutrition, University of Nairobi, Kenya

B. Daramola
Department of Food Technology Federal Polytechnic, Ado-Ekiti, Ekiti State, Nigeria

EJOH Shirley Isibhakhomen
Department of Human Nutrition, University of Ibadan, Ibadan, Oyo State, Nigeria

OBATOLU Veronica Adeoti
Institute of Agricultural Research and Training (IAR&T), Obafemi Awolowo University. P.M.B. 5029, Moor Plantation, Ibadan, Oyo State. Nigeria

OLANIPEKUN Oyeyoyin Taiwo
Institute of Agricultural Research and Training (IAR&T), Obafemi Awolowo University. P.M.B. 5029, Moor Plantation, Ibadan, Oyo State. Nigeria

FARINDE Elizabeth Oluremi
Institute of Agricultural Research and Training (IAR&T), Obafemi Awolowo University. P.M.B. 5029, Moor Plantation, Ibadan, Oyo State. Nigeria

A. Kameni
IRAD Nkolbisson, PO Box 2067, Yaounde, Cameroon

C. Kouebou
Food Technology Unit, IRAD Garoua, Box 415, Cameroon

D. A. K. Aboubakar
Food Technology Unit, IRAD Garoua, Box 415, Cameroon

C. The
IRAD Nkolbisson, PO Box 2067, Yaounde, Cameroon

E. Chiau
Department of Food Technology, Engineering and Nutrition, Lund University, PO Box 124, SE-221 00 Lund, Sweden
Chemical Engineering Department, Eduardo Mondlane University, PO Box 257, Maputo, Mozambique

J. Da Cruz Francisco
Chemical Engineering Department, Eduardo Mondlane University, PO Box 257, Maputo, Mozambique

B. Bergenståhl
Department of Food Technology, Engineering and Nutrition, Lund University, PO Box 124, SE-221 00 Lund, Sweden

I. Sjöholm
Department of Food Technology, Engineering and Nutrition, Lund University, PO Box 124, SE-221 00 Lund, Sweden

R Sadoudi
Département d'Agronomie. Faculté des Sciences Biologiques et Agronomiques. Université de Tizi-Ouzou, Algérie

A Ammouche
Ecole nationale supérieure agronomique (ENSA, Ex. INA). Département de Technologie alimentaire et de Nutrition Humaine. Hassen badi, El-Harrach (Alger). Algérie

D Ali Ahmed
Département d'Agronomie. Faculté des Sciences Biologiques et Agronomiques. Université de Tizi-Ouzou, Algérie

E. Zangoei
Plant Pathology, University of Tehran.Tehran, Iran

H. R. Etebarian
Plant Pathology, University of Tehran.Tehran, Iran

N. Sahebani
Plant Pathology, University of Tehran.Tehran, Iran

Getachew Addis
Ethiopian Health and Nutrition Research Institute, P.O. Box 1242, Addis Ababa, Ethiopia

Vasudeva Singh
Central Food Technological Research Institute, Mysore-570 020, India

Vishwas Pratape
Central Food Technological Research Institute, Mysore-570 020, India

Alok Srivastava
Central Food Technological Research Institute, Mysore-570 020, India

Lalitha Gowda
Central Food Technological Research Institute, Mysore-570 020, India

Madhavan Asha
Central Food Technological Research Institute, Mysore-570 020, India

Sila Bhattacharya
Central Food Technological Research Institute, Mysore-570 020, India

Khaled Meghawry El-Zahar
Food Science Department, Faculty of Agriculture, Zagazig University, 44511 Zagazig, Egypt

M. G. Umuhozariho
Department of Food Science and Technology, Faculty of Agriculture, Sokoine University of Agriculture, P.O. Box 3006, Morogoro, Tanzania
College of Agriculture, Animal Sciences and Veterinary Medicine, University of Rwanda, P.O. Box 117, Huye, Rwanda

N. B. Shayo
Department of Food Science and Technology, Faculty of Agriculture, Sokoine University of Agriculture, P.O. Box 3006, Morogoro, Tanzania

J. M. Msuya
Department of Food Science and Technology, Faculty of Agriculture, Sokoine University of Agriculture, P.O. Box 3006, Morogoro, Tanzania

P. Y. K. Sallah
College of Agriculture, Animal Sciences and Veterinary Medicine, University of Rwanda, P.O. Box 117, Huye, Rwanda

Ahmed A. Aly
Home Economic Department, Faculty of Specific Education, Benha University, Egypt

G.M.El-Aragi
Plasma Physics and Nuclear Fusion Department, Nuclear Research Center, AEA, PO 13759 Cairo, Egypt

Assia Amri
Department of Biochemistry, Faculty of Sciences, University of Annaba, BP 12 - 23000 - Annaba, Algeria

Ali Ladjama
Department of Biochemistry, Faculty of Sciences, University of Annaba, BP 12 - 23000 - Annaba, Algeria

Tadele Yohannes
Department of Chemistry, College of Natural Sciences, Jimma University, P.O. Box 378, Jimma, Ethiopia

Fekadu Melak
Department of Chemistry, College of Natural Sciences, Jimma University, P.O. Box 378, Jimma, Ethiopia

Khalid Siraj
Department of Chemistry, College of Natural Sciences, Jimma University, P.O. Box 378, Jimma, Ethiopia

Reyhaneh Sarani
Nutrition Faculty, Tabriz University of Medical Sciences, Tabriz, Iran

Javad MohtadiNia
Department of Food Science and Technology, Tabriz University of Medical Sciences, Tabriz, Iran

Mohammad Asghari Jafarabadi
Department of Statics and Epidemiology, Tabriz University of Medical Sciences, Tabriz, Iran

Reza Afshinpajouh
Zar Macaron Company, Iran

Soodabeh Heydarian
Zar Macaron Company, Iran

Mehdi Amini
Zar Macaron Company, Iran

Ehsan Saadatmand
Zar Macaron Company, Iran

Matin Yahyavi
Zar Macaron Company, Iran

Yonas Hailu
School of Animal and Range Sciences, Haramaya University, P.O. Box: 138, Haramaya, Ethiopia

Eyassu Seifu
Department of Food Science and Technology, Botswana College of Agriculture, Private Bag 0027, Gaborone, Botswana

Zelalem Yilma
Netherlands Development Organization (SNV), EDGET/EADD-ET Project, Country Project Mobilization Coordinator, Churchill Road, Tracon Building, 6th Floor, P.O. Box P.O. Box 40675, Addis Ababa, Ethiopia

Kamal Kishor
Department of Botany, D.S.B. Campus, Kumaun University, Nainital-263 001, Uttarakhand, India

Lalit M. Tewari
Department of Botany, D.S.B. Campus, Kumaun University, Nainital-263 001, Uttarakhand, India

Ashish Tewari
Department of Forestry and Environmental Science, D.S.B. Campus, Kumaun University, Nainital-263 001, Uttarakhand, India

Brij M. Upreti
Department of Botany, D.S.B. Campus, Kumaun University, Nainital-263 001, Uttarakhand, India

Muhammad Shahnawaz
Department of Agriculture and Food Technology, Karakoram International University, Gilgit

Saghir Ahmed Sheikhand
Institute of Food Sciences and Technology, Sindh Agriculture University, Tandojam

Zahid Ali Akbar
Department of Agriculture and Food Technology, Karakoram International University, Gilgit

R. Kezih
Institut de Nutrition, de l'Alimentation et des Technologies Agro-alimentaires, Université1, 7ème km, route de Sétif, Constantine, Algérie

F. Bekhouche
Institut de Nutrition, de l'Alimentation et des Technologies Agro-alimentaires, Université1, 7ème km, route de Sétif, Constantine, Algérie

A. Merazka
Institut de Nutrition, de l'Alimentation et des Technologies Agro-alimentaires, Université1, 7ème km, route de Sétif, Constantine, Algérie

M. O. Afolabi
Department of Food Science and Technology, Bowen University, Iwo, Osun State, Nigeria

G. O. Adegoke
Department of Food Technology, University of Ibadan, Ibadan, Oyo State, Nigeria